Advances in Oil and Gas Wellbore Integrity

Advances in Oil and Gas Wellbore Integrity

Topic Editors

Kai Wang
Jie Wu
Yongjun Deng
Lin Chen

Basel • Beijing • Wuhan • Barcelona • Belgrade • Novi Sad • Cluj • Manchester

Topic Editors

Kai Wang
School of Petroleum
Engineering
China University of
Petroleum (East China)
Qingdao
China

Jie Wu
Laboratory of Beam
Technology and Energy
Materials
Advanced Institute of Natural
Sciences
Beijing Normal University
Zhuhai
China

Yongjun Deng
School of Civil Engineering
and Architecture
Southwest University of
Science and Technology
Mianyang
China

Lin Chen
Institute of Radiation
Technology
Beijing Academy of Science
and Technology
Beijing
China

Editorial Office
MDPI AG
Grosspeteranlage 5
4052 Basel, Switzerland

This is a reprint of the Topic, published open access by the journals *Energies* (ISSN 1996-1073), *Geosciences* (ISSN 2076-3263) and *Applied Sciences* (ISSN 2076-3417), freely accessible at: https://www.mdpi.com/topics/UK45137ZV5.

For citation purposes, cite each article independently as indicated on the article page online and as indicated below:

Lastname, A.A.; Lastname, B.B. Article Title. *Journal Name* **Year**, *Volume Number*, Page Range.

ISBN 978-3-7258-3099-2 (Hbk)
ISBN 978-3-7258-3100-5 (PDF)
https://doi.org/10.3390/books978-3-7258-3100-5

© 2025 by the authors. Articles in this book are Open Access and distributed under the Creative Commons Attribution (CC BY) license. The book as a whole is distributed by MDPI under the terms and conditions of the Creative Commons Attribution-NonCommercial-NoDerivs (CC BY-NC-ND) license (https://creativecommons.org/licenses/by-nc-nd/4.0/).

Contents

Lev Eppelbaum and Youri Katz
A New Look to the Heletz–Ashdod Oil Field (Southern Israel): A Non-Conventional Hydrocarbon Deposit in the Easternmost Mediterranean
Reprinted from: *Geosciences* 2023, 13, 12, https://doi.org/10.3390/geosciences13010012 1

Alexandra Cedola and Runar Nygaard
Shale Cuttings Addition to Wellbore Cement and Their Effect on Unconfined Compressive Strength
Reprinted from: *Energies* 2023, 16, 4727, https://doi.org/10.3390/en16124727 26

Pengcheng Wu, Chentao Li, Zhen Zhang, Jingwei Yang, Yanzhe Gao, Xianbing Wang, et al.
Research on Cuttings Carrying Principle of New Aluminum Alloy Drill Pipe and Numerical Simulation Analysis
Reprinted from: *Energies* 2023, 16, 5618, https://doi.org/10.3390/en16155618 50

Tingting Jiang, Dongling Cao, Youqiang Liao, Dongzhou Xie, Tao He and Chaoyang Zhang
Leakage Monitoring and Quantitative Prediction Model of Injection–Production String in an Underground Gas Storage Salt Cavern
Reprinted from: *Energies* 2023, 16, 6173, https://doi.org/10.3390/en16176173 67

Evgeniya I. Lysakova, Andrey V. Minakov and Angelica D. Skorobogatova
Effect of Nanoparticle and Carbon Nanotube Additives on Thermal Stability of Hydrocarbon-Based Drilling Fluids
Reprinted from: *Energies* 2023, 16, 6875, https://doi.org/10.3390/en16196875 83

Weiguo Zhang, Deli Gao, Yijin Zeng and De Yan
Deepwater PDC Jetting Bit-Drilling Technology Based on Well Structure Slimming
Reprinted from: *Energies* 2023, 16, 7394, https://doi.org/10.3390/en16217394 103

Zhao Zheng, Jun Yang, Maoxuan Cui, Kui Yang, Hui Shang, Xue Ma and Yuxiang Li
Adsorption/Desorption Performances of Simulated Radioactive Nuclide Cs^+ on the Zeolite-Rich Geopolymer from the Hydrothermal Synthesis of Fly Ash
Reprinted from: *Energies* 2023, 16, 7815, https://doi.org/10.3390/en16237815 114

Anisa Noor Corina and Al Moghadam
The Sealing Performance of Cement Sheaths under Thermal Cycles for Low-Enthalpy Geothermal Wells
Reprinted from: *Energies* 2024, 17, 239, https://doi.org/10.3390/en17010239 128

Haifeng Zhu, Ming Xiang, Zhiqiang Lin, Jicheng Yang, Xuerui Wang, Xueqi Liu and Zhiyuan Wang
Study on the Mechanism of Gas Intrusion and Its Transportation in a Wellbore under Shut-in Conditions
Reprinted from: *Energies* 2024, 17, 242, https://doi.org/10.3390/en17010242 140

Anja Pfennig, Wencke Mohring and Marcus Wolf
The Insignificant Improvement of Corrosion and Corrosion Fatigue Behavior in Geothermal Environment Applying Boehmit Coatings on High Alloyed Steels
Reprinted from: *Appl. Sci.* 2024, 14, 1575, https://doi.org/10.3390/app14041575 157

Guolei He, Linqing Wang, Jiarui Wang, Kaixiang Shen, Hengfu Xiang, Jintang Wang, et al.
Load Calculation and Strength Analysis of the Deepwater Landing Drill Pipe-Lowering Operation
Reprinted from: *Energies* **2024**, *17*, 1258, https://doi.org/10.3390/en17051258 169

Tingting Jiang, Xiurui Shang, Dongzhou Xie, Dairong Yan, Mei Li and Chunyang Zhang
Study on the Mechanical and Permeability Characteristics of Gypsum Rock under the Condition of Crude Oil Immersion
Reprinted from: *Energies* **2024**, *17*, 1712, https://doi.org/10.3390/en17071712 184

Peng-Fei Yin, Sheng-Qi Yang and Pathegama Gamage Ranjith
Anisotropic Mechanical Behaviors of Shale Rock and Their Relation to Hydraulic Fracturing in a Shale Reservoir: A Review
Reprinted from: *Energies* **2024**, *17*, 1761, https://doi.org/10.3390/en17071761 200

Ning Zhang, Daiyin Yin, Guangsheng Cao and Tong Li
The Mechanism Study of Fracture Porosity in High-Water-Cut Reservoirs
Reprinted from: *Energies* **2024**, *17*, 1886, https://doi.org/10.3390/en17081886 235

Xiabin Wang, Shanpo Jia, Shaobo Gao, Long Zhao, Xianyin Qi and Haijun He
Wellbore Integrity Analysis of a Deviated Well Section of a CO_2 Sequestration Well under Anisotropic Geostress
Reprinted from: *Energies* **2024**, *17*, 3290, https://doi.org/10.3390/en17133290 252

Janaina Andrade de Lima Leon, Henrique Luiz de Barros Penteado, Geoffrey S. Ellis, Alexei Milkov and João Graciano Mendonça Filho
Recognition of Artificial Gases Formed during Drill-Bit Metamorphism Using Advanced Mud Gas
Reprinted from: *Energies* **2024**, *17*, 4383, https://doi.org/10.3390/en17174383 264

Renjun Xie and Laibin Zhang
A New Prediction Model of Annular Pressure Buildup for Offshore Wells
Reprinted from: *Appl. Sci.* **2024**, *14*, 9768, https://doi.org/10.3390/app14219768 279

Jingzhe Zhang, Rongrong Zhao, Hongyi An, Wenhao Li, Yuxin Geng, Xiangyu Fan and Qiangui Zhang
Numerical Simulation on the Influence of the Distribution Characteristics of Cracks and Solution Cavities on the Wellbore Stability in Carbonate Formation
Reprinted from: *Appl. Sci.* **2024**, *14*, 10099, https://doi.org/10.3390/app142210099 299

Zhengfeng Shan, Xiansi Wang, Zeqin Li, Zhenggang Gong, Nan Ma, Jianbo Zhang, et al.
Research on Key Parameters of Wellbore Stability for Horizontal Drilling in Offshore Hydrate Reservoirs
Reprinted from: *Appl. Sci.* **2024**, *14*, 10922, https://doi.org/10.3390/app142310922 321

Article

A New Look to the Heletz–Ashdod Oil Field (Southern Israel): A Non-Conventional Hydrocarbon Deposit in the Easternmost Mediterranean

Lev Eppelbaum [1,2,*] and Youri Katz [3]

1 Department of Geophysics, Faculty of Exact Sciences, Tel Aviv University, Ramat Aviv, Tel Aviv 6997801, Israel
2 Azerbaijan State Oil and Industry University, Azadlig Ave. 20, Baku AZ1010, Azerbaijan
3 Steinhardt Museum of Natural History & National Research Center, Faculty of Life Sciences, Tel Aviv University, Ramat Aviv, Tel Aviv 6997801, Israel
* Correspondence: levap@tauex.tau.ac.il

Abstract: Heletz–Ashdod oil field is the first oil deposit to have been discovered in the Eastern Mediterranean. This deposit has been exploited until the present despite its small reserves. However, this area's tectonic–geodynamic and structural features must be studied more. Based on the integrated regional geological–geophysical analysis, it was proposed that the Heletz terrane, which includes this deposit, is the nodal structure of the Eastern Mediterranean. This terrane is a composite part of the earlier identified Mesozoic terrane belt (MTB). The Heletz terrane's essential tectonic peculiarity is the MTB's impact, rotating counterclockwise. Analysis of local geophysical data in this area (gravity, magnetic, and seismic) and regional thermal data examination testifies to the complex mosaic composition of the Heletz structure. It is proved that the crystal basement below the Heletz terrane is characterized by specific properties that do not coincide with the adjacent areas. Finally, a series of structural–paleogeographic and thickness formation maps of the Heletz–Ashdod oil field has been compiled. This investigation shed light on this area's perspectives and searched other hydrocarbon deposits in the easternmost Mediterranean coastal and shelf zones.

Keywords: Heletz terrane; Mesozoic terrane belt; syn-sedimentary structural controls; structural–paleogeographic maps; thickness formation

Citation: Eppelbaum, L.; Katz, Y. A New Look to the Heletz–Ashdod Oil Field (Southern Israel): A Non-Conventional Hydrocarbon Deposit in the Easternmost Mediterranean. *Geosciences* **2023**, *13*, 12. https://doi.org/10.3390/geosciences13010012

Academic Editors: Kai Wang, Jie Wu, Yongjun Deng, Lin Chen and Jesus Martinez-Frias

Received: 6 December 2022
Revised: 27 December 2022
Accepted: 28 December 2022
Published: 29 December 2022

Copyright: © 2022 by the authors. Licensee MDPI, Basel, Switzerland. This article is an open access article distributed under the terms and conditions of the Creative Commons Attribution (CC BY) license (https://creativecommons.org/licenses/by/4.0/).

1. Introduction

Israel hosted its first oil drilling in 1947 and the completion of its first discovery in 1955 at the Heletz area. The Heletz–Ashdod oil field is located some 10 km from the Mediterranean Sea in southern Israel (Figure 1A,B). Interestingly, this area was identified as a perspective based on gravity survey analysis [1,2]. This deposit is associated with developing a Lower Cretaceous canyon-like incision, filled at the base with terrigenous continental oil source rocks, which are replaced by marine carbonates [3–8]. An extensive SSW–NNE trend of positive gravity anomalies $\Delta g_{Bouguer}$ [1] was found in the land area situated parallel to the Israel's coast, which served as one of the justifications for drilling test wells. Oil is produced mainly from Lower Cretaceous sandstones and, to a lesser extent, from dolomites, from depths ranging from 1.5 to 2.0 km. The source rock for the Heletz oil is either the Lower Cretaceous shales of the Gevar-Am formation [9] or the Jurassic fine-grained carbonates of the Barnea formation [4].

Analysis of the abovementioned geological sources points to a complex structure of the area. Some authors suggest that oil migrated into its present reservoirs during the Neogene following major tectonic movements [5,10,11]. Heletz gives a small commercial oil production from the mid-1950s until the present. At the same time, this field's concealed oil potential has yet to be entirely recognized.

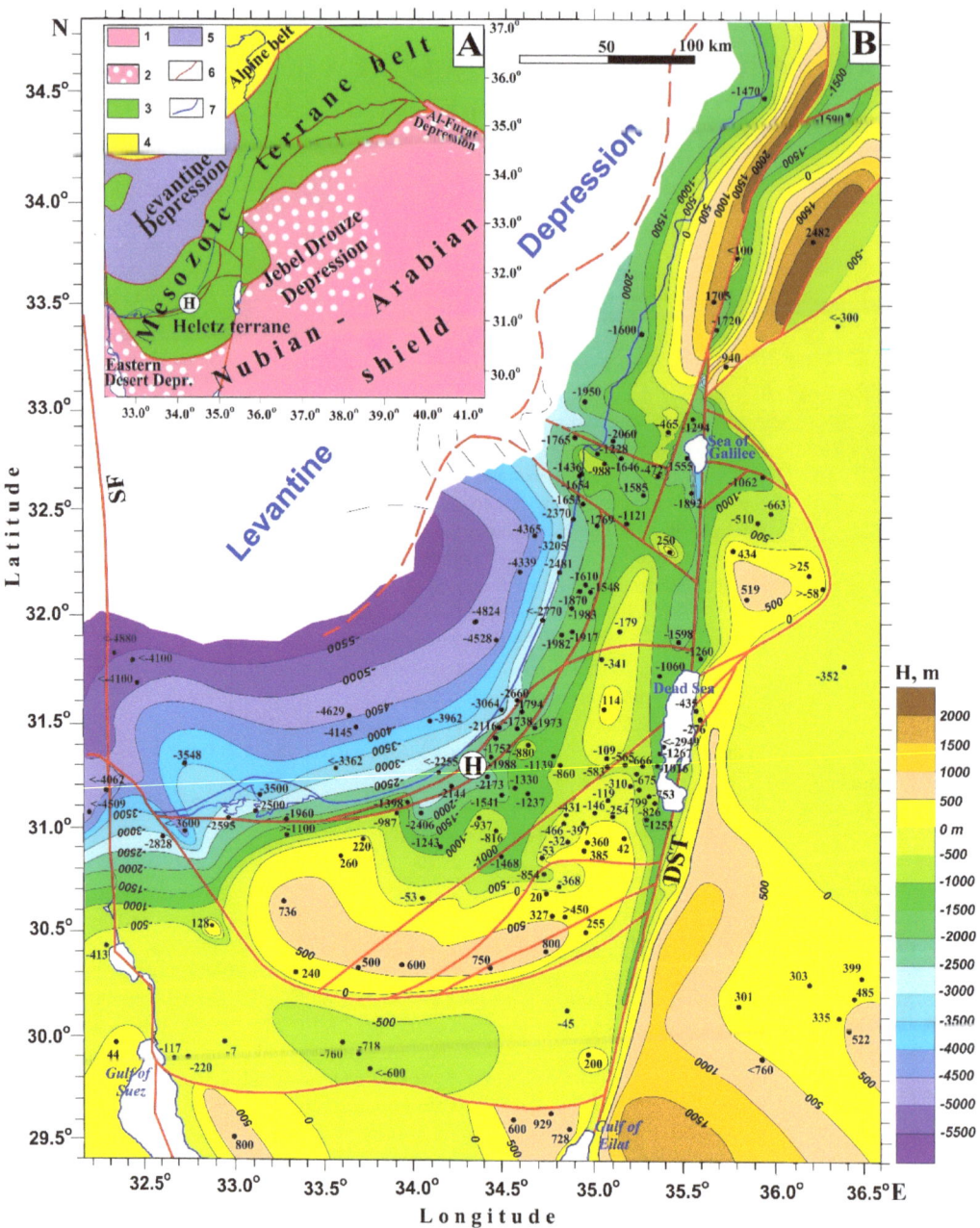

Figure 1. (**A**): Areal tectonic map, (**B**): Structural–neotectonic map of the base of the newest (post-Jurassic sediments) tectonic complex. (1) Precambrian Craton, (2) Precratonic Depression, (3) Mesozoic terrane belt, (4) Alpine Mediterranean Belt, (5) Tethyan oceanic crust, (6) faults, (7) coastline. SF, Sinai Fault, DST, Dead Sea Transform. (**B**) revised after [12].

A set of various geophysical studies was carried out here in sufficient detail, indicating a significant structural heterogeneity of the Heletz field [13]. At the same time, regional complex tectonic–geodynamic studies of the Eastern Mediterranean region were also performed here as in a model polygon for mapping complex structural elements of plate tectonics and geodynamics (e.g., [14–17]).

The need for such work was due to the development of active prospecting and exploration here and the discovery of large deposits of hydrocarbons, which could not be explained without applying the theoretical provisions of plate tectonics. The regional syn-sedimentary structural control method was used to precisely map the Heletz oil field and adjacent areas [7,17]. As a result, a series of structural maps of Mesozoic and Cenozoic sedimentary formations were obtained.

However, we postponed their publication because we got a picture of a rather complex structure formation, which could not be explained in light of the data of the regional tectonophysical analysis.

The long-term and comprehensively studied Heletz oil field is highly interested in structural morphology, evolution, and geomechanics. The latest detailed geological and geophysical surveys using various geophysical data (including satellite data) [18,19] have made it possible to build fundamentally new models of the deep structure of the Eastern Mediterranean. These physical-geological models can also substantiate local structural analysis for prospecting and contouring hydrocarbon deposits.

2. Tectonic–Geodynamic Pattern

2.1. Brief Regional Tectonic Background

The area under consideration is located near Israel's Mediterranean coast, west of the Dead Sea. Tectonically, it belongs to the Heletz terrane and adjacent structures of the Mesozoic terrane belt (MTB) located between the Nubian–Arabian shield of Gondwana and the Alpine mobile belt, alternating with the oceanic crust of the residual depressions of the Tethys Ocean (Figure 1A).

The terrane belt and the adjacent region of Gondwana, as well as part of the oceanic basin of the Eastern Mediterranean, are complicated by a submeridional system of deep faults—Sinai Fault (SF) (in the west) and Dead Sea Transform (DST) (in the east). These regional faults form the Sinai lithospheric plate [19,20]. Geodynamically, this plate consists of a series of uplifts and troughs associated with the instability of movements within the terrane belt [17]. This is reflected in the Jurassic top hypsometry map (Figure 1B).

To study the specifics of movements within the area under consideration, the data associated with the Heletz terrane (indicated by the letter 'H') are very critical. It is the most geodynamically unstable transition zone between the Negev terrane (where the top of the Jurassic is elevated up to +830 m above sea level) and the Pleshet terrane (where it descends to depths of about −5500 m). It should be noted that there is also a second linear zone of instability, discordantly crossing the Heletz terrane in the NW-SE direction, tracing through the southern part of the Dead Sea basin. The abovementioned geodynamic and structural features are significant for analyzing the cartographic material below.

2.2. The Heletz Terrane as a Nodal Regional Structure of the Eastern Mediterranean

The Heletz terrane is a regional nodal structure in terms of both structural analyses of syn-sedimentation sequences and the development of a methodology for prospecting hydrocarbon deposits, and in terms of understanding the geodynamics and history of the development of the Neotethys–Gondwana junction zone [6,7,17,21,22]. The instability of this terrane identified above also has a deeper geophysical substantiation since the granite-metamorphic layer underlying the carbonate platform of the MTB has a significant similarity in most terranes.

A mature granite-gneiss island-arc complex of the Neoproterozoic (island arc facies—813–610 Ma) is developed here, overlain by the Late Proterozoic arkose molasse of the Zenifim Formation. The narrow Heletz terrane is composed of the muscovite-chlorite Tashtit complex (Doroth Formation) consisting of the greenschist facies dating 636 ± 107 Ma [6]. Undoubtedly, this factor can serve as the basis for the difference in the structural differentiation of the Heletz terrane.

As for the use of the name of the terrane to designate the key structures of the Arabian Plate, where large-scale uplifts and troughs are distinguished, the latest regional tectonic-geophysical studies in the Middle East [19] revealed new plate-geodynamic criteria. The Heletz anticline cannot be part of the foreland tectonics, whose cover is composed of both the Meso-Cenozoic and the most ancient Paleozoic. However, the Heletz terrane does not belong to the Gondwana plate but to the MTB of the Neotethys basin, which experienced significant (up to 1000–1500 km) horizontal displacements with counterclockwise rotation of blocks at the Tethys–Gondwana boundary in the post-Paleozoic time [7,12,18,19]. Only such an active type of geodynamics formed nodal tectonic structures of the Heletz terrane type, which are indicators of regional movements of a deeper level. From this point of view, the presence of Early Permian igneous complexes in the Heletz terrane is significant, indicating the beginning of the initiation of the Neotethys Ocean and the fragmentation of the marginal zone of the thinned crust of the Neoproterozoic belt of Gondwana. It was then, probably, that the formation of the MTB began.

3. Materials and Methods

3.1. Materials

The following materials were used for the geological–geophysical data examination in the Heletz–Ashdod area: [1–3,5–11,13,17,21–42].

3.2. Methods

The following methods were applied for detailed analysis of geological data: an integrated analysis of geological (tectonic–geodynamic, syn-sedimentary structural controls, lithological-facial, event stratigraphy, formational, hydrospheric disturbances, structural-geological, cyclic stratigraphy, and palaeogeographical) and geophysical (gravity, magnetic, seismic, and thermal) data.

4. Results

With the aim to reveal the essence, typology, and nature of the structure and geodynamics of the considered oil and gas region, it is necessary to research historical geotectonics using the classical method of syn-sedimentary structural controls. At the regional level, we have done this in a series of previous works (e.g., [7,17]). This work was carried out to identify the cycles of structure genesis by using syn-sedimentary control of space-time sedimentary formations (complexes) within the framework of natural geological cycles of about 40 million years. The last of these cycles, the Neogene-Quaternary, is incomplete.

Such a methodology, linking classical and plate tectonics, is vital in the Eastern Mediterranean region, where the most diverse aspects of tectonic activity are present [6,19,43,44]. The series of structural maps with syn-sedimentary controls (Figures 2–6) and hiatuses and unconformities (Figures 7–12) presented in this paper includes three tectonic–geodynamic complexes formed during the Alpine Epoch in the Eastern Mediterranean:

(1) Post-collision—Neogene-Quaternary (Figure 2), Paleogene (Figure 3), and Late Cretaceous (Figure 4) structural stages;
(2) Collisional—Early Cretaceous structural stage (Figures 5–8);
(3) Pre-collisional—Jurassic (Figures 9–12) and earlier structural stages.

The constructed maps and their meaning will be discussed in detail below.

4.1. Post-Collisional Complex

The upper Neogene-Quaternary structural stage of the post-collision complex (Figure 2) has the simplest structure regarding the tectonic–geodynamic analysis. It includes two main structural zones: (1) a coastal uplift of the carbonate platform, including the Negev and Heletz terranes, and (2) a trough with strong tectonic subsidence at the boundary of the Heletz and Pleshet terranes. From the zone of uplift to the tectonic ledge along the diagonal fault, an erosional incision is stretched, which is a relic of the late Messinian crisis, which manifested itself as a unique catastrophe—the drying of the bottom of the Mediterranean Sea. The valley of the erosional canyon under consideration extended on the elevated part of the plateau at a distance of about 50 km from the eastern part of the Negev terrane to the northeastern part of the Heletz terrane [6,17,39].

The uplift and trough zones are not subject to any noticeable folding. This feature is very significant from the tectonic–geodynamic standpoint. It illustrates that the stage under consideration belongs to the final scene of the tectonic collision of the Alpine geodynamic cycle. At this stage, the contrast of uplifted coastal uplifts—consolidated marginal belts of tectonic plates—and powerful internal sedimentary troughs that make up the relict collision zone of the former ocean are clearly manifested. This provision is essential from the standpoint of the search for hydrocarbon deposits we will consider later.

The Paleogene and Upper Cretaceous structural maps (Figures 3 and 4) represented the middle and lower levels of the post-collision complex when the Mesozoic terrane belt (MTB) and the Neotethys Ocean geodynamically formed a single structure experiencing differentiated uplifts and subsidence [17]. The details of such differentiated structure formation are vital for the search for hydrocarbons. The Paleogene thickness distribution map of the Heletz oil exploration fields and adjacent areas (Figure 3) delineates the uplifted zone of the Negev and Heletz terranes and the relatively subsided zone of the Pleshet terrane. In this way, in general terms, it precedes the geodynamic stage of the Neogene-Quaternary stage. It is distinguished by a significant gently sloping folding—the development of elongated synclinal and anticlinal swell-like structures with an insignificant amplitude of uplifts and subsidence. The most abrupt zone of structural transition is the boundary of the Heletz and Pleshet terranes. A narrow ridge of this erosion zone of Paleogene deposits at a distance of 2 to 5 km sharply passes into a spot of increased thickness up to 600–800 m. To a certain extent, such an effect, as it were, precedes the subsequent subsidence of the basin of the Eastern Mediterranean in the Neogene-Quaternary time.

The lower stage of the post-collision complex (Figure 4) differs significantly from the Paleogene stage in significant differentiation of the structures of the terrane belt. A sharp zone of tectonic uplift is distinguished here, associated with the boundary of the Heletz and Pleshet terranes, where the Upper Cretaceous deposits are entirely eroded in the area of a relatively wide ridge of uplifts. According to the regional mapping data [17], such swells are developed throughout the Negev and Pleshet terranes field. Thus, the marked structural differentiation, as it were, inherits the previous (collisional) stage of arch-like uplift of blocky structures of the MTB. This structural and geodynamic phenomenon can be used for the directed search for hydrocarbons on the shelf and in the zone of development of the oceanic crust of the Eastern Mediterranean.

This phenomenon is all the more relevant because in the Upper Cretaceous carbonates of the Upper Campanian and Lower Maastrichtian of the Negev and Pleshet terranes, significant deposits of "black shales" [6,30]—upwelling facies that form hydrocarbon deposits in many parts of the world [45]. The northern part of the Pleshet terrane has been explored and created large commercial deposits prepared for development [8,12,46].

4.2. Collision Complex (Early Cretaceous)

Classical regional studies on the regional geology of the Near East [6,23,47] showed that in this area, in the transition zone of the Neotethys Ocean and the Gondwana Precambrian plate in the middle Mesozoic, between the thick carbonate the Jurassic and Upper Cretaceous strata formed an anomalous transitional sediment complex. Here, dominated terrigenous continental and partly marine formations and basalt traps [48], in places reaching 500 m in thickness [40]. The boundary deposits and structural complexes of the Jurassic and Lower Cretaceous of the Eastern Mediterranean, and especially the territory of Israel, are well and comprehensively studied geologically and contain a huge number of radiometric dates (e.g., [49]).

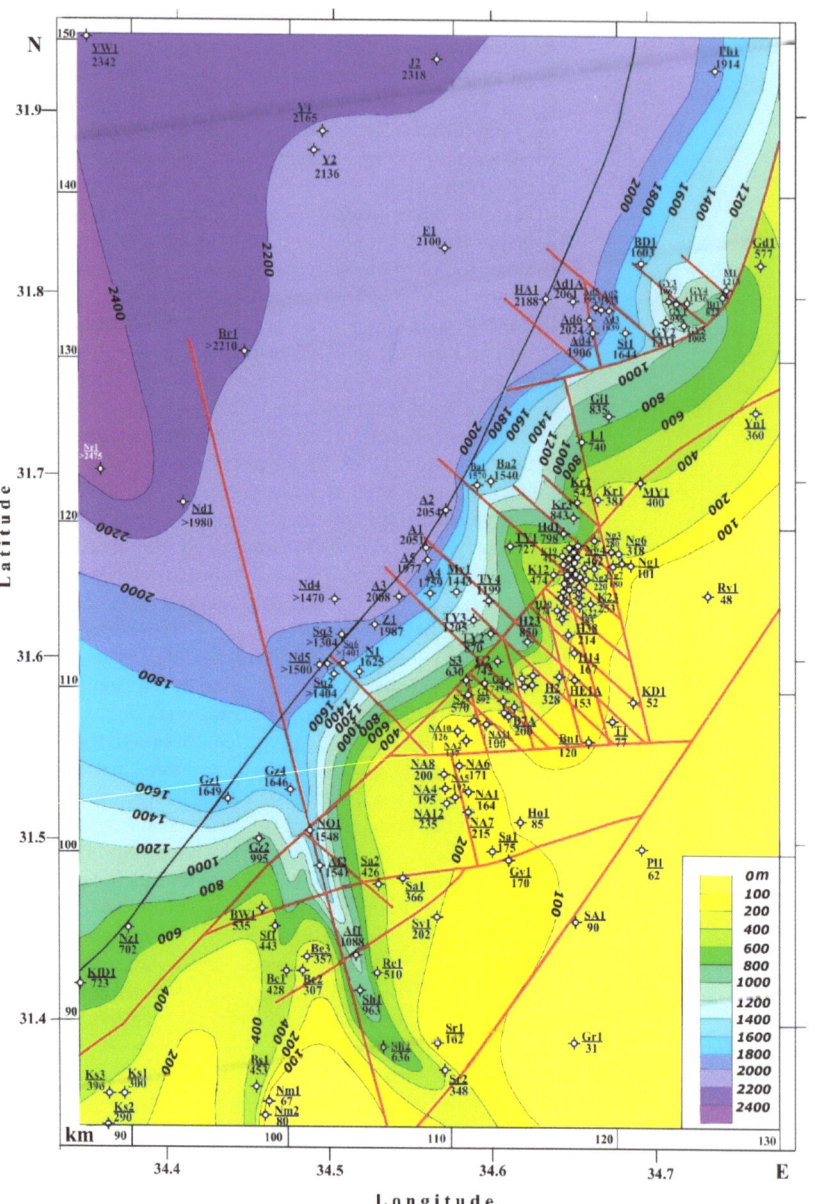

Figure 2. Neogene-Quaternary structural–paleogeographic map of the Heletz–Ashdod oil field and adjacent areas (isopach intervals are given in m). Borehole abbreviations: A, Ashkelon, Ad, Ashdod, Af, Afiq, B, Brur, Ba, Barnea, Be, Beeri, Bn, Binyamin, Br, Bravo, Bs, Bessor, Bt, Bitsaron, BD, Bene Darom, BW, Beeri West, E, Echo, G, Gevar Am, Gd, Gedera, Gi, Givati, Gr, Gerar, Gv, Gevim, Gz, Gaza, GY, Gan Yavne, H, Heletz, Hd, Hodiya, Ho, Hoga, HA, Hof Ashdod, HE, Heletz East, J, Joshua, K, Kokhav, Kr, Karmon, Ks, Kissufim, KD, King David, KfD, Kfar Darom, L, Lior, M, Mivtah, Mv, Mavkiim, MY, Massuot Yizhak, N, Nissanit, Nd, Nordan, Ng, Negba, Nm, Nirim, Nr, Nir, Nz, Nezarim, NA, Nir Am, NO, Nahal Oz, Ph, Palmahim, Pl, Pleshet, Rv, Revaha, S, Shimon, Sa, Saad, Sf, Sahaf, Sh, Shuva, Sq, Shiqma, Sr, Sharsheret, St, Shetulim, SA, Sedot Akiva, T, Telamim, TY, Talme Yafe, Y, Yam, Ya, Yakhini, Yn, YW, Yam West Yinnon, Z, Ziqim.

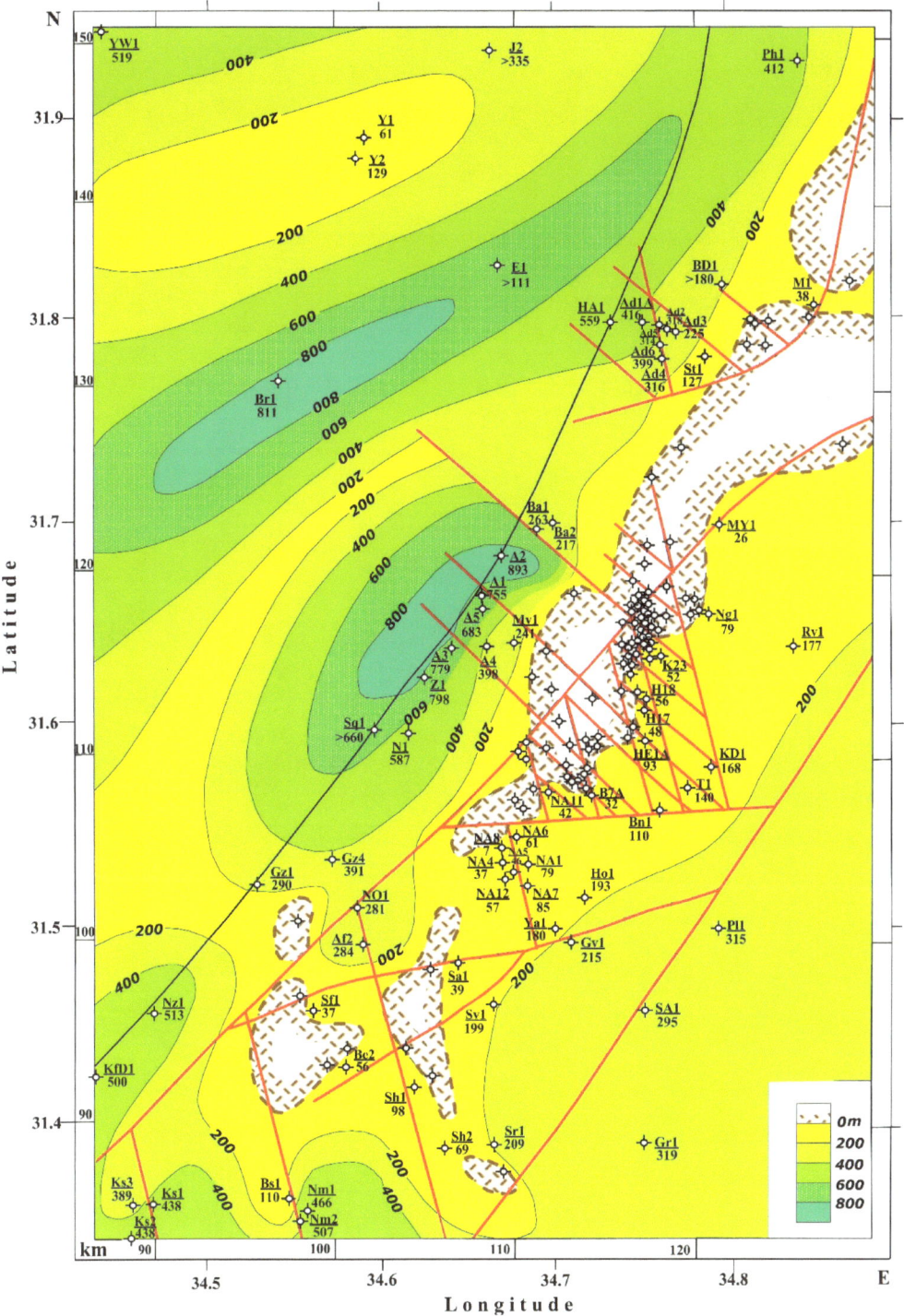

Figure 3. Paleogene structural–paleogeographic map of the Heletz–Ashdod oil field and adjacent areas. Borehole abbreviations: see captions for Figure 2.

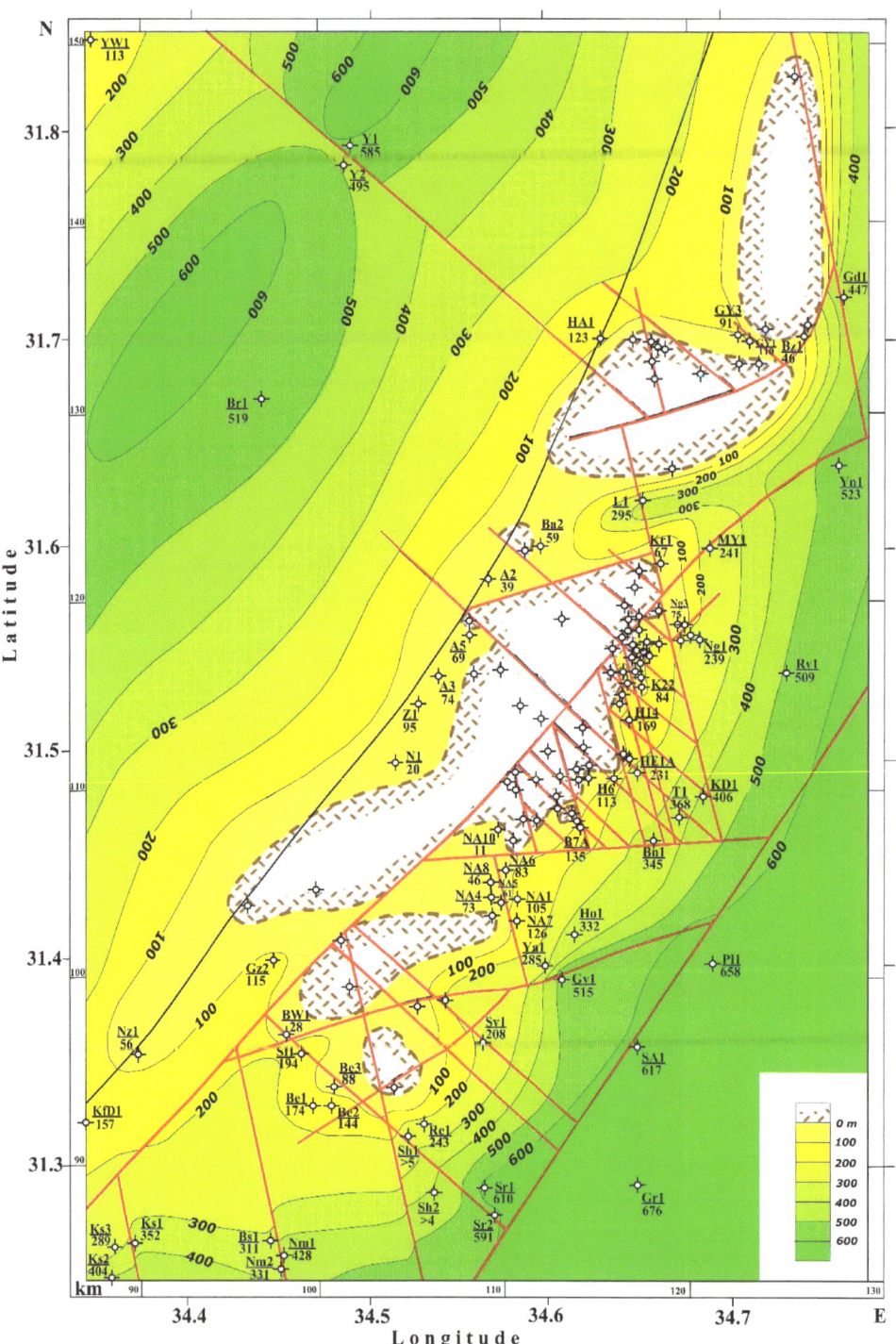

Figure 4. Late Cretaceous structural–paleogeographic map of the Heletz–Ashdod oil field and adjacent areas. Borehole abbreviations: see captions for Figure 2.

Regional structural-sedimentary studies [17] show that, in general, the collisional complex in the Early Cretaceous geodynamically differed sharply from the adjacent Precambrian shield of Gondwana by a significant thickness difference. Taking into account these data, as well as the phenomena of plate geodynamics in the transition zone of terrane massifs from the thinned continental crust and ophiolite sheets displaced from adjacent areas of the Tethys Ocean, the Levantine phase of the terrane belt consolidation was established, corresponding to the boundary of the Early and Late Hauterivian [50]. After the collision of blocks of continental and oceanic crust and the previous active erosion of the uplifted carbonate platform of this belt, its later movements could also occur, in the second half of the Early Cretaceous, when the thickest Aptian–Albian trap fields were formed to the north [36].

The northern part of the MTB was uplifted, and structural zones were distinguished in the Aleppo and Abdelaziz terranes, where the Lower Cretaceous was absent. The underlying formations were subjected to significant erosion. This erosion is especially clearly demonstrated by regional maps of the underlying Jurassic deposits, where the phenomenon of the Levantine uplift phase is reflected in the absence of Jurassic sediments in the Aleppo terrane and the western half of the Abdelaziz terrane.

Judging by the Jurassic thickness of the more southerly terranes displaced by 500–1000 km in the process of rotational motions counterclockwise in the southeast direction [18,19], the amplitude of the early secondary erosion of the Aleppo and Abdelaziz uplifts of the MTB could reach 1500–2500 m. In the Early Cretaceous, the southeastern part of the terrane belt moved to the Neotethys Ocean, on the contrary, was involved in active downward movements. Evidence of downward movements in the southeastern part of the terrane belt was obtained in the Makhtesh Ramon erosion-tectonic canyon, where more than 10 m blocks of Oxford limestone ("limestone blocks") with basaltoid and trachyte dikes of 135.1 Ma age were found brought to the surface [28,47]). The nearest Oxford formations are located 30 km to the north, in a different tectonic block, in the borehole Boqer-1 [31]. We explained this circumstance by underthrusting the subterrane block systems of the Negev terrane during their mutual displacement [51].

Thus, descending movements during terrane movements in the region's southeast-prevented significant phenomena of early secondary erosion of Jurassic deposits near the Neotethys Ocean. A reasonably distinct general subsidence in the Early Cretaceous occurred near this ocean. This is reflected in the structural map with the syn-sedimentary control of the Heletz oil-bearing basin and adjacent areas (Figures 5 and 6). Generally, the considered map is closest in type to the structural-sedimentary map of the Neogene-Quaternary stage (Figure 2).

Here, a clear tectonic step of the differential is distinguished from the uplifted Negev-Heletz terrane and the subsided Pleshet terrane. At the same time, an analogy is clearly seen between the erosional canyon of the Messinian age and the Early Cretaceous oil-bearing Heletz canyon. The difference lies in the development of local movements in the Early Cretaceous, which led to the formation of low-amplitude brachyanticlinal structures in the Heletz and Pleshet terranes. The most dislocated are the Lower Cretaceous rocks of the Heletz triangle (Figure 6).

Here, a unique fold-block system was formed in the era of the MTB collision with the Precambrian continental platform Gondwana and the Neotethys oceanic plate. In the center of the system is a brachysynclinal fold with the maximum thickness of the Lower Cretaceous deposits for the Eastern Mediterranean, exceeding 2359 m.

4.3. Paleogeographic and Palaeogeological Mapping of the Early Stages of the Collisional Complex

The initial stages of structure formation and geodynamics of the Mesozoic collisional complex of the Eastern Mediterranean cover the Jurassic-Cretaceous boundary and correspond to the final stages of the Late Jurassic and the first half of the Early Cretaceous-Neocomian—from the Berriasian to the Barremian. This is the stage of the equatorial contraction of the Earth's figure and the corresponding equatorial regression [52]. Therefore, this area, located south of the critical parallel of 35 degrees N, studied for more than 70 years by various scientific studies, is a reference in terms of structural, geodynamic and paleogeographic analysis of the Mesozoic history of Gondwana.

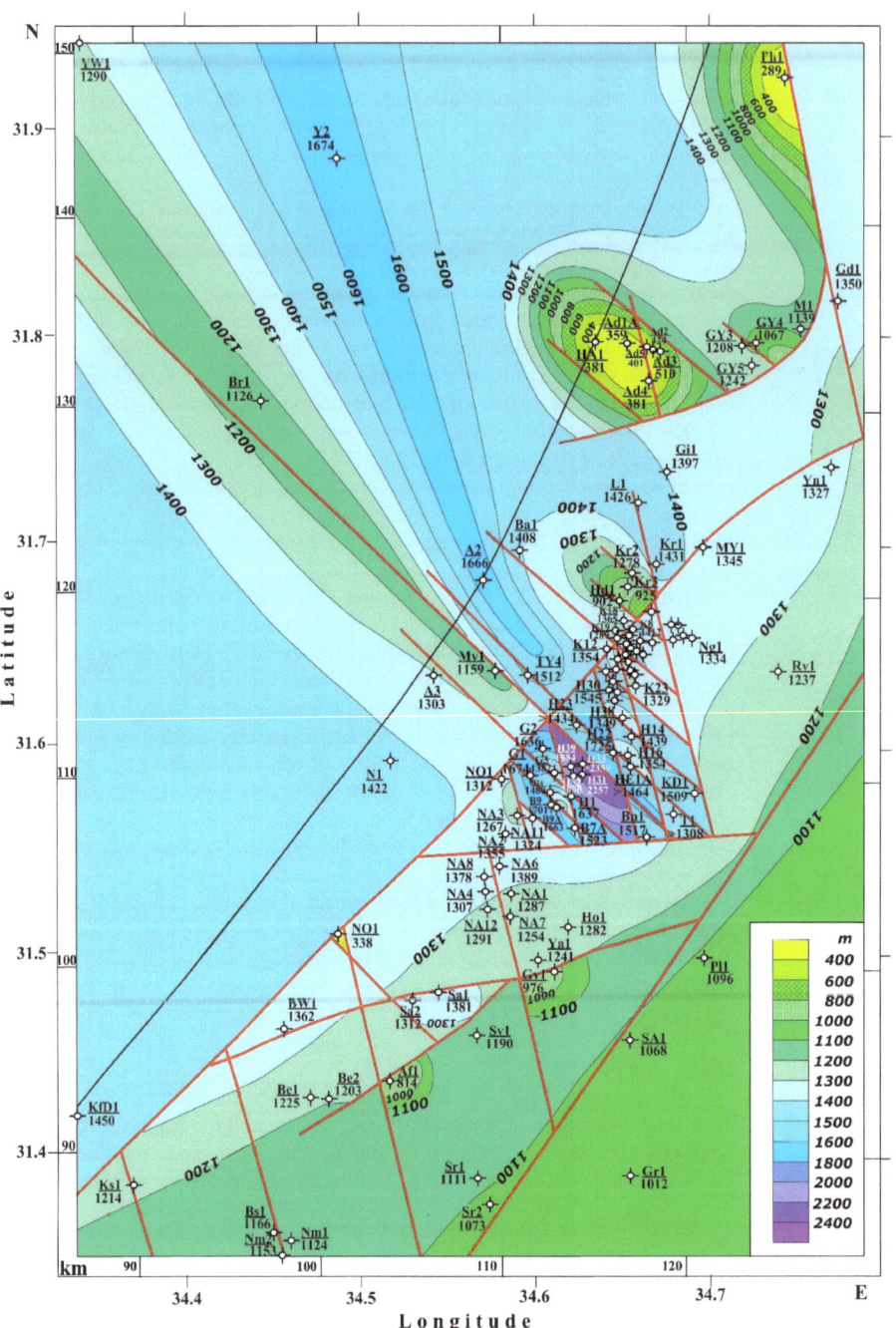

Figure 5. Early Cretaceous structural–paleogeographic map of the Heletz–Ashdod oil field and adjacent areas. Borehole abbreviations: see captions for Figure 2.

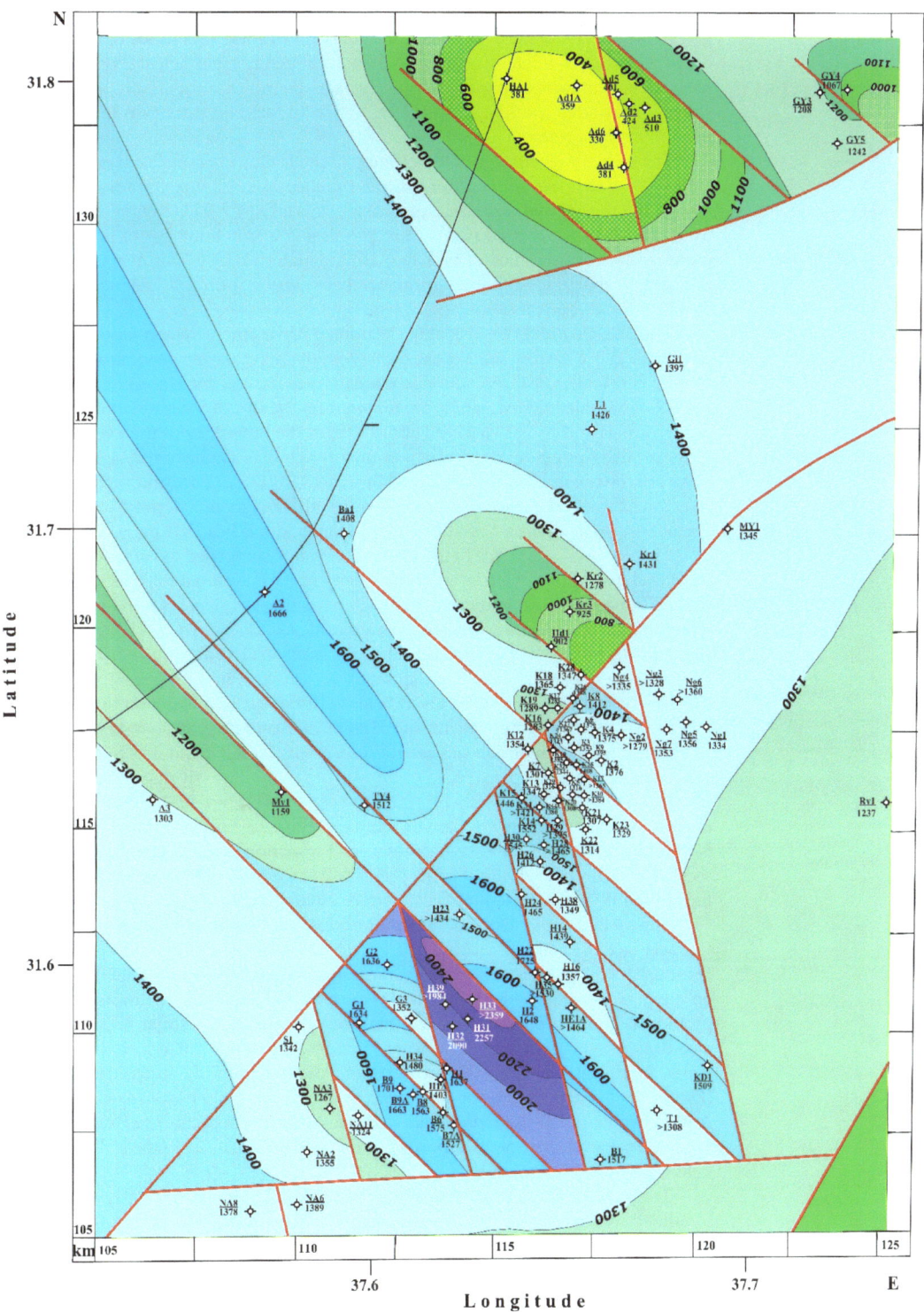

Figure 6. Early Cretaceous structural–paleogeographic map of the Heletz oil field. Borehole abbreviations: see captions for Figure 2.

To identify and analyze these processes, we relied on the classical methodology of geotectonics—the analysis of breaks and unconformities. It consists of the construction of special maps of the hiatus surface—structural-geomorphological (Figure 7), palaeogeological erosion maps (Figure 8), etc. Such maps indicate the amplitude of uplift and subsidence under conditions of differentiation of tectonic structures in the zone of the unconformity surface.

The structural geomorphological map of the Heletz erosional canyon was published earlier [7] and taken into account in the practice of specialized studies in the field of optimizing further hydrocarbon exploration in this area [42]. This circumstance contributed to the fact that we carried out further processing and refinement of the map due to a more careful consideration of available drilling materials [31]. This new variant (Figure 7) indicates a sharp geodynamic difference between the uplifted Heletz terrane and the subsided Pleshet terrane.

The map clearly shows that at the beginning of the Early Cretaceous, the Heletz terrane was essentially a narrow slope of the northwestern part of the elevated plateau of the Negev terrane carbonate platform. On this plateau, rare narrow incisions of Early Cretaceous river flows are developed, which are replaced on the slope by a more branched system of watercourses and initial vast canyons. Then, at the boundary of the Heletz and Pleshet terranes, this system abruptly changes into a delta-like expansion system. Finally, after 10 km to the northwest, the erosion valley turns into a canyon, and further to the north (the Yam-2 well), the thickness of predominantly terrigenous deposits of the Gevar Am Formation (Neocomian), which fills the canyon, and reaches 1304 m.

The underlying surface of the Heletz Canyon is composed of Jurassic deposits. Their structural map shows the relative homogeneity of the collisional stage tectonic movements in the Pleshet and Negev terranes and a sharp contrast of structures in the Heletz terrane, which is also complicated by a significant number of disjunctives. The triangular Heletz structure appears to be the most unstable, the main structural contrast o appearing on the Lower Cretaceous thickness map (Figure 6). A comparison of the Jurassic and Cretaceous stages of geodynamics will be given below. Therefore, it is essential to understand the inheritance of tectonic movements of the collisional and pre-collision stages in the MTB.

These unique materials of years of research by Israeli geologists made it possible to construct a palaeogeological map of the surface of the Jurassic deposits, confined to the Heletz Canyon and the surrounding land and sea areas (Figure 8). The stratigraphic volume of eroded deposits in Heletz Canyon from the Titonian Yam Formation to the Bathonian Sederot Formation and the Bajocian Barnea Formation is more than half of the Jurassic section. Geochronologically, it corresponds to a stage of 25 million years, and according to the eroded thickness of sediments, it reaches an amplitude of 1000–1400 m.

The total time of erosion of Jurassic carbonate sediments at the beginning of the Early Cretaceous erosional phase could be from 3.5 to 7.0 Ma. The subsequent erosion could be associated with sharp collapses of sedimentary masses of the Lower Cretaceous sediments on the canyon's steep sides. These issues are essential to solving hydrocarbon exploration problems, and the palaeogeological map can be used to optimize them. Attention is also drawn to another very significant phenomenon. They are associated with a Lower Cretaceous clastic complex (Figure 7) filling a deep canyon-like Jurassic erosional trough (Figure 8). Such thick erosion zones are absent in the higher latitude Galilee–Lebanon, Anti-Lebanon, Judea–Samaria terranes and in the northern part of the Pleshet terrane. It is favored by the absence of oil deposits north of the MTB, similar to the Heletz and Ashdod fields developed in the south.

The earliest stage of the considered collisional stage of the Early Cretaceous materialized in the form of premature secondary erosion of Jurassic deposits subjected to deformation. For this purpose, detailed information on the areal distribution and age of eroded and preserved rocks is needed. The Heletz oil field and adjacent prospecting and exploration areas are fine-drilled and thoroughly explored during prospecting and surveying, and the available base materials are in regional reports [6,31].

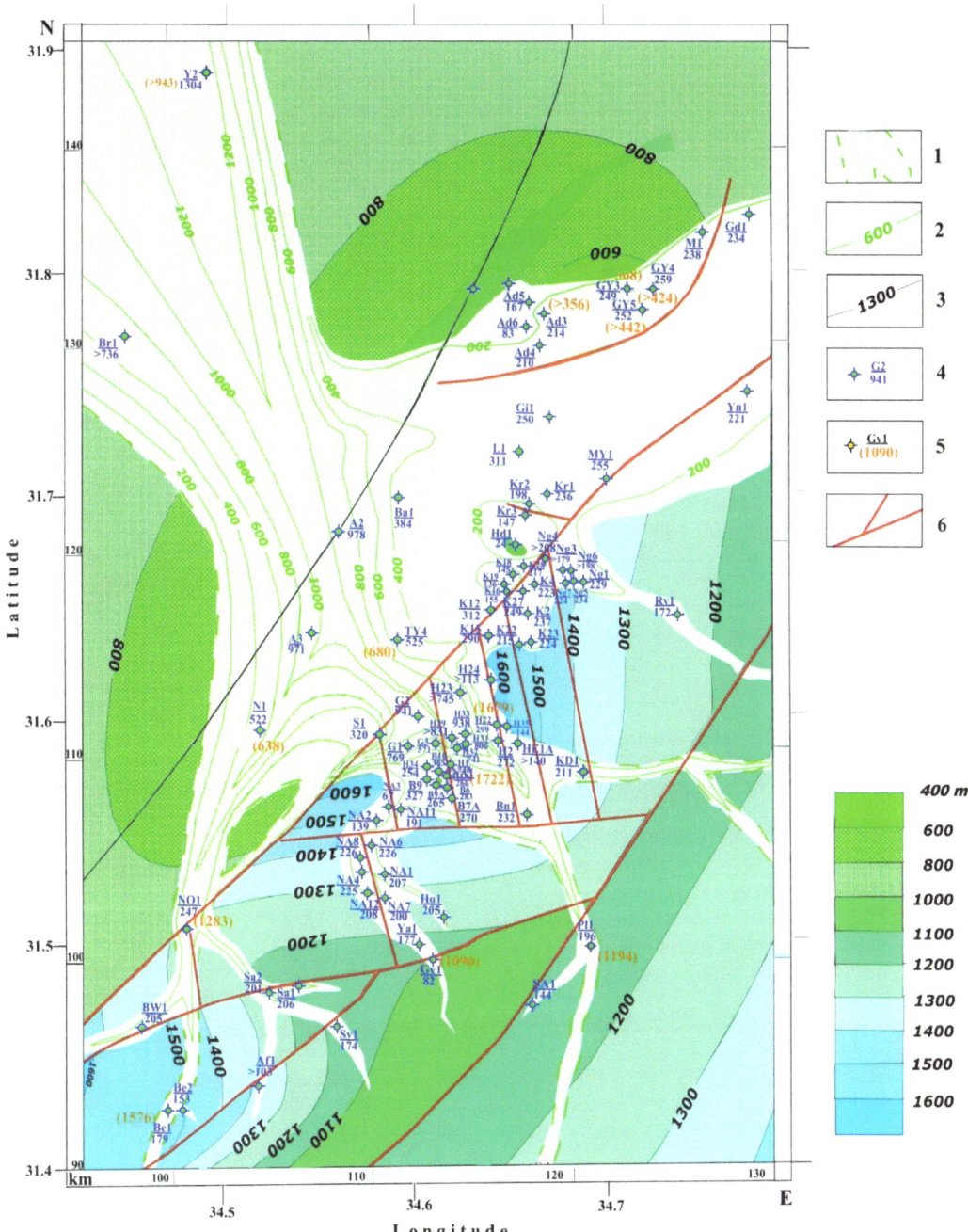

Figure 7. Paleogeomorphological map of the GevarAm erosional stage (Hauterivian–Barremian) of the Heletz–Ashdod oil field and adjacent areas. (1) The Early Cretaceous Gevar Am erosional zone, (2) isopachs of the Gevar Am Formation, (3) isopachs of the Upper—Middle Jurassic (Halutza—Upper Inmar formations) (in m), (4) boreholes penetrating Gevar Am Formation (nominator—borehole name and number, denominator—thickness in m, (5) boreholes penetrating Jurassic deposits (nominator—borehole name and number, denominator—thickness in m, (6) faults. Borehole abbreviations: see captions for Figure 2.

Our tectonic–paleogeographic and paleomagnetic mapping of these territories [38,40,51] made it possible to identify the cause of gender differences. It is due to the development of a vast field of thick Lower Cretaceous traps in the north. Mobile basaltic lavas quickly filled the erosional topography, preventing the development of canyons, which is clearly seen in the geological and paleomagnetic profiles [40,51].

It is significant that because of satellite geophysical data modeling and a thorough regional tectonic analysis, the underlying cause of the difference in geodynamics and tectonic-thermal processes in various parts of the Eastern Mediterranean has been established. This is due to the discovery of a deep mantle structure [18,19].

The northern terranes mentioned above were located in the axial zone of this structure, where hot spots formed. They are associated with fields of basalt traps, as well as deeper intrusions containing diamonds and their minerals-satellites [38,53]. A second significant circumstance related to the geodynamics of the counterclockwise rotating deep mantle structure. It led to the fact that the terranes' set moved along transform faults in the Mesozoic at the boundary of the Neotethys and Gondwana in the same direction from NE to SW. Therefore, the oil-bearing southwestern bend of the MTB is located outside the active zone of tectonic-thermal processes. In the Cretaceous period, significantly different geodynamics and the nature of tectonic-sedimentary processes were developed here, which is reflected in the presented maps (Figures 4–8). Accordingly, the forecast, searches, and prospects for discovering hydrocarbon deposits should be considered depending on the region's tectonic-geophysical zoning and geological history.

Evidence for filling erosion incisions with basalt—similar in age to the Lower Cretaceous Heletz Canyon—is reasonably well known and studied. In the Samaria region (Wadi Maliah), in the western block of the Anti-Lebanon terrane, a sequence of tuffaceous rocks and basalts was revealed up to 230 m thick (according to the age of the Gevar Am Formation (Neocomian)). These thicker formations are the apex tributary of an erosional valley that is deeper in trapping and extends EW, somewhat crossing the fault line separating the northern Galilee–Lebanon terranes from the more southerly Pleshet and Judea–Samaria terranes [40]. In the same work, using isopachs, the erosion valley was mapped. According to deep drilling data [31], to the east, in the Har Amir-1 borehole, the trap thickness is 401 m, and the total Neocomian thickness is 867 m. Further, about 25 km to the west along the line of depression of the erosion valley in the borehole Caesarea-2, the thickness of the traps reaches 463 m, and the total thickness of the Neocomian erosional complex is 1098 m. In the boreholes Caesarea-1 and 3, the thickness of the traps decreases to 184–268 m and the total Neocomian thickness to 591–803 m.

Comparing the total thickness of the Neocomian in this erosional valley with the Heletz Canyon (Figure 7), we are convinced of a significant difference in the intensity of erosion processes in the southern parts of the MTB compared to the northern ones. At the same time, the region of the Heletz terrane looks like the highest part of the collision system of block uplifts. According to drilling data, the magnitude of the erosional incision according to the Neocomian sediments filling the canyon in the Heletz 31, Heletz 32, and Heletz 33 wells is 1361 m, 1293 m, and 1514 m, respectively. These maximum values of the erosional incision directly in the Heletz triangle, in the source zone, contrast sharply with the incision data in the north, in Wadi Maliah, up to 230 m and near the mouth, in Caesarea, up to 1098 m.

4.4. Pre-Collision Complex

Structural analysis of the pre-collision complex, which was formed before the Early Cretaceous, is very important for understanding the structural evolution of the region under consideration and the strategy for searching for hydrocarbons. Three structural stages are clearly distinguished in this complex: Jurassic, Triassic, and Permian, separated by stratigraphic breaks and differing in the distribution of facies and thicknesses [6,7].

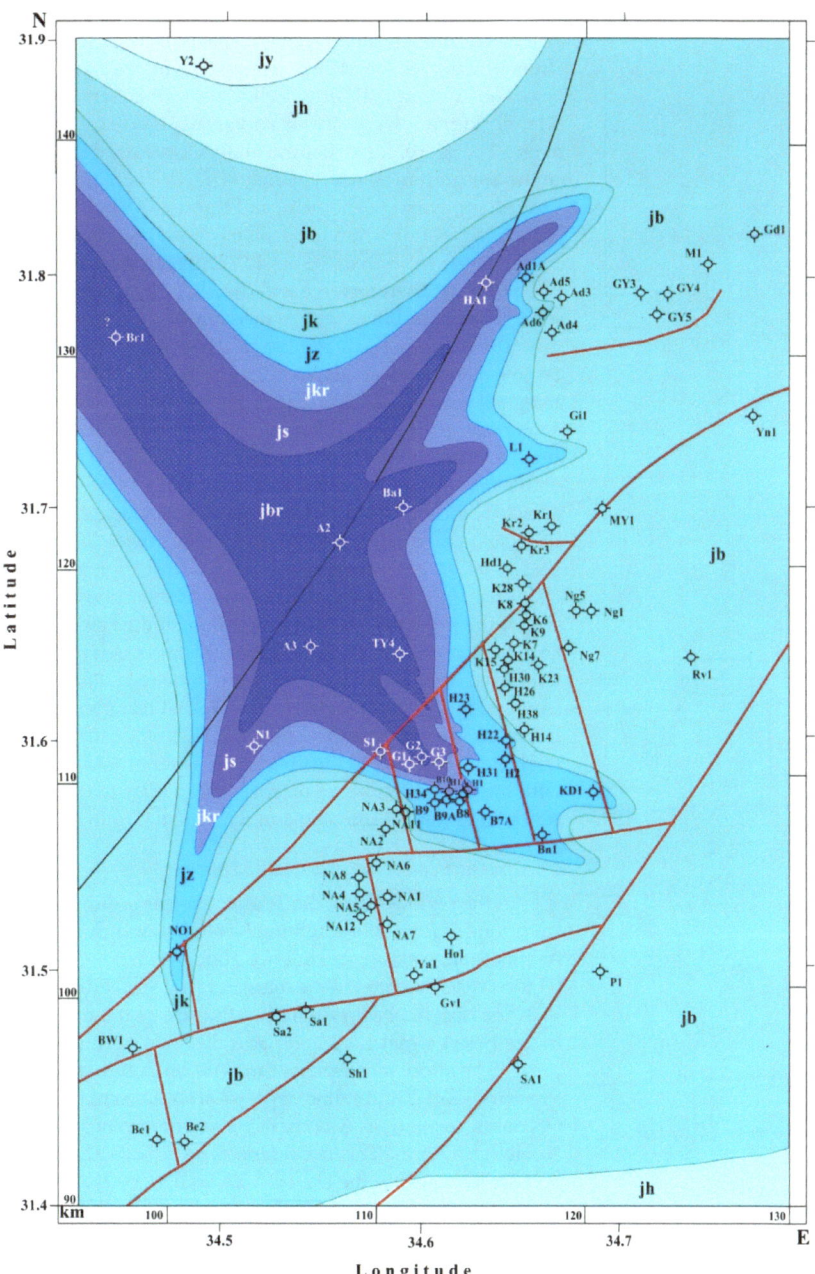

Figure 8. Palaeogeological map for the top Jurassic of the Heletz–Ashdod oil field and adjacent areas. Jurassic formations: jy—Yam, jh—Halutsa, jb—Beer-Sheva, jk—Kidod, Jz—Zohar, jkr—Karmon, js—Sederot, jbr—Barnea. Borehole abbreviations: see captions for Figure 2.

4.5. Syn-Sedimentation Sequence of the Late-Middle Jurassic of the Pre-Collision Complex

We were able to perform this analysis taking into account the consistent structural mapping with syn-sedimentary controls of the Middle-Upper Jurassic carbonate stratum fractionally dissected in the sections of numerous wells, which survived the early secondary erosion outside the erosional canyon (Figures 9–12). The sequence of structural maps covers half of the Jurassic and, in general, makes it possible to compare the nature of the structural differentiation of the MTB with subsequent stages—collisional and post-collisional.

The uppermost Jurassic horizons, the Yam and Halutsa formations (Tithonian-Kimmeridgian), were almost universally eroded, probably at the end of the pre-collision stage. Therefore, for structural analysis, we used formations that are relatively common in the area and reasonably widely drilled: Beer Sheba (Middle Oxfordian), Kidod (Lower Oxfordian), Zohar (Callovian), and Karmon (Upper Bathonian) (Figures 9–12, respectively). A detailed examination of each structural map indicates the unevenness of information related to the location of the erosion canyon zone, deep drilling areas, and maximum depths of boreholes. Therefore, we will dwell on the general trend that follows from the analysis of each structural map with syn-sedimentary controls.

The Negev terrane represents insignificant material at the boundary with the Heletz terrane. Nevertheless, generally, the amplitude of the zones of syn-sedimentary uplift of the marginal zones of both terranes does not differ significantly. The Heletz triangle depression subterrane is the most differentiated structurally and geodynamically. The most prominent troughs—up to 555 m in the Middle Oxfordian (Figure 9) are replaced by an uplift zone with an amplitude of up to 50 m in the Lower Oxfordian (Figure 10). At the same time, in adjacent blocks of the terrane Heletz, the subsidence amplitude was 150–275 m. This cannot be related to the erosion of the zone of the Lower Cretaceous erosional canyon present here since data for the overlying Beer Sheva formation are opposite.

Thus, we can fix the effect of oscillatory movements in the Heletz Triangle block at the end of the pre-collision stage. The data on the syn-sedimentary controls of the Callovian and Bathonian-Zohar and Karmon formations (Figures 11 and 12) concerning blocks of the Heletz terrane are different from in Oxford. However, there are still some differences between individual blocks. In the Pleshet terrane, the development of the Ashdod high is common to all structural maps, geodynamically most clearly manifested, similarly to the Heletz Triangle, in the Oxfordian, when the amplitude of its relative uplift compared with the adjacent troughs reached 150 m (Figures 9 and 10). These movements were less contrasting in the Callovian and Bathonian (Figures 11 and 12).

4.6. Triassic Syn-Sedimentation Sequence

In the regional plan, the Triassic stage is indicated [17] as the first complex of tectonic-stratigraphic formations, uniting Neotethys and Gondwana in a single transgressive phase. At this time, a stable subsidence zone, foreland, consisting of wide platform syneclises (Figure 1A) originated in the marginal Arabian-Nubian continental margin at the boundary with the oceanic basin located at the site of the present-day MTB. In the MTB, located in the ocean's marginal zone at a distance of more than 1000–1500 km ENE from the present position, a complex of marine predominantly carbonate formations and lagoon gypsum-bearing and saline depression strata were formed [6,33]. The maximum thickness of Triassic formations was characteristic of terranes located in the zone of contact with the Neotethys Ocean. This is evidenced by data on the Galilee–Lebanon terrane, where the Triassic thickness in the Deborah 2A well exceeds 2585 m [31].

The Triassic stage of the Heletz terrane, located between the relatively subsided Pleshet and Negev terranes, appears to be utterly anomalous in the composition of the entire terrane belt. This is especially evident in a significant gap between the Mesozoic and underlying Paleozoic formations with the fallout of the Lower Triassic and the development of the Middle Triassic Erez conglomerates up to 276 m thick, which follows from the data from the study of the Heletz deep 1A well [31]. Obviously, the well under consideration shows the highest thickness of the Triassic in this terrane (up to 926 m) [17].

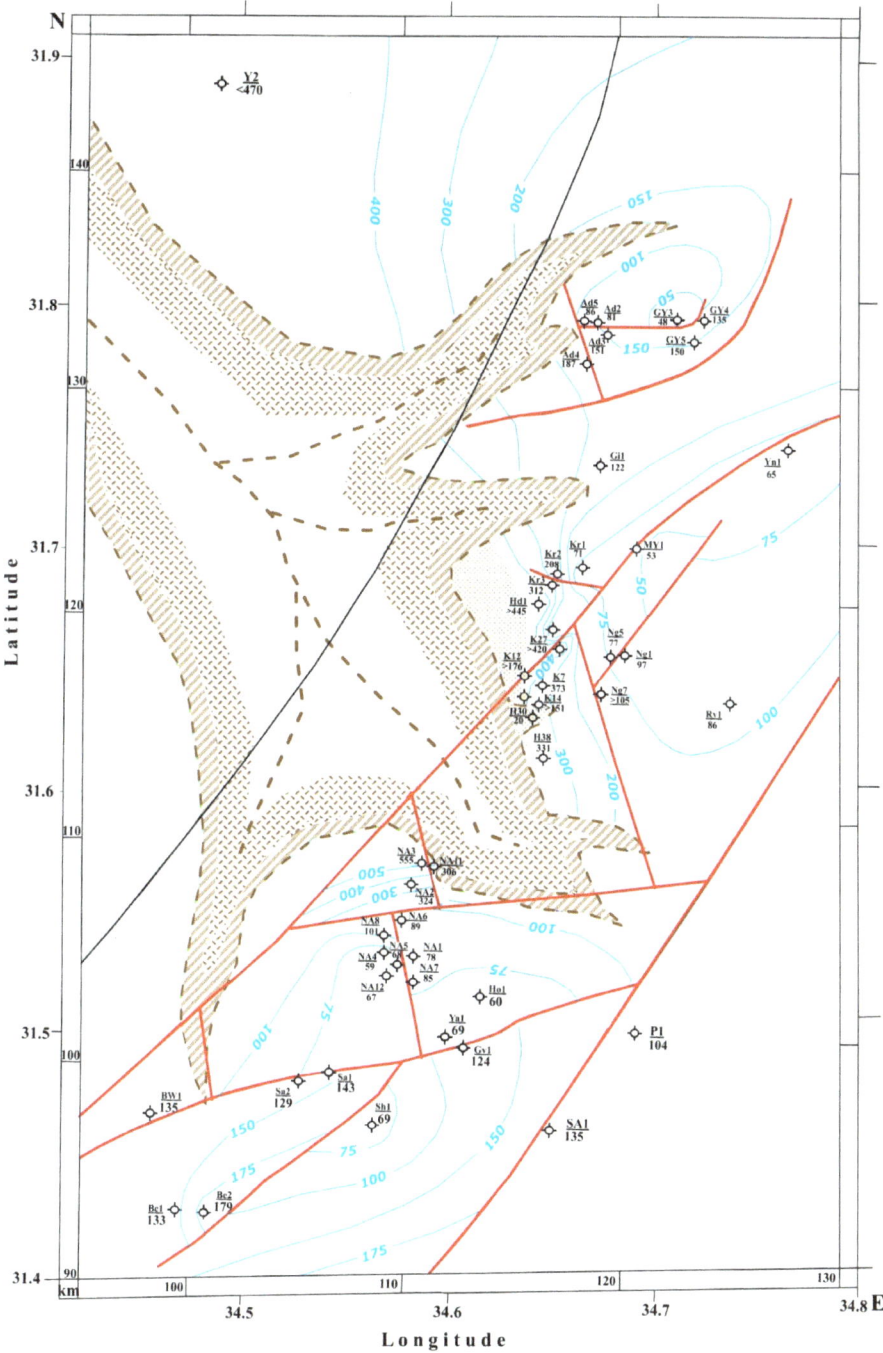

Figure 9. A map of the thicknesses of the Middle Oxfordian (Beer-Sheva Formation) of the Heletz–Ashdod oil field. The brown color shows the erosional canyon, and blue lines are the isopachs. Borehole abbreviations: see captions for Figure 2.

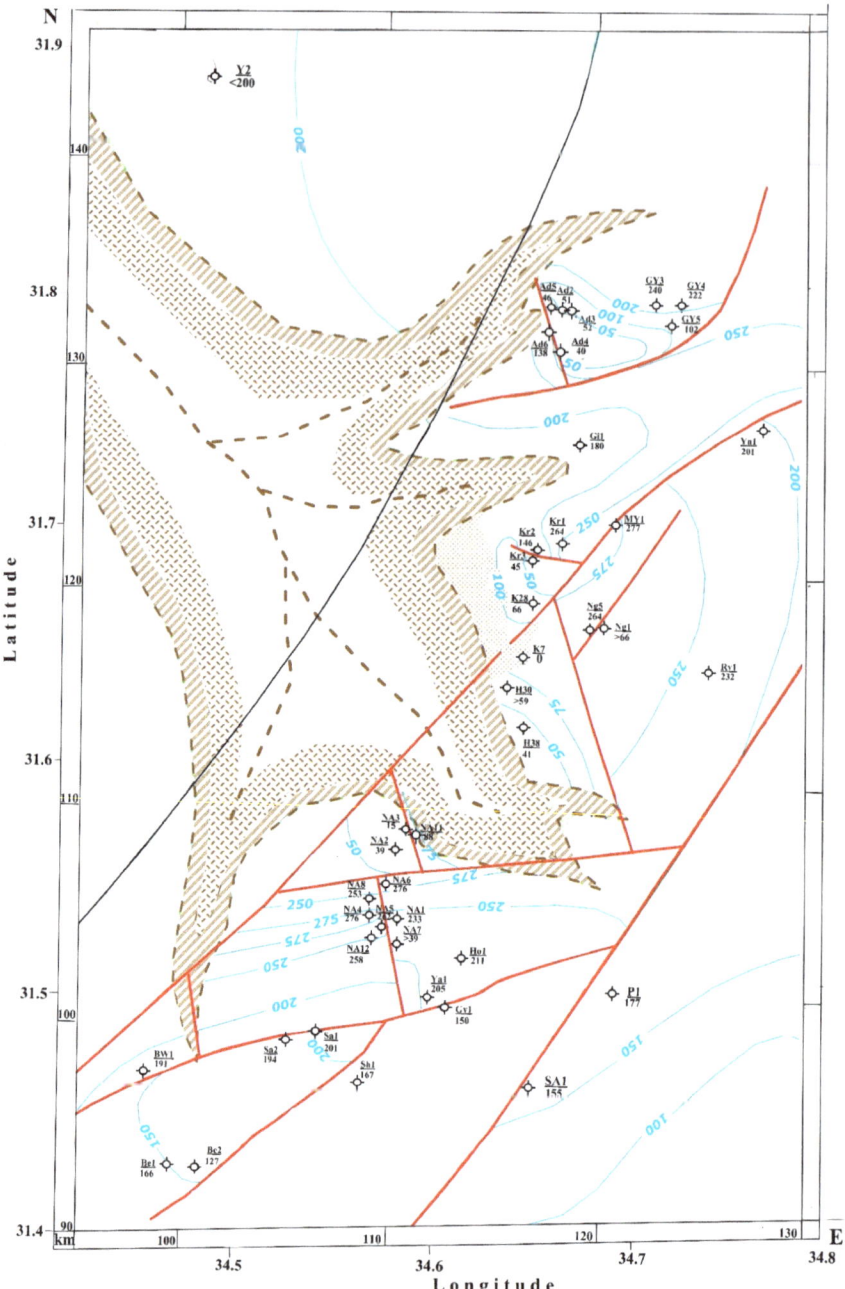

Figure 10. A map of the thicknesses of the Lower Oxfordian (Kidod Formation) of the Heletz–Ashdod oil field. The brown color shows the erosional canyon, and blue lines are the isopachs. Borehole abbreviations: see captions for Figure 2.

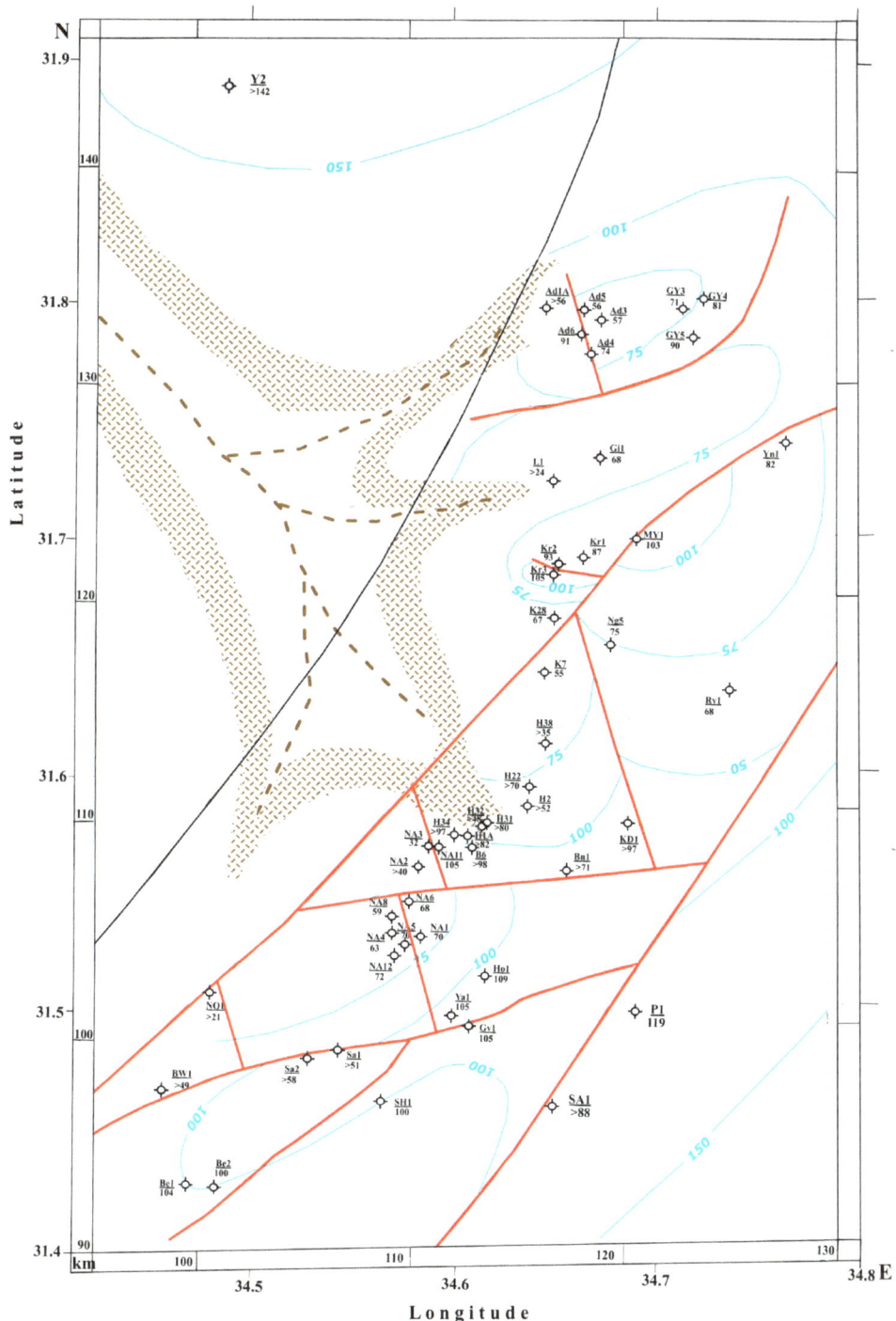

Figure 11. A map of the thicknesses of the Lower Callovian (Zohar Formation) of the Heletz–Ashdod oil field. The brown color shows the erosional canyon, and blue lines are the isopachs. Borehole abbreviations: see captions for Figure 2.

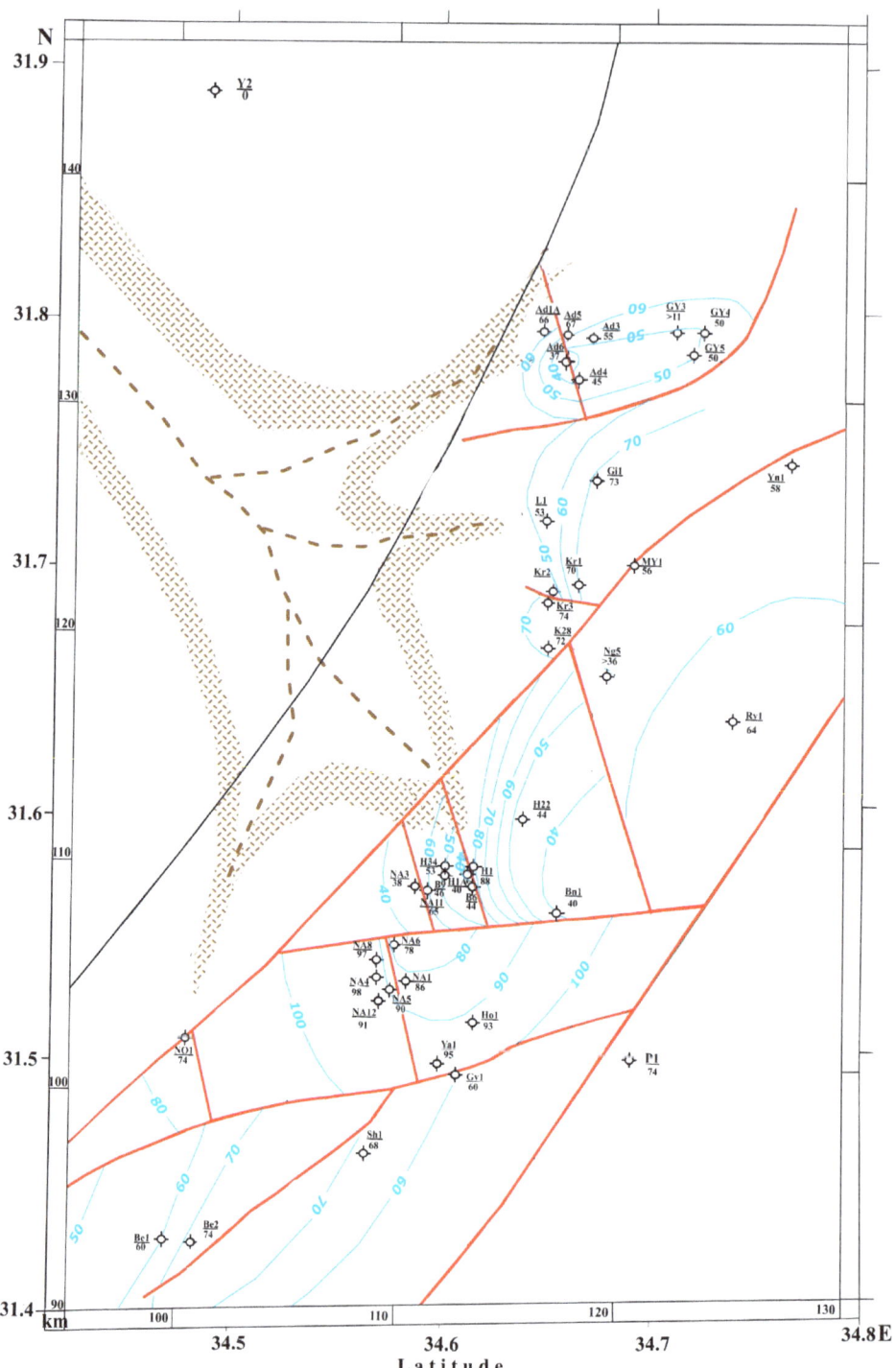

Figure 12. A map of the thicknesses of the Upper Bathonian (Karmon Formation) of the Heletz–Ashdod oil field. The brown color shows the erosional canyon, and blue lines are the isopachs. Borehole abbreviations: see captions for Figure 2.

Further, 10 km to the south in the Gevim-1 well, the thickness of the Triassic decreases to 208 m, and 20 km to the south-west in the section of the Bessor-1 well, the Triassic is almost completely eroded, and its thickness is 65 m. This situation of ascending Triassic movements The Heletz terrane is geodynamically paradoxical against the background of the general downward spreading movements of the Neotethys Ocean and the block massifs of the thinned continental crust of the terrane belt involved in them.

4.7. Upper Permian Syn-Sedimentation Sequence

The terrigenous-carbonate formations of the upper Permian, which, according to the study of foraminifers, correspond to the Julfian and Median stages of the Mediterranean scale, intermittently overlie the Precambrian basement or the Early Permian alkaline mafic (quartz porphyry) Gevim volcanics [6]. This stage does not extend to the foreland of Gondwana and is developed only in the terrane belt formed in the marginal part of the Neotethys Ocean. The thickness of marine sediments in the Upper Permian, according to the data from the deepest wells in the northern part of the Negev terrane, reaches 580 m (Agur1A deep) and 555 m (Pleshet 1). In the Heletz terrane this thickness varies from similar values—510 m (Gevim 1) to 235 m (Bessor 1) and 40 m (Heletz 1A deep) [6,31].

Thus, the structural analysis of the pre-collision syn-sedimentation sequence showed that the most active geodynamic processes were characteristic of the final stages immediately preceding the Early Cretaceous collision of the MTB.

At the same time, the Heletz terrane and, within its limits, the Heletz triangle subterrane, which encloses the main oil field, Heletz, had the most significant contrast of motions. The structural analysis confirmed that the second important structure of the collision complex, inherited in the collision and post-collision stages, is the Ashdod uplift, which also contains a complex of developed oil fields.

4.8. Geophysical Data Analysis

For clarifying the structural and geodynamic characteristics of the study area, results of different geophysical mapping were used [7,13,14,17,29,34,37].

The structural heterogeneity of hydrocarbon deposits is due to both tectonic-sedimentary and facies, causing the depth and physical nature to differ. This has been identified in and around this area by a variety of geological and geophysical survey methods [6,7,12–14,17,29,30,32,34,38,49,51]. As a result of combined structural analysis of syn-sedimentation sequence and gravity-magnetic mapping, a tectonic-geophysical scheme of the junction zone of the Pleshet, Heletz, and Negev terranes was compiled (Figure 13).

The Negev terrane is represented only by the northeastern margin adjacent to the narrow Heletz terrane. However, according to the gravity field distribution, the northwestern limb of the terrane is very gentle and undifferentiated, which is relatively consistent with the geological data. The magnetic field isolines, in general, also do not indicate any significant structural inhomogeneity in the considered Negev terrane margin. The gravity-magnetic data in the Heletz terrane shows an entirely different picture. They clearly trace all the tectonic blocks identified here: the Yinnon, Yakhini, and Gevim uplifts interspersed with submerged blocks—the Heletz triangle and Beeri.

In the zone of tectonic contact of the Heletz and Pleshet terranes, there is a sharp change in the nature of the distribution of isolines of gravity-magnetic changes, indicating a significant structural homogeneity of the Pleshet terrane, in which, according to gravity data, the Talme Yafo flat uplift is distinguished. The magnetic survey data located across the strike of the elongated axis of the Talme Jafe Rise contrast very interestingly with this pattern. The isolines of the magnetic field, branching off from the Heletz triangle, pass along the northeastern flank of this brachifold, crossing the isolines of the gravitational field at an angle of about 90 degrees.

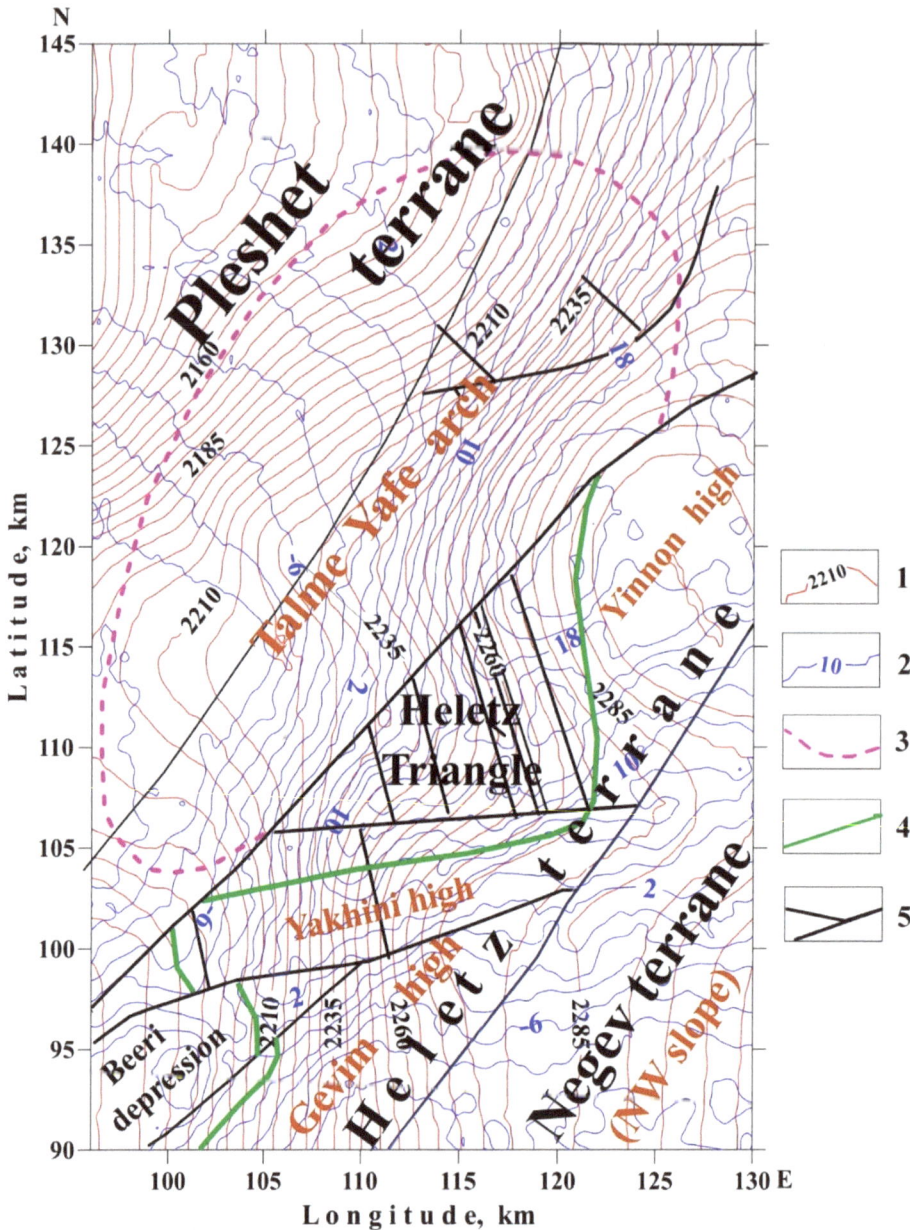

Figure 13. Combined gravity-magnetic map overlaid on the Late Jurassic-Early Cretaceous paleotectonic structure of the Heletz–Ashdod oil field and adjacent areas (revised after [7]). (1) Bouguer anomalies isolines (in mGal), (2) isolines of magnetic anomalies (in nT), (3) isopach 800 m of the Middle-Upper Jurassic rocks in the Pleshet terrane, (4) isopach 1300 m of the Middle-Upper Jurassic rocks in the Heletz terrane, (5) faults. The gravity and magnetic maps were constructed using the Gravity-Magnetic DB of Israel [37].

This situation is fully consistent with the Heletz erosional canyon map (Figure 7). It is very likely that the wide development of ferruginous minerals in the terrigenous sequence of the Lower Cretaceous canyon, revealed in the Heletz oil field [42], is even more developed than in the canyon itself in the Pleshet terrane. This factor could affect the nature of the distribution of magnetic field isolines.

According to thermal field analysis [34], the Curie surface depth is the most uplifted in the easternmost Mediterranean. This fact indicates that the Earth's crust under the Heletz structure is the more heated.

Seismic data analysis (preliminary interpretation was carried out using Multifocusing Technology [13]) also testifies the complex structure of the Heletz field. Of course, additional seismic studies are required here.

5. Conclusions

Thus, we can formulate the following conclusions:

1. For the first time, a comprehensive explanation has been proposed for understanding the nature of the Heletz regional structure for the Eastern Mediterranean.
2. The uniqueness of the Heletz uplift is also emphasized by the fact that only in this terrane and further on the Pleshet terrane adjacent to it from the west, an intense Early Cretaceous erosional canyon with hydrocarbon deposits was developed. It was not found on the surface of the Jurassic limestone both north and south of the Heletz region.
3. It is shown that the Heletz structure is geodynamically the most unstable part of the MTB, with a crystalline basement that differs from other terranes and is composed of metamorphic greenschist facies.
4. The geodynamic aspect of the Heletz uplift is considered a structure associated with the presence of industrial oil fields. For this purpose, a particular layer-by-layer tectonic analysis with the syn-sedimentary controls was applied in combination with constructing structural-geomorphological, paleo-geological, and tectonic-geophysical maps.

Author Contributions: Conceptualization, L.E. and Y.K.; methodology, L.E. and Y.K..; software, L.E.; validation, L.E. and Y.K.; formal analysis, L.E. and Y.K.; investigation, L.E. and Y.K.; resources, L.E. and Y.K.; data curation, L.E. and Y.K.; writing—original draft preparation, L.E. and Y.K.; writing—review and editing, L.E.; visualization, L.E. and Y.K.; project administration, L.E. All authors have read and agreed to the published version of the manuscript.

Funding: This research received no external funding.

Data Availability Statement: The data are not publicly available since they are partly from two databases for private use only.

Acknowledgments: The authors would like to thank three anonymous reviewers, who thoroughly reviewed the manuscript, and their valuable suggestions were helpful in preparing the final version of this paper.

Conflicts of Interest: The authors declare no conflict of interest.

References

1. Tschopp, H.J. The Oilfind of Heletz, Israel. *Bull. Swiss Ass. Pet.-Geol. Eng.* **1956**, *22*, 41–54.
2. Picard, L. Geology and oil exploration of Israel. In *5th World Petroleum Congress*; OnePetro: New York, NY, USA, 1959; pp. 311–336.
3. Cohen, Z. Early Cretaceous buried canyon: Influence on accumulation of hydrous in Helez Oil Field, Israel. *Am. Assoc. Petrol. Geol. Bull.* **1976**, *60*, 108–114.
4. Bein, A.; Sofer, Z. Origin of oils in the Heletz region, Israel–implication for exploration in the eastern Mediterranean. *Am. Assoc. Petrol. Geol. Bull.* **1987**, *71*, 65–75.
5. Gilboa, Y.; Fligelman, H.; Derin, B. Heletz-Brur-Kokhav field–Israel. In *Treatise of Petroleum Geology, Atlas of Oil and Gas Fields, Structural Traps*; Beaumont, E.A., Foster, N.H., Eds.; AAPG: Tulsa, OK, USA, 1990; Volume IV, pp. 319–341.
6. Hall, J.K.; Krasheninnikov, V.A.; Hirsch, F.; Benjamini, C.; Flexer, A. (Eds.) Volume II: The Levantine Basin and Israel. In *Geological Framework of the Levant*; Historical Productions-Hall: Jerusalem, Israel, 2005.
7. Eppelbaum, L.; Katz, Y. Tectonic-Geophysical Mapping of Israel and eastern Mediterranean: Implication for Hydrocarbon Prospecting. *Positioning* **2011**, *2*, 36–54. [CrossRef]

8. Gardosh, M.A.; Tannenbaum, E. The petroleum systems of Israel. In *Petroleum Systems of the Tethyan Region*; Marlow, L., Kendall, C., Yose, L., Eds.; AAPG Memoir: Tulsa, OK, USA, 2014; Volume 106, pp. 179–216.
9. Amit, O. Organochemical evaluation of the Gevar'am shales (Lower Cretaceous), Israel, as possible oil source rocks. *Am. Assoc. Petrol. Geol. Bull.* **1978**, *62*, 827–836.
10. Cohen, Z. The Geology of the Lower Cretaceous in the Heletz Field, Israel. Ph.D. Thesis, Hebrew University, Jerusalem, Israel, 1971. (in Hebrew, English Abstr.).
11. Gavrieli, I.; Starinsky, A.; Spiro, B.; Aizenshtat, Z.; Nielsen, H. Mechanisms of sulfate removal from subsurface calcium chloride brines: Heletz-Kokhav oilfields, Israel. *Geochim. Et Cosmochim. Acta* **1995**, *59*, 3525–3533. [CrossRef]
12. Eppelbaum, L.V.; Katz, Y.I.; Ben-Avraham, Z. Israel–Petroleum Geology and Prospective Provinces. *AAPG Eur. Newsl.* **2012**, *4*, 4–9.
13. Berkovitch, A.; Binkin, I.; Eppelbaum, L.; Scharff, N.; Guberman, E. Integration of advanced multifocusing seismics with potential field analysis: Heletz oil field (central Israel) example. *J. Balk. Geophys. Soc.* **2005**, *8*, 593–596. [CrossRef]
14. Ben-Avraham, Z.; Ginzburg AMakris, J.; Eppelbaum, L. Crustal structure of the Levant Basin, eastern Mediterranean. *Tectonophysics* **2002**, *346*, 23–43. [CrossRef]
15. Ben-Avraham, Z.; Schattner, U.; Lazar, M.; Hall, J.K.; Ben-Gai, Y.; Neev, D.; Reshef, M. Segmentation of the Levant continental margin, eastern Mediterranean. *Tectonics* **2006**, *25*, 1–17. [CrossRef]
16. Eppelbaum, L.V.; Katz, Y.; Ben-Avraham, Z. New data of the geodynamic evolution of the eastern Mediterranean. In Proceedings of the AAPG Conference Exploration and Development of Siliciclastic and Carbonate Reservoirs in the Eastern Mediterranean, Tel Aviv, Israel, 26–27 February 2019.
17. Eppelbaum, L.V.; Katz, Y.I. Eastern Mediterranean: Combined geological-geophysical zonation and paleogeodynamics of the Mesozoic and Cenozoic structural-sedimentation stages. *Mar. Pet. Geol.* **2015**, *65*, 198–216. [CrossRef]
18. Eppelbaum, L.V.; Ben-Avraham, Z.; Katz, Y.; Cloetingh, S.; Kaban, M. Combined Multifactor Evidence of a Giant Lower-Mantle Ring Structure below the Eastern Mediterranean. *Positioning* **2020**, *11*, 11–32. [CrossRef]
19. Eppelbaum, L.V.; Ben-Avraham, Z.; Katz, Y.; Cloetingh, S.; Kaban, M. Giant quasi-ring mantle structure in the African-Arabian junction: Results derived from the geological-geophysical data integration. *Geotectonics* **2021**, *55*, 67–93. [CrossRef]
20. Ben-Avraham, Z. The structure and tectonic setting of the Levant continental margin, Eastern Mediterranean. *Tectonophysics* **1978**, *46*, 313–331. [CrossRef]
21. Gvirtzman, G.; Weissbrod, T. The Hercynian geanticline of Heletz and the late Palaeozoic history of the Levant. In *The Geological Evolution of the Eastern Mediterranean*; Dixon, J.E., Robertson, A.H.F., Eds.; Special Publications: London, UK, 1984; pp. 177–186.
22. Gvirtzman, G.; Weissbrod, T.; Ginzburg, A. The Paleozoic Riyadh Geanticline in central Saudi Arabia: A structure similar to the Helez Geanticline. *Trans. Isr. Geol. Soc. Ann. Meet.* **1988**, 46–47.
23. Arkell, W.J. *Jurassic Geology of the World*; Olivier and Boyd: London, UK, 1956; p. 808.
24. Behr, M.A.; Ginzburg, A. Refraction survey in the Heletz-Negba area, Israel. In *Seismic Refraction Prospecting, Society of Exploration Geophysicists*; Musgrave, A.V., Ed.; Society of Exploration Geophysicists: Tulsa, OK, USA, 1967; pp. 505–521.
25. Lang, B.; Mimran, Y.A. An Early Cretaceous volcanic sequence in central Israel and its significance to the absolute date of the base of the Cretaceous. *J. Geol.* **1985**, *93*, 179–184. [CrossRef]
26. Levitte, D.; Olshina, A. *Isotherm and Geothermal Gradient Maps of Israel*; Geological Survey of Israel: Jerusalem, Israel, 1985.
27. Hirsch, F.; Picard, L. The Jurassic facies in the Levant. *J. Pet. Geol.* **1988**, *11*, 277–308. [CrossRef]
28. Lang, B.; Steinitz, G. K–Ar dating of Mesozoic magmatic rocks in Israel: A review. *Isr. J. Earth Sci.* **1989**, *38*, 89–103.
29. Cohen, Z.; Kaptsan, V.; Flexer, A. The Tectonic Mosaic of the Southern Levant: Implications for Hydrocarbon Prospects. *J. Pet. Geol.* **1990**, *13*, 437–462. [CrossRef]
30. Almogi-Labin, A.; Bein, A.; Sass, E. Late Cretaceous upwelling system along the southern Tethys margin (Israel): Interrelationship between productivity, bottom water environments and organic matter preservation. *Paleoceanography* **1993**, *8*, 671–690. [CrossRef]
31. Fleischer, L.; Varshavsky, A. *A Lithostratigraphic Data Base of Oil and Gas Wells Drilled in Israel, 2002–2014*; Ministry of National Infrastructure of Israel: Jerusalem, Israel, 2015.
32. Sneh, A.; Bartov, Y.; Rozensaft, M. Geological Map of Israel, Scale 1:200,000. In *Geological Survey of Israel*; Ministry of National Infrastructure: Jerusalem, Israel, 1998.
33. Krasheninnikov, V.A.; Hall, J.K.; Hirsch, F.; Benjamini, H.; Flexer, A. Volume 1: Cyprus and Syria. In *Geological Framework of the Levant*; Historical Productions-Hall: Jerusalem, Israel, 2005.
34. Eppelbaum, L.V.; Pilchin, A.N. Methodology of Curie discontinuity map development for regions with low thermal characteristics: An example from Israel. *Earth Planet. Sci. Lett.* **2006**, *243*, 536–551. [CrossRef]
35. Steinberg, J.; Gvirtzman, Z.; Gvirtzman, H.; Ben-Gai, Y. Late Tertiary faulting along the coastal plain of Israel. *Tectonics* **2008**, *27*, 1–22. [CrossRef]
36. Segev, A. ^{40}Ar/^{39}Ar and K-Ar geochronology of Berriasian-Hauterivian and Cenomanian tectonomagmatic events in northern Israel: Implications for regional stratigraphy. *Cretac. Res.* **2009**, *30*, 810–828. [CrossRef]
37. Institute of Geophysics of Israel. *Gravity-Magnetic Data Base of Israel*; Institute of Geophysics of Israel: Lod, Israel, 2010.
38. Eppelbaum, L.V.; Katz, Y.I. Mineral deposits in Israel: A contemporary view. In *Israel: Social, Economic and Political Developments*; Ya'ari, A., Zahavi, E.D., Eds.; Nova Science Publishers: Hauppauge, NY, USA, 2012; pp. 1–41.

39. Eppelbaum, L.; Katz, Y. Akchagylian Hydrospheric Phenomenon in Aspects of Deep Geodynamics. *Stratigr. Sediment. Oil-Gas Basins* **2021**, *2*, 8–26.
40. Eppelbaum, L.V.; Katz, Y.I. Paleomagnetic-geodynamic mapping of the transition zone from ocean to the continent: A review. *Appl. Sci.* **2022**, *12*, 5419. [CrossRef]
41. Katzir, Y.; Levin, A.; Calvo, R. Characterization of the Jurassic reservoir potential in the Eastern Levant margin using reconstruction of the depositional environment and the diagenetic history. In *The Geological Survey of Israel, Rep. MONI-ES-26-16*; Historical Productions-Hall: Jerusalem, Israel, 2016.
42. Edlmann, K.; Niemi, A.; Bensabat, J.; Haszeldin, R.S.; McDermott, C.I. Mineralogical properties of the caprock and reservoir sandstone of the Heletz field scale experimental CO_2 injection site, Israel; and their initial sensitivity to CO_2 injection. *Int. J. Greenh. Gas Control.* **2016**, *48*, 94–104. [CrossRef]
43. Ben-Avraham, Z.; Ginzburg, A. Displaced terranes and crustal evolution of the Levant and the eastern Mediterranean. *Tectonics* **1990**, *9*, 613–622. [CrossRef]
44. Eppelbaum, L.; Katz Yu Klokochnik, J.; Kosteletsky, J.; Zheludev, V.; Ben-Avraham, Z. Tectonic Insights into the Arabian-African Region inferred from a Comprehensive Examination of Satellite Gravity Big Data. *Glob. Planet. Chang.* **2018**, *171*, 65–87. [CrossRef]
45. Waples, D.W. *Geochemistry in Petroleum Exploration*; Springer: Dordrecht, The Netherlands, 1985; p. 232.
46. Oil and Natural Gas Exploration and Production in Israel. 2022. Available online: https://www.energy-sea.gov.il/English-Site/Pages/Oil%20And%20Gas%20in%20Israel/History-of-Oil--Gas-Exploration-and-Production-in-Israel.aspx (accessed on 27 December 2022).
47. Garfunkel, Z.; Derin, B. Reevaluation of the latest Jurassic-Early Cretaceous history of the Negev and the role of magmatic activity. *Isr. J. Earth Sci.* **1988**, *37*, 43–52.
48. Garfunkel, Z. Tectonic setting of Phanerozoic magmatism in Israel. *Isr. J. Earth Sci.* **1989**, *38*, 51–74.
49. Segev, A. Synchronous magmatic cycles during the fragmentation of Gondwana: Radiometric ages from the Levant and other provinces. *Tectonophysics* **2000**, *325*, 257–277. [CrossRef]
50. Katz, Y.I.; Eppelbaum, L.V. Levantine phase of tectonic-thermal activity in the Eastern Mediterranean. *Trans. Ann. Meet. Geol. Soc. Am. Sect Planet. Geol.* **1999**, *31*, A119.
51. Eppelbaum, L.V.; Katz, Y.I. Paleomagnetic Mapping in Various Areas of the Easternmost Mediterranean Based on an Integrated Geological-Geophysical Analysis. In *New Developments in Paleomagnetism Research*; Ser: Earth Sciences in the 21st Century; Eppelbaum, L., Ed.; Nova Science Publisher: Hauppauge, NY, USA, 2015; pp. 15–52.
52. Katz, Y.I. Cretaceous Thalassocratic Maximum and Planetary Movements of the Hydrosphere. In *Cretaceous Period. Paleogeography and Paleooceanology*; Doklady Earth Sciences; Naidin, D.P., Ed.; Nauka: Moscow, Russia, 1986; pp. 191–237. (in Russian)
53. Eppelbaum, L.V.; Vaksman, V.L.; Kouznetsov, S.V.; Sazonova, L.M.; Smirnov, S.A.; Surkov, A.V.; Bezlepkin, B.; Katz, Y.; Korotaeva, N.N.; Belovitskaya, G. Discovering of microdiamonds and minerals-satellites in Canyon Makhtesh Ramon (Negev desert, Israel). *Dokl. Earth Sci.* **2006**, *407*, 202–204. [CrossRef]

Disclaimer/Publisher's Note: The statements, opinions and data contained in all publications are solely those of the individual author(s) and contributor(s) and not of MDPI and/or the editor(s). MDPI and/or the editor(s) disclaim responsibility for any injury to people or property resulting from any ideas, methods, instructions or products referred to in the content.

Article

Shale Cuttings Addition to Wellbore Cement and Their Effect on Unconfined Compressive Strength

Alexandra Cedola * and Runar Nygaard

Mewbourne School of Petroleum and Geological Engineering, University of Oklahoma, Norman, OK 73019, USA; runar.nygaard@ou.edu
* Correspondence: aced@ou.edu

Abstract: Mitigation of greenhouse gas emissions is becoming a significant factor in all industries. Cement manufacturing is one of the industries responsible for greenhouse gas emissions, specifically carbon dioxide emissions. Pozzolanic materials have long been used as cement additives due to the pozzolanic reaction that occurs when hydrated and the formation a cementitious material similar to that of cement. In this study, shale, which is a common component found in wellbore drill cuttings, was used in various sizes and quantities to determine the effect it had on the mechanical properties of wellbore cement. The unconfined compressive strength of the cement containing shale was compared to the cement without shale to observe the effect that both the quantity and particle size had on this property. SEM–EDS microscopy was also performed to understand any notable variations in the cement microstructure or composition. The samples containing micron shale appeared to have the best results of all the samples containing shale, and some of the samples had a higher UCS than one or more of the base case samples. Utilization of cuttings as a cement additive is not just beneficial in that it minimizes the need for cuttings removal and recycling, but also in that it reduces the amount of greenhouse gas emissions associated with cement manufacturing.

Keywords: oil well cement; geologic additives; drill cuttings; mechanical properties; SEM; EDS

1. Introduction

Wellbore cement is the most used barrier material in hydrocarbon, carbon capture and storage, geothermal, and hydrocarbon wellbores, primarily due to the ability to manipulate the cement properties to adhere to the conditions at hand. For cement to mitigate hydrocarbon leakage, provide zonal isolation, and support the casing, the cement sheath must be able to withstand the overburden stress, remain intact during any subsequent wellbore operations, have the ability to securely bond to both the casing and formation, and be resistant to hydrocarbon migration. The cement must also be designed so that when it is in the liquid phase and being pumped downhole, it will be properly placed, remove any residual drilling mud, exhibit optimal rheological properties, and set in the appropriate period of time [1]. Cementitious materials are inflexible and susceptible to failures such as cracking and shrinking [2]. To enhance the mechanical properties of cement and prevent short- and long-term well integrity issues, additives are mixed with the cement slurry to procure the desired properties.

Additives can be divided into seven primary categories pertaining to the effect that they have on cement: densifiers, accelerators, retarders, viscosifiers, density reducers, friction reducers, and fluid loss prevention. Oftentimes, an additive will fall into more than one of these categories which makes finding a combination of additives to achieve each desired property difficult. Additive selection must take into consideration the pressure, temperature, chemical composition of the formation, mud type, and the presence of formation water, oil, and gas as well as the compatibility with other additives [3]. While the list of additive material that has been used to modify cement mechanical properties is

Citation: Cedola, A.; Nygaard, R. Shale Cuttings Addition to Wellbore Cement and Their Effect on Unconfined Compressive Strength. *Energies* **2023**, *16*, 4727. https://doi.org/10.3390/en16124727

Academic Editors: Kai Wang, Jie Wu, Yongjun Deng and Lin Chen

Received: 13 April 2023
Revised: 6 June 2023
Accepted: 9 June 2023
Published: 15 June 2023

Copyright: © 2023 by the authors. Licensee MDPI, Basel, Switzerland. This article is an open access article distributed under the terms and conditions of the Creative Commons Attribution (CC BY) license (https://creativecommons.org/licenses/by/4.0/).

ever-growing, some common additives are nanoparticles, cellulose material, polymers, and pozzolanic material.

Nanoparticles are often used to enhance the mechanical properties of cement because they have large surface areas and are thus able to be more reactive with surrounding cementitious material [4,5]. Nanoparticles used in wellbore cements include, but are not limited to, silica nanoparticles, magnesium oxide nanoparticles, alumina nanoparticles, iron nanoparticles, carbon nanotubes, and magnetic nanoparticles [6]. While nanoparticles have significant promise for enhancing cement mechanical properties, they are often expensive and significantly increase the cost of cementing [7].

Cellulose materials are often used in cement slurries to prevent fluid loss but have proved to be unreliable in both high- and low-temperatures as well as in areas with a high salt concentration [8]. Examples of cellulose materials commonly used in wellbore cement include hydroxyethyl cellulose (HEC), methyl hydroxyethyl cellulose (MHEC), and carboxymethyl hydroxyethyl cellulose (CMHEC), which when hydrated, immediately increase the slurry viscosity [9]. Cellulose materials can prove problematic in that this increased viscosity can lead to issues with pumping the slurry.

Polymers are often added to cement to enhance the elastic properties [10]. Encapsulation of polymers has also been investigated and it was found that such additives beneficially impact the cement mechanical properties and bonding, but due to the complex technique required to formulate such additives may prove costly [11]. The use of shape memory polymers (SMPs), which is a type of polymer that is composed of low-molecular weight pre-polymers and crosslinking agents, can alter its shape when stimulated [12–14]. The addition of SMPs would allow for a less brittle and more ductile cement which could prove beneficial for maintaining wellbore integrity for various aspects within the life of a well [15]. One of the drawbacks of using SMPs as sealing material is that they can be impacted by temperature fluctuations [16]. Another downside to this additive material is that little testing has been undertaken to determine how SMPs behave under downhole conditions [17]. Polymers, while improving elasticity in the short-term, could negatively impact the long-term well integrity in that the bond between the cement and polymers is often weak and the polymers could degrade under elevated temperatures and pressures [10].

There are a wide variety of pozzalanic materials that have been used as wellbore cement additives. Pozzolanic material works when the amorphous silica within the pozzolans react with calcium hydroxide that is formed during cement hydration [18]. This reaction leads to an increase in the compressive strength and durability by filling the effective pore space, thus, reducing the permeability, enhancing the calcium-silicate-hydrate (C-S-H) phases associated with the pozzolanic reaction, resulting in higher amounts of inert pozzolanic minerals, and increased nucleation sites for C-S-H formation [19,20]. Common pozzolanic materials used as wellbore cement additives are fly ash, metakaolin, blast furnace slag, and glass microspheres. Fly ash, metakaolin, and blast furnace slag are biproducts of industrial manufacturing processes and are readily available. The addition of fly ash created from the combustion of coal to cement has been shown to increase the compressive strength [21]. Metakaolin is made from calcined kaolin clay and when added to cement, decreases the porosity and permeability, minimizes shrinkage, and minimizes chemical degradation [22]. Blast furnace slag formed during the manufacturing of iron has been shown to aid in preventing gas migration, setting time, and bonding [23,24]. Glass microspheres are a type of pozzalanic material that is used to lower the density of cement but have many drawbacks such as their cost, separation tendencies, and the tendency to be crushed under pressure [25–27].

While the use of pozzolanic materials has proven beneficial for a variety of reasons, one of the key aspects that make them so desirable is the potential to minimize the amount of cement needed. Cement manufacturing is one of the largest producers of greenhouse gases, specifically, carbon dioxide (CO_2). Using pozzolanic material as a cement additive would reduce the amount of cement used in the petroleum industry which could lead to a significant decrease in CO_2 emissions.

While the aforementioned pozzolanic materials have been and continue to be investigated as cement additives, there is one type of pozzolanic material that is readily available in the petroleum industry yet has not been investigated: drill cuttings. Drill cuttings are fragments of formation coated in drilling mud that are brought to the surface during wellbore drilling operations. Cuttings disposal is dictated by the type of drilling fluid used and the disposal methods vary by geographic location.

The purpose of this research was to investigate the feasibility of using cuttings as a cement additive and to understand the effect on the cement's mechanical properties. Various particle sizes and quantities of shale were added to class H cement to understand how compressive strength was altered with the novel geologic additive. Woodford shale has a highly similar chemical composition to other pozzolanic additives, specifically class F fly ash and metakaolin, and its similarities can be seen in Figure 1 [28,29]. Class F fly ash and metakaolin have both been added to slurries that are pumped in the field, indicating the promise of shale additive potential.

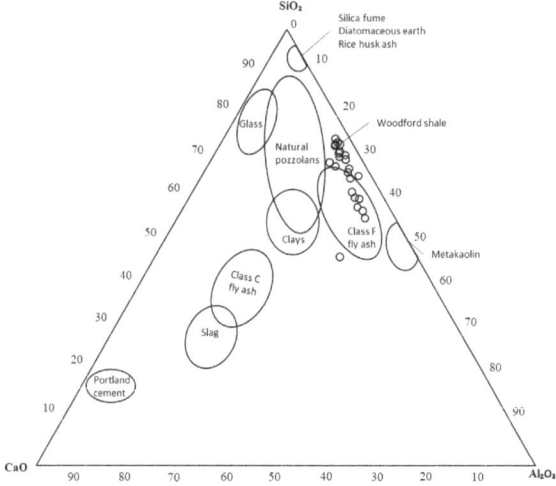

Figure 1. Ternary diagram showing various pozzolanic material chemical composition in comparison to Woodford shale; figure redrawn using information from [28] to include additional data obtained from [29].

While the purpose of this study was solely for feasibility purposes, it could prove to have significant beneficial effects if the work is furthered and various other properties of such slurries are tested and better understood. Annually, roughly 4.1 billion tons of cement is produced globally and is accountable for approximately 7.1% of global CO_2 emissions [30–32]. The use of cuttings within cement could lead to a reduction in the amount of cement used in wellbores, thus, reducing the amount of CO_2 emissions as well as decreasing the cost needed to recycle drill cuttings and any associated emissions.

2. Materials and Methods

To understand the feasibility of using cuttings as a cement additive, various sizes and quantities of shale were added to the Central Plains Cement Company class H wellbore cement. The class H cement was from the Central Plains Cement Company, and from the mill sheet the presence of calcium, oxygen, and silica were expected to be the primary constituents of the matrix with smaller quantities of aluminum, iron, magnesium, and sulfur as well as various elements attributed to the insoluble residue as shown in Table 1 [33].

Table 1. Central Plains class H cement chemical compositions [33].

Component	Amount in Cement (%)
SiO_2	21.40
Al_2O_3	3.20
Fe_2O_3	5.30
CaO	63.40
MgO	1.70
SO_3	2.40
Loss on Ignition	0.90
Na Eq.	0.40
Insoluble Residue	0.28
Free Lime	0.60

Woodford shale obtained from an outcrop in Murray County, Oklahoma was collected and ground so that there were equal amounts of millimeter (2–4.7 mm), micron (74–210 μm), and submicron (<1 μm) particles. While the millimeter and micron size shale were able to be obtained using conventional grinding techniques with a ring and puck grinder and sieved using a 10-mesh sieve, a 200-mesh sieve, and a Retsch sieve shaker, the submicron samples were obtained using a Retsch EMAX high energy ball mill and the particle size was verified using SEM techniques.

To make the samples, shale was added in 5%, 10%, 15%, 25%, 50%, and 75% quantities to a slurry composed of Central Plains class H cement, a defoamer, and deionized (DI) water. To ensure mixability, the density of each slurry was calculated to be 16.49 ppg and was obtained by changing the water content depending upon the amount of shale added to the system. To mix the slurry, a Chandler 3260 Constant Speed Mixer was used and mixing followed the American Petroleum Institute (API) 10 RB specifications [34]. The slurry was then poured into greased 1.5" by 4.0" stainless steel molds and allowed to set for 24 h. After this time, the samples were demolded and submerged in a sodium hydroxide (NaOH)-DI brine in a 150 °F oven for 28 days. After 28 days, the samples were removed from the oven, weighed wet, and placed in a 150 °F vacuum oven for 24 h, then removed, weighted dry and cut/ground to ensure both ends of the sample were smooth and free from any abnormalities, as shown in Table 2. Uniformity was confirmed using a caliper and rotating in at three locations with 90° spacing and the values were all within one-hundredth of a millimeter.

Table 2. Specimen density measurements.

Particle Size	% Shale	Avg. Dry Density (10^{-3} kg/m^3)	S. Dev. (10^{-3} kg/m^3)
Base	0	1.82	0.04
Millimeter	5	1.70	0.04
	10	1.78	0.01
	15	1.72	0.01
	25	1.69	0.05
	50	1.40	0.04
	75	1.63	0.05
Micron	5	1.78	0.03
	10	1.62	0.02
	15	1.57	0.02
	25	1.59	0.02
	50	1.43	0.04
	75	1.37	0.02
Submicron	5	1.69	0.02
	10	1.65	0.04
	15	1.64	0.03
	25	1.55	0.04

UCS testing was performed using a New England Research Autolab-500 uniaxial load frame with a strain rate of 0.04 mm/min. Once testing was initiated, samples were tested until axial force rapidly decreased to the initial axial force value, indicating that the sample had failed and UCS was reached indicating that the test was concluded (Figure 2).

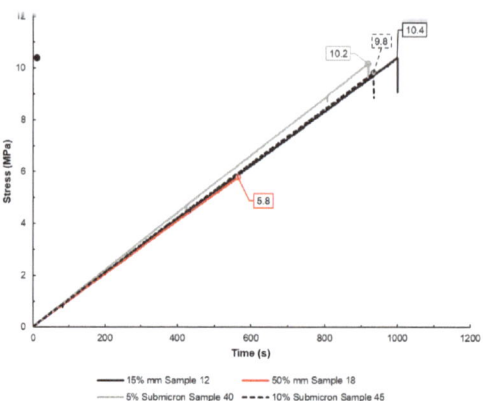

Figure 2. UCS test results for sample 12, 18, 40, and 45.

After the samples were tested, scanning electron microscopy (SEM) and energy dispersive X-ray spectroscopy (EDS) was performed on the millimeter and submicron sample sets that had the highest and lowest UCS values using a FEI Quanta 250 and Bruker XFlash 6130. Small pieces of the samples were sputter coated with gold and palladium using a Denton Vacuum Desk V Thin Film Deposition Solution. Once coated, SEM and EDS were performed on various locations at multiple magnifications on the sample surface.

3. Results

The mean UCS and standard deviation for all tests are given in Table 3 and shown in Figures 3–5. For the millimeter and micron sample sets, all samples were able to be tested regardless of the amount of shale. For the submicron sample set, the 50% and 75% samples broke during the curing process and could not be tested.

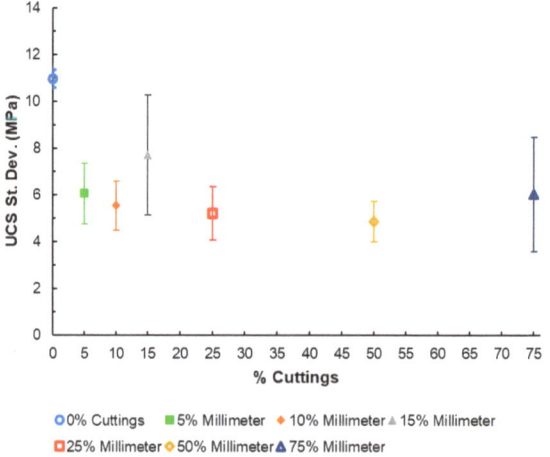

Figure 3. UCS and standard deviation results for the millimeter samples.

Table 3. UCS test results for the samples cured over 28 days.

Shale Size	% Shale	Sample Number	UCS (MPa)	Avg. UCS (MPa)	Std. Dev. (MPa)
Base	0%	1	10.52	10.97	0.38
		2	11.44		
		3	10.94		
Millimeter	5%	4	4.29	6.06	1.29
		5	7.33		
		6	6.57		
	10%	7	4.90	5.54	1.04
		8	4.71		
		9	7.01		
	15%	10	8.46	7.71	2.55
		11	4.28		
		12	10.39		
	25%	13	6.83	5.21	1.14
		14	4.34		
		15	4.47		
	50%	16	3.71	4.87	0.86
		17	5.13		
		18	5.77		
	75%	19	8.56	6.04	2.45
		20	2.72		
		21	6.85		
Micron	5%	22	12.06	9.79	2.02
		23	10.17		
		24	7.15		
	10%	25	10.83	8.85	2.61
		26	5.16		
		27	10.57		
	15%	28	11.60	9.26	2.55
		29	10.48		
		30	5.71		
	25%	31	10.93	7.96	2.25
		32	5.48		
		33	7.46		
	50%	34	5.63	5.75	0.64
		35	6.58		
		36	5.03		
	75%	37	7.29	6.07	1.11
		38	6.30		
		39	4.61		
Submicron	5%	40	10.16	9.58	0.63
		41	9.87		
		42	8.71		
	10%	43	5.14	7.00	2.03
		44	6.03		
		45	9.82		
	15%	46	6.00	6.08	0.90
		47	7.21		
		48	5.02		
	25%	49	6.76	4.88	1.34
		50	4.11		
		51	3.77		

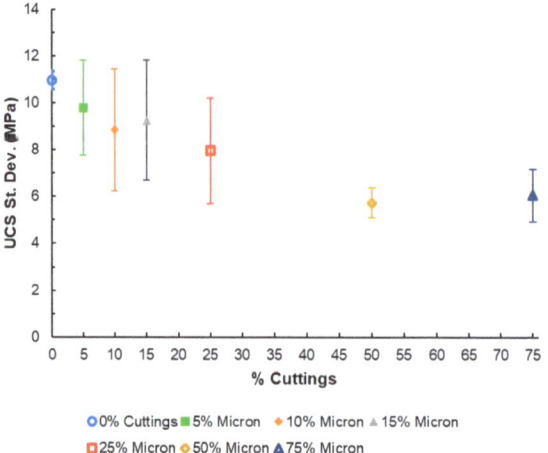

Figure 4. UCS and standard deviation results for the micron samples.

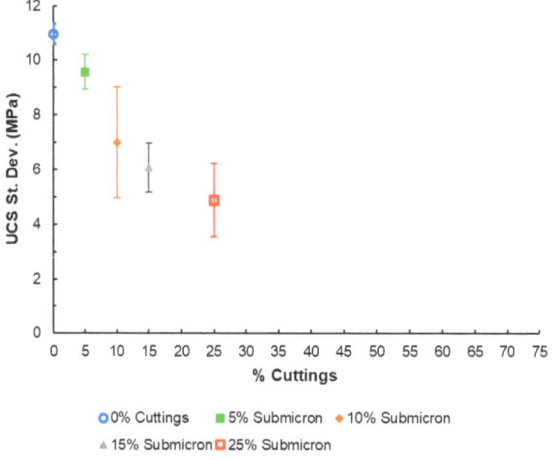

Figure 5. UCS and standard deviation results for the submicron samples.

Single-factor ANOVA tests were performed on the results in Table 3 to evaluate the statistical significance of the effect of shale size and quantity on UCS (Table A1).

The UCS test results fluctuated. From the UCS results, the base case achieved the highest average UCS of 10.97 MPa and the lowest standard deviation (Table 3). Adding the millimeter-sized cuttings lowered the UCS (Figure 3). For the millimeter sample sets, the 15% samples yielded the highest UCS while the 50%-millimeter shale samples had the lowest average UCS. From the 15%-millimeter sample set, sample 12 had the greatest UCS of 10.39 MPa, which was comparable to the base case results (Table 3). The standard deviation for the 15%-millimeter samples was the highest of all the millimeter sample sets with the difference between the highest and lowest samples being 6.11 MPa. Sample 11 achieved a UCS that was less than half of sample 12. The 50%-millimeter sample set had the lowest average UCS values but had the lowest standard deviation, indicating that the results for this sample set were fairly consistent. While the 75%-millimeter sample set had an average a UCS of 6.04 MPa, it also contained the sample with the lowest UCS of all the millimeter samples, sample 20, which had a UCS of 2.72 MPa. As shown in Figure 3, there was no trend in the reduction in strength with increasing cuttings volume.

Figure 4 shows that the micron-sized shale exhibited the most promising compressive strength properties. The samples containing the micron-sized shale appeared to achieve the highest strengths of all the samples containing shale. Of the micron sample set, the 5%, 10%, and 15% samples achieved the highest average UCS but also had high standard deviations. Sample 31, which contained 25%-micron shale, had a high UCS that was comparable to the samples containing lesser amounts of micron shale. The highest and lowest average UCS sample sets with micron shale addition were 5% and 50%, respectively. The 5%-micron sample set had an average UCS of 9.79 MPa with a standard deviation of 2.02 MPa. Sample 22 had the highest UCS of all the micron samples, 12.06 MPa, which was actually higher than any of the base case UCS values. Sample 28, which contained 15%-micron shale, had a UCS of 11.60 MPa, which was also higher than any of the base cases. The 50%-micron samples had both the lowest UCS and standard deviations, indicating that the samples all exhibited similar behavior and the UCS values showed little variation.

Figure 5 shows the submicron sample additives. The 5% and 10% submicron sample sets proved to have higher UCS values than those containing the same amounts of millimeter shale, yet the opposite was seen in the 15% and 25% sample sets with the same particle sizes. Of the submicron samples, the 5% samples had the highest average UCS of 9.58 MPa with a standard deviation of 0.63 MPa, while the 25% samples had the lowest average UCS of 4.88 MPa with a standard deviation of 1.34 MPa. The 10%-submicron samples had a high standard deviation primarily due to sample 45 having a significantly higher UCS than samples 43 and 44.

Figure 6 shows a comparison of the average UCS values for the base case, millimeter, micron, and submicron sample sets to the previously reported UCS values for class G and H cement [35].

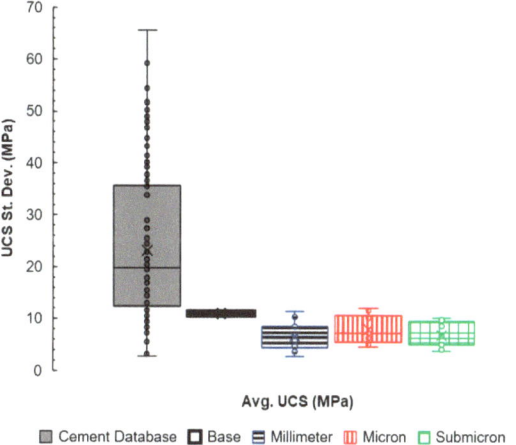

Figure 6. Comparison between various shale sizes on cement UCS.

SEM and EDS was performed on all the samples at low and high resolutions to understand the variations in both the chemical composition within the samples as well as the surface microstructure. Secondary electron and backscatter electron (BSE) images were created to analyze the sample topography and variations in the chemical composition, respectively. These dark areas were not wholly representative of the pores or voids within the sample. EDS was performed on a designated spot or a specified area on the surface of the sample to provide information about the sample at both a small scale and a macroscale. EDS was performed at low magnification at multiple locations to understand how the quantity of the shale could impact the overall chemical composition of the cement. Locations that appeared to contain shale particles were also selected for EDS analysis, to understand the compositional variations between the geologic material and cement. These locations were also observed as to whether any atypical reactions may have occurred at the cement–shale interface. SEM and EDS was performed on all of the 5%- and 10%-submicron samples as well as the 15%- and 50%-millimeter samples to investigate whether there were any differences in the microstructure, hydration, porosity, or chemical composition amongst the samples containing the same amounts of shale as well as the with the same particle sizes.

Areas and points were used in the EDS analyses; points were specifically selected when it appeared that the minerology differed. EDS analysis for sample 12 containing 15%-millimeter shale is shown in Figure 7.

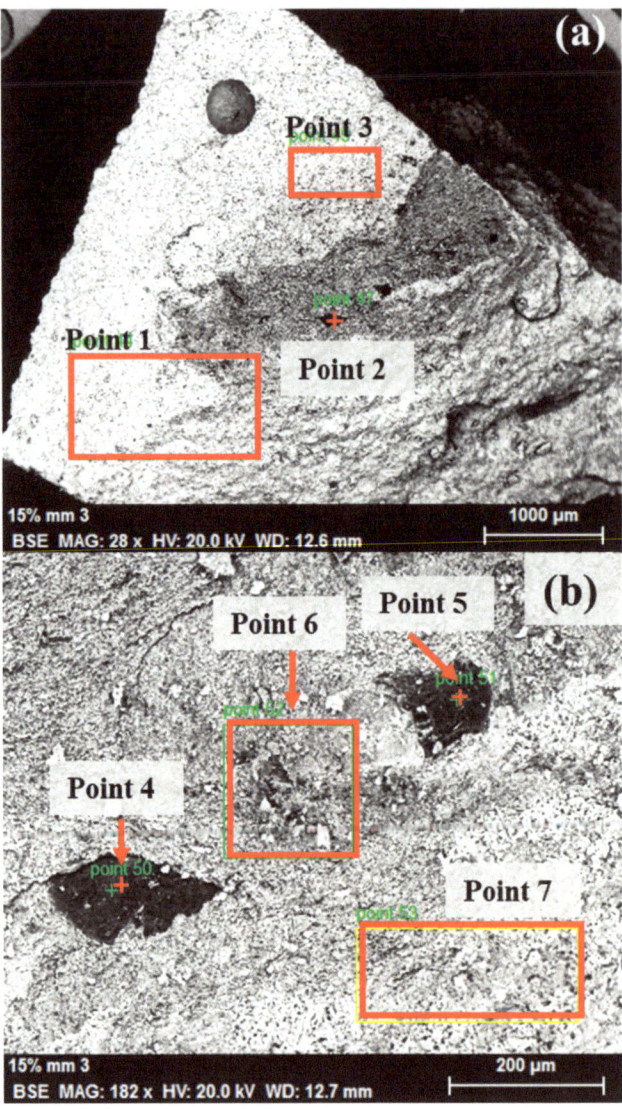

Figure 7. 15%-millimeter sample 12 EDS results at (**a**) 28× magnification; (**b**) 182× magnification.

SEM analysis taken at various surface locations on the 15%-millimeter sample 12 is shown in Figure 8.

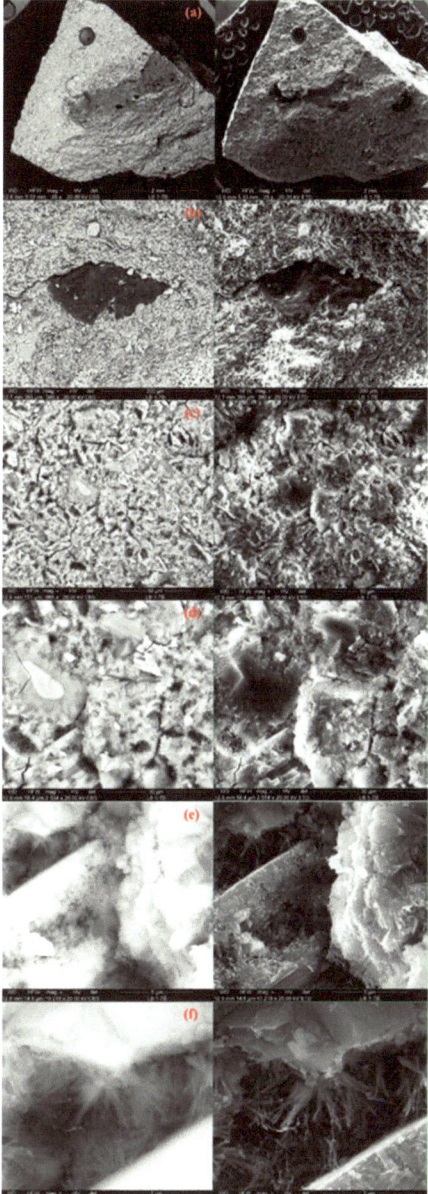

Figure 8. 15%-millimeter sample 12 at (**a**) 28× magnification; (**b**) 380× magnification; (**c**) 985× magnification; (**d**) 2554× magnification; (**e**) 10,218× magnification; (**f**) 31,516× magnification where the left images are the backscatter electron image and the right images are the secondary electron images.

EDS analysis to identify the elemental composition on the surface of the 50%-millimeter sample 18 is shown in Figure 9.

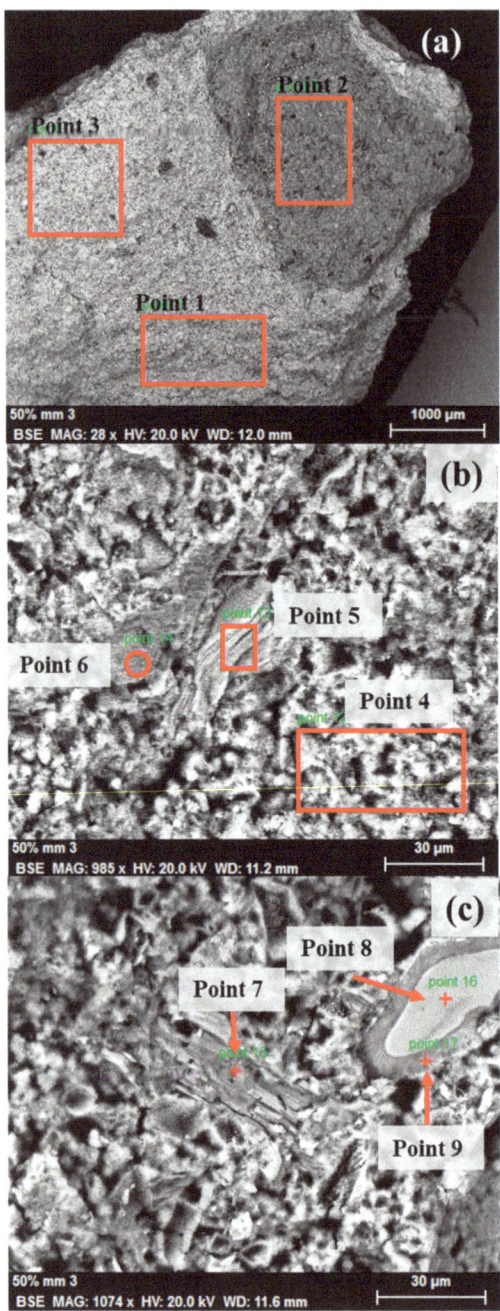

Figure 9. 50%-millimeter sample 18 EDS results at (**a**) 28× magnification; (**b**) 985× magnification; (**c**) 1074× magnification.

SEM analysis of the 50%-millimeter sample 18 is shown in Figure 10.

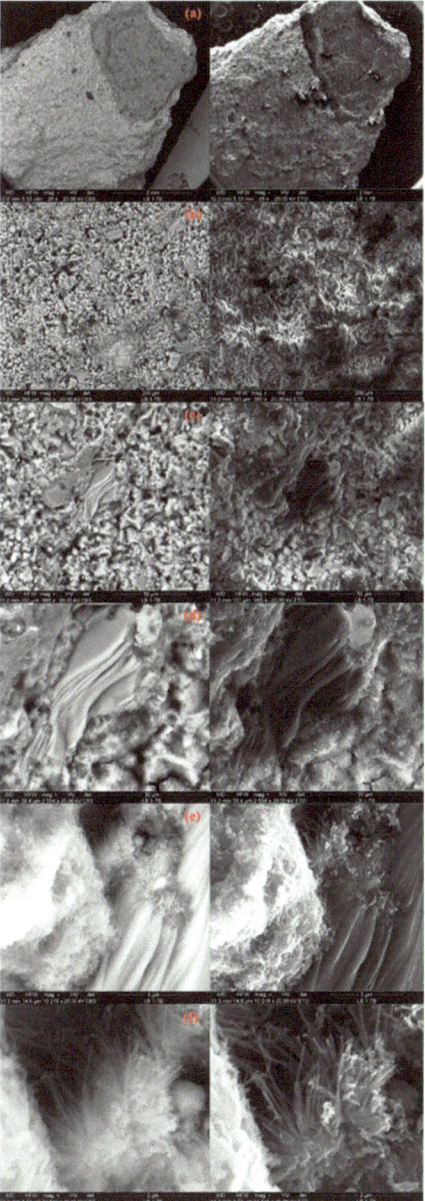

Figure 10. 50%-millimeter sample 18 at (**a**) 28× magnification; (**b**) 380× magnification; (**c**) 985× magnification; (**d**) 2554× magnification; (**e**) 10,218× magnification; (**f**) 31,516× magnification where the left images are the backscatter electron image and the right images are the secondary electron images.

EDS analysis to determine the compositional variations on the surface of the 5%-submicron sample 40 is shown in Figure 11.

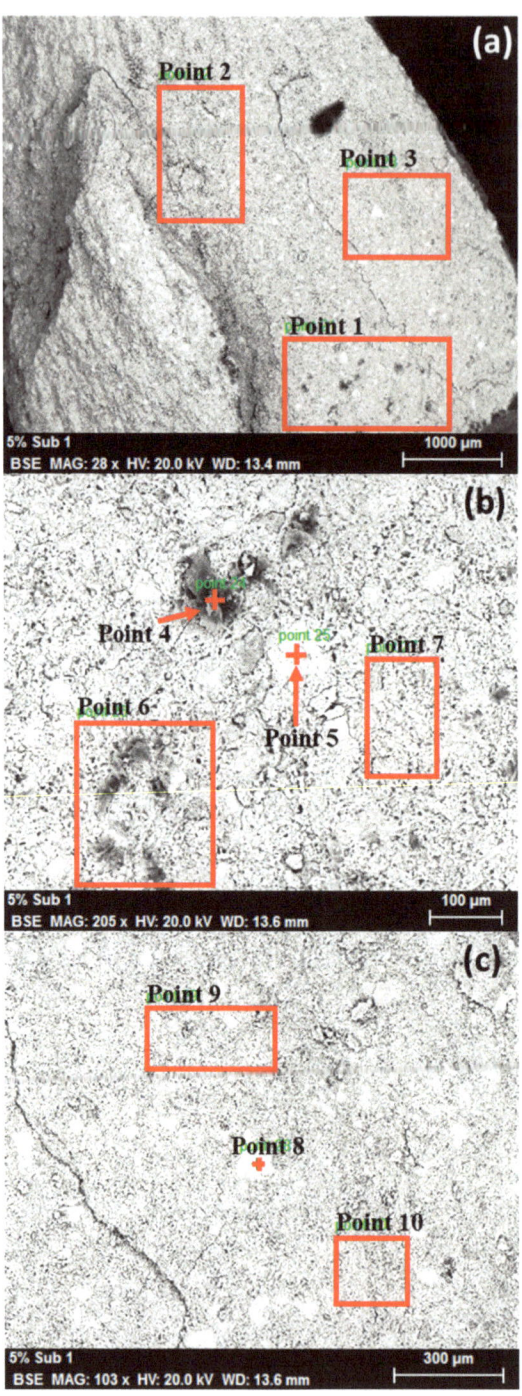

Figure 11. 5%-submicron sample 40 EDS results at (**a**) 28× magnification; (**b**) 209× magnification; (**c**) 103× magnification.

SEM analysis of samples 40 containing 5% submicron shale is shown in Figure 12.

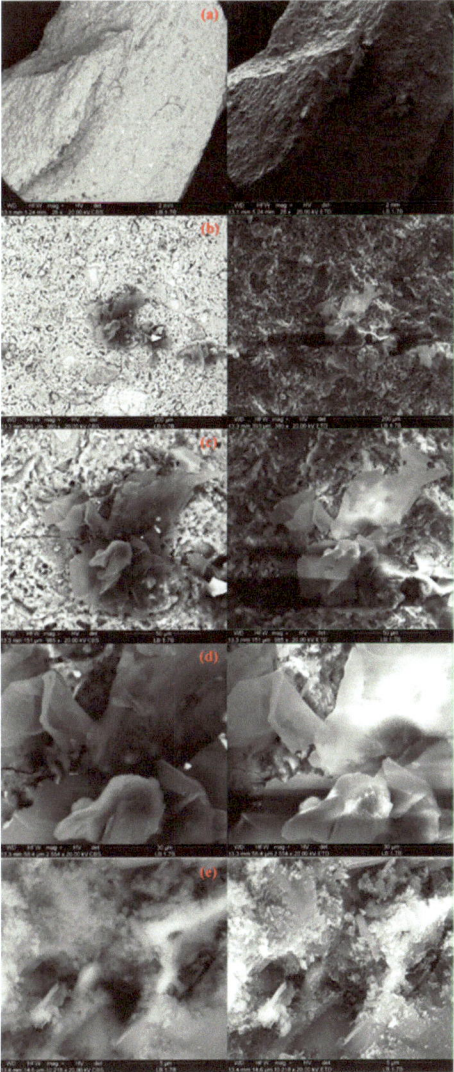

Figure 12. 5%-submicron sample 40 at (**a**) 28× magnification; (**b**) 380× magnification; (**c**) 985× magnification; (**d**) 2554× magnification; (**e**) 10,218× magnification where the left images are the backscatter electron images and the right images are the secondary electron images.

EDS analysis for the 10% submicron sample 45 is shown in Figure 13. EDS analysis was performed for the points taken at 28× magnification and 380× magnification.

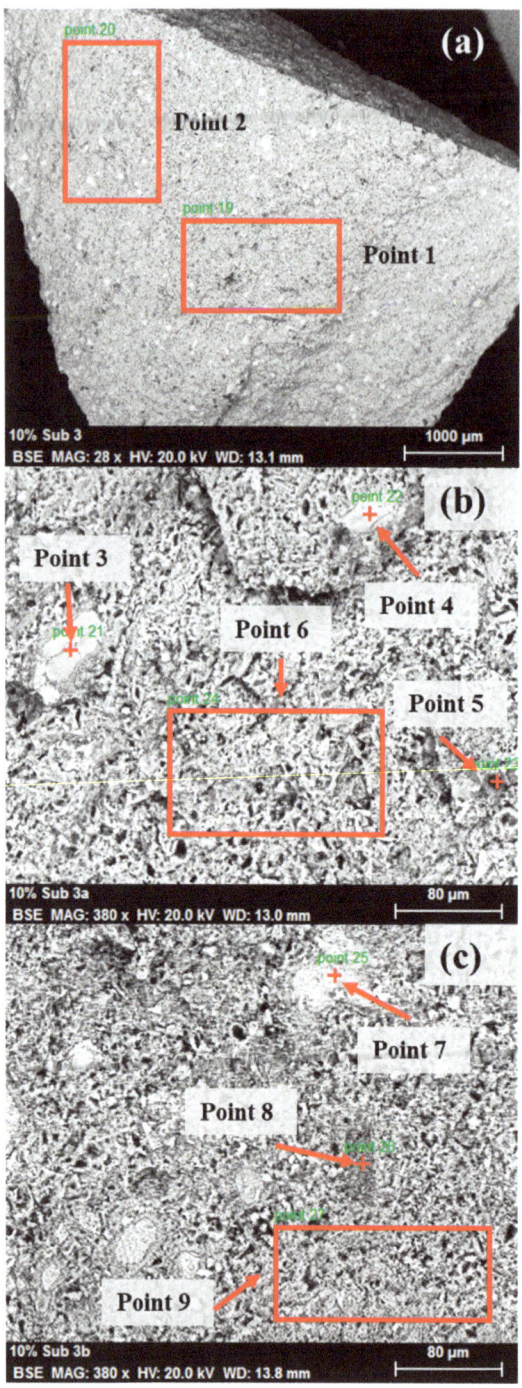

Figure 13. 10%-submicron sample 45 EDS results at (**a**) 28× magnification; (**b**) 380× magnification; (**c**) a separate location at 380× magnification.

SEM analysis of samples 45 containing 10% submicron shale is shown in Figure 14.

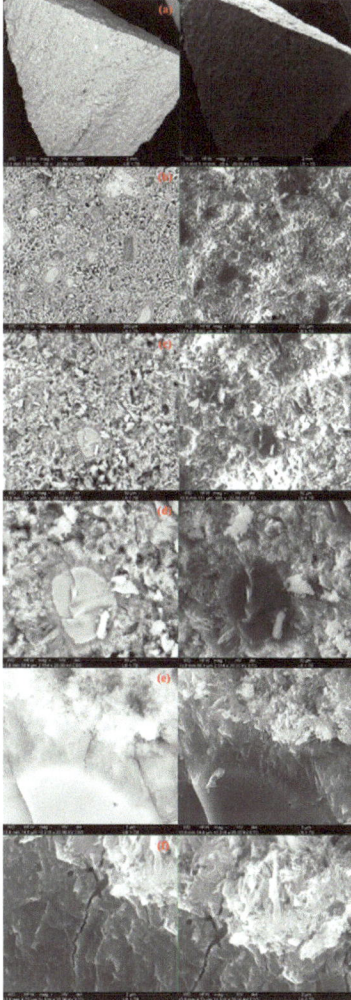

Figure 14. 10% submicron sample 45 at (**a**) 28× magnification; (**b**) 380× magnification; (**c**) 985× magnification; (**d**) 2554× magnification; (**e**) 10,218× magnification; (**f**) 31,516× magnification where the left images are the backscatter electron images and the right images are the secondary electron images.

4. Discussion

After the samples were crushed, a visual check was performed to determine whether the shale particles appeared to be evenly distributed within the samples. Visual inspection was not able to be carried out for the micron and submicron samples due to the particles being too small to see with the naked eye. For all the millimeter samples, particle distribution appeared homogenous; the shale did not settle to the bottom or float to the top and did not agglomerate within the sample and were evenly distributed throughout.

For all the millimeter sample sets, two samples were similar and one had a significantly higher or lower compressive strength. This indicated that the shale distribution varied between the samples yet upon visual inspection, this did not appear to be the case. From the results, the millimeter sample sets all achieved lower UCS values than the base case and the previously reported UCS values for class G and H cement. First, the addition of shale reduced the dry density of the samples (Table 2). These particles were much larger than the cement particle size and hindered the C-S-H nucleation and growth. This inhibition on hydration leads to weaker points within the cement where the cement is likely to experience failure. It also increases the porosity and permeability of the samples in that there is more free space where hydration products should be forming. In the SEM images of the 15%-shale sample 12, a large void and a number of large shale particles were apparent at low magnification. At higher magnification, the predominant microstructures present included ettringite needles and CH. Secondary electron images show that there were regions of the surface that appear to be slightly elevated. BSE images showed that there did appear to be some charging that occurred in sample 12, but from Figure 8d, it can be seen that the partially hydrated cement appeared darker in color than the surrounding structures. This was also true of the linear rod-like structures, which were CH. Sample 12 had the highest UCS of the three 15%-millimeter samples and showed that the shale was able to bond with the cement and the addition of shale to cement positively impacted compressive strength. The ettringite crystals seen in Figure 8f were similar in structure to those of a Portland cement [36]. EDS analysis for the 15%-millimeter sample locations all appeared highly similar. Figure 7 shows the seven locations where EDS was performed on sample 12. Points 2, 4, and 5 were darker than the other locations and appeared to be shale, but from the elemental analysis only points 6 and 7 show the presence of carbon, indicating that shale was present within the location areas. All sampled locations showed that oxygen was the most prevalent element followed by calcium, silicon, sodium, aluminum, and potassium.

From the SEM images of sample 18, which contained 50%-millimeter shale, an understanding of the interaction between the cement and shale was shown. Shale, which is the smooth, bedded structure, had begun to bond with the cement as shown in the high magnification secondary electron images. Ettringite and C-S-H appeared to be nucleating on the shale, thus, indicating that a reaction had occurred. Ettringite also surrounded a majority of the shale particles and at 2554× magnification, it appeared that the cement particles may actually have bonded to a portion of the shale. It was apparent that the shale was nearly the same brightness/color contrast as the surrounding matrix, meaning that the shale was of similar composition to the cement. Of the 50%-millimeter samples, sample 16 had the lowest UCS due to the occurrence of microcracking or tobermorite formation rather than other hydration products such as C-S-H and CH. While samples 17 and 18 had a similar UCS, sample 18 achieved a slightly higher UCS; this was indicative of the success of the shale bonding to the cement, as evidenced by the BSE and secondary electron images. Figure 10f had a similar microstructure to class G cement containing silica fume, which is a pozzolanic material [37]. The 50%-millimeter EDS results appeared to be consistent with one another and the presence of shale was more prevalent in these three samples. Sample 18 had the presence of carbon in five of the nine EDS locations but the carbon was present in small amounts ranging from 0.5% to 2.3%. Oxygen content appeared higher in the locations without carbon, which was due to the amount of oxygen within the class H cement. All the points also had high amounts of calcium except for point two, which had a higher silicon and aluminum content than the other sample locations. This could also be indicative of the presence of shale.

The micron sample sets had tests that either achieved a higher or comparable UCS than the base case but as a whole, achieved a lower UCS than the previously reported data for class G and H cement. Sample 22 (5%-micron shale) achieved the highest UCS of all the samples tested. Sample 28, which contained 15%-micron shale, had a higher UCS than any of the base case samples. This implied that at lower quantities of shale addition, there was a potential for increased compressive strength due to the pozzolanic reaction occurring between the pozzolanic material, the shale, and the cement. At lower quantities, the micron particles appeared to have provided enough surface area to allow for the hydration of the cementitious materials and allow for ample nucleation points for C-S-H to form and grow. While the 5%-micron shale had the highest average UCS, it also had a high standard deviation. This could have been due to one or more of the samples containing a higher amount of shale, problems during the debonding process, or increased porosity and permeability. The 50% and 75% sample sets all had UCS values that were considerably lower than the other micron sample sets. This indicated that higher amounts of shale hindered hydration and minimized C-S-H formation.

It appeared that as the amount of submicron shale was added to the cement, the UCS decreased linearly, indicating that additives or alterations to the water–cement (w/c) ratio were needed to increase the compressive strength. The addition of 5%-submicron shale particles yielded the highest compressive strength and lowest standard deviation of all the submicron sample sets. This showed that the shale was evenly distributed within the 5%-submicron samples, meaning that these particles aided the hydration and the formation of C-S-H.

The 10%-submicron sample 45 had a UCS similar to the 5%-submicron samples but the other two 10%-submicron samples had a much lower UCS. This indicated the potential for discrepancies in the composition, issues during debonding from the mold, or higher sample permeability. The 50%- and 75%-submicron samples were able to set after 24 h and were removed intact and heated in a brine bath. During the early hydration period in which the samples were being cured in the brine bath, the 50%- and 75%-submicron samples had all broken into smaller pieces and were unable to be tested. This occurrence could have been due to thermal degradation or poor bonding between the submicron particles and the cement. There are very few studies that have investigated the effect of such high concentrations of pozzolanic material being added to cement, specifically past the early stages of hydration. The 5%-submicron shale SEM images appeared to have more spherical particles than either the 15%- or 50%-millimeter shale samples. Sample 40 shows a submicron shale particle that had a thin, smooth, platy structure that was elevated within the cement. There was little to no presence of ettringite, implying that the shale did not reliably bond to the cement. A number of unhydrated and partially hydrated cement particles were also seen in the same area of the shale particle. This implied that there was a lack of water to appropriately hydrate the particles. From the secondary electron images, it was seen that the shale particle was raised within the cement with areas of the shale being higher than others. Figure 12e showed areas of dense spherical C-S-H on the edge of the smooth shale. From the UCS testing for the 5%-submicron shale addition to the cement, sample 40 had the highest UCS, and from the SEM it was seen that there appeared to be large amounts of CH and C-S-H, two of the primary cement hydration products within the sample, even around the shale particles. Secondary electron images showed that there were some shale particles that appeared to be raised above the matrix while others appeared to be within the cement. Sample 42 had the lowest UCS of the sample set and from the SEM images it appeared that there was less CH and C-S-H present but more C-S-H gel. For the three 5%-submicron samples, there did not appear to be a significant amount of ettringite. For the 10%-submicron samples, similar structures to those seen in both the millimeter and 5%-submicron samples were observed. The sample 45 secondary electron images showed that the topography of the surface was smooth; at high magnifications, there appeared to be a slightly raised area where the cement was bonding to the partially hydrated grain, but this elevation was minimal (Figure 14). The edges of the partially hydrated cement particle appeared to be more angular and there was an evident distinction between the smooth, unhydrated portion and the partially hydrated portion. C-S-H and ettringite were also present at the interface. At low magnification, BSE images showed a number of partially hydrated cement particles and a number of pores. The center of the partially hydrated cement was the cement that had not been hydrated and appeared much lighter than the surrounding material; while the outer area of these particles was somewhat darker than the unhydrated cement, it was not homogeneous to the matrix. The honeycomb-like structures seen around the unhydrated cement was C-S-H, which is a common crystalline structure in Portland cement [36]. For the 5%-submicron sample, shale was detected in two of the three samples as evidenced by the presence of carbon. The EDS weight percent results for the 5%-submicron shale sample locations were in good agreement with one another; calcium and oxygen were the two most prevalent elements with trace amounts of potassium, silicon, sulfur, iron, and sodium. Sample 40 showed that shale was present within the selected areas, and often portions of the sample locations appeared to be darker in areas. The cement material, which appeared lighter, was primarily composed of oxygen and calcium, which is in agreement with the Central Plains Cement Mill report. For the 10% submicron sample, oxygen and calcium appeared to be the most present elements at all the sample locations. This indicated that this piece of the sample had high amounts of cement and lower amounts of shale. For sample 45, only two of the nine locations where EDS was performed appeared to have any trace of shale, being locations five and eight. Along with having the presence of carbon, these two locations also had higher silicon content than the other locations. Silicon is a known constituent of Woodford shale but is also present in the cement. The two locations containing shale appeared to be darker than the surrounding areas on the sample surface.

From the UCS test results, it can be seen that the samples containing micron-sized shale exhibited the most promising compressive strength properties. The 5%- and 10%-submicron sample sets proved to have higher UCS values than those containing the same amounts of millimeter shale, yet the opposite was seen in the 15% and 25% sample sets with the same particle sizes. This indicated that the particle size played a significant role in the cement hydration and that there is an optimal particle size that should be used for future testing to ensure C-S-H formation.

The UCS results from a previous study by [38], which described the effect of 5%-, 10%-, and 15%-millimeter, micron, and submicron shale cuttings addition on the mechanical properties of cement after 7-days curing, were compared to the results found in this study due to the sizes and type of shale used being identical. For both short-term and long-term curing times, the average base case results achieved higher strengths than those with the shale. In both cases, it was seen that the millimeter samples had inferior strengths in comparison to the micron samples. For the long-term study, the average UCS for the 5%-submicron samples was closest to that of the base case. Altering the w/c ratio or using various additives would help to strengthen the systems and aid in achieving higher compressive strengths.

The SEM results were also indicative of the porosity and permeability of the cementitious material. From Figure 8, there were no large pores visible on the 15%-millimeter sample surface at high magnifications and at low magnifications, the pores were less than 10 microns in diameter, indicating that there was some porosity in the sample. Microcracks were not evident at low magnifications, but in Figure 8c,d, thin microcracks were seen on the material's surface. This was due to a combination of permeability and fracturing from the destructive UCS testing. Figure 10 showed that for sample 18 containing the 50%-millimeter shale, the presence of a long, thin microcrack could be seen at high magnification. At higher magnifications, there were microcracks in the matrix surrounding the dark shale particle indicating an increase in permeability near the shale. Porosity was also higher than the 15%-millimeter sample as the number of pores was greater throughout the matrix. The samples that contained the millimeter shale exhibited some permeability on the fractured sample surface, and this could have an impact on gas migration and fingering in a downhole environment. The increased porosity of the 50%-millimeter sample explained why the compressive strength decreased and indicated that the addition of millimeter shale increased the void space in the cementitious materials. For sample 40 shown in Figure 12, microcracks can be seen at every magnification. While the longer and wider microcracks seen in Figure 12a,b were due to the destructive testing that the sample underwent, the microcracks seen at higher magnifications were indicative of the sample permeability. It can be seen in Figure 12b that the matrix surrounding the shale particle had short, thin microcracks and few pores that were below 20 microns in diameter. Figure 14 shows that there were more pores in sample 45 but they were homogenous in size. The microcracks that were indicative of permeability within the sample are seen in Figure 14d–f; the cracks had penetrated the partially hydrated cement grains. This indicated that fluid flow within the matrix was possible and could lead to the formation of microannuli within the downhole environment. This phenomenon is responsible for oil and gas migration within the wellbore and possibly to the water table or surface. The high permeability needs to be remediated in future testing.

Overall, the micron sample sets appeared to have the best results of all the samples containing shale. This could be due to the fact that the micron shale particles were the most similar in size to that of the class H tested. Having additives with larger or smaller particle size in comparison to the cement can have an effect on the hydration and permeability of the system. Ideally, cement particles should be between 7–200 microns [39].

The results showed that cuttings as a cement additive reduced the strength of the cement, but the cement retained most of its strength (Figure 6) and showed promise as a method for cuttings upcycling. Additional work is needed to further understand the effect of shale on cement's mechanical properties. Determination of the Blaine fineness of the millimeter, micron, and submicron shale should be performed to understand the particle size of these additives in comparison to the cement; research on the rheology of cement with calcined clay showed that smaller particle size may lead to an increased yield point [40]. Another aspect that needs to be considered is the rheology, consistency, and thickening time of cement containing shale and how various sizes could impact pumpability [41]. An understanding of the ideal w/c ratio is needed to properly hydrate the slurries with this novel pozzolanic additive. It has been noted that pozzolanic materials such as fly ash require an increase in the amount of water needed for hydration, so using the conventional API water–cement ratio may not be ideal for shale or cuttings cements [42].

5. Conclusions

This paper addressed the effect of the use of an environmentally friendly, readily available, upcycled material on the mechanical properties of wellbore cement, to evaluate the feasibility of using the cuttings as a cement additive or replacement material. It is beneficial to reduce the need for cuttings removal and recycling reduces the amount of greenhouse gas emissions, specifically CO_2, associated with cement manufacturing. While this research is the first step in using cuttings as a wellbore cement additive, several insights were gained that may be beneficial for future applications.

1. While the base case samples, or samples containing no shale, achieved the highest average UCS, micron samples with 5%, 10%, 15% and 25% shale addition had at least one sample that had a comparable UCS.
2. The micron samples proved to have the highest average UCS for all sample sets, which also coincide with the particle sizes of the cement.
3. While the submicron samples containing 50% and 75% shale split during the curing process, the samples had a decreased UCS with an increased shale content.
4. For the millimeter and micron sample sets, the 75% average UCS values were similar and in both cases, were greater than the samples containing 50% shale, but the 75% sample sets had larger standard deviations. The 50%-shale sample sets for both the millimeter and micron shale had the lowest average UCS of all the sample sets indicating that this amount of shale had a negative impact on hydration and the formation of critical microstructure. The 75% sample sets had a higher UCS than the 50% millimeter and micron, but had a higher standard deviation due to variations in the shale distribution, debonding issues, or increased porosity and permeability. It was apparent from the decrease in density that the addition of shale increased the porosity of the samples with shale.
5. This paper shows promise for the upcycling of cuttings as a cement additive. Additional work is needed to further understand the effect of shale on cement's mechanical properties rheology, consistency, and thickening time. An understanding of the ideal *w/c* ratio is needed to properly hydrate the slurries with this novel pozzolanic additive.

Author Contributions: Conceptualization, A.C. and R.N.; methodology, A.C. and R.N.; formal analysis, A.C. and R.N.; writing—original draft preparation, A.C.; writing—review and editing, A.C. and R.N.; supervision, R.N. All authors have read and agreed to the published version of the manuscript.

Funding: This research received no external funding.

Data Availability Statement: All the data relevant to interpretation of results are available within the article.

Acknowledgments: The authors would like to thank Central Plains Cement Company for providing the cement used in this research. The authors would also like to thank Halliburton for providing the defoaming additive used in the cement systems. The authors would like to thank Gary Stowe at OU for his help with the equipment used in testing.

Conflicts of Interest: The authors declare no conflict of interest.

Abbreviations

API	American Petroleum Institute
BSE	Backscatter electron
CO_2	Carbon dioxide
C-S-H	Calcium-silicate-hydrate
CMHEC	Carboxymethyl hydroxyethyl cellulose
DI	Deionized
EDS	Energy dispersive X-ray spectroscopy
HEC	Hydroxyethyl cellulose
IEA	International Energy Agency
MHEC	Methyl hydroxyethyl cellulose
SEM	Scanning electron microscopy
SMP	Shape memory polymer
UCS	Unconfined compressive strength
w/c	Water–cement ratio

Appendix A

This appendix outlines the statistical analysis performed on the results given in Table 3.

Table A1. Single-factor ANOVA results for the class H cement samples containing various sizes and quantities of shale cured for 28 days.

Shale Size	Assumption	Groups	Count	Sum	Average	Variance	p-Value
Millimeter	Addition of millimeter shale has no effect on UCS.	0	3	32.8	10.9	0.2	0.02
		5	3	18.2	6.1	2.5	
		10	3	16.6	5.5	1.6	
		15	3	23.2	7.7	9.7	
		25	3	15.6	5.2	1.9	
		50	3	14.6	4.9	1.1	
		75	3	18.2	6.1	9.2	
	Concentration does not matter for millimeter samples.	5	3	18.2	6.1	2.5	0.6
		10	3	16.6	5.5	1.6	
		15	3	23.2	7.7	9.7	
		25	3	15.6	5.2	1.9	
		50	3	14.6	4.9	1.1	
		75	3	18.2	6.1	9.2	

Table A1. Cont.

Shale Size	Assumption	Groups	Count	Sum	Average	Variance	p-Value
Micron	Addition of micron shale has no effect on UCS.	0	3	32.8	10.9	0.2	0.1
		5	3	29.5	9.8	6.1	
		10	3	26.6	8.9	10.1	
		15	3	27.8	9.3	9.8	
		25	3	23.9	8.0	7.5	
		50	3	17.2	5.7	0.7	
		75	3	18.2	6.1	1.9	
	Concentration does not matter for micron samples.	5	3	29.5	9.8	6.1	0.3
		10	3	26.6	8.9	10.1	
		15	3	27.8	9.3	9.8	
		25	3	23.9	8.0	7.5	
		50	3	17.2	5.7	0.7	
		75	3	18.2	6.1	1.9	
Submicron	Addition of sub-micron shale has no effect on UCS.	0	3	32.8	10.9	0.2	0.008
		5	3	28.8	9.6	0.6	
		10	3	20.9	7.0	6.2	
		15	3	17.6	5.9	5.4	
		25	3	14.7	4.9	2.7	
	Concentration does not matter for submicron samples.	5	3	28.8	9.6	0.6	0.08
		10	3	20.9	7.0	6.2	
		15	3	17.6	5.9	5.4	
		25	3	14.7	4.9	2.7	

References

1. Kimanzi, R.; Wu, Y.; Salehi, S.; Mokhtari, M.; Khalifeh, M. Experimental Evaluation of Geopolymer, Nano-Modified, and Neat Class H Cement by Using Diametrically Compressive Tests. *J. Energy Resour. Technol.* **2020**, *142*, 092101. [CrossRef]
2. Jafariesfad, N.; Sangesland, S.; Gawel, K.; Toræter, M. New Materials and Technologies for Life-Lasting Cement Sheath: A Review of Recent Advances. *SPE Drill. Compl.* **2020**, *35*, 262–278. [CrossRef]
3. Boul, P.J.; Ellis, M.; Thaemlitz, C.J. Retarder Interactions in Oil Well Cements. In Proceedings of the 2016 AADE Fluids Technical Conference and Exhibition, Houston, TX, USA, 12–13 April 2016.
4. Lau, H.C.; Yu, M.; Nguyen, Q.P. Nanotechnology for Oilfield Applications: Challenges and Impact. In Proceedings of the Abu Dhabi International Petroleum Exhibition and Conference, Abu Dhabi, United Arab Emirates, 7–10 November 2016.
5. Deshpande, A.; Patil, R. Applications of Nanotechnology in Oilwell Cementing. In Proceedings of the SPE Middle East Oil & Gas Show and Conference, Manama, Kingdom of Saudi Arabia, 6–9 March 2017.
6. Alkhamis, M.; Imqam, A. New Cement Formulations Utilizing Graphene Nano Platelets to Improve Cement Properties and Long-term Reliability in Oil Wells. In Proceedings of the SPE Kingdom of Saudi Arabia Annual Technical Symposium and Exhibition, Dammam, Saudi Arabia, 23–26 April 2018.
7. Tabatabaei, M.; Santos, L.; Al Hassan, A.A.; Taleghani, A.D. Limiting Deteriorative Impacts of Oil-Based Mud Residuals on Cement Bonding. In Proceedings of the SPE Annual Technical Conference and Exhibition, Houston, TX, USA, 3–5 October 2022.
8. Dao, B.; Vijn, P. Environmentally Acceptable Cement Fluid Loss Additive. In Proceedings of the SPE International Conference on Health, Safety and Environment in Oil and Gas Exploration and Production, Kuala Lumpur, Malaysia, 20–22 March 2002.
9. Doan, A.; Brandl, A.; Vorderbruggen, M.; Leonard, D.D. Innovative Well Cementing Applications by Using Large Particle Sizes of Cement Additives. In Proceedings of the SPE/IATMI Asia Pacific Oil & Gas Conference and Exhibition, Nusa Dua, Indonesia, 20–22 October 2015.
10. Patel, H.; Johnson, K.; Martinez, R. Triazine Polymers for Improving Elastic Properties in Oil Well Cements. In Proceedings of the SPE International Conference on Oilfield Chemistry, The Woodlands, TX, USA, 6–7 December 2021.
11. Contreras, E.Q.; Santra, A. Wellbore Integrity and CO2 Sequestration Using Polyaramide Vesicles. In Proceedings of the SPE International Conference on Oilfield Cement Chemistry, The Woodlands, TX, USA, 6–7 December 2021.
12. Behl, M.; Lendlein, A. Shape-memory polymers. *Mater. Today* **2007**, *10*, 20–28. [CrossRef]
13. Li, J.; Duan, Q.; Zhang, E.; Wang, J. Applications of Shape Memory Polymers in Kinetic Buildings. *Adv. Mater. Sci. Eng.* **2018**, *2018*, 1–13. [CrossRef]
14. Ziashahabi, P.; Ravi, K.; Prohaska, M. Polymer-Based Sealing Materials as Cement Additives to Restore Wellbore Isolation. In Proceedings of the SPE Gas & Oil Technology Showcase and Conference, Dubai, United Arab Emirates, 21–23 October 2019.
15. Taleghani, A.D.; Li, G.; Moayeri, M. Smart Expandable Cement Additive to Achieve Better Wellbore Integrity. *J. Energy Resour. Technol.* **2017**, *139*, 062903–062911. [CrossRef]

16. Mansour, A.K.; Taleghani, A.D.; Li, G. Smart Lost Circulation Materials for Wellbore Strengthening. In Proceedings of the 51st US Rock Mechanics/Geomechanics Symposium, San Francisco, CA, USA, 25–28 June 2017.
17. Flores, J.C.; Patterson, D.J.; Wakefield, J.; Chace, D.; Malbrel, C. Acoustic and Nuclear Wireline Logging Validation of Shape Memory Polymer Screen Expansion. In Proceedings of the SPE Annual Technical Conference and Exhbition, Calgary, AB, Canada, 30 September–2 October 2019.
18. Hossain, K.M.A. Volcanic ash and pumice as cement additives: Pozzolanic, alkali-silica reaction and autoclave expansion characteristics. *Cement Concr. Res.* **2004**, *35*, 1141–1144. [CrossRef]
19. Barry, J.A.; Esafyana, E.; El Sayed, K.L.; El-Husseiny, M.A. Novel Applications of Pozzolans to Treat Wellbore prior to Cement, Casing and While Drilling to Prevent Overburden Stress Fractures. In Proceedings of the IADC/SPE International Drilling Conference and Exhibition, Galveston, TX, USA, 8–10 March 2022.
20. Brandl, A.; Bray, W.S.; Doherty, D.R. Technically and Economically Improved Cementing System with Sustainable Components. In Proceedings of the IADC/SPE Asia Pacific Drilling Technology Conference and Exhibition, Ho Chi Minh, Vietnam, 1–3 November 2010.
21. Ahdaya, M.S.; Imqam, A.; Jani, P.; Fakher, S.; ElGawady, M. New Formulation of Fly Ash Class C Based Geopolymer for Oil Well Cementing. In Proceedings of the International Petroleum Technology Conference, Beijing, China, 26–28 March 2019.
22. Foster, A.; Pollema, A.; Petitt, I.; Heathman, J.; Johnson, C.; Schlepers, R. Contemporary Approach Coupled with Traditional Techniques Tackles Extreme Wellbore Environment in Schoonebeek Heavy Oil Field. *SPE Drill. Compl.* **2012**, *27*, 516–530. [CrossRef]
23. Sweatman, R.E.; Nahm, J.J.; Loeb, D.A.; Porter, D.S. First High-Temperature Applications of Anti-Gas Migration Slag Cement and Settable Oil-Mud Removal Spacers in Deep South Texas Gas Wells. In Proceedings of the SPE Annual Technical Conference & Exhibition, Dallas, TX, USA, 22–25 October 1995.
24. Tare, U.A.; Growcock, F.B.; Takach, N.E.; Miska, S.Z.; Davis, N. Investigation of Blast Furnace Slag Addition to Water-Based Drilling Fluids for Reduction of Drilling Fluid Invasion into Permeable Formations. In Proceedings of the 1998 IADC/SPE Asia Pacific Drilling Conference, Jakarta, Indonesia, 7–9 September 1998.
25. Pang, X.; Hundt, G.; Lewis, S.; Vargo, R.; Tan, B. Storable Liquid-Bead System as Lightweight Additives for Oilwell Cementing. In Proceedings of the Offshore Technology Conference Asia, Kuala Lumpur, Malaysia, 20–23 March 2018.
26. Pernites, R.; Clark, J.; Padilla, F.; Jordan, A. New Advanced High-Performance Ultrafine Micromaterials for Providing Superior Properties to Cement Slurry and Set Cement in Horizontal Wells. In Proceedings of the SPE Annual Technical Conference and Exhibition, Dallas, TX, USA, 24–26 September 2018.
27. Anya, A. Lightweight and Ultra-Lightweight Cements for Well Cementing-A Review. In Proceedings of the SPE Western Regional Meeting, Garden Grove, CA, USA, 22–27 April 2018.
28. Aïtchin, P.-C. Supplementary cementitious materials and blended cements. In *Science and Technology of Concrete Admixtures*, 1st ed.; Aïtchin, P.-C., Flatt, R.J., Eds.; Woodhead Publishing: Cambridge, UK, 2016; Volume 1, pp. 53–73.
29. Mainali, P.; Crosby, Z.; Dix, M.; Yemidale, G.; Heard, S.; Austin, J.; Tilford, M. Advanced Well-Site Geochemistry While Drilling: Improved Wellbore Positioning and Formation Evaluation of Unconventional Reservoirs. In Proceedings of the AAPG International Convention and Exhibition, Cancun, Mexico, 6–9 September 2017.
30. Costa, F.N.; Ribeiro, D.V. Reduction in CO_2 emissions during production of cement, with partial replacement of traditional raw materials by civil construction waste (CCW). *J. Clean. Prod.* **2020**, *276*, 123302. [CrossRef]
31. Hellmann, J.R.; Scheetz, B.E.; Luscher, W.G.; Hartwich, D.G.; Koseki, R.P. Proppants for shale gas and oil recovery. *Am. Ceram. Soc. Bull.* **2014**, *93*, 28–35.
32. Poudyal, L.; Adhikari, K. Environmental sustainability in cement industry: An integrated approach for green and economical cement production. *Environ. Dev. Sustain.* **2021**, *4*, 100024. [CrossRef]
33. Central Plains Cement Company. 2019. Available online: http://www.centralplainscement.com/documents/2014/04/tulsa-class-h-mill-test-report.pdf (accessed on 31 May 2023).
34. American Petroleum Institute. *Recommended Practice for Testing Well Cements*; API RP 10B-2; American Petroleum Institute: Washington, DC, USA, 2019.
35. Wise, J.; Cedola, A.; Nygaard, R.; Hareland, G.; Arlid, Ø.; Lohne, H.P.; Ford, E.P. Wellbore characteristics that control debonding initiation and microannuli width in finite element simulations. *J. Pet. Sci. Eng.* **2020**, *191*, 107157. [CrossRef]
36. Franus, W.; Panek, R.; Wdowin, M. SEM Investigations of Microstructures in Hydration Products of Portland Cement. In *2nd International Multidisciplinary Microscopy and Microanalysis Congress*, 1st ed.; Polychroniadis, E.K., Oral, A.Y., Ozer, M., Eds.; Springer: Cham, Switzerland, 2014; pp. 105–112.
37. Tariq, Z.; Murtaza, M.; Mahmoud, M. Effects of Nanoclay and Silica Flour on the Mechanical Properties of Class G Cement. *ACS Omega* **2020**, *5*, 11643–11654. [CrossRef] [PubMed]
38. Cedola, A.E.; Nygaard, R.; Hareland, G. Cuttings Disposal in Cement: Investigation of the Effect on Mechanical Properties. In Proceedings of the 54th US Rock Mechanics/Geomechanics Symposium, Golden, CO, USA, 28 June–1 July 2020.
39. Zhang, H. *Building Materials in Civil Engineering*, 1st ed.; Woodhead Publishing Limited: Cambridge, UK, 2011; pp. 46–80.
40. Murtaza, M.; Rahman, M.K.; Al Majed, A.A. The Application and Comparative Study of Calcined Clay and Nanoclay Mixed Cement Slurries at HPHT Conditions. In Proceedings of the Offshore Technology Conference, Houston, TX, USA, 4–7 May 2020.

41. Mohamed, A.; Giovannetti, B.; Salehi, S.; Muhammed, F. A Novel Cement Additive to Prevent Gas Migration in Producing and Abandoned Oil and Gas Wells. In Proceedings of the ADIPEC, Abu Dhabi, United Arab Emirates, 31 October–3 November 2022.
42. Smith, D.K. *Cementing*, 2nd ed.; Society of Petroleum Engineers: New York, NY, USA, 1987.

Disclaimer/Publisher's Note: The statements, opinions and data contained in all publications are solely those of the individual author(s) and contributor(s) and not of MDPI and/or the editor(s). MDPI and/or the editor(s) disclaim responsibility for any injury to people or property resulting from any ideas, methods, instructions or products referred to in the content.

Article

Research on Cuttings Carrying Principle of New Aluminum Alloy Drill Pipe and Numerical Simulation Analysis

Pengcheng Wu [1,2,†], Chentao Li [2,†], Zhen Zhang [3], Jingwei Yang [2], Yanzhe Gao [4], Xianbing Wang [1], Xiumei Wan [1], Chengyu Xia [2,*,‡] and Qunying Guo [2,*,‡]

1. Engineering Technology Research Institute, PetroChina Southwest Oil & Gas Field Company, Chengdu 610017, China
2. School of Mechanical Engineering, Yangtze University, Jingzhou 434023, China
3. Shale Gas Research Institute, PetroChina Southwest Oil & Gas Field Company, Chengdu 610056, China
4. College of Electronic Information and Automation, Tianjin University of Science, Tianjin 300222, China
* Correspondence: xiachengyu2023@126.com (C.X.); qunyingguo1212@126.com (Q.G.)
† These authors contributed equally to this work and should be considered co-first authors.
‡ These authors also contributed equally to this work.

Abstract: In order to improve the cuttings transport ability and well hole purification effect of horizontal shale gas wells in the Sichuan and Chongqing area, a new type of aluminum alloy drill pipe is put forward, and the floatability is validated by theoretical analysis and actual parameter calculation. According to the cuttings migration mechanism, the cuttings cleaning simulation model of the new aluminum alloy drill pipe and the traditional steel drill pipe was established by Fluent, and the hexahedral mesh is used to divide the model. Thus, the mesh independence and convergence of the two models are verified. Then, the simulation model is verified by comparing the calculation results of the two simulation models with the experimental data of the indoor cuttings migration device. Finally, the rock-cleaning ability of the two drill pipes is analyzed under the conditions of changing the cuttings particle size, well inclination angle, displacement, and mechanical speed. Compared with the traditional steel drill pipe, the new aluminum alloy drill pipe can improve the borehole purification capacity by 13% on average. This research result is of great significance in reducing the quality of cuttings in the annulus and improving the borehole purification effect.

Keywords: new aluminum alloy drill pipe; cuttings migration mechanism; cuttings cleaning simulation model; cuttings particle size; well inclination angle

1. Introduction

With the deepening of oil and gas exploration and development resources, there are deeper and ultra-deep wells. Rock hardness, drillability level, and grinding ability are becoming higher and higher, which seriously affects the mechanical drilling speed and exploration and development cost of deep hard formation. The formation is hard and poor drillability and low mechanical drilling speed problems will lead to difficult debris return, poor drilling, blocking card, and other serious phenomena. Therefore, it is urgent to study the principle of cut bearing in the drilling process.

As clean unconventional fossil energy, shale gas has gradually drawn people's attention due to its long exploitation life and low pollution degree [1]. It has been proved that the shale gas reserves in the Sichuan-Chongqing area have reached 100 billion cubic meters [2,3] due to the emergence of new geological equipment such as aluminum alloy drill pipe [4] (after calculation, the drilling depth can reach 50,000 m). The drilling depth in the Sichuan-Chongqing region has gradually advanced from a deep well (4500–6000 m) to an ultra-deep well (6000–9000 m) [5]. With the increase in drilling depth, shale gas wells are facing problems such as high pressure, long string, large well diameter, and difficult migration of cuttings [6]. According to statistics, many stuck drilling accidents are caused by insufficient hole cleanliness [7]. Therefore, reducing the cuttings settlement in the hole, improving the cuttings migration speed, and reducing the formation of the cuttings bed are still urgent problems to be solved in drilling engineering [8].

Fully understanding the mechanism of cuttings migration and improving the cuttings carrying capacity of tools are the basis for ensuring underground safe operation. Therefore, researchers at home and abroad have done a lot of research. In terms of the study of cuttings migration mechanism, Wang et al. [9] studied cuttings migration in an eccentric annulus of a horizontal well from the microscopic point of view. When the annulus fluid reaction velocity is different, cuttings will move in the form of contact mass, saltation mass, layer mass, and suspended mass. Sorgun et al. [10] imported the cuttings migration model into Fluent to solve the flow field and concluded that the rotation of the drill pipe did not have a significant impact on the pressure gradient of Newtonian fluid in the eccentric annulus but could reduce the pressure gradient of non-Newtonian fluid. Amanna et al. [11] used Computational Fluid Dynamics (CFD) and experiments to study the effects of different flow rates, different drill pipe speeds, different cutting sizes, and inclination angles on cuttings migration and found that cuttings are the most difficult to clean when the well inclination angle is between 45° and 60°. In the research of cuttings cleaning device, Wu et al. [12] compared and analyzed the performance of two kinds of cuttings cleaning tools and concluded that the "V" type tool is suitable for cleaning cuttings with small particle size, while the spiral type tool is suitable for cleaning cuttings with large particle size. Puymbroeck et al. [13] proposed a borehole cleaning tool with a composite blade and verified that the tool could effectively improve cuttings transport efficiency through experiments. Pang et al. [14] studied the impact of pulsed jet drilling on cuttings migration. By analyzing the amplitude and frequency changes of drilling fluid inlet velocity, they compared the simulation results with traditional drilling and concluded that pulsed jet drilling was beneficial to improving cuttings migration velocity. In terms of the study of liquid–solid two-phase flow, Gao et al. [15], based on the gas–solid two-phase flow model, used CFD to simulate the migration characteristics of single-particle cuttings and cuttings particle groups under different working conditions from the perspective of microscopic movement of cuttings particles. It is found that the movement of single particle cuttings is mainly saltation, while the movement of particle groups is mainly creep and saltation. Mohammadreza et al. [16] used experiments and the Euler particle method to simulate the cuttings migration process of coiled tubing technology (CTT) and studied the influence of different parameters on cuttings migration. To sum up, the existing cuttings cleaning tools (pipe) with spiral wing structures mainly adopt mechanical cleaning and use a device around the rotation of the enhancement that is fluent, through stirring send cuttings at the bottom of the annular ring high-speed air flow area, but quite a number of debris particles did not reach the area to fall to the bottom hole, often needing tools to clean up many times. In the current simulation analysis of cuttings movement, most of the drilling fluid and cuttings particles in the annulus are considered as a liquid–solid two-phase flow, and the fluid force of drilling fluid on cuttings particles is ignored.

To sum up, a new type of aluminum alloy drill rod is proposed in this paper. The new structure drill pipe and the traditional steel drill pipe were simulated. Considering the interaction of drilling fluid and cuttings particles, the cuttings migration process was accurately simulated by using CFD-DEM (discrete element method) coupling. Finally, the rock-carrying (cuttings migration ratio) effect of the two drill pipes was compared. Consequently, the cuttings transport capacity and hole-cleaning effect of the aluminum alloy drill pipe of the drill pipe is improved.

2. Models and Equations

2.1. Structure and Rock-Carrying Principle of a New Aluminum Alloy Drill Pipe

2.1.1. Structure of a New Aluminum Alloy Drill Pipe

The new aluminum alloy drill pipe is a new rock-carrying tool (first proposed in this paper). Its structure is shown in Figure 1. The drill pipe is composed of a male joint, drill pipe body, hollow pipe (the hollow pipe is threaded with the chuck welded inside the male joint and female joint, respectively, during assembly), and female joint. The new aluminum alloy drill pipe is made of aluminum alloy as a whole, so its weight is relatively light. In order to prove the suspension of the drill pipe, the mechanical analysis of the new aluminum alloy drill pipe was carried out in the vertical direction.

$$F = F_{mjt} + F_{zg} + F_{gjt} + F_{kxg} \tag{1}$$

$$G = G_{mjt} + G_{zg} + G_{gjt} + G_{kxg} + G_{zjy} \tag{2}$$

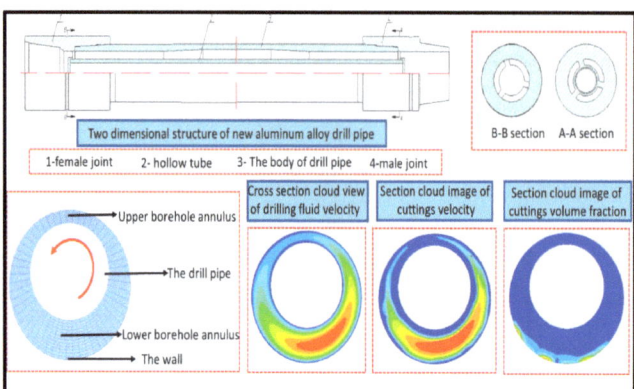

Figure 1. Structure and rock-carrying principle of new aluminum alloy drill pipe.

To make the total gravity of the new aluminum alloy drill pipe less than the buoyancy of the drilling fluid, simply:

$$F > G \tag{3}$$

In the formula, F is the total buoyancy of the new aluminum alloy drill pipe, kN; F_{mjt} is the buoyancy of the female joint in the drilling fluid, kN; F_{zg} is the buoyancy of the drill pipe body in the drilling fluid, kN; F_{gjt} is the buoyancy of the joint in drilling fluid, kN; F_{kxg} is the buoyancy of the hollow rod in drilling fluid (the fluid inside the drill pipe), kN; G is the total gravity of the suspended aluminum alloy drill pipe, kN; G_{mjt} is the self-gravity of the female joint, kN; G_{zg} is the self-gravity of the drill pipe body, kN; G_{gjt} is the self-gravity of the joint, kN; G_{kxg} is the self-gravity of the hollow tube, kN; G_{zjy} is the gravity of drilling fluid inside the drill pipe, kN.

Through the above theoretical analysis and actual parameter values (as shown in Table 1), it is seen that the gravity of the new aluminum alloy drill pipe is less than the buoyancy of the drilling fluid.

Table 1. Actual parameters of new aluminum alloy drill pipe.

The Structural Parameters	Unit	Numerical
Outer diameter of drill pipe	mm	140
Inner diameter of drill pipe	mm	114
Drill pipe length	m	8.7
Outer diameter of hollow pipe	mm	70
Inner diameter of hollow tube	mm	60
Length of hollow tube	m	9.1
Aluminum alloy density	kg/m^3	2780
Drilling fluid density	kg/m^3	2200
Male joint weight	kN	0.102
Female joint weight	kN	0.111

2.1.2. Rock-Carrying Principle of a New Aluminum Alloy Drill Pipe

Because the cuttings particle density is always greater than the drilling fluid density, the cuttings particles will always settle toward the bottom of the annulus. Because the conventional steel drill pipe is tilted lower in the annulus due to gravity, the high-velocity zone is located above the annulus, the low-velocity zone is located below the annulus, and the cuttings bed is located below the annulus, resulting in cuttings moving in three layers (contact, saltation, layering, and suspension). The new aluminum alloy drill pipe, due to buoyancy, is inclined to the upper part of the annulus, causing a relative shift in the position of the high-speed and low-speed fluid zones. This allows the high-speed fluid zone to be located in the lower part of the annulus, almost at the same height as the cuttings bed. In this case, the high-speed drilling fluid can easily carry cuttings toward the wellhead. Table 2 shows that in the high angle and horizontal well section, the new aluminum pipe under the relatively high-density drilling fluid, always in a state of suspension, under the effect of gravity, the deposited debris particles remain the fluid speed zone; we can see that cuttings particles suspended load movement, do not need an additional disturbed flow device, and at the same time can reduce friction, and reduce drilling hydraulic loss.

Table 2. Rock-carrying principle of new aluminum alloy drill pipe and traditional steel drill pipe.

Drill Pipe	Cloud Image of Axial Velocity of Drilling Fluid	Cloud Image of Debris Volume Fraction
Conventional steel drill pipe		
New aluminum alloy drill pipe		

2.2. Installation Position of New Aluminum Alloy Drill Pipe

In long horizontal sections, the traditional steel pipe is generally installed in the screw assembly, and after a bit, because the gravity of the traditional steel pipe is always greater than drilling fluid buoyancy, so under gravity, the axis of the traditional steel pipe to the lower part of the annulus, the high-speed area is located in the upper annulus fluid, and at the bottom of the low-speed region is located in the annulus, coupled with the cuttings particles (The orange dot in Figure 2) greater than the density of drilling fluid (The blue arrow in Figure 2) density. This causes debris to settle to the bottom of the annulus. This makes cuttings easy to settle and difficult to move. In order to improve the ability of cuttings migration in the horizontal section, replace the traditional steel pipe along the horizontal section with the proposed new aluminum pipe, as shown in Figure 2, a new type of aluminum alloy pipe, because it has the flotability of making the drill pipe in the drilling fluid to the upper wellbore annulus, at this point in the annulus fluid the high-speed area will change (high-speed area moves to the bottom of the annulus fluid). Cuttings settling at the bottom are affected by the high-velocity drilling fluid, which increases the axial velocity and improves cuttings transport. This, combined with the turbulence generated by the drill pipe rotation, further reduces cuttings settling at the bottom of the annulus.

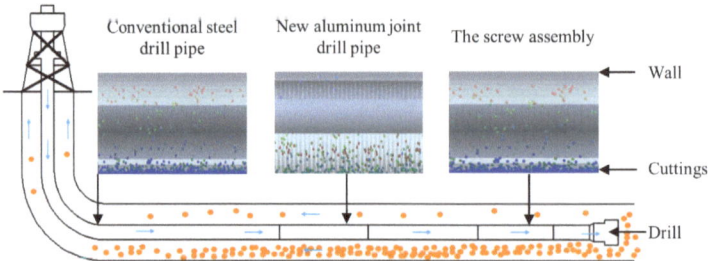

Figure 2. Installation position of new aluminum alloy drill pipe.

Because of the buoyancy of the new aluminum alloy drill pipe, the drill pipe is in the upper annulus, but it does not contact the upper borehole wall. This greatly reduces the wear of the drill pipe and also reduces the drilling hydraulic consumption, and avoids the supporting pressure. After calculating the maximum theoretical drilling depth of the new aluminum alloy drill pipe and the traditional steel drill pipe, respectively, it is seen that the maximum drilling depth of the traditional steel drill pipe is 6000 m, and the maximum drilling depth of the new aluminum alloy drill pipe is 3.3 times that of the traditional rigid drill pipe. Furthermore, the new aluminum alloy drill pipe has good extensibility.

2.3. Strength Check of New Aluminum Alloy Drill Pipe

As the new aluminum alloy drill pipe is lowered to the horizontal well section in actual working conditions, and in the process of drilling, the load on the drill pipe mainly includes dead weight of the drill pipe, rotary torque, static pressure of mud, buoyancy of drilling fluid, weight on bit, etc.; the stress on the drill pipe is shown in Figure 3. We assume that the pipe has been suspended in a certain position on the upper annulus, ignore the drilling fluids and drill pipe flow internal and external forces, the rod end joint as a fixed end constraint, drill pipe suspension buoyancy cavity provides buoyancy, its buoyancy cavity, drill pipe inner cavity, and drill pipe outside surface by the fluid pressure, buoyancy cavity is regarded as by upward buoyancy load and surface pressure load, the inner and outer surfaces of the drill pipe can be considered as surface pressure loads. The distribution of force load is shown in Table 3. L1 is the fixed end face of the female joint; L2, L3, and L4 are the internal pressure on the external surface of the drill pipe, the surface of the buoyancy chamber, and the internal cavity surface of the drill pipe, respectively. L5 is the gravity on the drill pipe, L6 is the buoyancy on the drill pipe, L7 is the torque on the drill pipe, and L8 is the bit pressure on the end face of the male joint.

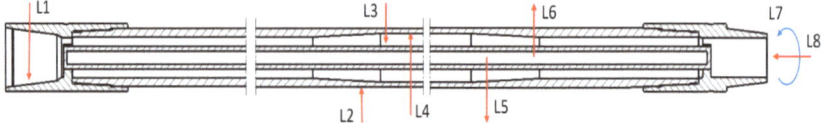

Figure 3. Schematic diagram of load application of new aluminum alloy drill pipe.

Table 3. Loading type and size of new aluminum alloy drill pipe.

Number	Location	Constraint Type	Load Type	Size
L1	End face of female joint	Fixed	Fixed Support	--
L2	Outer surface of drill pipe	Extrusion	Pressure	30 MPa

Table 3. *Cont.*

Number	Location	Constraint Type	Load Type	Size
L3	Outer surface of the buoyancy chamber	Extrusion	Pressure	30 MPa
L4	Drill pipe inner cavity surface	Extrusion	Pressure	30 MPa
L5	Drill rod torque	Gravity	Acceleration	9.81 m/s^2
L6	Buoyancy cavity	Buoyancy	Force	Buoyancy
L7	Drill pipe	Torque	Moment	5000 N·m
L8	Male joint face	Bit pressure	Force	100 KN

As shown in Figure 4, ANASYS 21.0 software was used for the mechanical analysis of the new aluminum alloy drill pipe. The model material was aluminum alloy material provided by the software. Mises equivalent stress and equivalent strain of the new aluminum alloy drill pipe were obtained under various composite loads. Finite element analysis shows that the maximum equivalent stress of the male and female joint is located at the joint between the stiffener and the buoyancy chamber, and the maximum equivalent stress of the buoyancy chamber is located in the middle part, where there is a large stress concentration, and compared with other parts, the maximum equivalent stress is received. This design strength check is based on the shape change specific energy theory (the fourth strength theory); shape change specific energy v_s is the main cause of material yield, no matter in what stress state, as long as the shape change specific energy v_s at the dangerous point reaches the limit v_{su} related to the material property, the material will yield. For plastic materials, such as steel, aluminum, copper, etc., this theory agrees with the experimental results, and the fourth strength theory is more consistent with the experimental results than the third strength theory. The yield stress of Al-Zn-Mg material σ_s = 350 MPa, and the allowable stress of al-Zn-Mg material [σ] = 291.67 MPa if the safety factor is set as 1.2. The maximum stress in the buoyancy chamber of the new aluminum alloy drill pipe is 228.87 MPa, which is less than the allowable stress [σ], so the strength of the modified aluminum alloy drill pipe can meet the requirements.

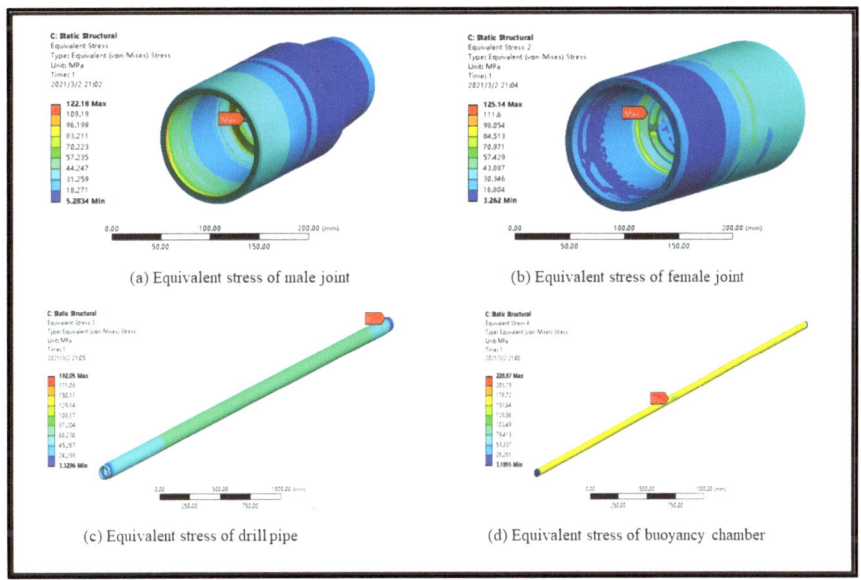

Figure 4. Mises equivalent stress pattern of new aluminum alloy drill pipe.

2.4. Calculation Method
2.4.1. Ordinary Rock-Carrying Drill Pipe

When the drilling fluid carries cuttings (generated by bit rotation) into the borehole annulus, the single liquid phase (drilling fluid) in the annulus becomes liquid–solid two-phase (drilling fluid and cuttings). The Eulerian multiphase flow model in Fluent is selected for solid–liquid two-phase flow simulation, and the annulus flow field is regarded as an incompressible turbulent flow field. The flow follows the Navier–Stokes equation (in fluid mechanics), so the following equation can be established in the Euler coordinate system [17].

Continuity equation:

$$\frac{\partial(\alpha_l \rho_l)}{\partial t} + \nabla \cdot (\alpha_l \rho_l v_l) = 0 \qquad (4)$$

Momentum equation:

$$\frac{\partial(\alpha_l \rho_l v_l)}{\partial t} + \nabla \cdot (\alpha_l \rho_l v_l) = \nabla \cdot (\alpha_l \tau_l) \alpha_l \nabla p + \alpha_l \rho_l g - f_{drag} \qquad (5)$$

In the above formula, the interaction force between the volume fraction of drilling fluid and fluid particles is:

$$\alpha_l = 1 - \sum_{i=1}^{m} \frac{V_i}{\Delta v} \qquad (6)$$

$$f_{drag} = \frac{1}{\Delta v} \sum_{i=1}^{m} F_{drag,i} \qquad (7)$$

In the formula, ρ_l is the drilling fluid density, kg/m^3; α_l is drilling fluid volume fraction, %; v_l is drilling fluid velocity, m/s; τ_l is shear stress of drilling fluid, Pa; f_{drag} is the interaction force of fluid particles, N; g is the acceleration of gravity, m/s^2; V_l is the volume of the i particle, m^3; Δ_v is the volume of the cell, m^3; F_{drag} is the drag force of a single particle, N; m is the total number of particles in the cell.

In order to improve the calculation accuracy of the flow field, the SST K-ω turbulence model (which has a good application effect in the rotating flow field) is selected, and its transport equation is as follows [11]:

$$\frac{\partial(\rho_l k)}{\partial t} + \frac{\partial}{\partial x_i}(k \rho_l v_l) = \frac{\partial}{\partial x_i}\left[\left(\mu + \frac{\mu_t}{\sigma_k}\right)\frac{\partial k}{\partial x_j}\right] + G_k - Y_k + S_k \qquad (8)$$

$$\frac{\partial(\rho_l \omega)}{\partial t} + \frac{\partial}{\partial x_i}(\omega \rho_l v_l) = \frac{\partial}{\partial x_i}\left[\left(\mu + \frac{\mu_t}{\sigma_\omega}\right)\frac{\partial \varepsilon}{\partial x_j}\right] + G_\omega + Y_\omega + D_\omega + S_\omega \qquad (9)$$

In the formula, k is turbulent kinetic energy, m^2/s^2; ω is the dissipation rate of turbulence, m^2/s^2; μ is the dynamic viscosity, kg/(m·s); G_k is the turbulent kinetic energy generated by laminar flow velocity gradient, J; G_w is the turbulent kinetic energy generated by ω equation, J; Y_k and Y_w are turbulence generated by diffusion, D_ω are orthogonal divergence terms, σ_k and σ_ω are turbulence Prandtl numbers of k equation and ω equation. S_k and S_ω are user-defined source entries.

Considering that the fluid in the rotating flow field meets the non-Newtonian rheological characteristics, the power-law fluid model is selected, and the rheological equation of the power-law fluid is:

$$\eta = k \gamma^n \qquad (10)$$

In the formula, η is shear stress, Pa; k is consistency coefficient, Pa·s; γ is shear strain rate, s^{-1}; n is the flow index.

2.4.2. DEM Governing Equation

The discrete element method is used to study the cuttings transport, which can improve the calculation accuracy of the rock-carrying flow field. Each moving cuttings particle is a discrete element. The corresponding momentum conservation equation is [18]:

$$m_i \frac{d^2}{d_t^2} x_i = F_{fluid} + m_i g \left(1 - \frac{\rho_l}{\rho_p}\right) + \sum_q F_c \quad (11)$$

In the formula, m_i is the mass of cuttings, kg; g is the acceleration of gravity, m/s^3; ρ_p is the cuttings density, kg/m^3; F_c is the contact force between cuttings particles (particles and wall surface); F_{fluid} refers to the interaction between a fluid and a particle.

The angular momentum conservation equation of debris particles is:

$$\frac{d}{d_t} I_i \omega_i = \sum_q (T_{t,j}^i + T_{n,j}^i) + T_D^i \quad (12)$$

In the formula, $T_{t,j}^i$ and $T_{n,j}^i$ are tangential and normal torques of cuttings particle I, respectively; I_i and ω_i are the inertia tensor and rotational moment of cuttings particle, respectively. T_D^i is the torque of the drag force in rotational motion.

The HertzMindlin (non-slip elastic contact model) model is selected in EDEM software, which can effectively simulate a series of processes such as collision, extrusion, and deformation between cuttings particles and wall surface so as to improve the accuracy of simulation results. In order to ensure the stability of the solution, the CFD time step is set to 20 times of DEM time step.

2.4.3. Cuttings Migration Rate Equation

The cuttings migration rate (CMR) is defined as the ratio of the total cuttings left in the annulus (the total cuttings stabilized in the annulus minus the remaining cuttings cleaned) to the total cuttings stabilized in the annulus.

$$CMR = \frac{m_w - m_s}{m_w} \cdot 100\% \quad (13)$$

In the formula, m_w is the mass of cuttings particle after stabilization in the annulus, kg; m_s is the residual cuttings mass after well cleaning, kg.

2.5. The Establishment of Simulation Model

According to the actual working conditions, the length of the drill pipe is 8.6 m, and the inner and outer diameters are 114 mm and 140 mm. Combined with the size of the drill pipe, the length of the hollow pipe is 9.1 m, and the inner and outer diameters are 60 mm and 70 mm. The aluminum alloy density was set as 2780 kg/m^3; the drilling fluid density was 2200 kg/m^3; the male joint weight was 0.102 kN; the female joint weight was 0.111 kN. The floatability of the new aluminum alloy drill pipe is verified by the calculation and theoretical analysis of the actual parameter values above.

Since the total gravity of the traditional steel drill pipe is greater than the buoyancy of the drilling fluid, while that of the new aluminum alloy drill pipe is less than the buoyancy of the drilling fluid, in order to more accurately simulate the borehole purification process, a borehole purification simulation model is established as shown in Figure 5. The simulation model contains a certain degree of eccentricity, and the calculation formula for eccentricity is as follows:

$$\varepsilon = \frac{2e}{D_w - D_n} \quad (14)$$

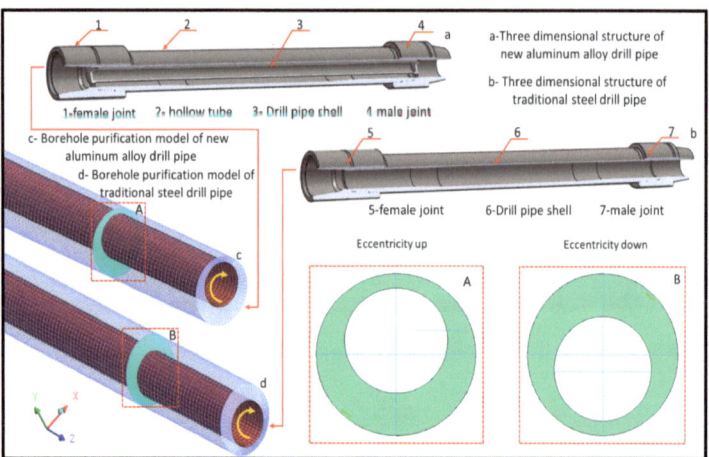

Figure 5. Three-dimensional structure of two kinds of drill pipe and borehole cleaning model.

In the formula, ε is eccentricity; e is the eccentric distance, mm; D_w is the diameter of shaft wall, mm; D_n is the outer diameter of drill pipe, mm.

2.6. Meshing and Convergence Analysis

Because the new aluminum alloy drill pipe model and the traditional steel drill pipe model are eccentric long straight pipes, in order to improve the accuracy of calculation and simulation, the hexahedral grid should be used in the selection of the mesh division method. It can be seen from Figure 6 that the convergence of the two models is verified to be independent of the grid by changing the numerical values of tangential, radial, and axial equal fractions of the two models. After changing the total number of grids, the total cuttings mass calculated by the traditional steel drill pipe model fluctuates between 7.18 kg and 7.45 kg. The total cuttings calculated with the new aluminum alloy drill pipe model fluctuated between 3.66 kg and 3.79 kg, both of which verified the mesh independence and showed good convergence. In the two enlarged images, since the total cuttings of #4 and #9 have little fluctuation, the #4 mesh method is used for the traditional steel drill pipe model, while the #9 mesh method is used for the new aluminum alloy drill pipe model.

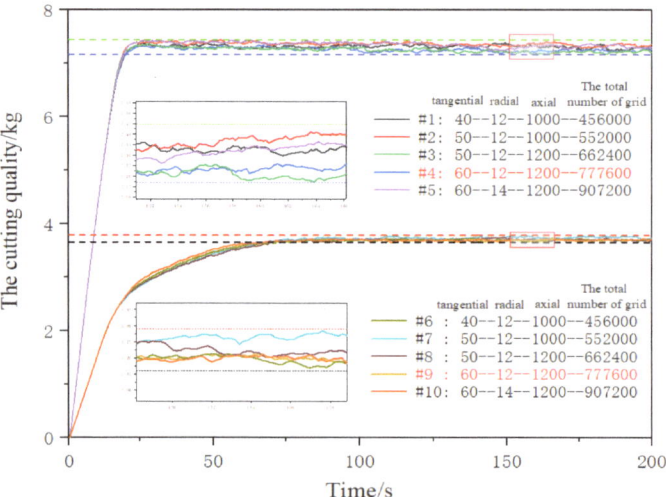

Figure 6. The mass of cuttings in annulus varies with the total mesh number.

2.7. Validation of Model

The cuttings migration experimental device shown in Figure 7 was set up indoors, which could simulate a 20 m long well section. The lifting hoist and lifting frame were used to change the inclination angle of the device so as to simulate cuttings migration in different well inclination sections. For intuitive analysis of cuttings migration, the outer layer uses thick transparent glass tubes. Since real drilling fluid easily pollutes the environment, xanthan gum and sodium formate powder are artificially added into the clean water (xanthan gum is used to increase the viscosity of the fluid, and sodium formate can improve the density of the fluid), and the mixed liquid is used to simulate the real drilling fluid, so as to improve the authenticity and accuracy of the experiment. Through the investigation of a shale gas well in the Sichuan Basin, the cuttings were sampled every 10 m to obtain the cuttings particle size distribution of the well section 4300–4500 m, and finally, it was concluded that the cuttings' particle size mainly concentrated in 0–4 mm. As shown in Figure 8, in order to analyze the cuttings transport capacity of different particle sizes, screens with different diameters were selected to screen the gravel. After screening different gravel sizes ranging from 1 mm to 5 mm, a dryer was used to dry the gravel. Finally, an electronic scale was used to weigh the same quality of gravel for each experiment.

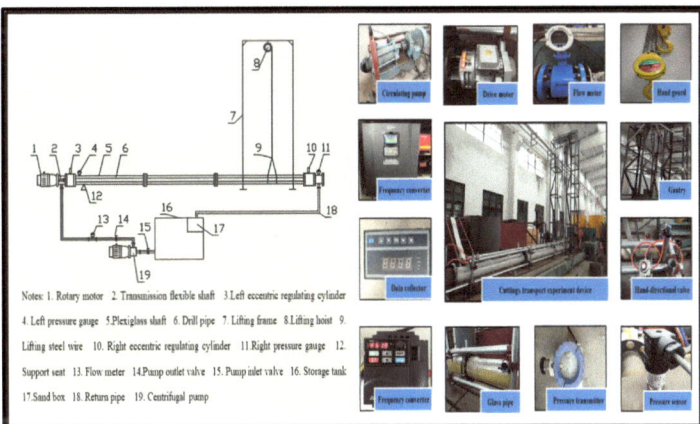

Figure 7. Experimental apparatus for cuttings migration.

Figure 8. Screening of sand and gravel.

The device can simulate the cuttings migration process with a hole diameter of 215.9 mm, drill pipe diameter of 139.7 mm, and drill pipe speed of 80 rpm. The diameter of the nozzle (new pulse jet drill pipe) is set as 12 mm, drilling fluid density of 2200 kg/m^3, displacement of 25 L/s, and mechanical speed of 5 m/h. Choose 3 mm to 4 mm gravel instead of debris particles, and take at regular intervals weighing and record the cutting quality of the tube method; the experimental results are shown in Table 4, then record the data and the results of finite element simulation, Figure 9 shows that when the tube cutting total quality gradually stabilized, the finite element simulation results and the indoor cuttings migration device recorded data close to the correctness of the simulation model are proved.

Table 4. Experimental data simulating rock-carrying capacity of two drill pipes.

Drill Pipe Position	Simulation of the Drill Pipe	Number of Experiments	Cuttings Quality after Stabilization(kg)	Average Cuttings Mass after Stabilization(kg)
	Conventional steel drill pipe	1 2 3	4.28 4.83 4.45	4.52
	New aluminum alloy drill pipe	1 2 3	3.96 3.82 3.47	3.75

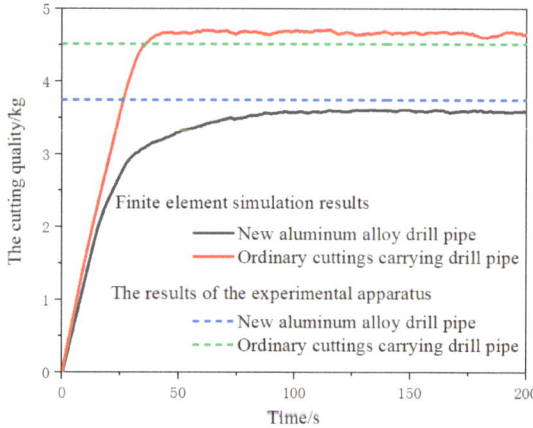

Figure 9. Verification of the model.

3. Results and Discussion

Taking a shale gas well as an example, its borehole diameter is 215.9 mm. The material density of the new aluminum alloy drill pipe (eccentricity is 0.5) is 2780 kg/m^3, the material density of the traditional steel drill pipe (eccentricity is −0.5) is 7850 kg/m^3, the outer diameter of the drill pipe is 139.7 mm, the length of the drill pipe is 20 m, the rotational speed of the drill pipe is 80 rpm, the cuttings density is 2600 kg/m^3, and the drilling fluid density is close. The viscosity coefficient was 0.45, and the fluidity index was 0.55. The borehole purification ability of new aluminum alloy drill pipe and traditional steel drill pipe is compared and analyzed.

3.1. Effect of Cuttings Particle Size on Well Cleaning

Due to the settlement of cuttings with cuttings particle size is directly related to how much, by setting model of the mechanical speed of 4 m/h, angle of 0°, drilling fluid inlet displacement of 25 L/s, in the case of cuttings particle size change, the traditional just drill pipe model and a new model of the aluminum alloy pipe impact the hole cleaning; Figure 10 shows that gradually increase with the diameter of cuttings, cuttings in the annulus are also increasing in quality. When the cuttings size is the same, the cuttings growth rate and total mass of stabilized cuttings in the annulus of the new aluminum alloy drill pipe model are lower than those of the traditional steel drill pipe model. When cuttings are stopped at 150 s, the new aluminum alloy drill pipe has a better well-cleaning effect than the traditional rigid drill pipe. When the diameter of the cutting is larger than 2 mm, the well-cleaning effect of the traditional steel drill pipe is significantly worse, resulting in a large amount of cuttings in the annulus, which is easy to form a cuttings bed. Figure 11 shows that the cuttings migration rate of the new aluminum alloy drill pipe is about 11% higher than that of the traditional steel drill pipe. As shown in Figure 12, the section cloud of cuttings volume fraction extracted from 13 m away from the inlet gradually increases as the cuttings particle size increases, but the maximum cuttings volume fraction of conventional steel drill pipe is always more than two to three times that of the new aluminum alloy drill pipe.

Figure 10. Verification of the model.

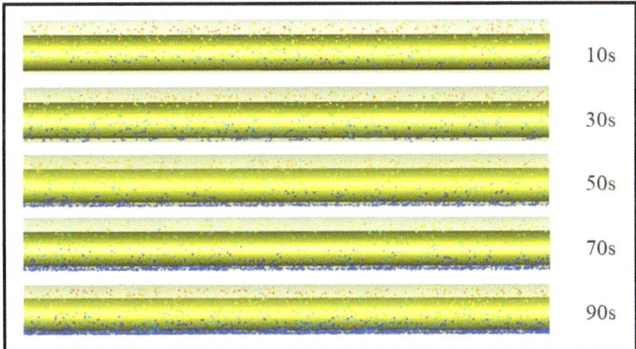

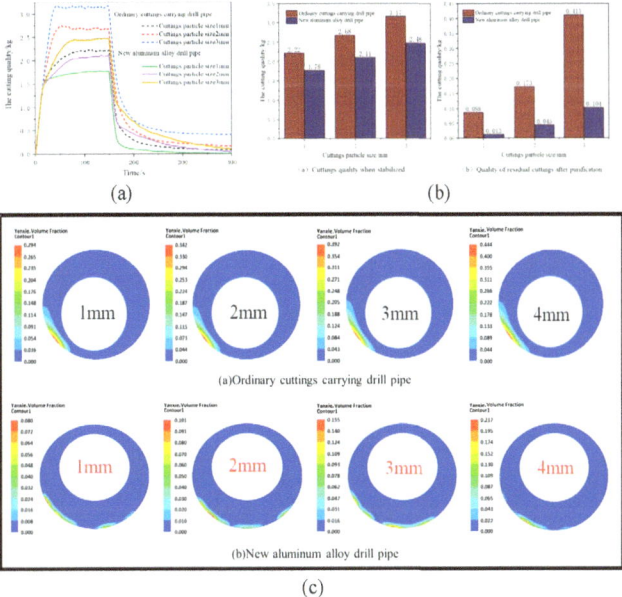

Figure 11. Effect of cuttings particle size on well cleaning: (**a**) effect of cuttings particle size on borehole cleaning; (**b**) effect of cuttings particle size on cuttings carrying capacity of drill pipe; (**c**) cross-section cloud map of cuttings volume fraction.

3.2. Effect of Hole Inclination Angle on Well Cleaning

Hole inclination angle changes often affect the size of the cuttings migration ability by setting a model of the mechanical speed of 4 m/h, cuttings particle size for 4 mm, drilling fluid inlet displacement of 25 L/s, in the case of hole inclination angle change, analyzes the traditional just drill pipe model and a new model of the aluminum alloy pipe impact the hole cleaning; Figure 12a shows that as the angle increases gradually, cuttings in the annulus are also increasing in quality. The cuttings growth rate and the total mass of stabilized cuttings in the annulus of the new aluminum alloy drill pipe model are lower than those of the traditional steel drill pipe model when the well inclination is the same. When cuttings are stopped at 150 s, the new aluminum drill pipe has a better hole-cleaning effect than the traditional rigid drill pipe, while the traditional steel drill pipe has a large amount of cuttings after the hole cleaning, which is easy to form a cuttings bed. Figure 12b shows that the cuttings migration rate of the new aluminum alloy drill pipe is about 10% higher than that of the traditional steel drill pipe. As shown in Figure 12c, the sectional cloud image of the cuttings volume fraction extracted 13 m from the inlet increases gradually as the well hole inclination angle increases, but the maximum cuttings volume fraction of conventional steel drill pipe is always more than two to four times that of the new aluminum alloy drill pipe.

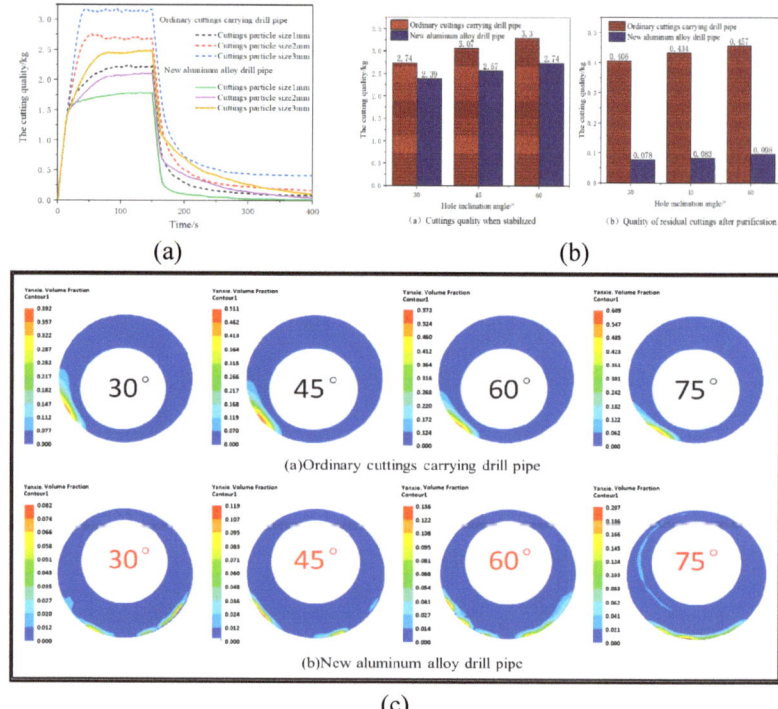

Figure 12. Effect of hole inclination angle on well cleaning: (**a**) effect of hole inclination angle on borehole cleaning; (**b**) effect of hole inclination angle on cuttings carrying capacity of drill pipe; (**c**) cross-section cloud map of cuttings volume fraction.

3.3. Effect of Displacement on Well Cleaning

Based on field feedback, the increased flow rate will improve cuttings migration and reduce the risk of stuck holes. By setting the model in the mechanical speed of 4 m/h, angle 60°, cuttings particle size for 4 mm under the condition of the drilling fluid displacement change, analyzes the traditional just drill pipe model and a new model of the aluminum alloy pipe impact the hole cleaning; Figure 13a shows that as the drilling fluid displacement increases gradually, the cutting quality is gradually reduced in the annulus. At the same displacement, the cuttings growth rate and total mass of stabilized cuttings in the annulus of the new aluminum alloy drill pipe model are lower than those of the traditional steel drill pipe model. When cuttings are stopped at 150 s, the new aluminum drill pipe has a better hole-cleaning effect than the traditional rigid drill pipe, while the traditional steel drill pipe has a large amount of cuttings after the hole cleaning, which is easy to form a cuttings bed. Figure 13b shows that the cuttings migration rate of the new aluminum alloy drill pipe is about 14% higher than that of the traditional steel drill pipe. As shown in Figure 13c, the sectional cloud of cuttings volume fraction extracted 13 m from the inlet gradually decreases as the drilling fluid displacement increases, but the maximum cuttings volume fraction of conventional steel drill pipe is always more than two to four times that of the new aluminum alloy drill pipe.

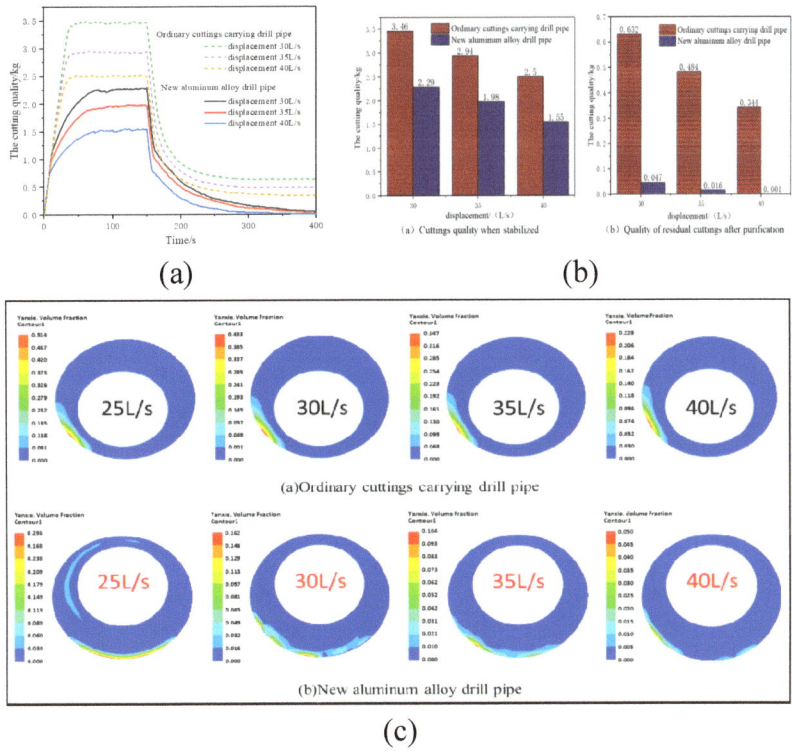

Figure 13. Effect of displacement on well cleaning: (**a**) effect of displacement rate on borehole cleaning; (**b**) effect of displacement on cuttings carrying capacity of drill pipe; (**c**) cross-section cloud map of cuttings volume fraction.

3.4. Effect of Rate of Penetration on Well Cleaning

The rate of penetration is one of the key parameters of drilling engineering. Its size directly affects the borehole debris sedimentation of how many by setting the model of angle as 55°, cuttings particle size for 4 mm, drilling fluid inlet capacity of 30 L/s, under the condition of the mechanical drilling rate change, analyzes the traditional just drill pipe model and a new model of the aluminum alloy pipe impact the hole cleaning; Figure 14a shows that with the increase of rate of penetration, the cuttings quality in annulus also increases gradually. When the rate of penetration is the same, the cuttings growth rate and total mass of stabilized cuttings in the annulus of the new aluminum alloy drill pipe model are lower than those of the traditional steel drill pipe model. When cuttings are stopped at

150 s, the new aluminum drill pipe has a better hole-cleaning effect than the traditional rigid drill pipe, while the traditional steel drill pipe has a large amount of cuttings after the hole cleaning, which is easy to form a cuttings bed. Figure 14b shows that the cuttings migration rate of the new aluminum alloy drill pipe is about 15% higher than that of the traditional steel drill pipe. As shown in Figure 14c, the sectional cloud of cuttings volume fraction extracted from 13 m away from the inlet gradually increases as the rate of penetration increases, but the maximum cuttings volume fraction of conventional steel drill pipe is always more than three to four times that of the new aluminum alloy drill pipe.

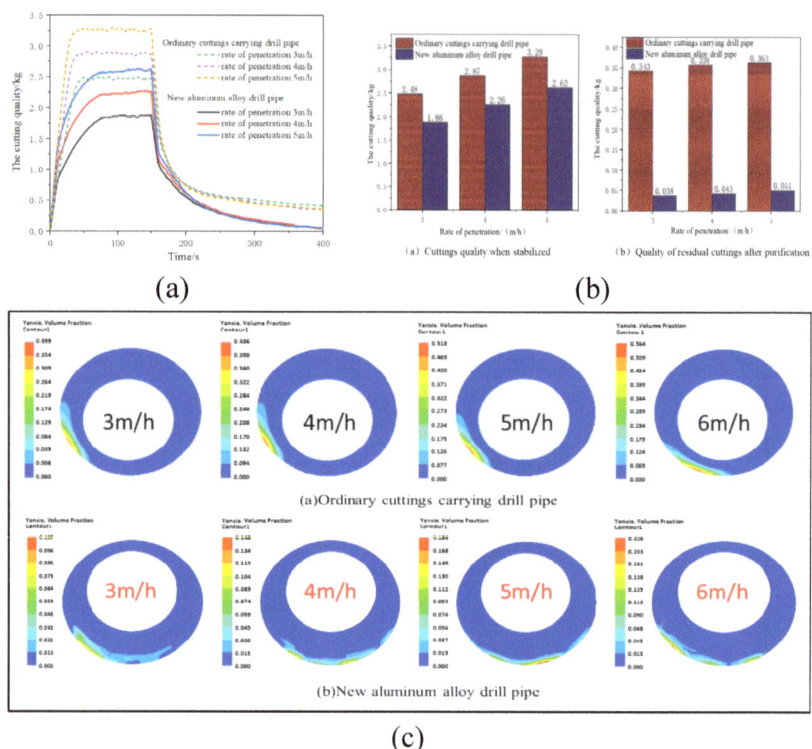

Figure 14. Effect of rate of penetration on well cleaning: (**a**) effect of rate of penetration on borehole cleaning; (**b**) effect of rate of penetration on cuttings carrying capacity of drill pipe; (**c**) cross-section cloud map of cuttings volume fraction.

4. Conclusions

In this paper, the cuttings migration process is accurately simulated by using CFD-DEM (discrete element method) coupling. The main conclusions are as follows:

1. To improve cuttings transport and wellbore cleaning, put forward the new aluminum alloy pipe to improve the structure of the aluminum alloy pipe (increase the drill pipe bottom annulus area, improve cuttings migration velocity, form a new type of structure of drill pipe), using theoretical analysis and actual parameters are calculated and proved that the new type aluminum alloy flotability of the drill pipe, and analyzes the installation location of the drill pipe and the intensity of its structure. Finally, according to the cuttings migration mechanism, the borehole purification simulation model of the new aluminum alloy drill pipe and the traditional rigid drill pipe was established in ANSYS. The hexahedral mesh division method was selected to verify the mesh independence and convergence of the two simulation models. The simulation model is verified by comparing the experimental data recorded by the indoor cuttings migration device with the finite element simulation results;

2. A just drill pipe will be a new aluminum pipe with the traditional simulation model of the import of fluent software, hole cleaning simulation analysis of the new aluminum pipe with the traditional just drill pipe to change the cutting size, angle, displacement, and mechanical speed under the factors such as hole cleaning ability, it is concluded that the new aluminum alloy pipe has a better hole-cleaning effect, compared with the traditional steel drill pipe, the cleanout capacity of the drill pipe is about 13% higher than that of the traditional steel drill pipe, which is of great significance to improve the cuttings transport capacity and the cleanout effect of the well;

3. Carbon fiber is 7 to 10 times stronger than steel, one-fourth as dense, and has high fatigue resistance and strength. With the development of new materials, a carbon fiber drill pipe may be developed. The research method of the aluminum alloy drill pipe in this paper is also applicable to the carbon fiber drill pipe. It can be inferred that the carbon fiber drill pipe will have better cuttings transport capacity and hole-cleaning effect than the aluminum alloy drill pipe.

Author Contributions: Methodology, P.W.; Formal analysis, P.W.; Investigation, J.Y.; Resources, X.W. (Xianbing Wang) and X.W. (Xiumei Wan); Data curation, Y.G.; Writing—original draft, C.L.; Visualization, Z.Z.; Supervision, C.X.; Project administration, Q.G. All authors have read and agreed to the published version of the manuscript.

Funding: This research was funded by the Key Research and development program of Hubei Province "Research on Intelligent High-Temperature Resistant Rotary Steerable Bit and Supporting Well Control Technology" (2021BAA053); National Key Laboratory of Shale Oil and Gas Enrichment Mechanism and Effective Development "Multi-dimensional Impact Rock Breaking Mechanism and Application of Deep Tight Hard Plastic Rock strata" (20-YYGZ-KF-GC-16); National Natural Science Foundation of China "Basic Research on Well Construction for Efficient Development of Shale and Tight Oil and Gas Fields" (U1762214).

Conflicts of Interest: The authors declare no conflict of interest.

References

1. Cao, H.; Wang, T.; Bao, T.; Sun, P.; Zhang, Z.; Wu, J. Effective Exploitation Potential of Shale Gas from Lower Cambrian Niutitang Formation, Northwestern Hunan, China. *Energies* **2018**, *11*, 3373.
2. Li, B.J.; He, Z.M.; Zhu, L.Y. Thoughts on the construction of China's natural gas market reform and development demonstration zones in Sichuan and Chongqing. *Nat. Gas Ind.* **2020**, *40*, 1–10.
3. Zhao, P.; Gao, B. Sinopec's shale gas development achievements during the "Thirteenth Five-Year Plan" period and outlook. *Oil Gas Geol.* **2021**, *42*, 16–27.
4. Li, K. Feasibility of extended drilling of aluminum alloy drill pipes in long horizontal wells. *Nat. Gas Ind.* **2020**, *40*, 88–96.
5. Su, Y.; Lu, B.; Liu, Y. Status and research suggestions on the drilling and completion technologies for onshore deep and ultra deep wells in China. *Oil Drill. Prod. Technol.* **2020**, *42*, 527–542.
6. Liang, J.; Zhang, J.; Yin, H.; Sun, J.H.; Feng, Q.Z. An efficient drilling tool aluminum alloy drill pipe. *Drill. Eng.* **2018**, *15*, 1–3.
7. Fan, Y.; Yang, H.; Fu, L.; Guo, K.; Wang, Y.; Liu, H. Sticking type of horizontal wells in Sichuan-Chongqing area and prevention measures. *West-China Explor. Eng.* **2020**, *32*, 87–90+93.
8. Fang, Y.; Wu, P.-c. Simulation analysis of cuttings migration in shale gas drilling. *Sci. Technol. Eng.* **2020**, *20*, 11532–11538.
9. Wang, H.; Liu, X.; Ding, G. Investigation of cutting transport mechanism in horizontal eccentric annulus. *Oil Drill. Prod. Technol.* **1993**, *15*, 8–17+34.
10. Sorgun, M.; Murat Ozbayoglu, A.; Evren Ozbayoglu, M. Support vector regression and computational fluid dynamics modeling of Newtonian and Non-Newtonian fluids in annulus with pipe rotation. *J. Energy Resour. Technol.* **2015**, *137*, 32901. [CrossRef]
11. Amanna, B.; Movaghar, M.R.K. Cuttings transport behavior in directional drilling using computational fluid dynamics (CFD). *J. Nat. Gas Sci. Eng.* **2016**, *34*, 670–679. [CrossRef]
12. Yu, W.-t.; Zhong, Z. Simulation analysis of cuttings cleaning tool in large displacement well. *Oil Field Mach.* **2019**, *48*, 73–77.
13. Van Puymbroeck, L.; Williams, H. Increasing drilling performance for ERD wells using new generation hydromechanical drill oipe. In Proceedings of the Unconventional Resources Technology Conference, Denver, Colorado, 12–14 August 2013; Volume 13, pp. 145–152.
14. Pang, B.; Wang, S.; Lu, C.; Cai, W.; Jiang, X.; Lu, H. Investigation of cuttings transport in directional and horizontal drilling wellbores injected with pulsed drilling fluid using CFD approach. *Tunn. Undergr. Space Technol.* **2019**, *90*, 183–193. [CrossRef]
15. Li, G.; Xiao, G.; Li, X.; Li, C. Numerical simulation of cuttings migration in horizontal gas drilling Well. *Pet. Drill. Tech.* **2015**, *43*, 66–72.

16. Kamyab, M.; Rasouli, V. Experimental and numerical simulation of cuttings transportation in coiled tubing drilling. *J. Nat. Gas Sci. Eng.* **2016**, *29*, 284–302. [CrossRef]
17. Guo, X.L.; Wang, Z.M.; Long, Z.H. Study on three-layer unsteady model of cuttings transport for extended-reach well. *J. Pet. Sci. Eng.* **2010**, *73*, 171–180. [CrossRef]
18. Oseh, J.O.; Norddin, M.M.; Ismail, I.; Gbadamosi, A.O.; Agi, A.; Ismail, A.R. Experimental investigation of cuttings transportation in deviated and horizontal wellbores using polypropylene–nanosilica composite drilling mud. *J. Pet. Sci. Eng.* **2020**, *189*, 106958. [CrossRef]

Disclaimer/Publisher's Note: The statements, opinions and data contained in all publications are solely those of the individual author(s) and contributor(s) and not of MDPI and/or the editor(s). MDPI and/or the editor(s) disclaim responsibility for any injury to people or property resulting from any ideas, methods, instructions or products referred to in the content.

Article

Leakage Monitoring and Quantitative Prediction Model of Injection–Production String in an Underground Gas Storage Salt Cavern

Tingting Jiang [1], Dongling Cao [1,*], Youqiang Liao [2], Dongzhou Xie [1], Tao He [2] and Chaoyang Zhang [3]

1. School of Resources and Environmental Engineering, Wuhan University of Technology, Wuhan 430070, China; jiangtingting104@163.com (T.J.); xiedongzhou20@163.com (D.X.)
2. Institute of Rock and Soil Mechanics, Chinese Academy of Sciences, Wuhan 430071, China; liaoyq2018@126.com (Y.L.); hetao18@mails.ucas.ac.cn (T.H.)
3. PetroChina Engineering Construction Co., Ltd., North China Branch, Cangzhou 062552, China; zhangchaoyang123@cpecc.com
* Correspondence: cdl912396152ncyb@163.com

Abstract: The leakage of the injection–production string is one of the important hidden dangers for the safe and efficient operation of underground salt cavern gas storage. Although distributed fiber optic temperature measurement system (DTS) can accurately locate the position of the string leakage port, how to establish the quantitative relationship between the temperature difference and leakage rate of the leakage port still needs further exploration. This paper proposes a new quantitative prediction model based on a DTS for the leakage monitoring of the injection–production string of salt cavern gas storage. The model takes into account the gas's physical parameters, unstable temperature conditions, and the Joule–Thomson effect. In order to verify the accuracy of the model, a simulation experiment of string leakage based on a DTS was carried out. The test results show that the relative deviation between the predicted leakage rate and the measured value is less than 5% compared with the calculated value. When the leakage rate drops to 0.16 m^3/h and the temperature range is less than 0.5 °C, it is difficult to accurately monitor the DTS. The results of this study help to improve the early warning time of underground salt cavern gas storage string leakage.

Keywords: salt cavern gas storage; string leakage; temperature field; leakage rate; Joule–Thomson effect

1. Introduction

The energy structure orientation of "carbon peak and carbon neutrality" has gradually promoted low-carbon energy, such as natural gas, to be favored by countries around the world. The imbalance between natural gas consumption and production regions has promoted the rapid development of large-scale energy storage [1,2]. The salt rock strata is an excellent geological body of deep gas storage, with low pores and good sealing advantages. Salt rock strata have been widely used for geological energy storage in the energy consumption of countries such as the United States, Germany, France, and China [3,4]. The sealing problem is one of the core issues in the operation of salt cavern gas storage [5,6]. Under the influence of high-frequency intense injection and mining, brine erosion, and corrosion for a long time, the injection–production string of gas storage often has small damage paths. During the operation of injection and production, high-pressure gas escapes from the path, and, once it leaks to the ground, it is highly likely to cause explosions and other accidents, resulting in inestimable economic and social impacts [7–9]. Therefore, the leakage monitoring of gas storage is of great significance to ensure the safe and efficient operation of gas storage.

The distributed optical fiber temperature measurement system, hereinafter referred to as the DTS, provides remote temperature monitoring through single-mode optical fibers. Due to its advantages of whole-well monitoring, permanent installation, low cost, accurate positioning, and visualization, the DTS is widely used in the field of the wellbore leakage monitoring of salt cave gas storage [10]. The latest DTS devices are capable of providing continuous temperature measurements with a high spatial and thermal resolution (up to 0.02 m spatial resolution and 0.1 °C) over distances of up to 20 km using optical cables [11]. Distributed optical fiber temperature sensing technology has become an excellent choice for string leakage monitoring due to its unique advantages [12]. The basic principle of the DTS is the use of the Joule–Thomson effect and multi-mode fiber optical time domain reflectometry (OTDR) technology [13]. OTDR technology is based on the optical propagation speed in the fiber and the backlight echo time to accurately locate the measured temperature points, which belong to the optical field [14]. The Joule–Thomson effect refers to the fact that, during the process of injection–production gas passing through the leakage path, the gas volume of compressible fluids, such as natural gas, is affected by the fluid nonlinear equation of state, and a sharp expansion occurs during the pressure drop process, which can cause a significant decrease in the temperature of the micro-leakage port [15]. According to these two characteristics, the DTS can achieve the minimum value of the peak fluctuation point of the temperature data of the string in real-time, to realize the accurate location of string micro-leakage [16]. However, for management decision making, it is far from enough to only grasp the location of the leakage port. We need to determine the leakage volume's specific value, evaluate the possible subsequent impact, and take remedial measures as soon as possible.

The calculation of the leakage of the injection–production string has always been the focus of experts and scholars. Because the downhole string is usually buried more than 1000 m deep, it is difficult to monitor the data of small leakage holes in the string. Most studies focus on the relationship between the rise of the annular pressure and wellbore casing temperature with the leakage rate. SCANNELL, a Norwegian company, designed and manufactured an annular leakage rate measurement sled to realize the real-time detection of multiple annular leakage points at the wellhead and the calculation of the annular leakage rate based on annular pressure changes, which has been successfully applied in the field [17]. Wu et al. proposed an algorithm for locating the leakage point of production strings based on the U-shaped tube principle and established a tool for predicting the location of tubing leakage under the continuous casing pressure of offshore gas wells based on the pressure difference and probability distribution [18]. Kabir et al. established the unsteady heat transfer wellbore model based on DTS temperature measurement data, the string, and other relevant data; proposed a method to estimate the total wellhead flow rate and distributed flow profile based on DTS temperature measurement; and verified the effectiveness of the method through field data [19]. Alan et al. accounted for the Joule–Thomson, isentropic expansion, conduction, and convection effects for predicting the transient temperature behavior and computing the wellbore temperature at different gauge depths [20]. Pan et al. made a semi analytical analysis of wellbore heat transfers and numerically solved the mass and heat balance equations using a finite difference scheme to describe the nonisothermal well opening flow of the co-brine mixture [21]. Based on distributed temperature sensing (DTS) and pressure data, Wiese determined the wellbore thermodynamics and heat transfer laws of the Ketzin CO_2 injection well in Germany and proposed two methods to measure heat flux. One is to determine the thermodynamic phase state along the well profile through the pressure data of the downhole point pressure gauge and the distributed temperature data. The other is to obtain heat flux based only on DTS data by analyzing spatial temperature differences [22]. Solima et al. obtained the relevant interface movement and gas flow in the string through the leakage profile developed by DTS technology and estimated the nitrogen leakage rate using the volume method and mass method [23]. Compared with the traditional temperature logging system, the estimation accuracy of the nitrogen leakage rate was improved. However, this method is still based on the ideal equation of state, is only applicable to the early sealing test stage, and does not apply to the unbalanced state during the operation of string injection and production. The above studies are all based on the temperature and pressure changes in the overall reservoir to predict wellbore leakage. In actual production, there are multiple leakage paths in the wellbore system at the same time, casing corrosion damage, casing cementing quality problems, and casing wire leakage in injection–production wells, which may cause an annular pressure rise and temperature change [24]. The monitoring of the temperature distribution of the injection–production string via a DTS makes it possible to calculate the leakage amount by using the transient temperature change of the leakage point.

Therefore, based on the Joule–Thomson effect, a new quantitative prediction model for the injection–production string leakage of salt cavern gas storage is proposed in this paper to solve the problem that the DTS can only be located but not quantified. The mathematical model takes into account the unsteady temperature condition, compression effect, and Joule–Thomson effect and combines the temperature data monitored by distributed optical fiber temperature sensing (DTS) to predict the gas leakage rate. In order to verify the accuracy of the model, a DTS monitoring string leakage test was carried out, and the relative deviation of pressure and pressure data measured via the DTS were compared with the predicted values. The prediction of the leakage rate was less than 5%, and the sensitivity of the leakage rate, ambient temperature, and related heat transfer coefficient were evaluated.

2. Model Development
2.1. Mathematical Model

As the salt cavern gas storage is buried thousands of meters underground, there is a large temperature gap between the fluid in the wellbore and the formation environment, resulting in a rapid increase in the fluid temperature and annulus pressure in the enclosed space of casing annulus in each layer during the injection and production process [25]. In particular, the fluid in the annulus between the string and casing is affected by the high-temperature formation environment, which disturbs the stability of the wellbore temperature field and forms additional radial unsteady heat transfer in the limited confined space. The physical model of the string leakage process is shown in Figure 1. In the process of gas injection and production, the gas escapes from the leakage port into the annular fluid of the pipe string and casing, and the primary forms of heat flow include convective heat transfer between the gas and the annular fluid, heat conduction between the pipe string and the annular fluid, the Joule–Johnson effect caused by the gas flow, and the internal coupling of these forms of heat transfer [13]. Figure 2 illustrates the flow and heat transfer characteristics of the leakage port of the string, and the string can be divided into three parts: the gas in the injection–production pipe string, the injection–production pipe string, and the annulus fluid between the pipe string and casing.

Before building the prediction model in this paper, the following assumptions need to be made:

(1) The temperature of the pipe column is symmetrical along the center line of the pipe column, and the pipe column is isotropic;
(2) There is only one case of string leakage in the casing;
(3) The annulus between the pipe string and the casing is sealed;
(4) The thermal physical properties of each material in the wellbore remain unchanged;
(5) No consideration is given to changes in borehole structure.

For the fluid in the string, there is a coupling effect between the fluid's pressure, temperature, flow rate, and density [26]. Assume a compressible fluid in the tubing. When there is a leak in the wellbore, the fluid flow variable will change with time. According to gas flow characteristics before equilibrium, it belongs to unsteady and nonuniform flow. Because the fluid flowing out of the leakage aperture at the instantaneous moment still conforms to the energy conservation law, the inconsistent term is equal to the sum of the diffusion and source terms. Since the external heat transfer to the wellbore can be ignored in the case of wellbore micro-leakage, the source term can be omitted, and the energy equation can be established to obtain the following:

$$\rho A\Delta z\left(U+\frac{v^2}{2}\right)\Big|_i^{t+\Delta t} - \rho A\Delta z\left(U+\frac{v^2}{2}\right)\Big|_i^{t} = \rho A v\Delta t\left(H+\frac{v^2}{2}\right)\Big|_i^{j+1} - \rho A v\Delta t\left(H+\frac{v^2}{2}\right)\Big|_i^{j} \quad (1)$$

where ρ is the fluid density, kg/m^3; A is the leakage port area, m^2; Δz is a micro-expression of the length of leakage path from the pipe column to the annulus; U is the internal energy of a molar gas at every moment, J/kg; v is the fluid leakage rate, m^2/s; t is for time, s; Δt is the increment of time, s; and H is the energy carried by the fluid, J/kg. As shown in Figure 2, calibrate the wellbore axial spatial discretization through i (i = 0, 1, 2... n); calibrate the wellbore axial spatial discretization through j (j = 0, 1, 2... n).

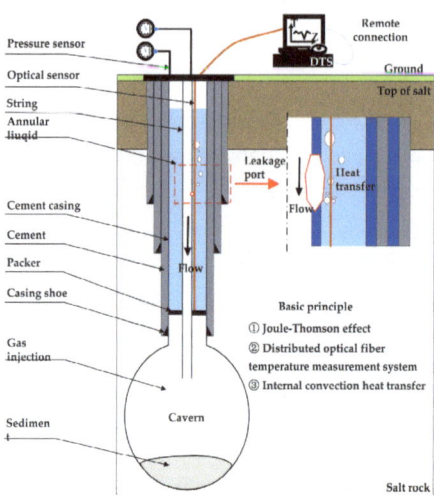

Figure 1. Schematic diagram of distributed temperature sensing (DTS) system used for leak monitoring of injection–production string of salt cavern gas storage.

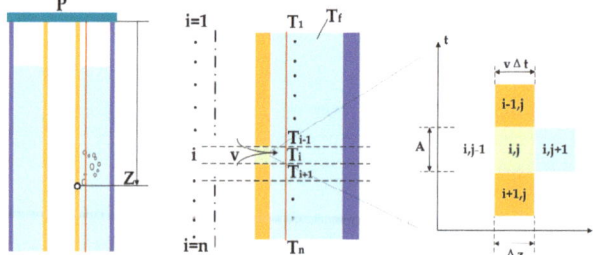

Figure 2. Schematic diagram of two-dimensional pipe leakage diameter to heat transfer.

Divide $A\Delta z\Delta t$ for both sides of the equation:

$$\frac{\rho(U+\frac{v^2}{2})\Big|_{t+\Delta t} - \rho(U+\frac{v^2}{2})\Big|_{t}}{\Delta t} = \frac{v\rho(H+\frac{v^2}{2})\Big|_{j+1} - v\rho(H+\frac{v^2}{2})\Big|_{j}}{\Delta z} \quad (2)$$

Take the partial derivative of both sides of the equation:

$$\frac{\partial\left[\rho(U+\frac{v^2}{2})\right]}{\partial t} = \frac{\partial\left[\rho v(H+\frac{v^2}{2})\right]}{\partial z} \quad (3)$$

According to the first law of thermodynamics,

$$U = H - \frac{p}{\rho} \quad (4)$$

Substitute Formula (4) into Formula (3):

$$\rho \frac{\partial H}{\partial t} + H \frac{\partial \rho}{\partial t} - \frac{\partial p}{\partial t} + \frac{v^2}{2}\frac{\partial \rho}{\partial t} + \rho v \frac{\partial v}{\partial t} \\ = vH\frac{\partial \rho}{\partial z} + \rho H \frac{\partial v}{\partial z} + \rho v \frac{\partial H}{\partial z} + \frac{v^3}{2}\frac{\partial \rho}{\partial z} + \frac{3\rho v^2}{2}\frac{\partial v}{\partial z} \quad (5)$$

The phenomenon that the temperature of a gas changes after irreversible adiabatic expansion when it passes through a porous medium is called the Joule–Thomson effect. The total derivative of enthalpy can be derived from thermodynamic equilibrium relations. For the gas phase, the total derivative of enthalpy can be obtained via the thermodynamic equilibrium as shown in Equation (6):

$$\begin{cases} dH = C_{v,g} dT - \mu_{jT} C_{v,g} d\rho \\ \frac{\partial H}{\partial t} = C_{v,g}\frac{\partial T_f}{\partial t} - \mu_{jT} C_{v,g} \frac{\partial p}{\partial t} \\ \nabla H = C_{v,g} \nabla T_f - \mu_{jT} C_{v,g} \nabla p \end{cases} \quad (6)$$

where p is fluid pressure, Pa; μ_{jT} is the Joule–Thomson effect coefficient, K/Pa; $C_{v,g}$ is the specific heat capacity of the fluid, J/(kg·°C); T is the temperature of the leakage point, °C; and T_f is the temperature of pore fluid, °C.

By substituting Formula (6) into Formula (5), the fluid energy equation in its final form can be obtained:

$$\rho C_{v,g} \frac{\partial T_f}{\partial t} - (\rho \mu_{jT} C_{v,g} + 1)\frac{\partial p}{\partial t} + (C_{v,g} T + \frac{v^2}{2})\frac{\partial \rho}{\partial t} + \rho v \frac{\partial v}{\partial t} \\ = \rho v C_{v,g} \frac{\partial T_f}{\partial z} - \rho v \mu_{jT} C_{v,g} \frac{\partial p}{\partial z} + (v C_{v,g} T + \frac{v^3}{2})\frac{\partial \rho}{\partial z} + (\rho C_{v,g} T + \frac{3\rho v^2}{2})\frac{\partial v}{\partial z} \quad (7)$$

2.2. Model Solving

The energy conservation, Equation (7), is solved numerically using the finite difference method, and the grid of the prediction model is first discretized. In this paper, the heterogeneous nodes of the prediction model are established according to the fluid in the pipe, the pipe, and the annulus fluid. For column axial space discretization I (i = 0, 1, 2...n) calibration and column radial space discretization j (0, 1, 2,...M) calibration, the node is represented by (i, j). The discrete time is calculated using n (n = 0, 1, 2...k). For calibration, Δt is a uniform time step. This prediction model considers the temperature, pressure, density, and velocity as variables and discretizes Formula (7) to obtain Formula (8), which represents the pore temperature value at the next time. The norehole unit division is shown in Figure 3:

$$\rho C_{v,g} \frac{T_f^{k+1} - T_f^k}{\Delta t} - (\rho \mu_{jT} C_{v,g} + 1)\frac{p^{k+1} - p^k}{\Delta t} + (C_{v,g} T + \frac{v^2}{2})\frac{\rho^{k+1} - \rho^k}{\Delta t} \\ + \rho v \frac{v^{k+1} - v^k}{\Delta t} = \rho v C_{v,g} \frac{T_f^{k+1} - T_f^k}{\Delta z} - \rho v \mu_{jT} C_{v,g} \frac{p^{k+1} - p^k}{\Delta z} \\ + (v C_{v,g} T + \frac{v^3}{2})\frac{\rho^{k+1} - \rho^k}{\Delta z} + (\rho C_{v,g} T + \frac{3\rho v^2}{2})\frac{v^{k+1} - v^k}{\Delta z} \quad (8)$$

To realize the coupling solution of fluid–string–annulus fluid, it is necessary not only to adopt the above fully implicit numerical discretization method but also to realize the simultaneous solution of the above equation under the same time step. The specific solution flow chart is shown in the figure below. The whole calculation process consists of two layers of iteration: the iterative solution of the pressure field and the iterative solution of the temperature field. Figure 4 is a detailed flowchart of the main program. The specific steps are as follows:

(1) Input the initial temperature and pressure distribution of the inner and outer tubes;
(2) Calculate the thermal physical property parameters of the gas according to the temperature and pressure field distribution calculated at time k or the last iteration;

(3) Assume the leakage rate of pipe leakage port Q at time k;
(4) The mass conservation equation is used to solve the pressure field distribution at k + 1 moment;
(5) The fluid–column–annulus heat transfer model established in this paper is used to solve the temperature field distribution at k + 1;
(6) Determine whether the temperature field distribution meets the convergence condition, and, if it does not, jump back to step (2);
(7) Perform the k + 2 moment solution until the solution time is complete.

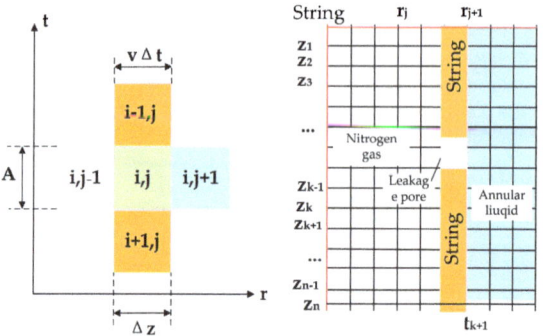

Figure 3. Discrete model of pipe leakage heat transfer.

The solution process is as follows:

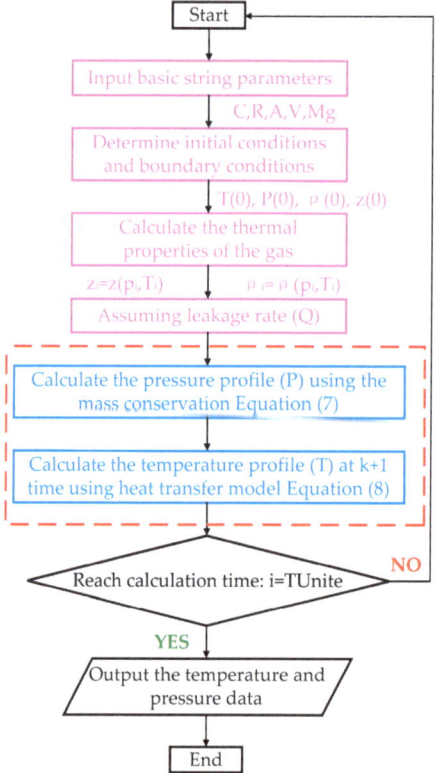

Figure 4. Detailed flow chart of the main program.

3. Experimental Verification

The second research work proposes a method and mathematical model for calculating the leakage rate of leakage ports in a string based on the temperature field. This section will verify the model by combining the DTS system and simulated wellbore installation. As shown in Figure 5a, the research group designed and built a set of full-size wellbore leakage indoor simulation devices with a double-layer string structure in the preliminary research work. The wall thickness of 304 stainless steel inner and outer strings is 0.01 m, the stainless steel outer string diameter d_{outer} is 0.246 m, the stainless steel inner string diameter d_{inner} is 0.178 m, the inner and outer strings are concentric, and the total length of the inner and outer strings L is 3 m. The side wall of the stainless steel inner string is provided with a leakage port with a diameter d of 0.025 m. The housing is connected with multiple flanges to facilitate the installation and replacement of leakage holes. There is a support structure between the internal line and the outer casing to ensure the concentricity of the inner and outer strings. The position of the leakage port is arranged on the inner wall of the string, and the leakage point is provided. The leakage valve is installed in the corresponding outer casing position to change the opening size of the leakage hole. The temperature-sensing cable is arranged to seal the inner casing wall close to the leakage valve. DTS measurements are made using multi-mode optical fibers. Table 1 summarizes the specifications of the DTS used. The spatial resolution is the minimum fiber distribution length required to characterize the DTS for accurate temperature measurement along the fiber length distribution. The high-spatial-resolution DTS equipment used in this paper can reach 0.02 m; the positioning accuracy is the maximum deviation between the measured length value of the optical fiber temperature measuring device and the measured value of the standard ruler, the positioning error value. The sampling interval is 0.02 m, depending on the accuracy of the hardware device. Figure 5b shows the temperature change curve along the optical fiber distance finally detected by the DTS. During the experiment, a 3 m long string can monitor about 150 temperature points, among which the lowest temperature point showing drastic fluctuations in the curve is regarded as the leakage point, that is, the location of the leakage port. For safety reasons, gas is injected through a nitrogen simulated gas reservoir, and annular protection fluid is affected through water. In the simulation test device, a nitrogen cylinder is used to fill high-purity nitrogen into the oil pipe to manufacture natural gas, and water is set outside the casing to cram into the annulus to affect the annular protection fluid. Plugs are formed at both ends of the pipe string combination to seal, and the pipe string combination is lifted off the ground and kept vertically placed during the experiment. During the investigation, the compressed gas enters the annulus from the leak hole of the inner tube string. The temperature signal generated by the gas leak is received by the flexible optical cable (temperature-measuring multi-mode fiber) installed on the wall of the inner tube string and transmitted to the signal demodulation instrument, which converts the optical signal into an electrical signal and transfers it to the monitoring system in the computer, showing the graph between the visual fiber distance on the wall of the inner tube string and the temperature.

Compressed nitrogen was selected as the experimental gas because the preparation of nitrogen is simple and safe. At the same time, the main factors that affect the temperature change characteristics of the pipe string after pipe leakage are the pressure difference between the string and casing and the size of the leakage hole. Gas properties have little effect on optical fiber monitoring. In the DTS monitoring string leak test, it is first necessary to ensure the tightness of the device. After 2~6 MPa nitrogen is injected into the inner pipe string and the pressure of the inner and outer tanks is observed to remain unchanged for a while, it is considered that the device is qualified for sealing. After the sealing test, the leakage hole parameters are set according to the test plan. The leak aperture is 25 mm, and the opening angle of the leak opening can be set to 180° (245.313 mm^2) by changing the valve size. Then, open the gas injection valve to inject nitrogen into the inner string at a preset pressure and stop the nitrogen injection when the pressure in the internal string reaches a specific value. Nitrogen in the inner string can enter the annulus through the leak hole, increasing the annular pressure. When the tension between the inner tube string and the annulus at the leak point is equal, when the pressure balance is reached, a set of temperature data of the internal tube string is collected. After one experiment, perform the following steps:

(1) Open the bolts and flanges on the wall of the jacket tube.
(2) Release the pressure in the annulus.
(3) Replace the leak holes of other sizes and leak points of different depths.
(4) Repeat the above experiment process.

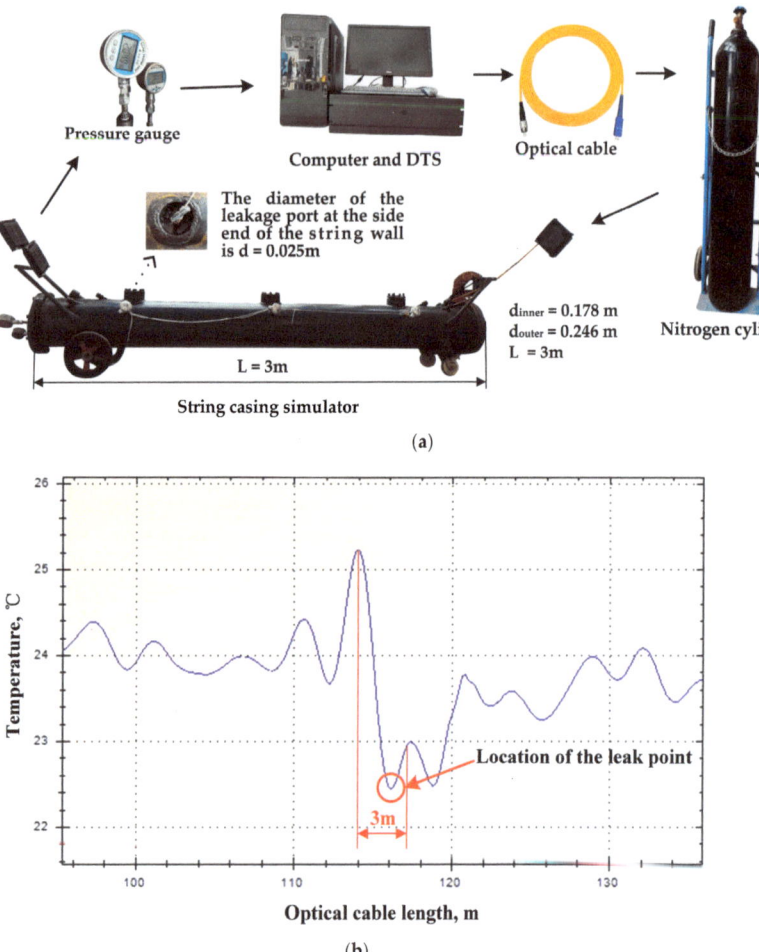

Figure 5. Schematic diagram of the test device. (**a**) The connection of the test device and the expression of the position of the leakage port; (**b**) Temperature changes along the length of the cable and the specific display of the leak point location monitored by DTS.

Table 1. Measurement specifications of DTS.

DTS	Index
Detection unit length (spatial resolution)	0.02 m
Positioning accuracy	±1 m
Measuring time	<30 s
Temperature resolution	0.1 °C
Sampling interval	0.02 m
Temperature variation accuracy	0.5 °C (2000 m)

The basic principle of DTS positioning technology based on the micro-leakage monitoring of the injection–production pipe string is as follows: the gas on both sides of the pipe leakage port flows out at the same time, and the gas accumulation on both sides causes the heat release phenomenon, resulting in the temperature increase of the pipe string on both sides of the leak port. When the gas passes through the hole plug, it absorbs heat due to the rapid expansion in the pressure drop process. The Joule–Thomson effect results in a significantly lower peak temperature at the leakage port. Based on this feature, the position of the string leakage port can be observed. As shown in Figure 6a,b, the temperature and pressure data at this point of the pipe string were obtained through the leakage test. The leakage rates at different times were calculated using Formula (9), and Formula (10) was fitted. The change law of the annular pressure $p(t)$ and pipe temperature with the leakage time t of the leakage port under the condition of local micro-leakage at the leakage port was calculated to calculate the leakage rate prediction at the leakage port of the injectional production pipe string. A comparison was made with the experimental simulation results, as shown in Figure 6c.

$$v = \frac{VM}{\rho ZRT}\frac{dP_A}{dt} \tag{9}$$

$$v = 0.0022 - 0.000002t \tag{10}$$

where P_A is the wellhead annulus pressure, Pa; V is the annulus volume, m^3; M is the molar mass of gas, kg/kmol; Z is the compression factor; and R is the molar gas constant, J/(mol·K).

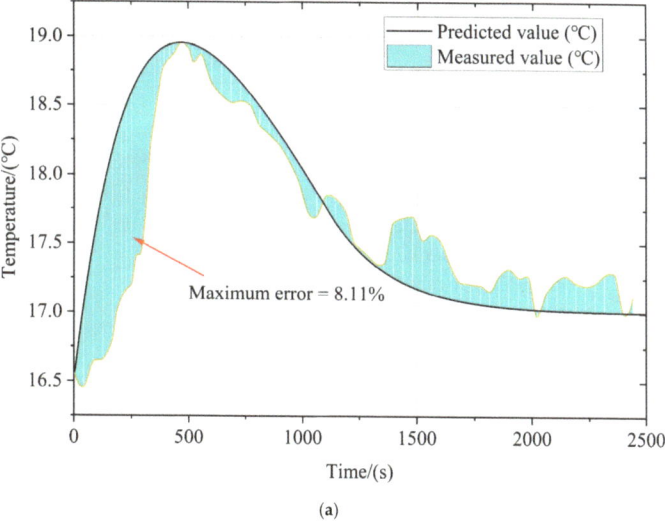

(a)

Figure 6. *Cont.*

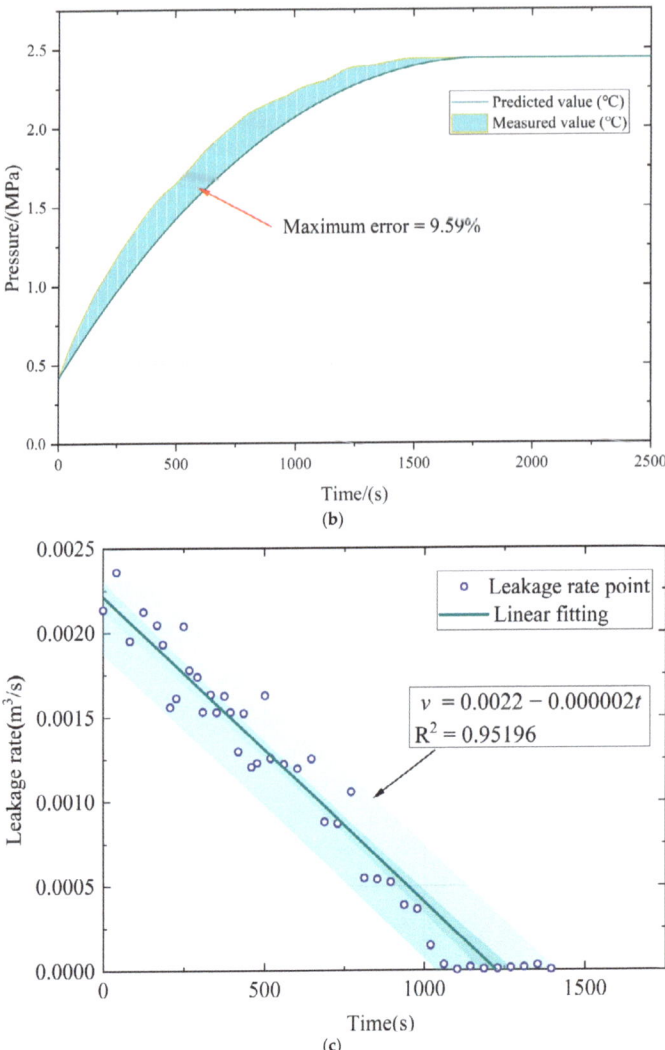

Figure 6. Comparison between the predicted value and the measured value when the leakage opening is 180°, and the internal and external pressure difference is 2 MPa: (**a**) Time–annular pressure curve at the leak point; (**b**) description of time–temperature curve at the leak point; (**c**) description of time–leakage rate curve at the leak point.

As shown in Figure 6a, due to leakage, nitrogen in the pipe column floods into the annulus liquid, bubbles are generated due to the pressure difference, bubble diffusion is dispersed, and, after hitting the optical cable, it breaks back and constantly releases heat. However, this behavior is intermittent, and the temperature curve shows that the measured value's temperature curve is an uninterrupted fluctuating rise. However, the peak value of the predicted leakage point temperature change calculated via software simulation and the estimated weight of the actual device coincides with the curve shape, and the maximum error is 8.11%.

As shown in Figure 6b, due to nitrogen injection into the annulus liquid at the leakage port, bubbles formed by pressure difference will spread around and surge to the place with low pressure. The pressure gauge at the side of the device will monitor the pressure value of the entire annulus, and the pressure curve will show slight fluctuations and not smooth due to the continuous rupture of bubbles. However, on the whole, when the pressure difference is 2 MPa, the predicted value of the leakage point pressure change calculated via software simulation and the final stabilizing value measured by the actual device coincide with the curve shape, and the maximum error is 9.59%.

As shown in Figure 6c, the initial leakage rate is the largest, and the pressure difference gradually decreases with the leakage process. On the whole, the leakage rate of the leakage point decreases linearly with time. The temperature and pressure values of the leakage point at different times were obtained through the test, and the time interval was 1 s. The leakage rate values of the leakage point at other times were obtained by calculating the temperature and pressure data combined with Formula (9). The error between the fitting curve of the leakage rate and the actual value is less than 5%, indicating that the prediction model proposed in this paper can accurately predict the leakage rate of the leakage point of the pipe string.

4. Sensitivity Analysis

4.1. Leakage Rate

According to the test results of the wellbore tightness of an injection–production well in Jintan Salt Cave storage (Table 2) [27], the maximum leakage rate can reach 9.66 m^3/h. In addition, the leakage rate of each wellbore in the second stage decreased significantly compared with that in the first stage: W1 decreased by 76.8%, and W3 decreased by 81.2%. The leakage rate decreases gradually with increases in time, which indicates that the leakage of the string is first rapid then slow and finally tends to be stable. Its basic parameters are shown in Table 2. To explore the change in temperature with time under different gas leakage rates, eight cases are explored in this section. The leakage rate values of the leakage point are 9.66 m^3/h, 3.06 m^3/h, 0.71 m^3/h, 0.16 m^3/h, 0.08 m^3/h, 0.04 m^3/h, 0.01 m^3/h, and 0.001 m^3/h. As shown in Figure 7, with the gradual decrease in the leakage rate, the fluctuation trend of the temperature curve at the leak point gradually weakens. When the leakage rate is 0.16 m^3/h, the peak fluctuation of the temperature curve is less than 0.5 °C, while the leakage rate is reduced to 0.01 m^3/h and 0.001 m^3/h, and the temperature curve is close to the 0.16 m^3/h curve. When the leakage rate is reduced to 0.16 m^3/h, it is difficult for the DTS to accurately monitor when the temperature range is less than 0.5 °C.

Table 2. Tightness test data of wellbores, including leakage location, average leakage rate, and interface displacement.

No.	Leakage Position (m)	Test Phase	Average Leakage Rate (m^3/h)	Interface Displacement (m/h)
W1#	892.1~894.5	Phase 1	3.06	4.33
		Phase 2	0.71	4.11
W2#	893.2~895.4	Phase 1	0.10	0.40
		Phase 2	0.08	0.08
W3#	894.3~896.2	Phase 1	0.16	1.05
		Phase 2	0.03	0.42
W4#	903.2~904.3	Phase 1	0.12	0.05
		Phase 2	0.04	0.18
W5#	888.9~889.7	Phase 1	0.14	0.24
		Phase 2	0.09	0.03

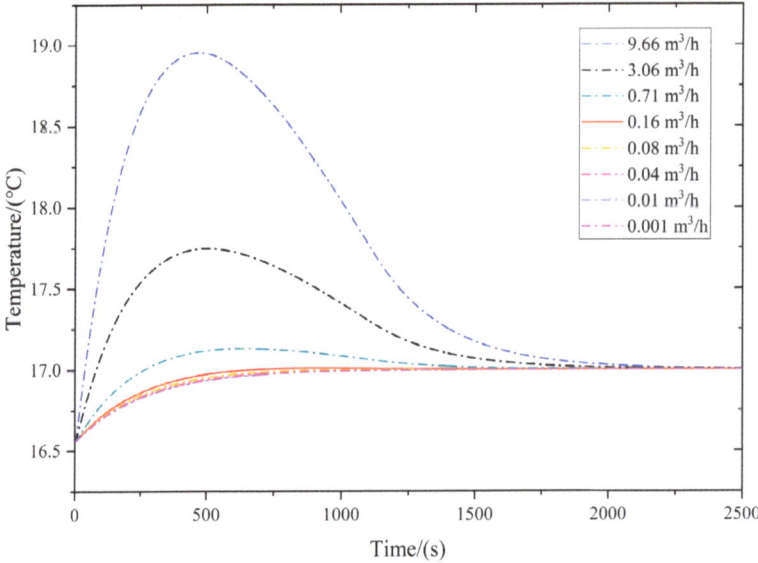

Figure 7. Temperature change with time at different leakage rates.

4.2. Ambient Temperature

The buried depth of salt cavern gas storage is generally 600–2000 m, and the string distribution is 0–1200 m. Due to the existence of a geothermal gradient, the influence of different ambient temperatures at different depths and gas densities on the evolution of the temperature field still needs to be supported by more detailed numerical simulation. In this section, sensitivity analysis under different ambient temperatures will be carried out to further reveal the temperature evolution law after gas leakage and provide theoretical support for further leakage prediction. Taking the basic parameters of Jintan injection production as an example (Table 3), the initial surface temperature is 20 °C, and the ground temperature gradient is 0.03 °C/m. The temperature variation rule of the well depth is 200 m (26 °C), 400 m (32 °C), 600 m (38 °C), 800 m (44 °C), and 1000 m (50 °C), and six conditions are studied under this rule. As can be seen from Figure 8, with the increase in depth and the growth in ambient temperature, the variation range of the temperature gradually decreases. In general, the variation trend of temperature curves at leakage points under different geothermal gradients is similar, and the temperature fluctuation difference between 0 and 1000 m depth is 0.45 °C, indicating that the geothermal gradient has a significant impact on the temperature monitoring of the DTS, which should be considered in the design of the DTS.

Table 3. Summary of the basic parameters used for model performance.

Parameter	Value	Parameter	Value
Wellbore diameter	0.254 m	Ground temperature	20 °C
Gas recovery rate	35 m^3/s	Geothermal gradient	0.03 °C/m
Gas production duration	8 d	Resting time	8 d
Initial gas storage pressure	16.5 MPa	The density of the surrounding rock	2650 kg/m^3
Specific heat of gas	2347 J/(kg·K)	Thermal conductivity of gas	0.15 W/(m·K)
Specific heat of surrounding rock	999 J/(kg·K)	Thermal conductivity of surrounding rock	2.09 W/(m·K)

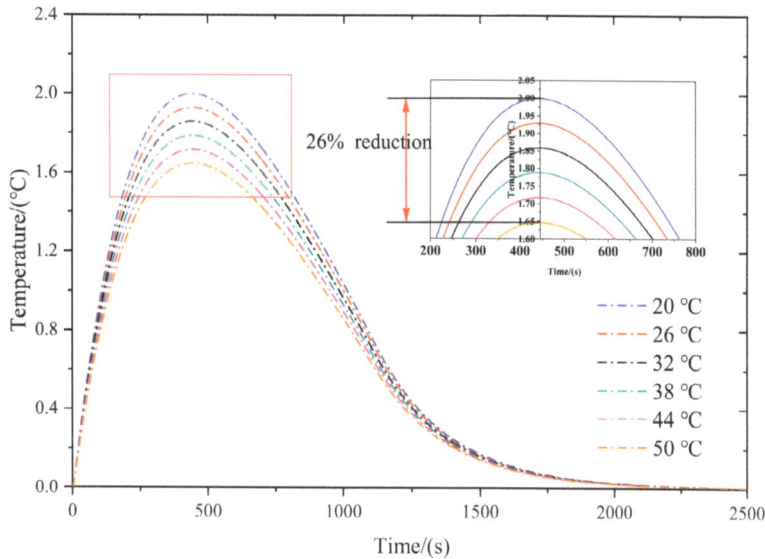

Figure 8. Temperature fluctuation range of leakage points at different depths with time.

4.3. Heat Transfer Coefficient

The convective heat transfer coefficient is closely related to the physical characteristics, depth, and leakage rate of injection–production and annulus fluids. The convective heat transfer coefficient, meaning the convective heat transfer rate, is numerically equal to the convective heat transfer rate per unit heat transfer area under the unit temperature difference. The heat transfer coefficient of the string wall is closely related to the heat transfer coefficient of the gas and annulus liquid in the string, wall thickness, and fouling thermal resistance. This section will conduct sensitivity analysis under different convective heat transfer coefficients and column wall surface heat transfer coefficients and explore the influence law of different convective heat transfer coefficients and column wall heat surface transfer coefficients on the temperature change after gas leakage when the shape, depth, and leakage rate are unchanged. The values of the convective heat transfer coefficient were selected as 222 W/(m^2·K), 422 W/(m^2·K), 522 W/(m^2·K), and 822 W/(m^2·K). It can be seen from Figure 9 that the higher the convective heat transfer coefficient and the stronger the heat transfer capacity between the fluid and the solid, the more pronounced the temperature change. The column wall surface heat transfer coefficient refers to the ability of a string wall to transfer heat per unit area in unit time. The values of the string wall surface heat transfer coefficient are 17 W/(m·K), 27 W/(m·K), 37 W/(m·K), 47 W/(m·K), and 67 W/(m·K), respectively. It can be seen from Figure 10 that, with the increase in the wall heat transfer coefficient, the temperature range gradually increases, and the speed of rising to the peak point gradually increases. However, with the gradual decrease in the leakage rate, the temperature of the leak point with a significant wall heat transfer coefficient decreases faster, which indicates that DTS monitoring equipment is more suitable for the injection production string with a significant wall heat transfer coefficient, the string diameter is more extensive, and the wall thickness is thinner. This study will help analyze the feasibility of DTS monitoring string leaks.

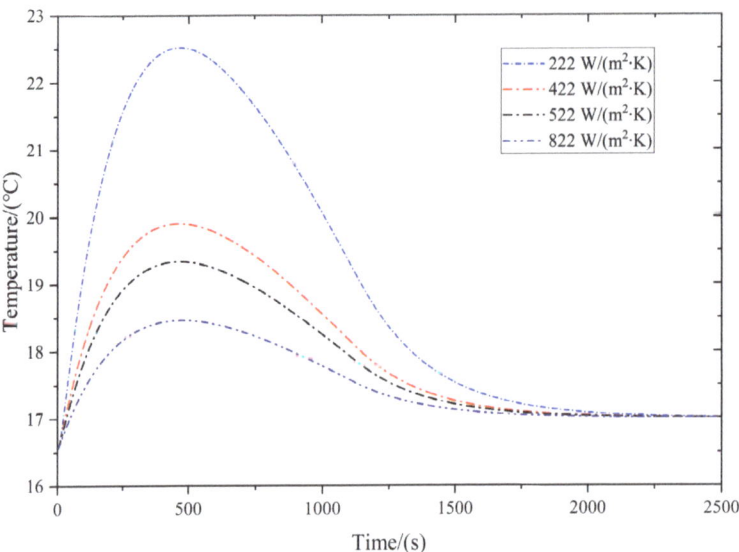

Figure 9. Variation in temperature fluctuation range of leakage point with time under different convective heat transfer coefficients.

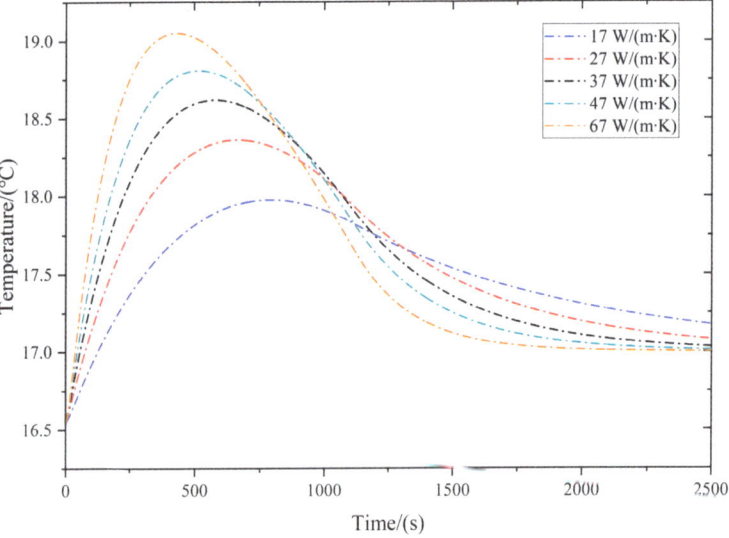

Figure 10. Temperature fluctuation range of leakage point under different string wall heat transfer coefficient changes with time.

5. Conclusions

In this paper, a new mathematical formula for predicting the gas leakage rate in an injection–production string of salt cavern gas storage using distributed temperature sensing technology (DTS) is presented, and the following conclusions are obtained:

1. The formula for calculating the heat transfer at the leakage port under unsteady temperature conditions was established, and the quantitative relationship between the temperature difference and leakage rate was set by considering the Joule–Thomson

effect. The predicted value of the gas leakage rate was obtained by inverting the temperature data of the string.
2. A simulation test of leakage monitoring for the injection–production string of salt cavern gas storage was carried out. Combined with DTS monitoring technology, the measured temperature value of string leakage under pressure was obtained. By comparing the calculated value with the predicted value, the prediction of the leakage rate was realized, and the maximum error was less than 5%, which verified the accuracy of the mathematical model.
3. The sensitivity of the leakage rate, ambient temperature, and related heat transfer coefficient was evaluated. The results showed that when the leakage rate value was reduced to 0.16 m^3/h, it would be difficult to accurately monitor the DTS when the temperature change range was less than 0.5 °C; the ambient temperature significantly influences temperature monitoring, and the temperature fluctuation error at 20 to 50 °C is 26%. This influence factor should be considered when designing the DTS.

Author Contributions: Conceptualization, D.C.; Methodology, D.C., Y.L., T.H. and C.Z.; Software, Y.L. and C.Z.; Validation, D.C., Y.L. and C.Z.; Formal analysis, D.X.; Investigation, T.J., Y.L., D.X. and C.Z.; Resources, D.X.; Writing—original draft, D.C.; Writing—review & editing, T.J., D.X. and T.H.; Supervision, Y.L., T.H. and C.Z.; Project administration, Y.L. and C.Z. All authors have read and agreed to the published version of the manuscript.

Funding: Supported by the CRSRI Open Research Program (Program SN: CKWV2019736/KY).

Data Availability Statement: No new data were created or analyzed in this study. Data sharing is not applicable to this article.

Acknowledgments: We would like to sincerely thank all of our previous and current teachers and classmates who laid the basis for this research.

Conflicts of Interest: The authors declare no conflict of interest.

Nomenclature

A	leakage port area, m^2
$C_{v,g}$	specific heat capacity of the fluid, J/(kg·°C)
H	energy carried by the fluid, J/kg
M	molar mass of gas, kg/kmol
p	fluid pressure, Pa
P_A	wellhead annulus pressure, Pa
R	molar gas constant, J/(mol·K)
t	leakage time, s
T	temperature of leakage point, °C
T_f	temperature of pore fluid, °C
U	internal energy of a molar gas at every moment, J/kg
v	fluid leakage rate, m^2/s
Z	compression factor
ρ	fluid density, kg/m^3
μ_{jT}	Joule–Thomson effect coefficient, K/Pa
Δt	increment of time, s
Δz	a micro-expression of the length of the leakage path from the pipe column to the annulus

References

1. Yang, C.; Wang, T. Deep Underground Energy Storage: Aiming for Carbon Neutrality and Its Challenges. In *Engineering*; Elsevier: Amsterdam, The Netherlands, 2023.
2. Haldorsen, H.H. Invited Perspective: The Outlook for Energy: A View to 2040. *J. Pet. Technol.* **2015**, *67*, 14–19.
3. Yang, C.; Wang, T.; Chen, H. Theoretical and Technological Challenges of Deep Underground Energy Storage in China. *Engineering* **2022**. [CrossRef]
4. Al-Shafi, M.; Massarweh, O.; Abushaikha, A.S.; Bicer, Y. A review on underground gas storage systems: Natural gas, hydrogen and carbon sequestration. *Energy Rep.* **2023**, *9*, 6251–6266. [CrossRef]

5. Zhang, Y.; Oldenburg, C.M.; Zhou, Q.; Pan, L.; Freifeld, B.M.; Jeanne, P.; Tribaldos, V.R.; Vasco, D.W. Advanced monitoring and simulation for underground gas storage risk management. *J. Pet. Sci. Eng.* **2022**, *208*, 109763. [CrossRef]
6. He, T.; Wang, T.; Wang, D.; Xie, D.; Dong, Z.; Zhang, H.; Ma, T.; Daemen, J. Integrity analysis of wellbores in the bedded salt cavern for energy storage. *Energy* **2023**, *263*, 125841. [CrossRef]
7. Evans, D.J. A review of underground fuel storage events and putting risk into perspective with other areas of the energy supply chain. *Geol. Soc. Lond. Spec. Publ.* **2009**, *313*, 173–216. [CrossRef]
8. Li, H.Z.; Saint-Vincent, P.M.B.; Mundia-Howe, M.; Pekney, N.J. A national estimate of U.S. underground natural gas storage incident emissions. *Environ. Res. Lett.* **2022**, *17*, 084013. [CrossRef]
9. Zhao, L.; Yan, Y.; Wang, P.; Yan, X. A risk analysis model for underground gas storage well integrity failure (Article). *J. Loss Prev. Process Ind.* **2019**, *62*, 103951. [CrossRef]
10. Jayawickrema, U.M.N.; Herath, H.; Hettiarachchi, N.; Sooriyaarachchi, H.; Epaarachchi, J. Fibre-optic sensor and deep learning-based structural health monitoring systems for civil structures: A review. *Measurement* **2022**, *199*, 111543. [CrossRef]
11. Wang, J.; Han, Y.; Cao, Z.; Xu, X.; Zhang, J.; Xiao, F. Applications of optical fiber sensor in pavement Engineering: A review. *Constr. Build. Mater.* **2023**, *400*, 132713. [CrossRef]
12. Sun, Y.; Xue, Z.; Hashimoto, T. Fiber optic distributed sensing technology for real-time monitoring water jet tests: Implications for wellbore integrity diagnostics. *J. Nat. Gas Sci. Eng.* **2018**, *58*, 241–250. [CrossRef]
13. Tabjula, J.L.; Wei, C.; Sharma, J.; Santos, O.L.; Chen, Y.; Kunju, M.; Almeida, M.; Upchurch, E.R.; Samdani, G.A.; Moganaradjou, Y.; et al. Well-scale experimental and numerical modeling studies of gas bullheading using fiber-optic DAS and DTS. *Geoenergy Sci. Eng.* **2023**, *225*, 211662. [CrossRef]
14. Leone, M. (INVITED) Advances in fiber optic sensors for soil moisture monitoring: A review. *Results Opt.* **2022**, *7*, 100213. [CrossRef]
15. Liu, Y.H.; Ai, G. Numerical Study on the Heat Transfer in the Leakage of Pressure Vessels Considering the Joule-Thomson Cooling Effect. *Procedia Eng.* **2015**, *130*, 232–249. [CrossRef]
16. Nuñez-Lopez, V.; Muñoz-Torres, J.; Zeidouni, M. Temperature monitoring using Distributed Temperature Sensing (DTS) technology. *Energy Procedia* **2014**, *63*, 3984–3991. [CrossRef]
17. Tarmoom, I.; Thabet, H.B.; Samad, S.; Chishti, K.; Hussain, A.; Arafat, M. A Comprehensive Approach to Well-Integrity Management in Adma-Opco. In *SPE Middle East Oil and Gas Show and Conference*; SPE: Kuala Lumpur, Malaysia, 2007.
18. Wu, S.; Zhang, L.; Fan, J.; Zhang, X.; Zhou, Y.; Wang, D. Prediction analysis of downhole tubing leakage location for offshore gas production wells. *Measurement* **2018**, *127*, 546–553. [CrossRef]
19. Kabir, C.; Izgec, B.; Hasan, A.; Wang, X. Computing flow profiles and total flow rate with temperature surveys in gas wells. *J. Nat. Gas Sci. Eng.* **2012**, *4*, 1–7. [CrossRef]
20. Alan, C.; Cinar, M. Interpretation of temperature transient data from coupled reservoir and wellbore model for single phase fluids. *J. Pet. Sci. Eng.* **2022**, *209*, 109913. [CrossRef]
21. Pan, L.; Oldenburg, C.M.; Wu, Y.-S.; Pruess, K. Wellbore flow model for carbon dioxide and brine. *Energy Procedia* **2009**, *1*, 71–78. [CrossRef]
22. Wiese, B. Thermodynamics and heat transfer in a CO2 injection well using distributed temperature sensing (DTS) and pressure data. *Int. J. Greenh. Gas Control* **2014**, *21*, 232–242. [CrossRef]
23. Soliman, E.M.; Lee, D.; Stormont, J.C.; Taha, M.M.R. New mathematical formulations for accurate estimate of nitrogen leakage rate using distributed temperature sensing in Mechanical Integrity Tests. *J. Pet. Sci. Eng.* **2022**, *215*, 110710. [CrossRef]
24. Trudel, E.; Frigaard, I.A. Stochastic modelling of wellbore leakage in British Columbia. *J. Pet. Sci. Eng.* **2023**, *220*, 111199. [CrossRef]
25. Wu, S.; Zhang, L.; Fan, J.; Zhang, X.; Liu, D.; Wang, D. A leakage diagnosis testing model for gas wells with sustained casing pressure from offshore platform. *J. Nat. Gas Sci. Eng.* **2018**, *55*, 276–287. [CrossRef]
26. Hasan, A.R.; Kabir, C.S. Wellbore heat-transfer modeling and applications (Review). *J. Pet. Sci. Eng.* **2012**, *86–87*, 127–136. [CrossRef]
27. Wang, T.; Chai, G.; Cen, X.; Yang, J.; Daemen, J. Safe distance between debrining tubing inlet and sediment in a gas storage salt cavern. *J. Pet. Sci. Eng.* **2021**, *196*, 107707. [CrossRef]

Disclaimer/Publisher's Note: The statements, opinions and data contained in all publications are solely those of the individual author(s) and contributor(s) and not of MDPI and/or the editor(s). MDPI and/or the editor(s) disclaim responsibility for any injury to people or property resulting from any ideas, methods, instructions or products referred to in the content.

Article

Effect of Nanoparticle and Carbon Nanotube Additives on Thermal Stability of Hydrocarbon-Based Drilling Fluids

Evgeniya I. Lysakova, Andrey V. Minakov * and Angelica D. Skorobogatova

Laboratory of Physical and Chemical Technologies for the Development of Hard-to-Recover Hydrocarbon Reserves, Siberian Federal University, Krasnoyarsk 660041, Russia; mihienkova_evgeniya@mail.ru (E.I.L.); taisvivat@mail.ru (A.D.S.)
* Correspondence: tov-andrey@yandex.ru

Abstract: The article presents the results of experimental study on the effect of additives of silicon oxide nanoparticles, as well as single-walled and multi-walled carbon nanotubes on the colloidal stability and thermal degradation process of hydrocarbon-based drilling fluids. Such a comprehensive study of hydrocarbon-based drilling fluids was carried out for the first time. The effect of the concentration and size of silicon oxide nanoparticles, as well as the type and concentration of nanotubes on the colloidal stability of drilling fluids during thermal aging tests at different temperatures, was investigated. The nanoparticle size varied from 18 to 70 nm, and the concentration ranged from 0.25 to 2 wt.%. Single-walled and multi-walled nanotubes were studied, whose concentration varied from 0.01 to 0.5 wt.%. The thermal aging temperature varied from 30 to 150 °C. According to the results of the investigation, it was shown that the temperature stability of hydrocarbon-based drilling fluids can be significantly improved by adding the above substances. At the same time, it was shown that the use of single-walled nanotubes for thermal stabilization of drilling fluids was several times more effective than the use of multi-walled nanotubes, and tens of times more effective than the use of spherical silicon oxide nanoparticles.

Keywords: hydrocarbon-based drilling fluid; thermal degradation; colloidal stability; thermal stability; single-walled and multi-walled carbon nanotubes; nanoparticles

Citation: Lysakova, E.I.; Minakov, A.V.; Skorobogatova, A.D. Effect of Nanoparticle and Carbon Nanotube Additives on Thermal Stability of Hydrocarbon-Based Drilling Fluids. *Energies* **2023**, *16*, 6875. https://doi.org/10.3390/en16196875

Academic Editors: Yongjun Deng, Lin Chen, Kai Wang and Jie Wu

Received: 7 September 2023
Revised: 21 September 2023
Accepted: 26 September 2023
Published: 29 September 2023

Copyright: © 2023 by the authors. Licensee MDPI, Basel, Switzerland. This article is an open access article distributed under the terms and conditions of the Creative Commons Attribution (CC BY) license (https://creativecommons.org/licenses/by/4.0/).

1. Introduction

Despite the tremendous development and widespread implementation of green technologies and the gradual transition to renewable energy, the consumption of hydrocarbon (HC) fuels is currently increasing [1]. According to various estimates, the demand for oil, in particular, is projected to grow until at least 2040, after which only a gradual, slow decline is likely due to the increasing share of alternative energy [2]. The growing demand for hydrocarbons and simultaneous depletion of conventional oil and gas fields forces oil companies to develop fields in difficult mining and geological conditions or to move to deep and ultra-deep drilling horizons. In addition to increasing the cost of drilling, this causes certain related fundamental challenges. The bottomhole temperature (BHT) increases with increasing depth of the well. In some fields, the BHT of deep wells exceeds 250 °C. High BHT imposes increased requirements on the thermal stability of drilling fluids used for drilling wells. Therefore, the development of drilling fluid formulations applicable at high BHT is a very urgent task, which is dealt with by a large number of researchers [3–6]. As a rule, clay–polymer-based and hydrocarbon-based drilling fluids are used for drilling. However, the main disadvantage of traditional polymer-based and hydrocarbon-based drilling fluids is their low thermal stability due to degradation of the polymers and emulsifiers used in their formulations. In practice, this means that after the operation of such drilling fluids in the well for several hours at high BHT (above 90 °C), they completely or partially lose their initial functional characteristics. To overcome this problem, high-temperature polymer modifications, based on polyacrylamides, sulfonic acid copolymers, amphoteric polymers (PEX), etc., are being developed [7,8].

Another direction in the field of temperature stabilization of drilling fluids is the use of nanoparticles (NPs) [9]. Obviously, NPs are not susceptible to thermal degradation, and are stable in environmental conditions. A large number of studies have been carried out recently in this area. These studies are mainly focused on the stabilization of water-based drilling fluids by NPs. Thus, Al-Malkli et al. [10] studied the effect of sepiolite NP additives on the functional characteristics of water-based bentonite drilling fluid at high temperatures, ranging from 50 to 180 °C, and elevated pressures. The drilling fluid with sepiolite additive showed a generally higher stability of rheological properties over a wide range of temperatures and pressures. However, at temperatures above 120 °C, the rheological parameters, such as plastic viscosity and yield stress, decreased sharply. This indicates irreversible destruction of the drilling fluid due to thermal degradation. Moreover, according to the authors, the considered nano-additives turned out to be ineffective at low pressure and low temperature. The effect of temperature on the rheological properties of water-based bentonite drilling fluid, modified by Fe_2O_3 NPs with an average size of about 30 nm, was investigated in [11]. The temperature ranged from 25 to 85 °C. The concentration of NPs was about 1%. It was shown that the addition of NPs increased the yield stress and plastic viscosity by 45–200% and 20–105%, respectively. The plastic viscosity and yield stress of all the solutions decreased with the increase in temperature. However, for the solutions containing Fe_2O_3 NPs, this decrease was much less. The authors [12] investigated the effect of adding 40 nm silicon and graphite NPs to a water-based complex clay–polymer drilling fluid with bentonite, barite, and starch additives. The temperature varied within the range of 25 to 85 °C. It was shown that the drilling fluid, modified by the NPs, better retained its rheological properties at high temperatures.

A systematic experimental study of the temperature dependence of the viscosity of drilling fluids, modified by NPs of different size and composition, was performed in our recent work [13]. It was shown that the addition of NPs significantly influences the temperature dependence of the viscosity of clay and clay–polymer dispersions. It was found that the effect of NPs on the viscosity of drilling fluids decreased with increasing temperature. Thus, while at room temperature, the addition of 3 wt.% of NPs increased the viscosity of clay slurry by about eight times, and at a temperature of 80 °C, this increase was about three times. Moreover, it was found that the addition of NPs makes the viscosity of the drilling fluid less sensitive to temperature changes. For example, for a drilling fluid with 3 wt% of 10 nm silicon oxide NPs, increasing the temperature from 25 to 80 °C caused an increase in the effective viscosity by about 2.5 times, whereas, for the base drilling fluid, the increase in viscosity, caused by a similar increase in temperature, was about ten times. In addition, it has been shown that the size of NPs also affects the temperature dependence of the viscosity of drilling fluids. In general, a decrease in NP size leads to an increase in the temperature dependence of drilling fluid viscosity. Therefore, when increasing temperature from 25 to 55 °C, the viscosity of mortar slurry with 150 nm aluminum oxide NPs increased by 1.4 times, and, with 50 nm particles, it increased by 1.7 times. Moreover, it was shown that NPs of different composition at close average sizes and the same concentration have a different effect on the temperature dependence of the viscosity of clay and clay–polymer suspensions. In this respect, solutions with aluminum oxide particles are the most sensitive to temperature.

Recently, to modify the thermal properties of drilling fluids, carbon nanotube additives have also been investigated. For this purpose, mainly multi-walled carbon nanotubes (MWCNTs) are used. Thus, Ahmad et al. [14] attempted to improve the properties of a water-based bentonite drilling fluid using carbon nanotubes of different modifications. In this study, the temperature varied from 25 to 85 °C. It was demonstrated that the addition of 0.25 wt.% of polymer and the same amount of nanotubes improved the rheological properties of the drilling fluid. With increasing temperature, plastic viscosity decreased and yield stress increased, indicating the thermal degradation of the drilling fluid. The authors of the study [15] investigated the effect of MWCNT concentration on a drilling fluid also based on water. In this study, the temperature varied from 25 to 180 °C. It was shown that the fluids were viscoplastic non-Newtonian at all pressures and temperatures. The addition of MWCNTs led to an increase in drilling fluid viscosity with increasing concentration. A non-monotonic dependence of the effective viscosity of the fluid on temperature was established for the basic and modified drilling fluids. At temperatures above 120 °C, the effective viscosity of the drilling fluid decreased sharply. This may indicate fluid thinning down due to thermal degradation. Increasing the concentration of nanotubes made the dependence of viscosity on temperature more monotonic. The research goal of work [16] was to improve the properties of water-based drilling fluids using a novel nanocomposite based on MWCNTs and polyethylene glycol. In this study, a relatively small temperature variation range (from 30 to 40 °C) was considered. It was shown that the addition of a relatively small amount of nanocomposite (1 wt.%) to the drilling fluid led to a significant improvement in the rheological performance. However, while analyzing the temperature dependence of the rheological characteristics, it was found that even this very insignificant change in temperature has led to a significant change in the rheological characteristics. While increasing the temperature from 30 to 40 °C, the plastic viscosity of the drilling fluid has decreased almost twofold, and the yield stress increased almost six times. This indicates a very strong loss of colloidal stability of the fluid. Ahmad et al. [17] evaluated the effect of CNT additives on the thermal stability of polymer solutions. They showed that, at a nanotube concentration of 2 vol.%, the rheological characteristics of the solution remained stable up to 150 °C. It was also found that the introduction of CNTs reduced filtration loss and fostered a reduction in the hydration of clayey rocks.

Nanocellulose is also widely used to stabilize the rheological properties of drilling fluids. A comprehensive study of using nanocellulose in various forms for these purposes was carried out in [18], where it was used to modify water-based bentonite drilling fluid. As a result, the advantage of nano-structured cellulose was demonstrated, specifically in the form of NPs. In this study, the temperature varied from 25 to 80 °C. It was shown that the nanocellulose-modified solutions had better rheological characteristics and higher thermal stability.

There have also been recent studies using boron nitride-based nanomaterials as additives to improve the performance of oil-based drilling fluids (OBDF). For example, in [19], the authors developed a hybrid invert-emulsion drilling fluid based on graphene nanoplates and boron nitride nanoparticles. An extensive and comprehensive experimental investigation was conducted to study and quantify the effect of these nanomaterials on all functional properties of oil-based drilling fluid at different temperatures (25–125 °C). The developed hybrid nanodrilling fluid demonstrated improved rheological properties (it increased the plastic viscosity and yield stress by 17% and 36%, respectively, with the introduction of 0.2 ppb nanomaterials) and filtration characteristics (the maximum reduction in filtrate losses was 60% at a concentration of 0.6 ppb). It was also shown that the high thermal stability, uniform dispersion, and large surface area of nanomaterials significantly improved the performance of OBDF at all temperatures considered.

The authors of [20] studied the thermal and rheological behavior of two-dimensional nanostructures of oxidized hexagonal boron nitride dispersed directly in xanthan gum and obtained similar results on the influence of these nanoparticles on the properties of the drilling fluid. The results of thermal conductivity measurements showed that the drilling fluid prepared with 6 wt % nano-additives had a thermal conductivity value 7% higher than the sample without nanoparticles. A rheological study of the nanofluids showed that the viscosity values of all samples increased with increasing concentration of boron nitride nanoparticles, and suspensions with higher concentrations exhibited yield stress, whereas suspensions with lower concentrations were shear thinning fluids. Thus, the work showed that these systems may have advantages for use in deeper fields due to their increased thermal conductivity and viscosity.

Thermal degradation of drilling fluids refers to changes in their properties caused by exposure to high temperatures. When drilling fluids are exposed to heat, they can undergo various processes of degradation and compositional changes, which can negatively affect their efficiency and productivity. High temperatures can lead to changes in the physicochemical properties of fluids, such as viscosity, rheological characteristics, and electrostatic and emulsion stability, as well as to changes in the chemical composition of the drilling fluid. This may require adjustment of the solution composition or the use of stabilizing additives. That is why it is necessary to select drilling fluid formulations that will be stable over a wide temperature range and, at the same time, maintain all other functional characteristics for the quality of the drilling process. For this purpose, this work proposes modification of drilling fluids with nanomaterials, which, as shown, meet these requirements.

The literature review has shown that most of the studies on the modification of the high-temperature properties of drilling fluids using various nanomaterials were performed for water-based drilling fluids, whereas, from the practical standpoint, the most promising fluids for drilling in difficult conditions are hydrocarbon-based drilling fluids, which are invert emulsions. There is very little data on the effect of NP and, especially, carbon nanotube additives on the thermal stability of hydrocarbon-based drilling emulsions, and there are no works at all on thermal stabilization of drilling fluids by means of single-walled carbon nanotubes (SWCNTs). At the same time, SWCNTs, as additives for the modification of various properties of drilling fluids, are much more effective compared to other nanomaterials. Thus, this research area looks very promising.

The aim of the present work was to carry out a comparative experimental analysis of the effectiveness of additives of silicon oxide NPs, SWCNTs, and MWCNTs on colloidal stability during the thermal degradation of hydrocarbon-based drilling fluids. Such a comprehensive and systematic study of high-temperature stability for hydrocarbon-based drilling fluids has been performed for the first time.

2. Drilling Fluid Preparation Technique and Experimental Facility for Studying Their Thermal Degradation

Emulsions are unstable thermodynamic dispersed systems consisting of two or more mutually insoluble liquids stabilized by chemical reagents. Proper selection of reagents influences the stability, as well as the rheological, filtration, and antifriction parameters of emulsion drilling fluids.

Hydrocarbon-based drilling fluid is a water-in-oil invert emulsion, i.e., an emulsion in which water (W) is dispersed into tiny droplets, while the HC liquid is a dispersion medium. In this work, we investigated emulsions whose HC/W ratio was 65/35. This ratio was chosen based on the fact that the emulsion loses stability with the increase in the water concentration. In addition, the increase in oil content leads to an increase in the cost of the drilling fluid. Therefore, the chosen ratio is optimal.

The composition of the reagents that were used when preparing the drilling emulsion is described below.

REBASE PC-230 low-viscosity base oils (produced by OOO NPO REASIB, Tomsk, Russia) were used as the base hydrocarbon fluid (64.9 vol.%). REBASE PC-230 base oil with viscosity of 3.9 cP and density of 850 kg/m^3 (hereinafter referred to as Oil-1) was used as a base HC fluid in solutions modified by spherical NPs. REBASE low-viscosity oil with viscosity of 3.3 cP and density of 815 kg/m^3 (Oil-2) was used as a base HC fluid when preparing drilling fluids with carbon nanotubes (CNTs). This oil is a mixture of synthetic hydrocarbons of narrow fractional composition characterized by low kinematic viscosity, low content of aromatic substances, high flash-point, and low congelation temperature. This oil is used as an HC base in the preparation technology of heat-resistant invert emulsion drilling fluids. Due to highly-stable viscous properties, RS-230 allows preparing invert drilling fluids with a so-called flat rheological flow profile; flat rheology invert drilling fluids provide high technical and technological drilling performance in difficult geological conditions, especially in the case of a narrow regulation range of equivalent circulation density (ECD). They also prevent the loss of stability when drilling in shale rocks, as well as drilling fluid absorption due to hydraulic fractures of weakly cemented formations.

Calcium chloride ($CaCl_2$) brine (30 vol.%) reduces interface surface tension. By increasing the concentration of electrolytes in the aqueous phase, the rate of positive adsorption at the interface increases. Electrolyte ions cause dehydration of the hydrophilic groups of surfactants forming the adsorption layer, which promotes their convergence and more compact arrangement at the interface, reducing the size of molecule landing pad and increasing the condensation of the adsorption layer, which strengthens the protective film of globules. The charges of electrolyte ions attract opposite charges of ionic surfactants and contribute to the formation of a stronger and more condensed adsorption shell.

The non-ionogenic REBASE PC-510 emulsifier (2 vol.%) was used to stabilize the emulsion. The PS-510 emulsifier is a solution of compound ethers of higher unsaturated fatty acids based on mineral oils with the addition of high molecular weight organic compounds. PS-510 is designed as the main emulsifier when preparing heat-resistant invert emulsion drilling fluids.

The ABR hydrophobizer (0.4 vol.%) is used as a co-emulsifier of invert emulsion solutions and as a wetting agent for the solid phase of hydrocarbon-based drilling fluids. ABR is a surfactant that reduces surface tension at the oil/solid interface. It is used for hydrophobization of barite, marble, colmatant, drilled cuttings, and drilling tools. ABR stabilizes the solution in the case of stratum water inflow. In the case of solid phase development and emulsion weighting, the hydrophobizer envelops particles and prevents them from wetting with water and, as a consequence, from aggregating and settling. ABR also serves to reduce the filtration rate.

Organophilic clay (0.8 vol.%) is a structure-forming agent representing montmorillonite clay powder treated with amines with a certain length of HC radical, forming complex aggregates on the surface, which give the clay powder hydrophobic properties and the ability to swell in HC media. Organophilic clays are used as a structure-forming and wall-building component of oil-based solutions intended for use at high temperatures. They form structured solutions if their concentration in the solution exceeds 5 wt.%. Bitumen solutions, structured with organophilic clay, have a high retention capacity with respect to the weighting agent, and remain stable.

Oil-soluble polymer NRP-20M (0.5 vol.%) is a 20 wt.% solution of polyisobutylene in I-20A industrial oil. The introduction of NRP into the emulsion allows increasing the thermal stability of the solution due to the fact that polyisobutylene molecules, being in the interfacial layer, contribute to increasing the density of protective shells on water droplets and contribute to increasing the aggregative stability of the emulsion. Moreover, by thickening the dispersion medium of emulsion drilling fluid, polyisobutylene increases the structural and rheological properties of the solution, reduces its filtration, and increases the viscosity of the filtrate, which has a favorable effect on the preservation of porosity and permeability properties of the productive formation.

Micro calcite was used as LCM (lost circulation material) (4.4 vol.%). It is a fine-dispersed product of the carbonate group, which is produced by the mechanical grinding of marble. The following grades of calcium carbonate ($CaCO_3$) having different average particle sizes were used: MK-100, with an average particle size of 32 µm, MK-60—23 µm, and MK-10—10 µm.

Hydrophobic nanosized silicon oxide (SiO_2) particles were used to modify the drilling emulsion; they are fumed silica, treated with polydimethylsiloxane and/or cetyltrimethylammonium chloride. Silicon dioxide powder has high chemical resistance to many reagents and is characterized by high adsorption properties. The mass fraction of the main substance (SiO_2), in terms of the calcined substance, is not less than 99.6%. The average size of the NPs ranged from 18 to 70 nm. The weight fraction of NPs was set to 0.25, 0.5, 1, and 2 wt.%. Electron photographs of SiO_2 NPs are shown in Figure 1a. Electron microscopy was performed using a JSM-7001F scanning microscope (JEOL, Japan) and an S-5500 super-resolution scanning electron microscope (Hitachi, Japan). Electron microscopic examination was carried out using a base fluid pre-evaporation technique.

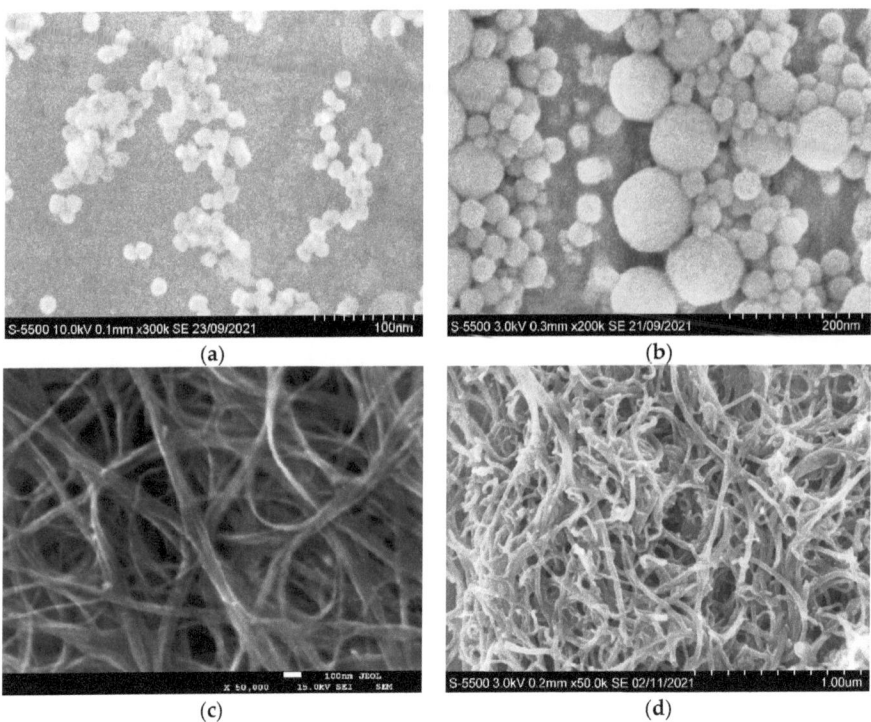

Figure 1. Electron microscopy of SiO$_2$ NPs with an average size of 18 nm (**a**), and 70 nm (**b**), SWCNTs (**c**), and MWCNTs (**d**).

SWCNTs and MWCNTs were used to modify the drilling emulsions. TUBALLTM; powder (OOO OCSiAl.ru, Novosibirsk, Russia) was used as SWCNTs [21]. This material is a loose black powder. The average diameter of the SWCNTs was 1.6 ± 0.4 nm, and the specific surface area, according to the BET method, was 510 m^2/g. On the evidence from atomic force microscopy data, the length of the SWCNTs exceeded 4 µm, and their average and bulk densities were 1.8 and 0.1 g/cm^3. Moreover, to improve the thermal stability of the drilling fluid, the effect of MWCNT additives, namely, Taunit-MD, produced by Nanotech Center (Tambov, Russia), was investigated [22]. The inner diameter of these MWCNTs was 5–15 nm, and the outer diameter was 8–30 nm. The specific surface area was higher than 270 m^2, and the length exceeded 5 µm. The number of carbon layers amounted to 30–40. The electron photographs of the MWCNTs are shown in Figure 1b. The weight fraction of nanotubes in the solutions was set equal to 0.1, 0.25, and 0.5 wt.%. Methodological experiments have shown that using higher concentrations of MWCNTs for the modification of drilling fluids is impractical due to a very significant increase in viscosity.

The methodology for preparing drilling fluids with NPs and nanotubes was slightly different. NPs were introduced into the HC phase. This was performed to avoid the fragmentation of the dispersed phase into nanoglobules after addition of CaCl$_2$ brine. To break the NP conglomerates, the solution was subjected to intensive ultrasonic treatment after their addition to the HC medium. Ultrasonic treatment was carried out by the Volna UZTA-0.4/22-OM ultrasonic device. The treatment was performed for 30 min at maximum performance.

When preparing nanotube-containing solutions, highly concentrated aqueous calcium chloride brine with a density of 1.1 g/cm³ was prepared first, and nanotubes were introduced into the aqueous phase and also treated by ultrasound. The ultrasonic treatment was carried out for 60 min at maximum performance, providing water cooling to prevent the dispersion from heating.

The dispersions, prepared according to the above technology, were used further for preparing emulsions. Emulsions were prepared using the coarse-droplet crushing technique, whose essence consists in drop-by-drop introduction of components in the course of simultaneous intensive mechanical dispergation by means of a high-speed stirrer. When preparing drilling fluid, HC base, emulsifier, hydrophobizer, oil-soluble polymer, microcolmatants, and calcium chloride brine were introduced sequentially. Intensive mixing was carried out using the Hamilton Beach triple-spindle mixer for 30 min at 20,000 rpm.

To evaluate the effect of nano-additives on the thermal stability of the drilling fluid, a comprehensive study of the hydrodynamic properties of the obtained emulsions was carried out. The colloidal stability of the drilling fluid was investigated in the course of thermal aging.

Thermal aging of drilling fluids was investigated using the OFITE roller oven (U.S. Patent No. 4,677,843), which allowed us to determine the temperature effect on drilling fluids circulating through the borehole (Figure 2a). Pressurizing drilling fluid in containers clearly demonstrates the effect of heat on viscosity, as well as the behavior of various additives at elevated temperatures. The aging was performed in conditions ranging from static to dynamic, and from ambient to elevated temperatures. Thermal aging cells inside the oven were continuously rotated to simulate drilling fluid circulation. The operating temperatures ranged from 38 to 232 °C. The temperature was monitored by an electronic semiconductor thermostat. The temperature varied from 30 to 150 °C, and the thermal aging time varied from 2 to 16 h.

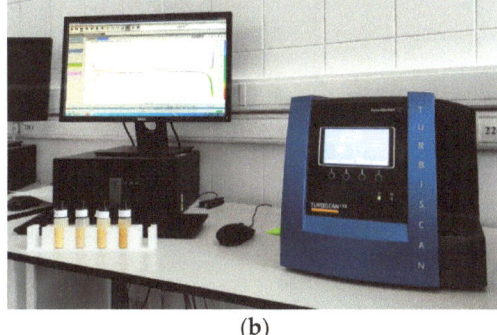

(a) (b)

Figure 2. The OFITE roller oven with cells for drilling fluids (**a**), the Turbiscan LAB colloidal stability analyzer (**b**).

Colloidal stability of emulsions was monitored using the Turbiscan LAB colloidal stability analyzer (Figure 2b), which allowed the analysis of the long-term stability of drilling fluid samples, as well as the calculation of the average particle/droplet size and volume fraction in the case of simple destabilization mechanisms. The measurement error did not exceed 0.1%.

3. Results of the Study of Thermal Stability of Drilling Fluids with Nano-Additives

One of the most important requirements for any emulsion is its tolerance to coalescence. For the drilling emulsion, this requirement is of fundamental importance. The rate of the destabilization processes increases when the drilling fluid is heated. When drilling wells, drilling fluid often has to operate at very high temperatures; therefore, the thermal stability problem is given great consideration. There are many different ways to stabilize the emulsion. The ability of fine particles to stabilize emulsions has been known for more than a hundred years. For example, the Pickering effect [23] is well-known. However, disperse systems, stabilized by NPs, have been intensively studied only in the last ten years. Currently, stabilization technologies of direct (oil-in-water) emulsions are being actively investigated due to their promising application in the pharmaceutical, cosmetic, food, and oil and gas industries [24–26]. Works are known in which the effect of MWCNTs on the stabilization of emulsions were studied [27,28]. It follows from these works that it is possible to reduce the droplet size in the emulsion several times by reducing the surface tension by adding MWCNTs. Drilling emulsions are much more complex colloidal systems; however, at present, no studies on the effect of CNTs on drilling fluids stability are available.

At the same time, there are practically no studies on the thermal stabilization of invert drilling emulsions by NPs, and there are even fewer studies on thermal stabilization through the use of carbon nanotubes. In this respect, the present work investigates the effect of nanosized silicon oxide and MWCNT additives on the stability of drilling fluids in the thermal stabilization process.

For this purpose, the measurement series of colloidal stability of drilling emulsions was carried out using Turbiscan LAB colloidal stability analyzer. The operation principle of Turbiscan LAB is based on the Multiple Light Scattering (MLS) method. The 20 mL sample to be analyzed was carefully placed in a cylindrical glass vial so that the meniscus was clearly visible. A near-infrared (880 nm) laser diode was used as the radiation source. Two synchronous optical sensors were used to measure the intensity of the radiation that passed through the sample (transmission detector, measuring 180° of the incident beam) and the radiation scattered by the sample in the opposite direction (backscatter detector, measuring 45° from the incident beam). The Turbiscan LAB scanned the sample along the vial height (up to 55 mm), recording transmission (T) and backscatter (BS) signals at every 40 µm. The transmission signal T was used when analyzing transparent or translucent samples, and the backscatter signal BS was used when analyzing translucent and opaque samples (concentrated dispersed systems). Measurements of repeated profiles at different time points t allows characterizing the changes occurring in the sample under study. The Turbiscan LAB uses a Turbiscan Stability Index (TSI) to quantitatively compare the stability of multiple samples. The TSI is calculated as the difference in the backscattering profiles of radiation at neighboring time points, averaged over the height of the vial. The lower this parameter, the more stable the emulsion is. In the beginning, the effect of the nano-additive concentration was investigated. To analyze the effect of nano-additives on the high-temperature stability of drilling emulsions, transmission and backscatter profiles were plotted for tested samples before and after 8 h thermal aging tests at 150 °C. Figure 3 shows the test results for the solution with 18 nm silicon oxide NPs. The changes in the samples over time are clearly visible from the backscatter profile. The emulsion without the addition of NPs demonstrates the most striking example. It can be seen that, after thermal aging, the segregation of the emulsion occurs almost up to half of the vial height, whereas the addition of NPs causes the backscattering profile to become more homogeneous, and the addition of 2 wt.% makes the emulsion stable and unchanged over time for a much longer period, indicating an enhancement in the colloidal stability of the emulsion.

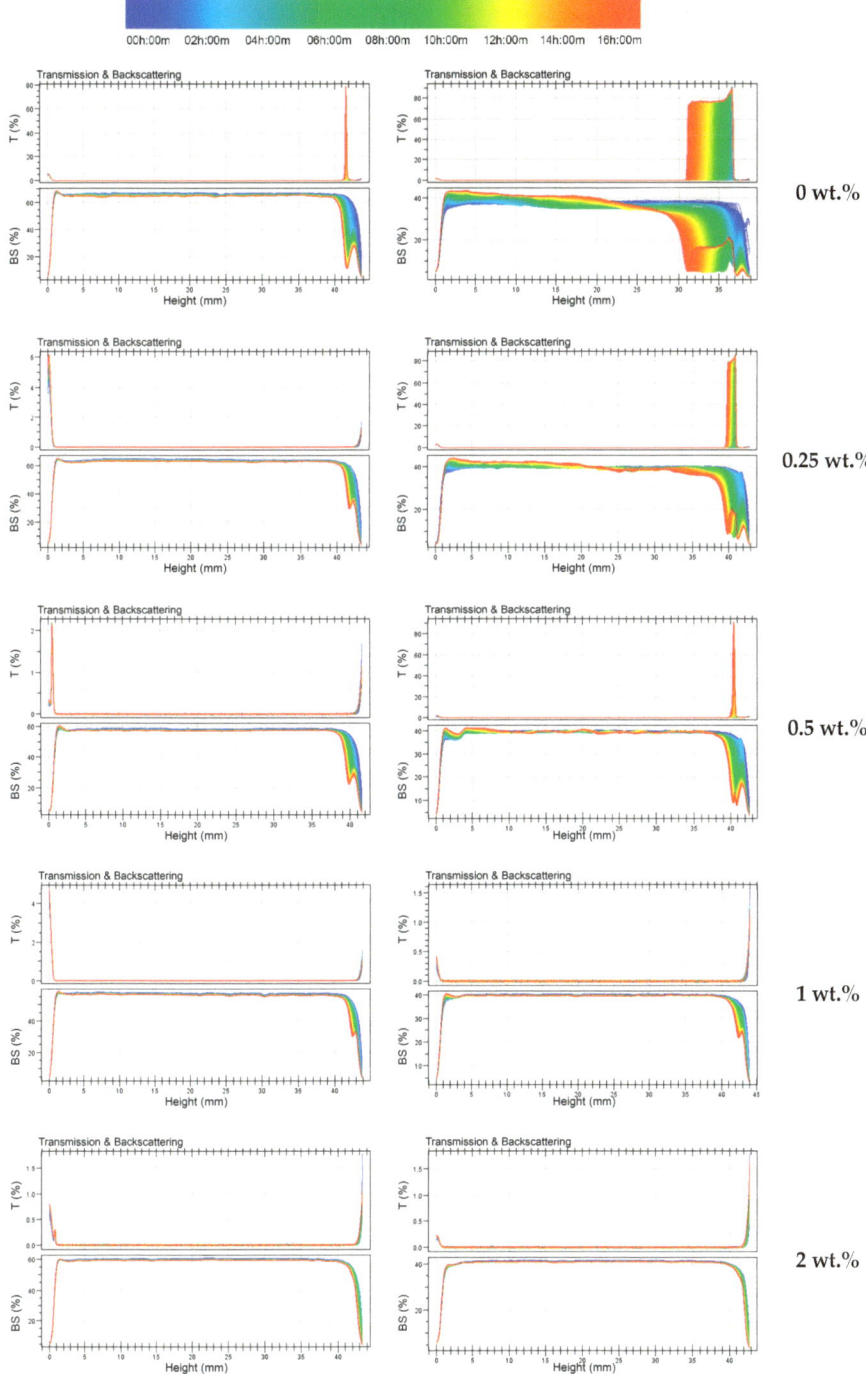

Figure 3. Evolution of transmission and backscattering profiles with increasing concentration of 18 nm SiO$_2$ nanoparticles for Oil-1-based drilling fluids before (**left**) and after 8 h thermal aging at 150 °C (**right**).

Quantitative comparison of the results on the effect of different NP concentrations on the thermal stability of drilling fluids is shown in Figure 4. The study of the effect of NP concentration on the colloidal stability of drilling fluids has shown that the colloidal stability of the basic drilling fluid before the thermal aging test has improved significantly after adding NPs, even at a minimum concentration of 0.25 wt.% (see Figure 4). Compared to the base fluid, the TSI has decreased 2.7 times. Solutions not subjected to thermal aging weakly depend on NP concentration. Further increase in the NP concentration has no significant effect on the stabilization of drilling emulsions at room temperature.

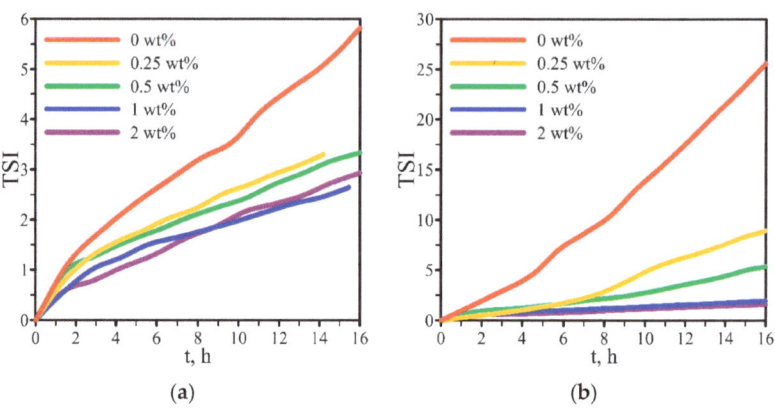

Figure 4. Evolution of the TSI over time depending on the concentration of 18 nm NPs for Oil-1-based drilling fluid before (**a**) and after 8 h thermal aging at 150 °C (**b**).

After being kept in the oven at high temperature during the thermal aging test, the base drilling fluid was significantly destabilized. This is clearly visible from both the backscattering and transmission profiles (Figure 3 (left)) and the TSI, measured 16 h after fluid preparation (see Figure 4b). However, the addition of NPs significantly enhances the colloidal stability of the drilling emulsion after thermal aging. Therefore, after adding 0.25 wt.% of NPs, the TSI of the emulsion after thermal aging had decreased by 2.9 times. A further increase in the concentration of NPs enhanced the stability of the samples, and, at a NP concentration of 2 wt.%, achieved a 16-fold decrease in TSI compared to the basic solution. It has also been shown that, at higher concentrations of NPs (more than 0.5 wt.%), the samples were even more stable than before the thermal aging test. However, increasing the concentration over 1 wt.% is unreasonable, because, in this case, the effect of enhancement is insignificant, and there is no need to increase the optimal cost of drilling fluid.

Similar studies were carried out for drilling fluids modified by carbon nanotubes.

Figure 5 shows the time history of the TSI for drilling emulsion with MWCNT additives before and after thermal aging. In this case, the lower viscosity Oil-2 was used to prepare the CNT-containing drilling fluids, which are more stable in the initial state compared to the above-discussed solutions. After 16 h of observation, the TSI of the base fluid did not exceed 2, while, for the Oil-2-based drilling fluid, the TSI was about 5. Nevertheless, the addition of CNTs to this drilling fluid prior to thermal aging test further stabilizes it. Unlike in the previous case, the TSI decreased monotonically with increasing NT concentration (see Figure 5a). At the concentration of MWCNTs of 0.5 wt.%, the rate of destabilization of the emulsion decreased about four times. Figure 5b shows the destabilization rate of emulsions with CNTs for 16 h of observation, pre-aged at 150 °C for 8 h. In this case, the HC base for the drilling fluid was chosen to be optimal. As can be seen, in this case, thermal aging practically did not affect colloidal stability of the studied emulsion. Nevertheless, adding CNTs to a greater extent contributed to its stabilization. Therefore, it was demonstrated that the addition of MWCNTs significantly improves the stability of drilling fluid, both in the initial state and after being subjected to thermal aging.

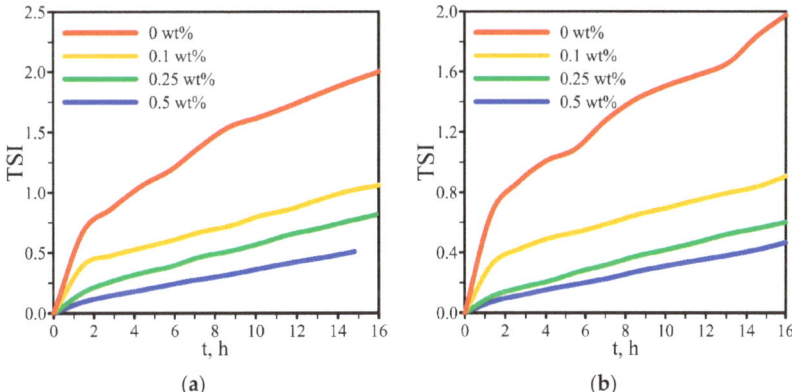

Figure 5. Evolution of the TSI over time depending on the concentration of MWCNTs for Oil-2-based drilling fluid before (**a**) and after 8 h thermal aging at 150 °C (**b**).

Generally, similar effects are manifested by SWCNT additives. Figure 6 shows the destabilization kinetics of drilling fluids with SWCNTs before and after thermal aging depending on the concentration of SWCNTs. It was shown that the TSI decreases monotonically with increasing SWCNT concentration. SWCNTs enhance the stability of hydrocarbon-based drilling fluids more effectively. Therefore, the addition of 0.1 wt.% of SWCNTs reduces the TSI by 3.1 times. The analysis of the effect of the SWCNT on the destabilization kinetics of the solutions subjected to thermal aging (see Figures 6b and 7b) has shown that their addition effectively enhances the colloidal stability of the emulsion. Thus, the addition of 0.1 wt.% of SWCNTs reduces the TSI by almost six times compared to that of the base solution. In our opinion, the main reason for this effect of CNTs on the stability of drilling emulsions is the formation of nanotube network, which, like polymer molecules, forms a developed structure inside the liquid that limits the mobility of water droplets and solid particles in the oil, thereby contributing to greater colloidal stability. Thus, it was found that the addition of CNTs significantly reduces the destabilization kinetics of hydrocarbon-based drilling fluids, both in the initial state and in the thermal degradation process.

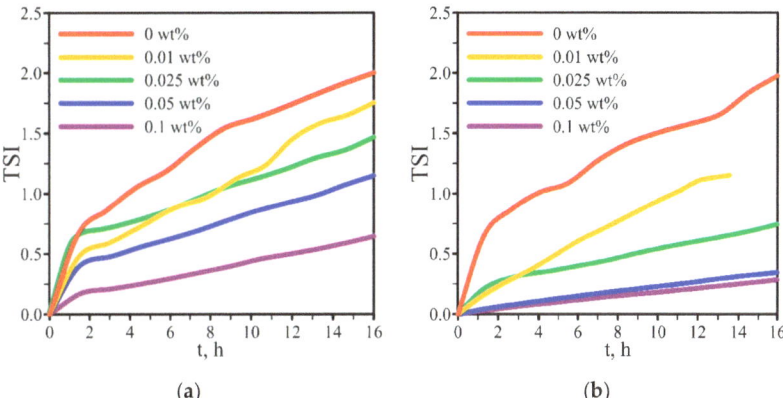

Figure 6. Evolution of the TSI over time depending on the concentration of SWCNTs for Oil-2-based drilling fluid before (**a**) and after 8 h thermal aging at 150 °C (**b**).

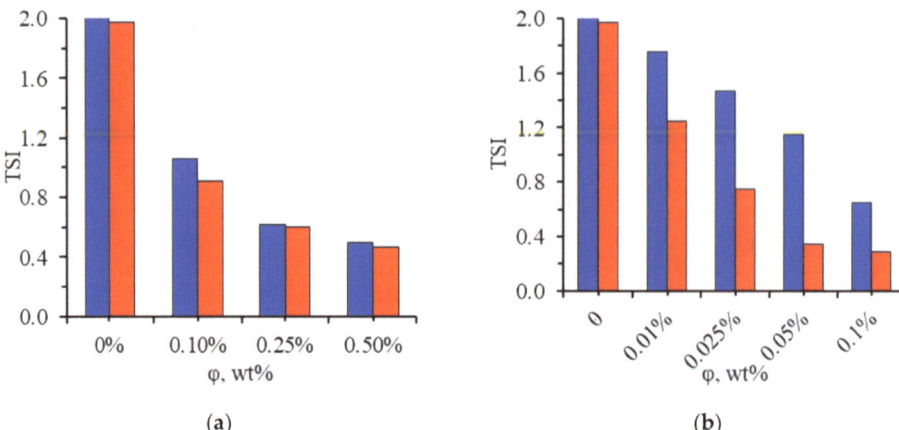

Figure 7. Dependence of the TSI on the concentration of MWCNTs (**a**) and SWCNTs (**b**) for Oil-2-based drilling fluid before (blue) and after 8 h thermal aging at 150 °C (red).

Due to coalescence processes, the droplets increase in size, and the emulsion begins to gradually segregate. As a result, after a while, a thin layer of lighter weight mineral oil is formed at the surface of the drilling fluid cell. Figure 8 shows the thickness (a) and growth rate (b) of the segregated layer of drilling fluids with SWCNTs and MWCNTs (0.1 wt.%) before and after 8 h exposure at 150 °C. As is obvious, the addition of nanotubes, especially SWCNTs, significantly reduces the segregation rate compared to the base drilling fluid.

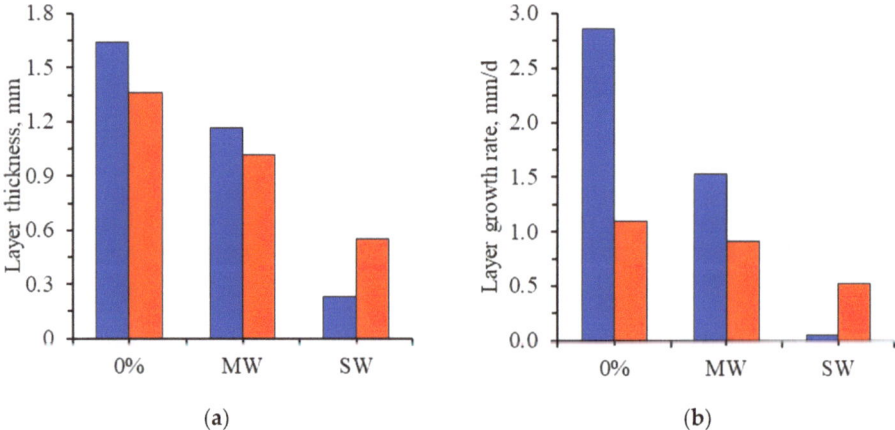

Figure 8. Dependence of layer thickness (**a**) and segregation rate (**b**) of drilling fluid without additives and with 0.1 wt.% of SWCNTs and MWCNTs before (blue) and after 8 h thermal aging at 150 °C (red).

Next, the effect of the NP size on the colloidal stability of hydrocarbon-based drilling fluids containing SiO_2 additive was analyzed directly in the thermal aging process. For this purpose, NP additives of three different average particle sizes, namely 18, 50, and 70 nm, were considered. Such a study was performed for the first time. The NP size distributions are shown in Figure 9. The particle size distributions directly in the liquid were measured using an acoustic and electroacoustic analyzer DT1202. The acoustic method is based on the measurement of the attenuation degree of the ultrasonic signal, and is used for determining particle size, including in opaque and concentrated media.

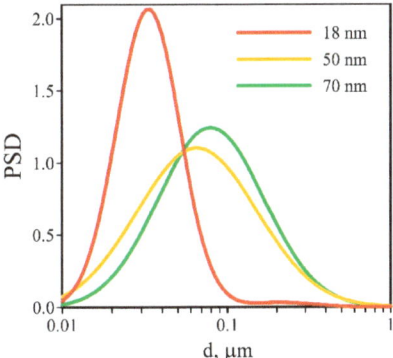

Figure 9. Weight basis particle size distribution of SiO$_2$ nanoparticles in the drilling fluid.

Determining the size of nanoparticles in a suspension is a very important task [29,30] since the properties of the suspension depend significantly on the size of the nanoparticles. In this work, the average size of the nanoparticles was determined directly in the liquid, which is fundamentally important. The DT-1200 series acoustic and electroacoustic analyzer for characterizing dispersions is designed for characterizing dispersions in a wide range of particle concentrations (0.1 ÷ 60 vol.%) and measuring average particle sizes, dynamic viscosity, and specific electrical conductivity of various dispersions. The acoustic sensor of the device measures the ultrasonic attenuation coefficient (0 ÷ 20 ± 0.01 dB cm^{-1} MHz^{-1}) in a wide dynamic frequency range (from 1 to 100 MHz), as well as the speed of sound, the values of which are subsequently used to determine the average particle size, dynamic viscosity, and specific electrical conductivity of the dispersions under study.

For drilling emulsions with 50 and 70 nm NPs, studies were performed similar to those described above for 18 nm NPs. For samples with larger NPs, a significant improvement in colloidal stability of drilling emulsions, not subjected to thermal aging, was observed already at minimal concentrations (0.25 wt.%). Dependences of the resulting TSI on the concentration of NPs of different sizes, obtained for 16 h of observation, are shown in Figure 10. Data analysis shows that, for the solutions not subjected to thermal aging, the TSI monotonically decreases with the decreasing size of NPs (see Figure 11). Reduction in NP size from 70 to 18 nm resulted in enhancement of the colloidal stability of the initial solution by about 12%.

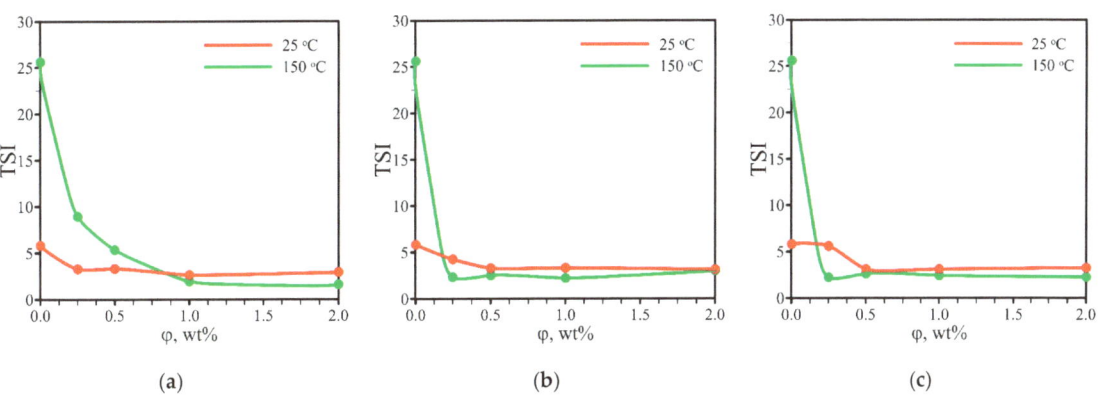

Figure 10. The TSI of the drilling fluid with NPs of 18 nm (**a**), 50 (**b**), and 70 (**c**), measured 16 h after the preparation, depending on the NP concentration.

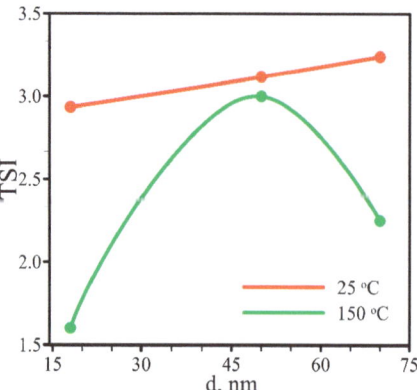

Figure 11. The TSI of drilling fluid with 2 wt.% of SiO_2 additive, measured 16 h after the preparation, depending on the NP size.

Significantly, a greater influence of coarse-sized NPs was observed in the solutions after their thermal aging test. It was found that, for all NP concentrations, the TSI, measured after 16 h, was lower after the exposure of the solutions to high temperature, comparing to that at room temperature (see Figures 10 and 11). This is very important for practical applications, since, usually, standard additives, used to maintain the thermal stability of drilling fluids, keep the drilling emulsions stable in the best case, while NPs improve it.

These results are clearly demonstrated in the graphs, showing the effect of NP size on the TSI (see Figure 10). It can be seen that the most significant influence on the colloidal stability of drilling emulsions is exerted by 18 nm NPs, i.e., the smallest of all considered sizes. Moreover, this influence is the most significant both before and after thermal aging tests. It is seen that with the increase in the size of the particles, the stability of fluid samples, not subjected to thermal aging, varies within 12%. Therefore, although all the NPs reduce the TSI after 8 h thermal treatment of the solutions at 150 °C, it is most appropriate to use particles with the smallest size. Thus, it is shown that, as the NP size decreases, the stability of hydrocarbon-based drilling fluids to thermal degradation generally increases.

Next, the effect of temperature on the colloidal stability of drilling emulsions with NPs and nanotubes was analyzed for the first time. The main difference from the thermal aging tests of drilling fluids was that, in this case, the change in stability of the fluid samples was studied in static conditions, i.e., without fluid circulation (as was the case of in the thermal aging cells, which were continuously rotating throughout the measurement). The colloidal stability of the drilling fluid in a static state, depending on temperature change, is a very important indicator in the case of forced cessation of drilling, which may occur in case of an emergency or the need to seal the well. In this context, to avoid complications when resuming circulation, the drilling fluid should be maximally stable for a long period at different temperatures (the average temperature gradient is 3 °C per every 100 m). This is important, because if the so-called segregation behavior of the drilling fluid occurs, the hydrostatic pressure will change abruptly over the well depth, which may lead to spontaneous uncontrolled flow of reservoir fluid into the well, creating a risk of drilling fluid release, and, consequently, its loss, as well reservoir fracturing, etc., which is a serious violation of drilling technology.

For this test, a drilling fluid sample containing 50 nm NPs at concentration of 2 wt.% was compared with the base fluid without NP additive. The temperature varied from 30 to 60 °C. The destabilization evolution of the samples was investigated for 8 h.

The transmission and backscattering profiles clearly show the positive effect of NPs on the stability of uncirculated drilling fluids with increasing temperature (see Figure 12). At all temperatures considered, with the addition of NPs into the emulsions, the profiles became more homogeneous, indicating an enhancement of colloidal stability.

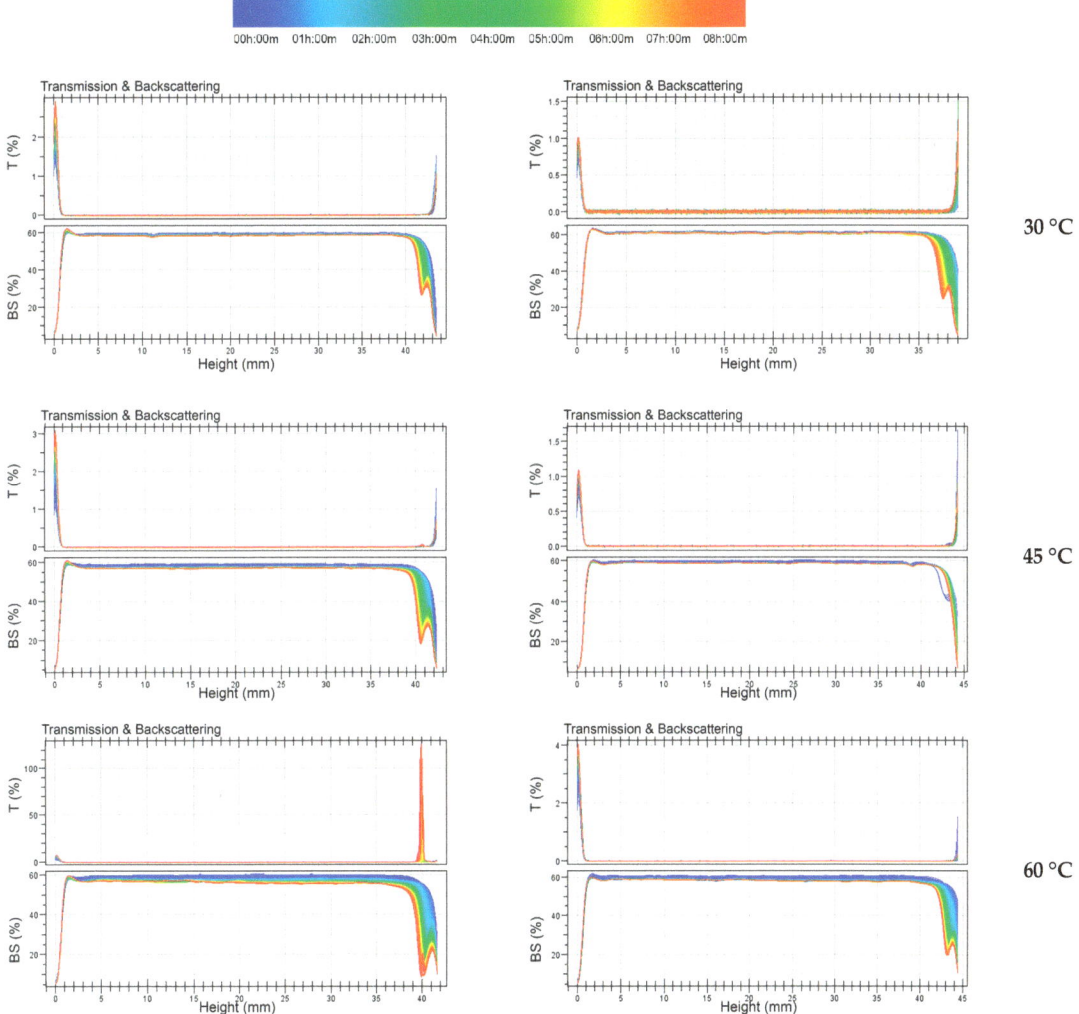

Figure 12. Evolution of transmission and backscattering profiles with increasing temperature for drilling fluids without NPs (**left**) and with 55 nm NPs at a concentration of 2 wt.% (**right**).

Figure 13 demonstrates that the colloidal stability decreases with increasing temperature for all the samples considered. With an increase in temperature by 30 °C, the TSI for the basic solution increased almost three times. However, the addition of NPs stabilizes the solutions. Moreover, the higher the temperature, the stronger the effect of NPs on the TSI. This is clearly seen in Figure 13. Thus, for a solution at 45 °C, the NPs reduce the TSI by a factor of 1.5, and, for a solution at 60 °C, by a factor of 1.8.

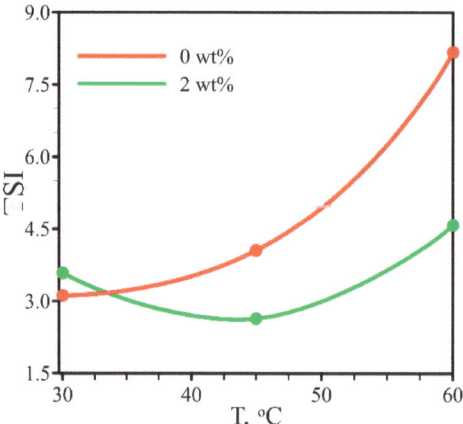

Figure 13. The TSI, measured 8 h after preparation of the solution without NPs (blue) and with SiO$_2$ NPs (red) at a concentration of 2 wt.% depending on temperature.

A similar study on the effect of temperature on destabilization kinetics was carried out for drilling fluids with additives of SWCNTs and MWCNTs. The main results of this study are presented in Figures 14 and 15. The evolution of the increment of the relative radiation backscattering for drilling fluid samples, held at 60 °C for 16 h, is shown in Figure 14. Such a change in backscattering indicates the droplet coalescence processes occurring in the fluid volumes. At the same time, the intensity of this destabilization process in solutions with nanotubes is several times lower.

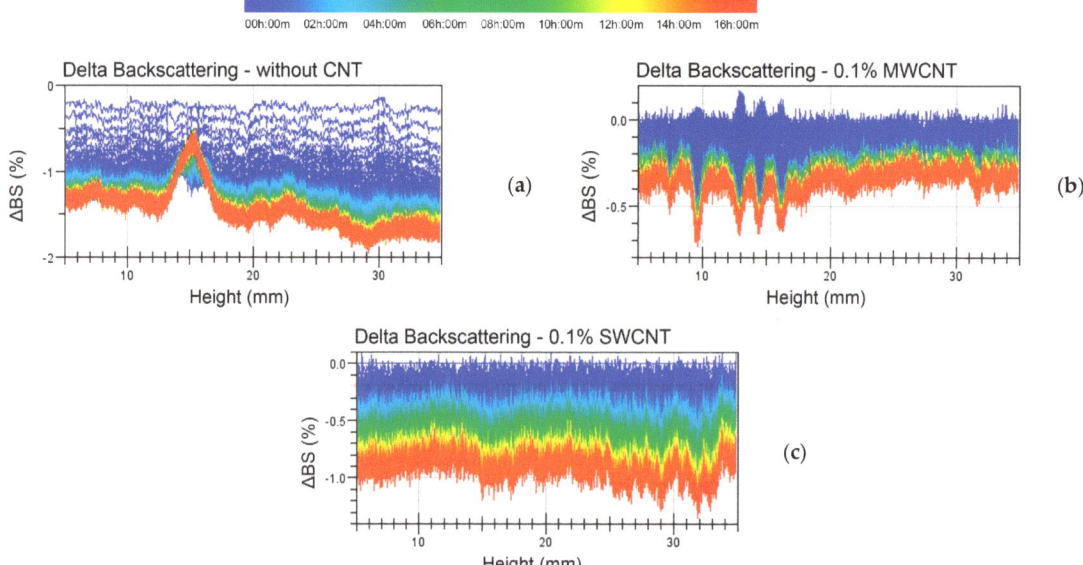

Figure 14. Evolution of the backscattering increment profiles for drilling fluids at 60 °C: without NPs (**a**), with MWCNTs (**b**), and with SWCNTs (**c**), both at concentrations of 0.1 wt.%.

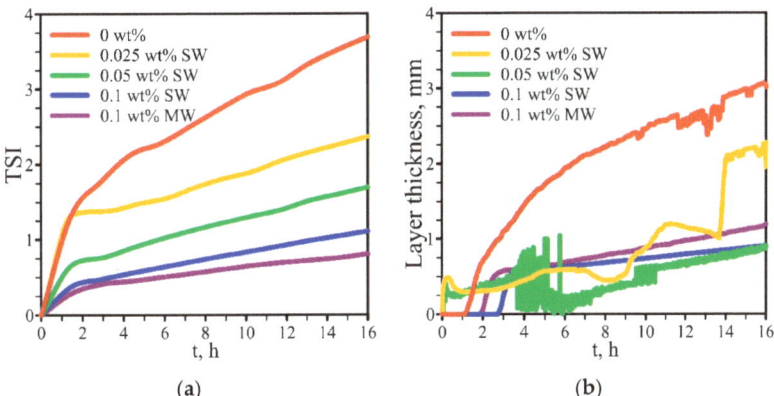

Figure 15. Evolution of the TSI (**a**) and segregation layer thickness (**b**) over time depending on CNT concentration for Oil-2-based drilling fluid at 60 °C.

This statement is proved by the destabilization kinetics of drilling fluids with CNTs directly in the process of their exposure at high temperature, shown in Figure 15a. Increasing temperature significantly destabilizes drilling emulsions. While comparing similar results obtained at room temperature, shown in Figures 5 and 6, it can be noted that the destabilization rate of the drilling fluids at 60 °C increases by a factor of about 1.8 for all samples. The addition of nanotubes decreases the destabilization rate both at room and high temperatures. With increasing temperature, the effect of CNT addition increases slightly. Therefore, at 30 °C, the addition of 0.1 wt.% of SWCNTs reduces the TSI by 3.1 times, while, at 60 °C, it reduces by 3.4 times. The addition of MWCNTs gives quite similar results.

Figure 15b shows the thickness of this segregation layer at the surface of the drilling fluid over time when aging the fluid at 60 °C. For the base drilling fluid without additives, this layer starts growing after about one and a half hours of exposure to the set temperature, and it reaches about 3 mm after 16 h of observation. For drilling fluids, modified by 0.1 wt.% of MWCNTs, this layer appears after about two hours, and, for the drilling fluid with SWCNTs at the same concentration, the effect is achieved after three hours, while the layer thickness reaches 1 mm. At the same time, the growth rate of the segregation layer for drilling fluids, modified by nanotubes, is also significantly lower than that for the basic solution. This all indicates a significant stabilization of the drilling fluid samples when adding nanotubes. With increasing nanotube concentration, the stabilizing effect enhances.

In the end, it is necessary to discuss the possible mechanisms of the influence of nanotubes on the processes of colloidal stability of drilling emulsions. As the results of this work showed, with the introduction of nanotubes, the colloidal stability of the emulsion increased significantly. Nanotubes have interfacial wettability and are embedded in the interface between the aqueous and hydrocarbon phases, thereby preventing droplet coalescence. In addition, there are works that studied the effect of multi-walled nanotubes on the size of emulsion droplets [28]. It has been shown that, with the MWCNT additives, by reducing surface tension, it is possible to reduce the size of the droplets in the emulsion by several times. Reducing the size of emulsion droplets can contribute to the emulsion stabilization process. Another important circumstance that contributes to the stabilization of emulsions using nanotube additives is the fact that their addition significantly increases the viscosity of liquids and changes their rheology [31]. The addition of nanotubes significantly increases the yield strength of liquids. This indicates that, in emulsions with nanotubes, structure formation processes associated with tube percolation processes occur more intensively. As the length of the tubes increases, the concentration at which percolation occurs, decreases. Nanotubes inside a liquid, like polymer molecules, form a stable network-like structure, which impedes the movement of droplets, thereby preventing their coalescence. We are confident that this is not a complete list of the mechanisms of the influence of nanotube additives on the colloidal stability of drilling emulsions, and this issue requires further research.

In addition to increasing the thermal stability of drilling fluids when adding nanomaterials, one more important fact should be noted. The addition of nanoparticles and carbon nanotubes changes the thermophysical characteristics of the medium. In particular, it increases the coefficient of thermal conductivity and increases heat transfer. It should be noted that the problem of enhancing the thermal conductivity coefficient of fluids using nanotubes is not a new issue. This technology has been widely investigated as a method of enhancing heat transfer using nanofluids. Several dozens of studies, described in review articles, have been carried out in this area [31–34]. It has been shown that, indeed, adding nanotubes can enhance the thermal conductivity of fluids by tens of percent. Similar effects are observed when adding MWCNTs to drilling fluids. However, while enhancing thermal conductivity of the heat transfer fluid is beneficial, enhancing that of the drilling fluids is not so unambiguous. On the one hand, it is a beneficial factor, because it contributes to more effective cooling of the drilling bit; on the other hand, it can have a negative impact on thawing rate while drilling in permafrost conditions. Therefore, studying the effect of MWCNT additives on the thermal conductivity of drilling fluids requires a deeper understanding.

4. Conclusions

A systematic experimental study of the effects of silicon oxide NPs, as well as SWCNTs and MWCNTs on colloidal stability during thermal degradation of hydrocarbon-based drilling fluids, has been carried out, as well as a study of the effects of NP and nanotube concentration and size on the stability of drilling fluids during thermal aging tests. The size of the NPs varied from 18 to 70 nm, and the concentration ranged from 0.25 to 2 wt.%. The concentration of CNTs varied from 0.01 to 0.5 wt.%. The temperature in the thermal aging tests varied from 30 to 150 °C. The thermal aging time ranged from 2 to 16 h. During the experiments, the effect of temperature on the destabilization kinetics of drilling hydrocarbon-based emulsions stabilized by NPs and CNTs was investigated.

The following conclusions have been drawn.

1. It has been shown that the addition of NPs significantly enhances the colloidal stability of the drilling emulsion, both in the initial state and when subjected to thermal degradation. Thus, with the addition of 2 wt.% of silicon oxide NPs with an average particle size of 18 nm, the TSI of the drilling fluid after thermal aging has decreased by 16 times compared to the basic drilling fluid.
2. The analysis of the effect of NP size on the colloidal stability of hydrocarbon-based drilling fluids with the addition of NPs in the course of thermal aging has shown that, with decreasing NP size, the stability of hydrocarbon-based drilling fluids subjected to thermal degradation processes enhances significantly.
3. It is established that CNT additives also significantly stabilize hydrocarbon-based emulsions, enhancing their thermal stability. With increasing CNT concentration, the TSI of solutions before and after thermal aging monotonically decreases. In this regard, SWCNTs proved to be the most effective. Compared to the solution without additives, addition of 0.1 wt.% of SWCNTs reduces the destabilization rate of the initial solution by 3.1 times, and almost by 6 times for the solution aged for 8 h at 150 °C.
4. MWCNTs also enhance the temperature stability of drilling fluids. However, achieving the same effects requires significantly higher concentrations of MWCNTs compared to SWCNTs. The addition of 0.5 wt.% of MWCNTs would be required to reduce the stabilization rate of drilling fluids aged at high temperature.
5. The effect of temperature on the colloidal stability of drilling emulsions with NPs and nanotubes was analyzed. It was revealed that, at all considered temperatures, nano-additives increase the thermal stability of drilling fluids. Moreover, the higher the temperature, the greater is the effect of nanomaterial additives on thermal stability.

Thus, based on the results of this study, it has been shown that it is possible to significantly enhance the temperature stability of hydrocarbon-based drilling fluids by adding various nanomaterials. At the same time, it was shown that the use of SWCNTs for thermal stability was several times more effective than using MWCNTs and tens of times more effective than adding spherically shaped silicon oxide nanoparticles.

Another important feature of NP and carbon nanotube additives is the fact that they, unlike standard additives used for maintaining the thermal stability of drilling fluids, are, themselves, stable, and do not degrade at high temperatures. Therefore, nano-additives can be recommended for thermal stabilization of hydrocarbon-based fluids, to be used when drilling wells in formations with elevated temperatures.

A logical and reasonable continuation of this work will be research on the effect of nanotube additives on the thermal destruction of water-based drilling fluids, which are essentially suspensions. Despite the fact that these solutions are less effective when drilling unstable and permafrost rocks, they have a significant advantage: they do not have such a negative impact on the environment due to a significantly lower amount of toxic components. Therefore, improving the formulations of water-based drilling fluids to increase their thermal stability with the help of nanoparticles also seems to be an urgent task.

Author Contributions: Conceptualization, E.I.L. and A.V.M.; methodology, E.I.L. and A.V.M.; validation, A.V.M.; investigation, E.I.L. and A.D.S.; data curation, A.V.M.; writing—original draft preparation, E.I.L. and A.D.S.; writing—review and editing, E.I.L. and A.V.M.; visualization, E.I.L.; supervision, A.V.M.; project administration, A.V.M.; funding acquisition, A.V.M. All authors have read and agreed to the published version of the manuscript.

Funding: The research was funded by a grant from the Russian Science Foundation No. 23-79-30022, https://rscf.ru/project/23-79-30022/, accessed on 1 September 2023.

Data Availability Statement: Not applicable.

Conflicts of Interest: The authors declare no conflict of interest.

References

1. Available online: https://ourworldindata.org/grapher/fossil-fuel-production (accessed on 1 September 2023).
2. Holechek, J.L.; Geli, H.M.E. A global assessment: Can renewable energy replace fossil fuels by 2050? *Sustainability* **2022**, *14*, 4792. [CrossRef]
3. Zhao, X.; Qiu, Z.; Huang, W.; Wang, M. Mechanism and method for controlling low-temperature rheology of water-based drilling fluids in deepwater drilling. *JPSE* **2017**, *154*, 405–416. [CrossRef]
4. Beg, M.; Kumar, P.; Choudhary, P.; Sharma, S. Effect of high temperature ageing on TiO2 nanoparticles enhanced drilling fluids: A rheological and filtration study. *Upstream Oil Gas Technol.* **2020**, *5*, 100019. [CrossRef]
5. Huang, X.; Lv, K.; Sun, J.; Lu, Z.; Bai, Y. Enhancement of thermal stability of drilling fluid using laponite nanoparticles under extreme temperature conditions. *Mater. Lett.* **2019**, *248*, 146–149. [CrossRef]
6. Mao, H.; Qiu, Z.; Shen, Z.; Huang, W. Hydrophobic associated polymer based silica nanoparticles composite with core–shell structure as a filtrate reducer for drilling fluid at utra-high temperature. *JSPE* **2015**, *129*, 1–14. [CrossRef]
7. Dai, Z.; Sun, J.; Wang, Y. A polymer-based drilling fluid with high temperature, salt and calcium resistance property. *IOP Conf. Ser. EES* **2019**, *237*, 052058. [CrossRef]
8. Hamad, B.A.; He, M.; Xu, M.; Liu, W.; Mpelwa, M. A novel amphoteric polymer as a rheology enhancer and fluid-loss control agent for water-based drilling muds at elevated temperatures. *ACS Omega* **2020**, *5*, 8483–8495. [CrossRef]
9. Oseh, J.O.; Mohd, N.M.N.A.; Gbadamosi, A.O.; Agi, A.; Blkoor, S.O. Polymer nanocomposites application in drilling fluids: A review. *Geoenergy Sci. Eng.* **2023**, *222*, 211416. [CrossRef]
10. Al-Malkli, N.; Pourafshary, P.; Al-Hadrami, H.; Abdo, J. Controlling bentonite-based drilling mud properties using sepiolite nanoparticles. *Pet. Explor. Dev.* **2016**, *43*, 717–723. [CrossRef]
11. Mohammed, A.S. Effect of temperature on the rheological properties with shear stress limit of iron oxide nanoparticle modified bentonite drilling muds. *Egypt. J. Pet.* **2017**, *26*, 791–802. [CrossRef]
12. Devi, G.; Al-Ruqeishi, M.; Mohiuddin, T. Experimental investigation of drilling fluid performance as nanoparticles. *WJNSE* **2013**, *3*, 53–61.
13. Minakov, A.V.; Pryazhnikov, M.I.; Mikhienkova, E.I.; Voronenkova, Y.O. Systematic experimental study of the temperature dependence of viscosity and rheological behavior of water-based drilling fluids with nano-additives. *Petroleum*, **2022**; *in press*.

14. Ahmad, H.M.; Kamal, M.S.; Murtaza, M.; Al-Harthi, M.A.; Fahd, K. Improving the drilling fluid properties using nanoparticles and water-soluble polymers. In Proceedings of the SPE Kingdom of Saudi Arabia Annual Technical Symposium and Exhibition, Dammam, Saudi Arabia, 24–27 April 2017. SPE-188140-MS.
15. Anoop, K.; Sadr, R.; Yrac, R.; Amani, M. Rheology of a colloidal suspension of carbon nanotube particles in a water-based drilling fluid. *Powder Technol.* **2019**, *342*, 585–593. [CrossRef]
16. Kazemi-Beydokhti, A.; Hajiabadi, S.H. Rheological investigation of smart polymer/carbon nanotube complex on properties of water-based drilling fluids. *Colloids Surf. A Physicochem. Eng. Asp.* **2018**, *556*, 23–29. [CrossRef]
17. Ahmad, M.H.; Kamal, S.M.; Al-Harthi, M.A. Synthesis and experimental investigation of novel CNT-polymer nanocomposite to enhance borehole stability at high temperature drilling applications. In Proceedings of the SPE Kingdom of Saudi Arabia Annual Technical Symposium and Exhibition, Dammam, Saudi Arabia, 23–26 April 2018. SPE-192352-MS.
18. Li, M.-C.; Wu, Q.; Song, K.; Qing, Y.; Wu, Y. Cellulose nanoparticles as modifiers for rheology and fluid loss in bentonite water-based fluids. *ACS Appl. Mater. Interfaces* **2015**, *7*, 5006–5016. [CrossRef]
19. Arain, A.H.; Ridha, S.; Mohyaldinn, M.E.; Suppiah, R.R. Improving the performance of invert emulsion drilling fluid using boron nitride and graphene nanoplatelets for drilling of unconventional high-temperature shale formations. *J. Mol. Liq.* **2022**, *363*, 119806. [CrossRef]
20. Soares, Y.C.F.; Yokoyama, D.D.; Costa, L.C.; de Oliveira Cremonezzi, J.M.; Ribeiro, H.; Naccache, M.F.; Andrade, R.J.E. Multifunctional hexagonal boron nitride dispersions based in xanthan gum for use in drilling fluids. *Geoenergy Sci. Eng.* **2023**, *221*, 111311. [CrossRef]
21. Predtechenskiya, M.R.; Khasina, A.A.; Bezrodnya, A.E.; Bobrenoka, O.F.; Dubova, D.Y.; Muradyana, V.E.; Saika, V.O.; Smirnov, S.N. New perspectives in SWCNT applications: Tuball SWCNTs. Part 1. Tuball by itself—All you need to know about it. *Carbon Trends* **2022**, *8*, 100175. [CrossRef]
22. Korusenko, P.M.; Nesov, S.N. Composite based on multi-walled carbon nanotubes and manganese oxide with rhenium additive for supercapacitors: Structural and electrochemical studies. *Appl. Sci.* **2022**, *12*, 12827. [CrossRef]
23. Pickering, S.U. Emulsions. *J. Chem. Soc.* **1907**, *91*, 2001–2002. [CrossRef]
24. Arab, D.; Kantzas, A.; Bryant, S.L. Nanoparticle stabilized oil in water emulsions: A critical review. *J. Pet. Sci. Eng.* **2018**, *163*, 217–242. [CrossRef]
25. Ashaolu, T.J. Nanoemulsions for health, food, and cosmetics: A review. *Environ. Chem. Lett.* **2021**, *19*, 3381–3395. [CrossRef] [PubMed]
26. Mariyate, J.; Bera, A. A critical review on selection of microemulsions or nanoemulsions for enhanced oil recovery. *J. Mol. Liq.* **2022**, *353*, 118791. [CrossRef]
27. Briggs, N.M.; Weston, J.S.; Li, B.; Venkataramani, D.; Aichele, C.P. Multiwalled carbon nanotubes at the interface of pickering emulsions. *Langmuir* **2015**, *31*, 13077–13084. [CrossRef]
28. Briggs, N.; Raman, A.K.Y.; Barrett, L.; Brown, C.; Li, B. Stable pickering emulsions using multi-walled carbon nanotubes of varying wettability. *Colloids Surf. A Physicochem. Eng. Asp.* **2018**, *537*, 227–235. [CrossRef]
29. Singh, S.B.; De, M. Scope of doped mesoporous (<10 nm) surfactant-modified alumina templated carbons for hydrogen storage applications. *Int. J. Energy Res.* **2019**, *43*, 4264–4280.
30. Alumina based doped templated carbons: A comparative study with zeolite and silica gel templates. *Microporous Mesoporous Mater.* **2018**, *257*, 241–252. [CrossRef]
31. Rudyak, V.Y.; Dashapilov, G.R.; Minakov, A.V.; Pryazhnikov, M.I. Comparative characteristics of viscosity and rheology of nanofluids with multi-walled and single-walled carbon nanotubes. *Diam. Relat. Mater.* **2023**, *132*, 109616. [CrossRef]
32. Selvam, C.; Mohan Lal, D.; Harish, S. Thermal conductivity and specific heat capacity of water–ethylene glycol mixture-based nanofluids with graphene nanoplatelets. *J. Therm. Anal. Calorim.* **2017**, *129*, 947–955. [CrossRef]
33. Kode, V.R.; Stuckenberg, D.J.; Went, E.K.; Erickson, O.M.; Plumer, E. Techno-economic analysis of atmospheric water generation by hybrid nanofluids to mitigate global water scarcity. *Liquids* **2022**, *2*, 183–195. [CrossRef]
34. Angayarkanni, S.A.; Sunny, V.; Philip, J. Effect of nanoparticle size, morphology and concentration on specific heat capacity and thermal conductivity of nanofluids. *J. Nanofluids* **2015**, *4*, 302–309. [CrossRef]

Disclaimer/Publisher's Note: The statements, opinions and data contained in all publications are solely those of the individual author(s) and contributor(s) and not of MDPI and/or the editor(s). MDPI and/or the editor(s) disclaim responsibility for any injury to people or property resulting from any ideas, methods, instructions or products referred to in the content.

Article

Deepwater PDC Jetting Bit-Drilling Technology Based on Well Structure Slimming

Weiguo Zhang [1,2], Deli Gao [1], Yijin Zeng [1,3] and De Yan [2,*]

1 College of Petroleum Engineering, China University of Petroleum (Beijing), Beijing 102249, China; zhangwg@cnooc.com.cn (W.Z.)
2 CNOOC China Ltd. Shenzhen Branch, Shenzhen 518000, China
3 SINOPEC Research Institute of Petroleum Engineering Co., Ltd., Beijing 102206, China
* Correspondence: yande@cnooc.com.cn

Abstract: Growing global energy demand and limited reserves of traditional energy resources are causing a growing energy shortage. In order to meet future energy needs, new energy resources must be continuously explored. Deepwater drilling research has emerged as one of the key ways to address this issue, and well structure slimming is an effective way to increase drilling speed and reduce costs. The hole size of the second section of deepwater wells decreases from a conventional 660.4 mm to 444.5 mm and increases from 500–800 m to 800–1200 m, creating problems where the conventional 660.4 mm cone bit cannot be used, the rate of penetration (ROP) of the cone bit is low, and the service life is short. To solve these problems, a 444.5 mm artificial polycrystalline diamond compact (PDC) was designed for the first time for use at home or abroad, and according to the characteristics and operation requirements of the 914.4 mm conductor jetting process, a unique anti-collision gauge protector was designed, an innovative bypass nozzle was configured, and a hydraulic design to prevent bit balling in shallow soft mudstone was formulated. PDC jetting bit-drilling technology based on well structure slimming was successfully applied to eight deepwater wells in the eastern South China Sea, which successfully jetted a 914 mm conductor and greatly improved the ROP of their second-section holes. When the below-mudline depth of the second-section hole increased by 37.01%, the average ROP increased by 227.84%. These technical achievements have successfully realized deep drilling with seawater, increased speed and efficiency, achieved good application results, and accumulated valuable experiences that can be used for reducing the cost and increasing the efficiency of offshore drilling operations.

Keywords: well structure; slimming; deep drilling with seawater; PDC jetting bit; bypass nozzle; jetting

1. Introduction

In the contemporary era, energy is the vital lifeblood of our global society, serving as the foundational pillar upon which the multifarious sectors of the world's economy pivot and flourish. Nevertheless, the inexorable rise in global energy demand—driven by burgeoning population dynamics, the relentless march of industrialization, and the ascendancy of emerging economies—portends a looming specter of depletion of conventional energy resources. This precipitates a pressing inquiry: how can we satiate future energy exigencies while concurrently alleviating our reliance on the finite reservoirs of traditional energy sources? The realm of deepwater drilling emerges as a beacon of hope in grappling with this exigency [1,2]. Deepwater regions enshroud vast, untapped reservoirs of petroleum and natural gas. Through the meticulous inquiry and extraction endeavors of deepwater drilling, we are poised to unearth and exploit these reservoirs, potentially extending the temporal bounds of conventional energy resource provisioning. Deepwater drilling, therefore, proffers a conduit toward diversifying our energy supply matrix, ameliorating the proclivity toward dependence on specific regions or sources, and bolstering the global energy supply's resilience and security.

Citation: Zhang, W.; Gao, D.; Zeng, Y.; Yan, D. Deepwater PDC Jetting Bit-Drilling Technology Based on Well Structure Slimming. *Energies* 2023, *16*, 7394. https://doi.org/10.3390/en16217394

Academic Editors: Antonio Zuorro and Hossein Hamidi

Received: 5 September 2023
Revised: 26 September 2023
Accepted: 25 October 2023
Published: 1 November 2023

Copyright: © 2023 by the authors. Licensee MDPI, Basel, Switzerland. This article is an open access article distributed under the terms and conditions of the Creative Commons Attribution (CC BY) license (https://creativecommons.org/licenses/by/4.0/).

One of the main reasons for the rapid spread and wide application of surface conductor jetting technology, as an advanced method for operating the upper well section of deepwater wells, is its excellent adaptability and efficiency [3]. Oil and gas exploration and development in deepwater environments urgently require innovative technologies, and studies by scholars such as Jeanjean [4] and Yan [5] have shown that the introduction of surface conductor jetting technology provides a practical solution to meet this need. Wang et al. [6–8] revealed the effect of fluid phase transition on multiphase flow in deepwater wellbore and proposed a multiphase flow model by considering fluid phase transition, which lays an important foundation for the safety of deepwater drilling. Domestic and international deepwater surface conductor jetting technology utilizes well-designed combinations of drilling tools. It is common for these combinations to use a cone bit to drill a 914 mm diameter conduit and then move to a 660.4 mm cone bit to jet and complete the lowering of the second-opening 660.4 mm borehole, which is ultimately combined with a 508 mm surface casing [2,9]. This well structure design has been widely verified in the international deepwater well development field and has achieved remarkable success. Similarly, China's self-operated deepwater drilling project in the South China Sea also adopted this proven form of wellbore structure in its initial stage to ensure stable downhole operations in the complex deepwater environment [10].

The control of the surface conductor in deepwater wells in terms of mud depth is crucial. Usually, based on the geological conditions of the deepwater seabed and the stability requirements of the submerged wellhead, the mud depth of the surface casing is controlled to be around 80 m. Wei et al. [11] found that the mud depth of the 508 mm surface casing for the design of a second-opening 660.4 mm wellbore is generally between 500 and 800 m. This strategy is designed to effectively seal off the shallow, unformed formations in deepwater from the complex third-opening borehole. The purpose of this well-designed mud entry depth strategy is to effectively seal the deepwater shallow, unformed formation in order to meet the requirements of complex three-opening borehole operations. Applying surface conduit injection technology to the upper section of deepwater wells is a well-established technique and plays a vital role in international and domestic deepwater oil and gas exploration and development projects. Its high adaptability, operational efficiency, and strong suitability for deepwater environments make it one of the formidable tools that can be used in addressing future challenges in deepwater exploration and development.

Jin and other scholars [12] have shown that streamlining and slimming the design of the wellbore structure is a key initiative in realizing cost savings and efficiency improvements in oil and gas well drilling operations. With the continuous advancement in self-operated deepwater drilling projects in the South China Sea, a useful way to reduce the cost of deepwater well drilling is carefully designing and slimming down the wellbore structure of the upper well section of deepwater wells to improve speed and efficiency. In deepwater well drilling, the high cost has always been a challenge, and Lukawski [13] found that the cost mainly covers several aspects, such as equipment, manpower, operating time, and resources. Therefore, by slimming down the design of the well structure, the time and resources required for drilling operations can be effectively reduced, thus increasing productivity while reducing the overall cost. In this process, the wellbore structure of the upper well section of a deepwater well plays a key role. By optimizing the wellbore design, not only can the efficiency of the drilling operation be improved, but unnecessary resistance and friction during the drilling process can also be reduced, equipment wear and tear can be reduced, and equipment life can be extended, which further reduces the burden of operating costs. This design concept has been widely used in self-operated deepwater drilling projects in the South China Sea, providing solid support for continued advancement in the project and the realization of economic benefits. Slimming down the well structure of the upper section of deepwater wells is one of the key strategies that can be used to reduce costs and improve efficiency in deepwater drilling operations. This useful technique not only helps to meet challenges in deepwater oil and gas exploration and development, but also contributes positively to improving the competitiveness and sustainability of deepwater drilling projects.

Currently, deepwater well drilling technology research mainly focuses on the analysis of the risk of dynamic positioning of deepwater drilling platforms [14], subsea casing technology [15], design of the surface casing entry depth [16], verification of subsea wellhead stability [17], drill string mechanics analysis [18], low-temperature/low-density cement slurries [19], and low-temperature drilling fluid systems [20]. A series of studies by Wang et al. [6,21] and Zhang et al. [22,23] showed that the problem of hydrate formation control is also very important for the deepwater drilling process. However, research has not covered the slimming down of wellbore structures and polycrystalline diamond compact (PDC) jetting bit-drilling technology [24,25]. Therefore, conducting an analysis on the slimming design of wellbore structures in the upper wellbore section of deepwater wells and innovatively researching matching small-sized PDC jetting bit-drilling technology is of significant importance. This endeavor will promote the practical application of the technology, contributing to the acceleration and cost reduction of deepwater drilling. Simultaneously, this research will continue to play a crucial role in addressing global energy shortages and contributing to the sustainability of future energy resources.

2. Materials and Methods

2.1. Deepwater Well Structure Slimming Technology

With advancements in deepwater self-operated drilling operations and the deepening of the understanding of the geological conditions of the deepwater seabed in the South China Sea and deepwater drilling technology, in order to reduce the operating costs of deepwater wells and improve operational efficiency, starting in 2020, the deepwater wells' second-section borehole design was optimized. A comparison of the slimmed-down upper-section well structure of the deepwater well with the conventional upper-section well structure of the deepwater well is shown in Figure 1, which is slimmed from a conventional 660.4 mm borehole to 444.5 mm (with the use of a 339.7 mm casing instead of the original 508 mm surface casing). The penetration depth of the second section was deepened from the conventional 500–800 m to 800–1200 m for seawater deep drilling. In order to meet the requirement of a 476.25 mm high-pressure wellhead head (HPWH) sitting on a 914 mm low-pressure wellhead head (LPWH), the top of the 339.7 mm casing was connected with 1–2 pcs of 508 mm surface casing via a fitter casing, and the top of the 508 mm surface casing was welded with a 476.25 mm HPWH. This method of connecting a 508 mm casing at the top of a fitter casing and a 339.7 mm casing at the bottom is called a composite casing string. Because of this well structure design, the 508 mm surface casing is only installed in 1–2 pcs, the length of which is within 24 m. In comparison, the depth of the first opening of the 914 mm surface conductor is ±80 m, so the 508 mm surface casing is located inside of the 914 mm surface conductor after it is installed in the well. The size of the 444.5 mm wellbore's second section will not affect its installation into the well.

The slimmed-down design of the second-section borehole in the upper section of deepwater wells has multiple technical advantages: (1) The size of the borehole is reduced from the conventional 660.4 mm borehole to 444.5 mm, which reduces the volume of rock breakage and rock cuttings. According to the calculation, under the same penetration depth, the cuttings volume of 444.5 mm borehole is only 45% of the 660.4 mm borehole, which reduces the energy required for drilling in the second section and is conducive to speeding up the drilling process and reducing the wear and tear on the drill bit. (2) The operation of the upper section of the deepwater wells is carried out via seawater drilling and the cuttings are returned from the seabed to be discharged into the sea, so after the structure of the second section has been slimmed and the penetration has been deepened, it can conduct deep seawater drilling. After slimming down the structure of the second section and increasing the penetration depth, deep drilling with seawater is realized, and the amount of drilling fluid system required for the lower well section is reduced. Deepwater wells are affected by low temperatures in the seabed, the polyammonium drilling fluid system has good rheology, and hydrate inhibition is generally used [15], which is expensive.

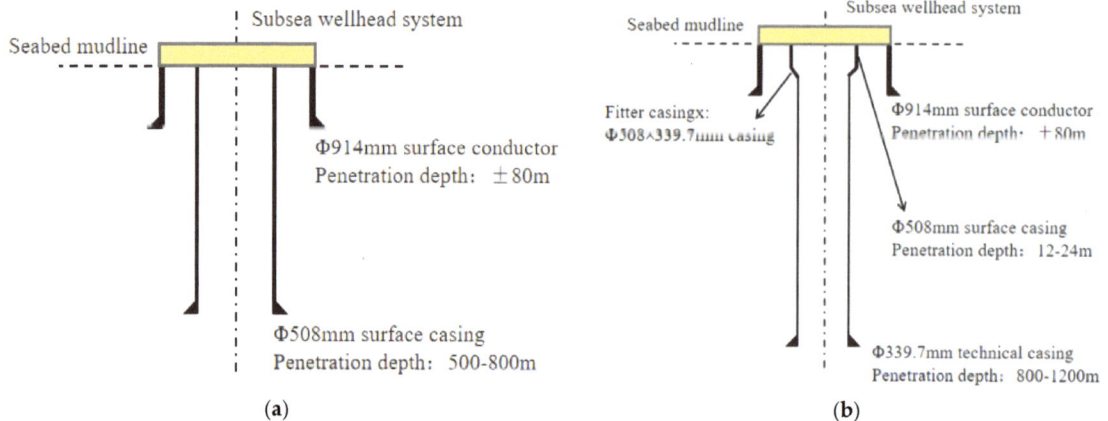

Figure 1. Slimmed-down upper-well section of the well structure of deepwater wells and conventional upper-well section structure of deepwater wells. (**a**) Upper-section well structure of conventional deepwater well; (**b**) Upper-section well structure of slimmed-down deepwater well.

(3) After the well structure is slimmed down, the conventional 508 mm surface casing is replaced with 508 mm + 339.7 mm composite casing strings, which saves the cost of casing tubing, and the small-size casing connection is more efficient. Compared with the conventional cementing method of a 508 mm surface casing in a 660.4 mm borehole, a 339.7 mm casing in a 444.5 mm borehole can significantly reduce the cost of cementing slurry and additives and reduce the time of cementing operation. After calculations, we determined that the second section reaches a penetration depth of 800 m; for example, the use of slimming after the well structure compared to the conventional design of the pair is shown in Figure 2. This design can save casing tubing costs up to 1.55 million CNY, 1.5 h of casing time, cementing slurry costs up to 2 million CNY, and up to 1.5 h of cementing time.
(4) The reduction in the size of the second section is conducive to correcting the deviation correction operation, and in the early stage in a number of deepwater wells, a second open 660.4 mm borehole drilling well deviation occurs. In the previous period, there were many deepwater wells with 660.4 mm boreholes where well deviation occurred, and it was difficult and inefficient to deviate correction using the motor and large-size cone bit. (5) Due to the reduction in the borehole size, there is no need to use the conventional 241.3 mm drill collars and measuring tools in the bottom drilling tools, and it is possible to use the 203.2 mm collars and measuring tools, which can be used directly in the drilling derrick after the completion of the operation of the upper section and in the subsequent lower section of the 311.3 mm collars and measuring tools. After completing the upper section, these drilling tools can stand directly on the drilling platform derrick and be reused in the subsequent 311.15 borehole of the lower section, saving the time required for dismantling and reassembling the drilling tools.

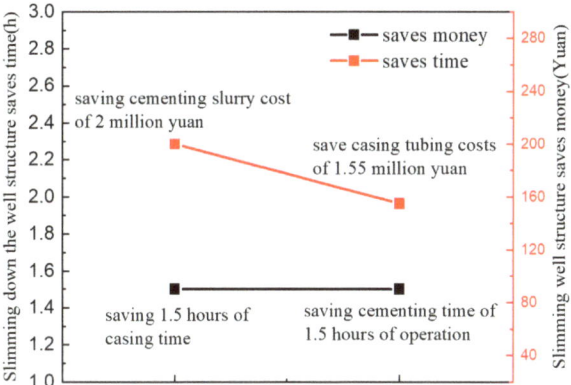

Figure 2. Comparison of the structure of the slimmed-down borehole with the conventional design after the second section at a penetration depth of 800 m.

2.2. Innovative Design of PDC Jetting Bit

The first problem introduced by the slimming of the size of the second-section borehole in the upper section of the deepwater well is that the conventional 660.4 mm cone bit is unable to perform continuous drilling of the second-section 444.5 mm borehole after the conductor jetting is in place. It is necessary to design a 444.5 mm drill bit to drill the second-section 444.5 mm borehole and perform the first-opening conduit jetting operation simultaneously. Subsequently, there is an increase in the penetration depth of the second section, which leads to an increase in the formation strength and a decrease in drilling ability, and the use of a cone bit will result in a slow rate of penetration (ROP) and low drilling efficiency, which has been demonstrated in the second open drilling operation using a conventional 660.4 mm cone bit. For example, in a deepwater well in Liwan, South China Sea, with a water depth of 1583.6 m, a 914 mm conductor depth of 80.4 m, and a 660.4 mm borehole depth of 853.4 m (starting depth of 1690 m, ending depth of 2463 m, and footage of 773 m), a conventional 660.4 mm cone bit is used to perform the conductor jetting and the second section of the borehole drilling. As can be seen in Figure 3, the mechanical drilling speed of the second section of the well can be reduced from an average of about 100 m/h to an average of about 50 m/h 550 m below the mudline. In the last 100 m of drilling, it is even reduced to about 30 m/h, a significant reduction, resulting in an average ROP of 61.2 m/h in the second section of the borehole, which is less than the expected operational efficiency.

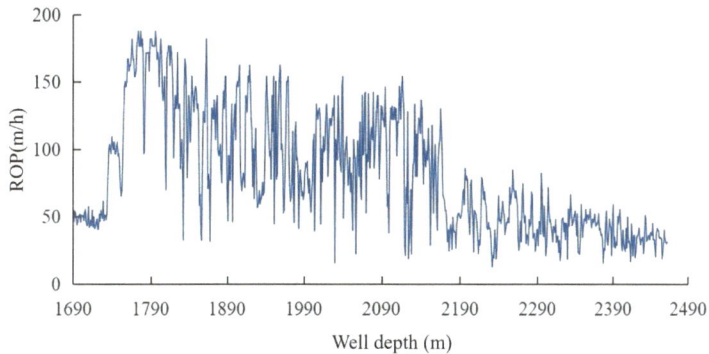

Figure 3. ROP diagram of a deepwater well with second section.

In addition, since the cone bit consists of movable parts such as wheel carriers and bearings, it has a limited service life under deep seawater drilling operations, with increased drilling pressure, risk of dropping the parts into the well, and subsequent complications.

Based on the above-mentioned problems of slimming the well structure in the upper section of deepwater wells, combined with the geological characteristics wherein the shallow stratum of deepwater wells in the eastern part of the South China Sea is poorly formed and the lithology is mainly a large set of mudstones, as well as the mechanism of cutting and breaking the rock by the teeth of the PDC bit—an innovative 444.5 mm PDC jetting bit design was developed. The advantages of the innovatively designed PDC jetting drill bit over the conventional cone bits used at home and abroad are shown in Table 1 below.

Table 1. Comparison of advantages of PDC jetting drill bits over conventional cone bits.

Drill Type	PDC Jetting Drill Bit	Conventional Cone Bits
Applicable borehole size	Can drill 444.5 mm boreholes	The 660.4 mm cone bit fails to drill 444.5 mm borehole
Rock-breaking efficiency	High breaking efficiency in shallow mudstone formations	High breaking efficiency in hard and brittle formations
Structural characteristics	No moving parts	Has moving parts and is at risk of objects falling down hole

The structure of the innovatively designed 444.5 mm PDC jetting drill bit is shown in Figure 4, and the hole is designed with five cutter blades and a single row of teeth. In order to support smooth jetting operation in a 914 mm large-size conductor, this bit has two obvious design features compared with the conventional PDC bit:

(1) Unique anti-collision design. Due to the smaller size of the drill bit and the increased gap between the drill bit and the surface conductor ring, the drill bit is prone to swinging and collision at the conductor's shoe position during the jetting operation, resulting in damage to the drill bit. In order to solve this problem, an anti-collision retainer is designed at the conventional retainer, and anti-collision retainer vertical bars are arranged. In order to take into account the abrasion resistance of the drill bit and the high feed rate, combined with the theoretical analysis of the length of the drill bit extending from the conductor shoe [26,27], the length of the drill bit extending from the bottom end of the conductor shoe is designed to be in the range of 150–200 mm, ensuring that the lower edge of the conductor shoe is within the length of the drill bit's anti-collision retaining diameter.

(2) Innovative bypass nozzle configuration. The increase in annular space between the drill bit and the conductor will lead to two problems: one is that it is more difficult to return the cuttings in the annular space of the conductor, and the other is that a stratigraphic step surface can easily form at the conductor shoe, which may lead to a low injection efficiency or the accumulation of cuttings to block the annular space of the conductor, resulting in a failed operation. Therefore, a bypass nozzle with a 45° inclination angle toward the bottom of the drill bit and an inner diameter of 11.1 mm was designed to hydrodynamically flush the stratum step between the drill bit and the conductor shoe and assist in the hydrodynamic return of cuttings through the design of a reasonable bit space.

In addition to the above two technical features, in terms of the characteristics of deepwater shallow soft mudstone, in order to prevent bit balling, the height of the cutter blade at the crown of the bit and the space of the drill flute were increased, and the depth of the drill flute was designed to be 95–130 mm. Different from conventional drilling, where rock breaking is mainly accomplished by the mechanical cutting or milling action of the teeth of the bit [28], the conductor jetting process relies on the water jets in the water eye of the bit to flush the soft mud to form a hole, so the hydrodynamic design of the jetting bit is the key. Based on the hydrodynamic parameter design, the innovatively designed PDC jetting drill bit has nine Φ11.11 mm nozzles in the crown of the drill bit, of which four small-angle nozzles are arranged near the center of the drill bit and five large-angle inclined nozzles are positioned near the shoulder of the drill bit. As measured by the mud pump displacement of 5200 m/L on the offshore platform, the total hydraulic force of the drill bit reaches 490 hp, and the impact strength of the water jet is over 60 kpa, which meets the requirements of deepwater surface conductor jetting operation.

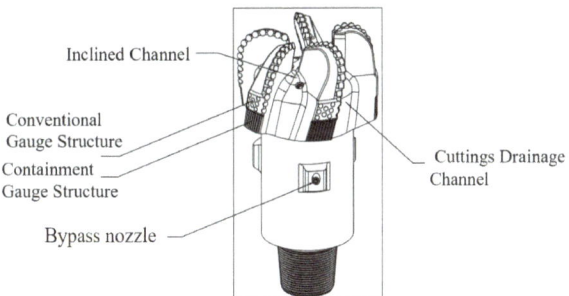

Figure 4. Schematic structure of 444.5 mm PDC jetting drill bit.

3. Field Applications

The PDC jetting bit-drilling technology based on the slimming of well structures has been implemented and applied in eight deepwater wells in the eastern South China Sea; these eight wells all use the innovatively designed 444.5 mm PDC jetting bit to jet down the 914 mm conductor, and the second-section drilling operation process is safe and smooth.

The operating water depth of the technology application well is shown in Figure 5, and the operating data of the first-section borehole and the second-section borehole of the technology application well are shown in Figures 6 and 7. The operation data of the eight wells are shown in Table 2; the water depth range is 572–1277.2 m, the penetration depth of the 914 mm surface conductor is 67.5–83.06 m, and the penetration depth of the 444.5 mm borehole is 738.27–1190.5 m. The maximum penetration depth of the second borehole is 1190.50 m, the average mechanical drilling speed of the second borehole exceeds 200 m/h, and the maximum is 214.51 m/h, which is the record for the highest mechanical drilling speed of the second borehole in the deepwater drilling operation.

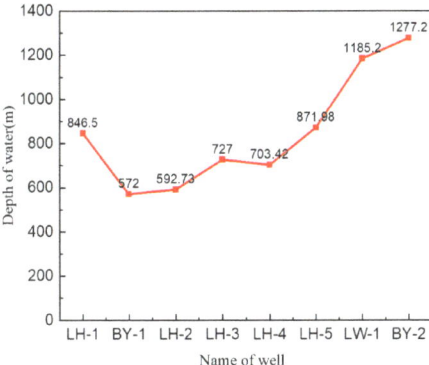

Figure 5. Operational water depth of technical application wells.

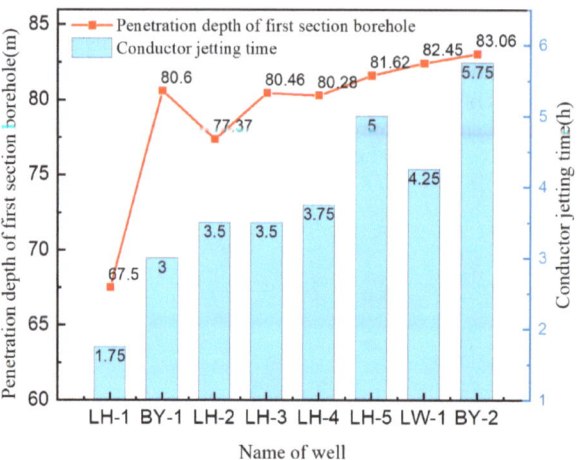

Figure 6. Operational data of the technical application well's first-section borehole.

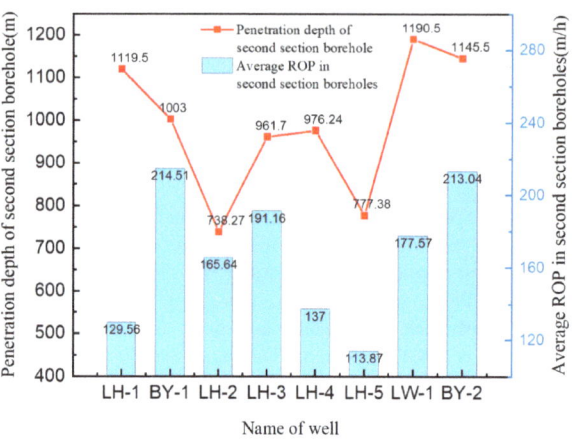

Figure 7. Operational data of the technical application well's second-section borehole.

Table 2. Statistical table of operational data of wells where the technology is applied.

Serial Number	Name of Well	Depth of Water (m)	Penetration Depth of First-Section Borehole (m)	Conductor Jetting Time (h)	Penetration Depth of Second-Section Borehole (m)	Average ROP in Second-Section Boreholes (m/h)
1	LH-1	846.5	67.50	1.75	1119.50	129.56
2	BY-1	572	80.60	3.0	1003.00	214.51
3	LH-2	592.73	77.37	3.5	738.27	165.64
4	LH-3	727.00	80.46	3.50	961.70	191.16
5	LH-4	703.42	80.28	3.75	976.24	137.00
6	LH-5	871.98	81.62	5.00	777.38	113.87
7	LW-1	1185.20	82.45	4.25	1190.50	177.57
8	BY-2	1277.20	83.06	5.75	1145.50	213.04

The average time for conductor jetting in the eight wells is 3.81 h, compared with the average of 5 h in the previous operation. This result is 23.8% faster, among which the time for conductor jetting of LH-1 well is only 1.75 h, which sets a new record for the shortest time for deepwater conductor jetting in China. As shown in Figure 8 below, after the deepwater well structure is slimmed down, the average depth below the mudline in the eight wells' second-section 444.5 mm boreholes reaches 989.01 m, compared with 721.67 m, the average depth below the mudline in the conventional 660.4 mm boreholes in the previous operation; overall, this is an increase of 37.01%. After adopting the innovatively designed 444.5 mm PDC jetting bit, the average mechanical drilling speed of the eight wells' second section was 162.15 m/h, compared with the average mechanical drilling speed of 49.46 m/h of the conventional 660.4 mm bit in the previous operation. The average mechanical drilling speed was increased by 227.84%, with an increase in penetration depth of 37.01%, realizing the increase in speed and efficiency.

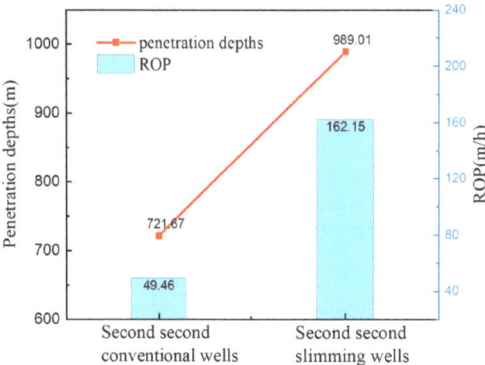

Figure 8. Comparison of conventional and slimmed-down technologies for deepwater well body structures.

On site, one innovatively designed 444.5 mm PDC jetting drill bit was used to complete the drilling of the upper section of all eight deepwater wells, and the drill bit was only slightly worn and could be used again. If a cone bit was used, only one bit could be used for a maximum of three wells, and three cone bits were needed to complete the eight wells. Considering the operating time saved by speeding up drilling, the cost of cementing slurry and casing materials saved by the slimming of the well body, and the cost of drilling fluid materials saved by the deep drilling of seawater, the operating cost saved for a single well exceeded 4 million CNY, and the cumulative economic benefit of the eight deepwater wells exceeded 30 million CNY, a great reduction in the cost of deepwater operations.

4. Conclusions

Deepwater drilling research is critical to solving the world's energy shortages. Deepwater drilling provides critical support to meet the growing global energy demand by discovering new hydrocarbon reserves, diversifying energy supplies, and supporting technological innovation and economic development. Research on deepwater PDC jetting bit-drilling technology with lean well structures can contribute to future energy sustainability.

The innovatively designed 444.5 mm PDC jetting bit substantially increased the ROP of the second section in deepwater wells compared with the conventional cone bit, met the design requirements of slimming the structure of deepwater wells, realized deep drilling with seawater, and saved time in the operation period and drilling material costs.

The small-size 444.5 mm PDC jetting bit and large-size 914.4 mm surface conductor have the problem of a small bit in the conductor shoe position swinging back and forth, as well as low jetting efficiency and other problems. Unique anti-collision to protect the diameter of the design, innovative configuration of the bypass nozzle, and improving the jetting displacement and other methods can effectively solve these problems to ensure that the conductor's surface is safe and smooth jetting operations are performed.

The deepwater well's second-section borehole size is reduced after slimming, reducing the stability of the subsea wellhead. The need for target well field investigation information, as well as operating conditions and strict calibration of the stability of the subsea wellhead, can ensure that the energy extraction of deepwater wells is safe and reliable and provides the necessary guarantee for a sustainable energy supply.

Author Contributions: Conceptualization, W.Z.; methodology, D.G.; formal analysis, W.Z.; resources, Y.Z.; data curation, D.Y.; writing—original draft, W.Z., D.G., Y.Z. and D.Y. All authors have read and agreed to the published version of the manuscript.

Funding: This research was funded by the Major Scientific and Technological Project "Deepwater Drilling and Completion Engineering Technology" (2016ZX0528-001) in China, as well as the National Natural Science Foundation of China (NSFC) Key Project "Basic Theory Research on Key Technology of Deepwater and Shallow Drilling" (51434009).

Data Availability Statement: Not applicable.

Conflicts of Interest: The authors declare no conflict of interest.

References

1. Amadi, K.; Iyalla, I. Application of mechanical specific energy techniques in reducing drilling cost in deepwater development. In Proceedings of the SPE Deepwater Drilling and Completions Conference, OnePetro, Galveston, TX, USA, 20–21 June 2012.
2. Akers, T.J. Jetting of structural casing in deepwater environments: Job design and operational practices. *SPE Drill. Complet.* **2008**, *23*, 29–40. [CrossRef]
3. D'Ambrosio, P.; Prochaska, E.; Bouska, R.; Tinsley, D.; Hart, S. Cost Effective Ultra-Large Diameter PDC Bit Drilling in Deepwater Gulf of Mexico. In Proceedings of the SPE/IADC Drilling Conference, Amsterdam, The Netherlands, 5–7 March 2013.
4. Jeanjean, P. Innovative design method for deepwater surface casings. In Proceedings of the SPE Annual Technical Conference and Exhibition, San Antonio, TX, USA, 29 September–2 October 2002; p. 77357.
5. De, Y.A.; Bin, C.H.; Ruirui, T.I.; Lingan, S.O.; Biao, W.A.; Baobo, L.I. Research on Increasing Drilling Speed for Deepwater conductor Jetting. *Drill. Prod. Technol.* **2021**, *44*, 36.
6. Wang, Z.; Sun, B. Annular multiphase flow behavior during deep water drilling and the effect of hydrate phase transition. *Pet. Sci.* **2009**, *6*, 57–63. [CrossRef]
7. Wang, Z.; Sun, B.; Sun, X.; Li, H.; Wang, J. Phase state variations for supercritical carbon dioxide drilling. *Greenh. Gases Sci. Technol.* **2016**, *6*, 83–93. [CrossRef]
8. Wang, Z.; Lou, W.; Sun, B.; Shaowei, P.; Xinxin, Z.; Hui, L. A model for predicting bubble velocity in yield stress fluid at low Reynolds number. *Chem. Eng. Sci.* **2019**, *201*, 14815. [CrossRef]
9. Syazwan, M.; Shafei, A.; Paimin, M.R.; Su'if, M.Z.; Abidin, H.Z.; Ismail, Y.; Natarajan, S. Enhancement of Structural conductor Design and Execution of Jetting Operations for Ultra-Deepwater Wells in Brunei. In Proceedings of the SPE Asia Pacific Oil and Gas Conference and Exhibition, Perth, Australia, 25–27 October 2016; p. 182427.
10. Liu, S.; Xie, R.; Tong, G.; Wu, Y.; Xu, G. Progress and prospect of deepwater well drilling and completion technique of CNOOC. *Acta Pet. Sin.* **2019**, *40*, 168.
11. Wei, H.; Yang, J.; Liu, Z.; Ye, J.; Zhang, J.; Yan, D.; Li, S.; Zhang, K.; Wang, W. Study on safety control technology of surface conductor jetting penetration in ultra-deep water soft formation drilling. In Proceedings of the ISOPE International Ocean and Polar Engineering Conference, ISOPE, Honolulu, HI, USA, 16–21 June 2019.
12. Jin, X.; Wen, J.; Li, L.; Guo, W.; Zhou, G. Analysis of the emergency disconnection range of dynamically positioned platforms in deepwater drilling operations. *China Offshore Oil Gas* **2017**, *29*, 116–121.
13. Lukawski, M.Z.; Anderson, B.J.; Augustine, C.; Capuano, L.E., Jr.; Beckers, K.F.; Livesay, B.; Tester, J.W. Cost analysis of oil, gas, and geothermal well drilling. *J. Pet. Sci. Eng.* **2014**, *118*, 1–14. [CrossRef]
14. Shen, Y.; Zhao, L. Safety emergency response measures for out of control drift of offshore dynamic positioning drilling platform. *Drill. Prod. Technol.* **2013**, *36*, 118–120.
15. Zhou, J.L.; Xu, L.B. Research progress of key technologies for deepwater drilling risers. *China Offshore Oil Gas* **2018**, *30*, 135–143.
16. Jin, Y. Calculation method of surface conductor setting depth in deepwater oil and gas wells. *Acta Pet. Sin.* **2019**, *40*, 1396.
17. Wang, L.; Yang, J.; Li, L.; Hu, Z.; Ke, K.; Zang, Y.; Sun, T. Wellhead stability in gas hydrate formation during deep-water drilling. *Chin. J. Geotech. Eng.* **2022**, *44*, 2312–2318.
18. Gao, D.L.; Wang, Y.B. Progress in tubular mechanics and design control techniques for deepwater drilling. *Pet. Sci. Bull.* **2016**, *1*, 61–80.
19. Feng, Y.; Song, M.; Zhang, H.; Li, H. Liquid cementing fluid reduced lightened deepwater light-weight cement slurry. *Drill. Fluid Complet. Fluid* **2017**, *34*, 80–84.
20. Geng, T.; Ziu, Z.; Tang, Z.; Zhao, X.; Miao, H. The development and application of high temperature resistant and strong inhibitive water-based drilling fluid for deepwater drilling. *Pet. Drill. Tech.* **2019**, *47*, 82–88.
21. Wang, Z.; Sun, B. Deepwater gas kick simulation with consideration of the gas hydrate phase transition. *J. Hydrodyn. Ser. B* **2014**, *26*, 94–103. [CrossRef]

22. Zhang, J.; Wang, Z.; Liu, S.; Meng, W.; Sun, B.; Sun, J.; Wang, J. A method for preventing hydrates from blocking flow during deep-water gas well testing. *Pet. Explor. Dev.* **2020**, *47*, 1256–1264. [CrossRef]
23. Zhang, J.; Wang, Z.; Sun, B.; Sun, X.; Liao, Y. An integrated prediction model of hydrate blockage formation in deep-water gas wells. *Int. J. Heat Mass Transf.* **2019**, *140*, 187–202. [CrossRef]
24. Hu, H.; Guan, Z.; Wang, B.; Liang, D.; Sun, M.; Wang, X.; Chen, W. Weight-on-bit self-adjusting dual-diameter bit leads a step change in drilling bit technology. In Proceedings of the International Petroleum Technology Conference, IPTC, Beijing, China, 26–28 March 2019.
25. Eustes, A.; Bourdon, N.; Joshi, D.; McKenna, K.; Uzun, O.; Zody, Z.; Alhaidari, S.; Amer, A. 4 Onshore Drilling. In *Advances in Terrestrial and Extraterrestrial Drilling: Ground, Ice, and Underwater*; CRC Press: Boca Raton, FL, USA, 2021; Volume 63.
26. Yang, J.; Yan, D.; Tian, R.; Zhou, B.; Liu, S.; Zhou, J.; Tang, H. Bit stick-out calculation for the deepwater conductor jetting technique. *Pet. Explor. Dev.* **2013**, *40*, 394–397. [CrossRef]
27. Zhou, B.; Yang, J.; Liu, Z.; Zhou, R. Model and experimental study on jetting flow rate for installing surface conductor in deep-water. *Appl. Ocean. Res.* **2016**, *60*, 155–163. [CrossRef]
28. Zhang, Z.; Zhao, D.; Zhao, Y.; Zhou, Y.; Tang, Q.; Han, J. Simulation and experimental study on temperature and stress field of full-sized PDC bits in rock breaking process. *J. Pet. Sci. Eng.* **2020**, *186*, 106679. [CrossRef]

Disclaimer/Publisher's Note: The statements, opinions and data contained in all publications are solely those of the individual author(s) and contributor(s) and not of MDPI and/or the editor(s). MDPI and/or the editor(s) disclaim responsibility for any injury to people or property resulting from any ideas, methods, instructions or products referred to in the content.

Article

Adsorption/Desorption Performances of Simulated Radioactive Nuclide Cs$^+$ on the Zeolite-Rich Geopolymer from the Hydrothermal Synthesis of Fly Ash

Zhao Zheng [1,2,*], Jun Yang [3], Maoxuan Cui [3], Kui Yang [4], Hui Shang [2], Xue Ma [3] and Yuxiang Li [3]

[1] Shock and Vibration of Engineering Materials and Structures Key Laboratory of Sichuan Province, Southwest University of Science and Technology, Mianyang 621010, China
[2] School of Civil Engineering and Architecture, Southwest University of Science and Technology, Mianyang 621010, China; shanghui@swust.edu.cn
[3] School of Materials and Chemistry, Southwest University of Science and Technology, Mianyang 621010, China; yangjun_9908@163.com (J.Y.); c1722847383@163.com (M.C.); maxue@swust.edu.cn (X.M.); liyuxiang@swust.edu.cn (Y.L.)
[4] Sino Shaanxi Nuclear Industry Group, Xi'an 710100, China; yangkui@ssn-hs.com
* Correspondence: zhengzhao@swust.edu.cn

Abstract: The operation of nuclear power plants generates a large amount of low- and intermediate-level radioactive waste liquid. Zeolite-rich geopolymers, which are synthesized under hydrothermal conditions from industrial waste fly ash, can effectively immobilize radioactive nuclides. In this study, the synthesis law of zeolite-rich geopolymers and the adsorption/desorption performances of radioactive nuclide Cs$^+$ were researched using XRD, SEM and ICP. The results show that the increase in curing temperatures and NaOH concentrations leads to the transformation of Y-type zeolite to chabazite and cancrinite at low NaNO$_3$ concentrations. However, at high NaNO$_3$ concentrations, NaOH above 2 M has no obvious effect on the phase transformation of the main zeolite of chabazite and cancrinite. In the adsorption and desorption experiment of Cs$^+$ on the chabazite/garronite-rich geopolymer, it was found that the adsorption of Cs$^+$ in the low initial concentration range is more suitable for the Freundlich equation, while the Langmuir equation fits in the adsorption process at the high initial concentration range. Moreover, the desorption kinetics of Cs$^+$ are in good agreement with the pseudo-second-order rate equation. Thus, the adsorption of Cs$^+$ on chabazite/garronite-rich geopolymers is controlled by both physical and chemical reactions, while desorption is a chemical process.

Keywords: low- and intermediate-level radioactive wastes; adsorption/desorption performances; zeolite-rich geopolymer; radioactive nuclide Cs$^+$; NaNO$_3$ concentration

1. Introduction

With the rapid growth of global energy demands, nuclear energy is considered a clean and reliable energy source in the world [1]. Subsequently, a large amount of radioactive waste liquid will be generated. ^{137}Cs is one of the most important fission products of low- and intermediate-level radioactive waste liquid (LILW) due to its short half-life ($t_{1/2}$ = 30.2 a) and high yield. Compared with ^{90}Sr, ^{137}Cs is an alkali metal element, and its high solubility in water gives it strong fluidity [2]. Geopolymers are often used to solidify radionuclide ions, which are usually prepared from fly ash activated by alkaline, sometimes called alkali-activated materials [3]. The radionuclide ions are solidified in the geopolymer in the form of physical adsorption or chemical bonding [4]. Geopolymers are superior to Portland cement in mechanical strength, acid/salt resistance, and thermal stability; they are conducive to the sustainable development of the construction industry and also have a great application prospect in the solidification of hazardous wastes [5]. Furthermore, a large amount of fly ash will be produced every year in the world [6], and geopolymers prepared from fly ash do not need to emit CO_2, which alleviates some of the greenhouse effect [7,8]. Thus, the utilization of fly ash is also of great significance for the disposal of LILW [9].

At present, there are many technologies to produce zeolite, such as precipitation, hydrothermal, microwave, and sol–gel synthesis [10,11]. On the basis of traditional alkali melting–hydrothermal synthesis of zeolites, the steps of the calcination activation pretreatment of fly ash, water-soluble desiliconization, alkali-soluble aluminosilicate, filtration, and other steps to synthesize various types of high-purity zeolite products have emerged. Many researchers [12–14] have investigated the effects of preparation methods on the synthesis of fly-ash-based geopolymers and discussed the effects of preparation conditions such as the alkalinity, liquid/solid ratio, and crystallization time on the synthesis products. He et al. [5] systematically studied the effect of hydrothermal treatment parameters on the phase composition and microstructure of hydrothermal products of the amorphous sodium-based geopolymer (NaGP), and they concluded that the best hydrothermal method for preparing NaGP was treated with a 1M NaOH solution at 160 °C for 6 h. The products of fly-ash-based geopolymers are various, with different raw materials and curing conditions. Wang et al. [15] and Wdowin et al. [16] found that the products of geopolymers include amorphous geopolymer gel and zeolite N (C)-A-S-H, NaA-X zeolite, NaP zeolite, NaP1 zeolite, cesium garnet, A zeolite, and so on.

Lei et al. [17] investigated the adsorption characteristics of zeolite microspheres in metakaolin/slag-based geopolymers for Cs^+ and Sr^{2+}, and the results showed that the adsorption kinetic process of Cs^+ and Sr^{2+} on zeolite microspheres was in good agreement with the pseudo-second-order kinetic equation. Lee et al. [18] prepared fly ash/slag-based mesoporous geopolymers using the hydrothermal method, and the results showed that when the initial concentration of Cs^+ was 100 mg/L, the maximum adsorption capacity of mesoporous geopolymers was 15.24 mg/g. Hui et al. [19] prepared 4A zeolite from fly ash and investigated the adsorption kinetics of different metal ions on fly-ash-based 4A zeolite using pseudo-first-order and pseudo-second-order kinetic equations; it was found that the pseudo-first-order model fits in Ni^{2+} well, and the pseudo-second-order model fits in Co^{2+} and Cr^{3+} well.

Large-volume geopolymer solidified matrices have been used in the treatment of LILW with high levels of alkali (NaOH) and salts ($NaNO_3$), while the high center temperature of the matrices has prompted the generation of many crystal products. However, the transformation law of the fly-ash-based geopolymer products formed under different curing temperatures and sodium salt concentrations is under researched, and the adsorption/desorption mechanism of the simulated nuclide Cs^+ is not clear. This study aims to clarify the relationship between the nuclides and a single phase of fly-ash-based geopolymers. The influence of the curing temperature, NaOH concentrations, and $NaNO_3$ concentrations on the transformation of fly-ash-based geopolymer products in the system was explored. Moreover, the isothermal adsorption characteristics and desorption kinetics of Cs^+ by the main zeolites were studied. The Langmuir and Freundlich isotherm adsorption equations were used for the fitting analysis, while the pseudo-first-order and pseudo-second-order kinetic models were used to fit the desorption test data.

2. Materials and Methods

2.1. Materials

Fly ash, mainly composed of SiO_2 and Al_2O_3, was used as the raw material for the synthesis of zeolite materials [20]. The oxide composition of fly ash (wt.%) was analyzed by an X-ray fluorescence spectrometer, and the results are listed in Table 1. The D_{50} of the fly ash was 18.75 μm, determined by a laser particle analyzer (Mastersizer2000, Malvern, UK). The fly ash used in this study was obtained from the Jiangyou Power Station, China. The analytically pure reagents, such as CsCl, NaOH, $NaNO_3$, etc., are all from Kelong Chemical Co., Ltd., Chengdu, China. The Cs^+ solution used in the experiment refers to the CsCl solution, and its concentration refers to the concentration (mg/L) of the cesium element in the solution.

Table 1. Oxide composition of fly ash.

Oxide Composition	SiO_2	Al_2O_3	CaO	Fe_2O_3	K_2O	MgO	Other
Content (%)	59.79	21.99	5.55	5.13	3.21	1.35	2.98

2.2. Sample Preparation Methods

A fly-ash-based geopolymer was prepared using the hydrothermal synthesis method. Fly ash was fixed at 5 g, and the water/fly ash ratio was 8. The mass concentrations of $NaNO_3$ solution (100~500 g/L), NaOH molar concentrations (0.66~8 M), and curing temperatures (60~150 °C) were varied to prepare the fly-ash-based geopolymer. The curing time was constant at 24 h. The specific design ratios are shown in Tables 2 and 3.

Table 2. Design ratios with different NaOH concentrations and different $NaNO_3$ concentrations.

60 °C Water Bath for 8 h; Curing at 90 °C for 24 h					
Sample Number	$NaNO_3$ (g/L)	$NaNO_3$ (g)	W/F	NaOH (g)	Fly Ash (g)
100N-0.66H	100	4.15	8	1.06	5
100N-2H	100	4.15	8	3.23	5
100N-4H	100	4.15	8	6.45	5
100N-6H	100	4.15	8	9.68	5
100N-8H	100	4.15	8	12.90	5
300N-0.66H	300	13.55	8	1.06	5
300N-2H	300	13.55	8	3.23	5
300N-4H	300	13.55	8	6.45	5
300N-6H	300	13.55	8	9.68	5
300N-8H	300	13.55	8	12.90	5
500N-0.66H	500	24.90	8	1.06	5
500N-2H	500	24.90	8	3.23	5
500N-4H	500	24.90	8	6.45	5
500N-6H	500	24.90	8	9.68	5
500N-8H	500	24.90	8	12.90	5

Notes: N means $NaNO_3$ solution concentration, and H means NaOH molar concentration.

Table 3. The ratios of sample 100N-2H at different curing temperatures.

Stirring in Water Bath at 60 °C for 8 h, Curing at Different Temperatures for 24 h					
Sample Number	$NaNO_3$ (g/L)	$NaNO_3$/g	W/F	NaOH/g	Fly Ash/g
100N-2H-60 °C	100	4.15	8	3.23	5
100N-2H-90 °C	100	4.15	8	3.23	5
100N-2H-120 °C	100	4.15	8	3.23	5
100N-2H-150 °C	100	4.15	8	3.23	5

2.3. Isothermal Adsorption Experiment

The initial concentration of the adsorbent is a key parameter that determines the final adsorption behavior of the adsorbent, and the concentration gradient between the adsorbent surface and the adsorbate solution controls the mass transfer rate [18]. Thus, a 50 mL CsCl solution with mass concentrations of 50, 100, 150, 200, 300, 400, 500, 600, and 800 mg/L (Cs^+) was added with 0.2 g products after washing and drying in several centrifuge tubes and then shaken in a constant temperature shaker (300 r/min) for 6 h. The supernatant was filtered with a 0.45 um needle filter, and the mass concentration of the remaining Cs$^+$ was detected to calculate the adsorption rate (Equation (1)) and adsorption capacity (Equation (2)) [21]. The adsorbed product was dried for further tests.

$$R = \frac{C_0 - C_e}{C_0} \times 100 \quad (1)$$

$$Q_e = \frac{(C_0 - C_e)}{m} \times V \quad (2)$$

where R is the adsorption rate, %; C_0 is the initial mass concentration of Cs$^+$, mg/L; C_e is the residual mass concentration of Cs$^+$, mg/L; Q_e is the equilibrium adsorption capacity, mg/g; m is the amount of the products added, g; V is the volume of Cs$^+$ solution, L.

2.4. Desorption Kinetics Test

A total of 0.2 g of the test products was added in multiple centrifuge tubes, then 50 mL of the Cs$^+$ solution with a mass concentration of 150 mg/L was added. The tubes were placed in a constant temperature oscillation box (300 r/min) to conduct an oscillation adsorption for 6 h, and finally, they were moved to a centrifuge. The remaining solid samples in the centrifuge tube were washed with deionized water to remove non-specifically adsorbed metal ions on the products, and then they were placed in a new centrifuge tube. A total of 50 mL of a hydrochloric acid solution (pH = 1) was added in the centrifugal tube as the eluate, and the tubes were put into a constant temperature oscillation box (300 r/min). At different times (5, 10, 15, 30, 45, and 60 min), the centrifuge tube was taken out, and the supernatant was filtered with a 0.45 μm needle filter. The remaining Cs$^+$ concentration was detected to calculate the desorption capacity Q_j (Equation (3)) and desorption rate R_j (Equation (4)), and the results were used to fit the subsequent desorption kinetic curve.

$$Q_j = C_{je} V_j / m \quad (3)$$

$$R_j = \frac{Q_j}{Q_e} \times 100 \quad (4)$$

where R_j is the desorption rate, %; Q_e is the equilibrium adsorption capacity in 150 mg/L Cs$^+$ solution, mg/g; m is the additional amount of the products, g; Q_j is the desorption capacity of Cs$^+$ from the products, mg/g; C_{je} is the desorption equilibrium mass concentration of Cs$^+$, mg/L; V_j is the volume of desorption solution, L.

2.5. Characterization

The composition of oxides in fly ash was determined by X-ray Fluorescence (XRF) (Axios, PANalytical, Almelo, The Netherlands). The products of fly-ash-based geopolymers were detected by X-ray Diffractometry (XRD) (Smartlab, Rigaku, Tokyo, Japan) from 5° to 80° of 2θ with a rate of 20°/min using Cu K radiation. The morphology of fly-ash-based geopolymers was analyzed by Scanning Electron Microscope (SEM) (TM4000, Hitachi, Tokyo, Japan), and the surfaces of the fly-ash-based geopolymers were sprayed with a gold conductive layer and pasted onto conductive double-sided tape. The Fourier Transform Infrared Spectrometer (FTIR) (IS5, Thermo Fisher Scientific, Waltham, MA, USA) was used to detect the evolution of the H–O, Si–O–T (T = Si and Al), N–O, and C–O bond, and the fly-ash-based geopolymers were soaked in anhydrous ethanol and then dried. Subsequently, the dried fly-ash-based zeolite was mixed with KBr crystal and ground into a powder, which was then compressed in a mold (12–14 MPa) for 15 s to form an FTIR sample. The concentration of Cs$^+$ was detected by an Atomic Absorption Spectrometer (AAS) (AA700, PerkinElmer, Waltham, MA, USA).

3. Results and Discussion
3.1. Products Analysis
3.1.1. XRD

The XRD diffraction patterns of the products of the fly-ash-based geopolymers synthesized based on Tables 2 and 3 are shown in Figures 1 and 2. It can be seen from Figure 1a that in the system with the $NaNO_3$ concentration of 100 g/L and NaOH concentration of 2 M, the diffraction peaks of the Y-type zeolite (PDF # 89-1629) and chabazite (PDF # 44-0248) appear, while the diffraction peak intensity of chabazite is rather low. At the same time, the diffraction peaks of faujasite (PDF # 28-1036) are observed. With the increase in NaOH concentrations (4~8 M), the diffraction peaks of faujasite increase, and the diffraction peaks of cancrinite (PDF # 46-1332) also occur, while the characteristic peaks of cancrinite become stronger and stronger. It can be seen from Figure 1b that in the system with a $NaNO_3$ concentration of 300 g/L and a low NaOH concentration (0.66 M), a small number of diffraction peaks of chabazite emerge, but the intensity of the diffraction peaks is low. As the NaOH concentrations varied from 2 M to 8 M, the diffraction peaks of SiO_2 (PDF # 85-0796) show a decreasing trend. The crystal phase types in the fly-ash-based geopolymer products increase obviously, and the diffraction peaks of cancrinite and chabazite rise. Indeed, the characteristic peak of $NaNO_3$ (PDF # 72-0025) with a high diffraction peak intensity also appear, which indicates that there is too much unreacted $NaNO_3$ in the system. It can be seen from Figure 1c that in the system with the $NaNO_3$ concentration of 500 g/L, the diffraction peak of SiO_2 has disappeared, and the main phase is $NaNO_3$. Meanwhile, the diffraction peaks of cancrinite, chabazite, and faujasite zeolites still exist in the system, indicating that high $NaNO_3$ concentrations have no inhibition effect on the growth of zeolites. However, chabazite is more suitable for growth in high $NaNO_3$ (300~500 g/L) environments compared with cancrinite. When NaOH concentrations are in the range of 6~8 M, an increase in $NaNO_3$ will promote the transformation of cancrinite into chabazite.

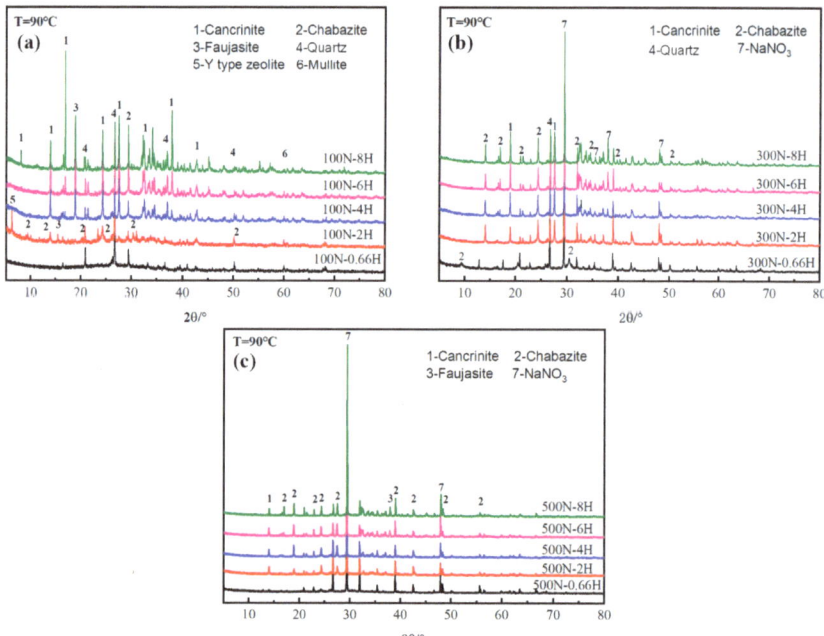

Figure 1. XRD patterns of products with different NaOH contents at different $NaNO_3$ concentrations: (**a**) 100 g/L $NaNO_3$; (**b**) 300 g/L $NaNO_3$; (**c**) 500 g/L $NaNO_3$.

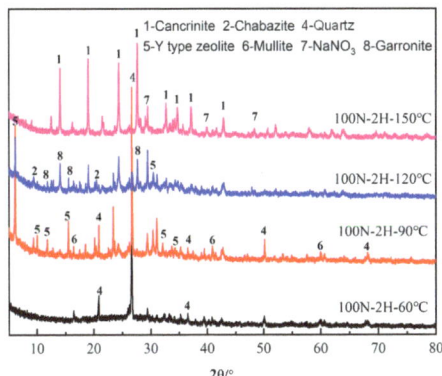

Figure 2. XRD patterns of products 100N-2H at different curing temperatures.

In the following experiment, the concentration of the NaOH solution in the system was controlled at 2 M as no obvious zeolite formed at 0.66 M, and the NaNO$_3$ concentration was set as 100 g/L, which is near to the normal concentration of NaNO$_3$ in the LILW. The phase transformation of alkali-activated fly-ash-based geopolymer products at different curing temperatures from 60 °C to 150 °C was studied. High curing temperatures are another key parameter affecting hydrothermal conversion that promote the grains to crystallize and grow during hydrothermal treatment [5]. From the analysis of Figure 2, it can be seen that there is an obvious crystallization peak of silica near 2θ = 26.5° in the sample after the curing treatment of low temperatures (60 °C), while no characteristic diffraction peaks of other crystal phases are found, which indicates that no crystal phase emerges under this condition. When the curing temperature rises to 90 °C, the diffraction peaks of silica decrease obviously, and the characteristic diffraction peak of Y-type zeolites occurs. However, the crystallinity is low, and there are still unreacted mullite and silica. When the temperature reaches 120 °C, the characteristic diffraction peaks of Y-type zeolites in the product are weakened, and the characteristic diffraction peaks of chabazite and garronite (PDF # 39-1374) appear. However, the crystallinity of chabazite is not high, and the silica diffraction peaks weaken. This indicates that a curing temperature of 120 °C is beneficial to the crystal growth and transformation into chabazite and garronite, but the reaction under hydrothermal conditions is still incomplete. The weakening of the characteristic diffraction peak of silica was caused by the dissolution of silica by a high concentration of the NaOH solution [2]. By further increasing the curing temperature to 150 °C, the characteristic diffraction peaks of cancrinite emerge, and the crystallinity is high. Moreover, the characteristic diffraction peaks of silica disappear. Thus, it can be concluded that the increase in temperature is beneficial to the crystallization and phase transformation of zeolites. Also, Palomo et al. [22] reported the promoting effect of high temperatures on the formation of the zeolite phase. Zheng et al. [23] studied the effect of curing temperatures on the evolution of the crystalline phase of geopolymers, and the results showed that a relatively high curing temperature (>60 °C) was a necessary condition for the phase transition of fly-ash-based geopolymers.

3.1.2. SEM

XRD diffraction pattern analysis has roughly determined the zeolite generated at the NaOH concentration of 2 M and NaNO$_3$ concentration of 100 g/L, as Y-type zeolite, chabazite/garronite, and cancrinite have been obtained at the curing temperatures of 90 °C, 120 °C, and 150 °C, respectively. Figures 3–5 are the SEM images of the fly-ash-based geopolymer products prepared by the above three conditions.

Figure 3. SEM diagrams of product 100N-2H-90°C.

Figure 4. SEM diagrams of product 100N-2H-120°C.

Figure 5. SEM diagrams of product 100N-2H-150°C.

In Figure 3, it can be seen that the fly-ash-based geopolymer has a polyhedron-shaped product under the NaOH concentration of 2 M and 100 g/L of NaNO$_3$ at the curing temperature of 90 °C, which is the characteristic morphology of Y-type zeolites [24]. The particle size and shape of Y-type zeolites are relatively uniform. Indeed, it can also be observed that the surface presents a wrinkled spherical particle, which is the characteristic morphology of the chabazite. Moreover, the spherical particles are larger than the polyhedron-shaped particles. At the same time, dissolved and broken spherical particles can be observed. Combined with the results of XRD, it can be concluded that Y-type zeolites and chabazite were formed in the alkali-activated fly-ash-based geopolymer products under this condition. From Figure 4, it can be seen that the fly-ash-based geopolymer has well-crystallized particles under a NaOH concentration of 2 M and a $NaNO_3$ concentration of 100 g/L at a curing temperature of 120 °C. Most of the particles are regularly spherical, and the surface has obvious folds. The particle size is small, but the size is uniform, and the typical chabazite crystal shows an aggregation growth morphology [25]. There is also a small part of the particles with a cross-shaped columnar structure, which is larger than chabazite. The well-crystallized phase indicates that new crystals have been generated, and it is also proven that the fly ash in the raw material is fully involved in the reaction. Combined with the XRD results, it was determined that chabazite and garronite were formed under this condition. From Figure 5, it can be seen that the fly-ash-based geopolymer has thorny spherical particles under the NaOH concentration of 2 M and $NaNO_3$ concentration of 100 g/L at the curing temperature of 150 °C, which are the characteristic morphology of cancrinite. The particle size is uniform, the crystallization is good, and the small irregular particles are grown around the spherical particles. Thus, it can be concluded that the alkali-activated fly-ash-based geopolymer generates cancrinite under this condition. Above all, the fly-ash-based geopolymer product of chabazite and garronite formed at 120 °C (sample T120: chabazite/garronite-rich geopolymers) was selected for the subsequent isothermal adsorption test and desorption kinetic test to explore the adsorption and desorption characteristics of Cs^+, as the main crystal phases of chabazite and garronite obtained excellent directional adsorption ability to Cs^+ [26,27].

3.1.3. FTIR

Figure 6 shows the FTIR results of T120 products before and after the adsorption of Cs^+. The sharp absorption peak at 462.76 cm^{-1} belongs to the internal bending vibration of the silicon aluminum oxide tetrahedron, while its symmetrical tensile vibration is at 572.99 cm^{-1} [24]. The absorption peak near 1635.74 cm^{-1} is generated by the stretching and bending vibrations of O–H and H–O–H. As the hydration of fly-ash-based geopolymer was terminated by soaking in absolute ethanol and the test product was dried sufficiently to eliminate the influence of free water on the absorption peak, the wider peaks in these two places are related to the hydration gel generated during the reaction process, such as N–A–S–H gel. The absorption peak at 1423.67 cm^{-1} is caused by the asymmetric stretching vibration of O–C–O [28,29]. An amount of 1384.70 cm^{-1} is the stretching vibration of N–O, and the source of nitrogen in the experiment is only $NaNO_3$. It can be concluded that there is still unreacted $NaNO_3$ in the fly-ash-based geopolymer under the condition of alkali excitation [25], which is consistent with the phenomenon that a strong $NaNO_3$ diffraction peak is found in the XRD pattern of the fly-ash-based geopolymer products generated at high concentrations of $NaNO_3$. The small peak at 685.61 cm^{-1} is the bending vibration of T–O–T (T = Si, Al). Corresponding to the Si-O bond at 995.97 cm^{-1}, it can be inferred that the T120 products contain a small amount of cancrinite [30]. The infrared spectra of chabazite/garronite-rich geopolymers before and after adsorption show no obvious change in the peak. No evidence has shown the chemical bond fracture and new bond formation in the adsorption process of nuclide Cs^+ on the products; therefore, it can be deduced that the adsorption of Cs on chabazite/garronite-rich geopolymer is physical adsorption.

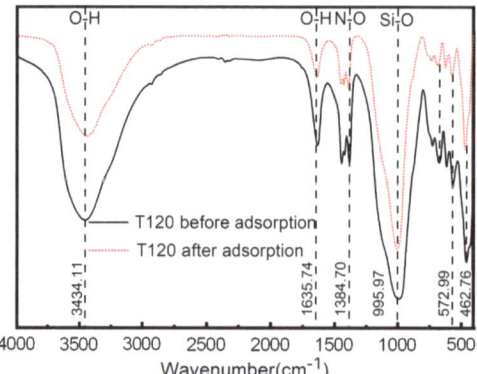

Figure 6. FTIR spectra of T120 products before and after adsorption of Cs^+.

3.2. Adsorption Performances

3.2.1. Adsorption Capacity and Adsorption Rate

Figure 7 outlines the adsorption rate curves of T120 products with different initial concentrations of the Cs^+ solution, and the adsorption time is 6 h. From Figure 7, it can be seen that as the concentration of Cs^+ in the system varies from 50 mg/L to 800 mg/L, the adsorption capacity of chabazite/garronite-rich geopolymers to Cs^+ increases from 11.73 mg/g to 82.95 mg/g. The increasing initial mass concentration of Cs^+ elevates the ion concentration in the unit volume solution and extends the contact time between the adsorbate and the adsorbent to a certain extent. From Figure 7, it can also be seen that the adsorption rate of Cs^+ by chabazite/garronite-rich geopolymers decreases with the increase in the initial concentration of Cs^+, and the adsorption rate of Cs^+ by chabazite/garronite-rich geopolymers decreases from 93.80% to 41.48%. When the concentration of Cs^+ is 50 mg/L, the adsorption removal rate of Cs^+ by chabazite/garronite-rich geopolymers is the highest at 93.80%.

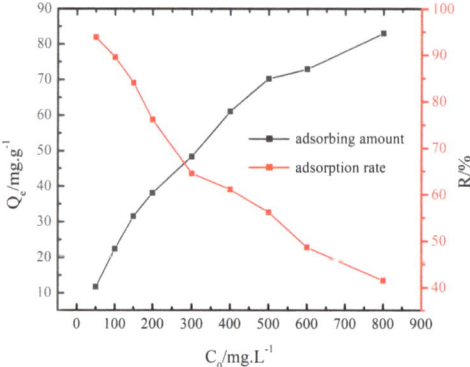

Figure 7. Adsorption rate of Cs^+ on T120 products with different initial concentrations.

3.2.2. Adsorption Isotherm Model

The adsorption data were analyzed, and the fitting parameters were obtained to infer the adsorption reaction mechanism. For the adsorption of heavy metals in a solution, the Langmuir isothermal adsorption model [31] and the Freundlich isothermal adsorption model [32] were used to fit the experimental data. The Langmuir equation is a model based on monolayer adsorption (chemical adsorption), while the Freundlich equation is suitable for describing the adsorption process of uneven surfaces (physical adsorption). Figure 8 demonstrates the Langmuir and Freundlich isothermal adsorption equation fitting curve of Cs$^+$ on chabazite/garronite-rich geopolymers. The fitting parameters of Langmuir and Freundlich's isothermal adsorption equations of Cs$^+$ on the chabazite/garronite-rich geopolymer are shown in Table 4.

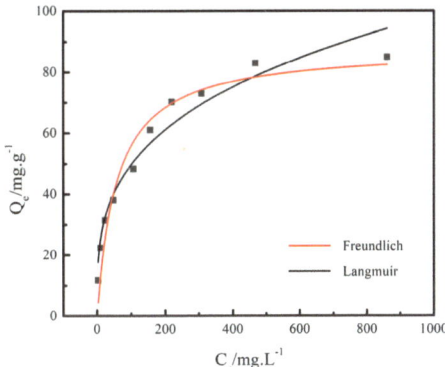

Figure 8. The fitting curves of isothermal adsorption of Cs$^+$ on T120 products.

Table 4. Isothermal adsorption equation and fitting parameters of Cs$^+$ on T120 products (25 °C).

	Langmuir			Freundlich		
	Q_m (mg/g)	K_L	R^2	n	K_F (mg/g)	R^2
T120	87.91	1.75×10^{-2}	0.95	3.36	12.64	0.96

According to the fitting parameters of the isothermal adsorption curve, the isothermal adsorption curve of Cs$^+$ is fitted well with the Freundlich equation and Langmuir equation, both of which are suitable for describing the isothermal adsorption behavior of Cs$^+$ on the chabazite/garronite-rich geopolymer. It indicates that the adsorption of Cs$^+$ on the chabazite/garronite-rich geopolymer is dominated by monolayer adsorption and the adsorption of uneven surfaces at the same time. Moreover, the isothermal adsorption fitting curve shows that the isothermal adsorption data of Cs$^+$ in the low initial concentration range (C_o at 50~300 mg/L) are more suitable for the Freundlich equation, while the Langmuir equation fits the adsorption data at the high initial concentration range (C_o at 400~800 mg/L). When the residue concentration of Cs$^+$ is in the range of 50~800 mg/L, the maximum theoretical equilibrium adsorption capacity of chabazite/garronite-rich geopolymers to Cs$^+$ is 87.91 mg/g. The calculated separation coefficients (K_L) using the Langmuir model fall within the range of 0–1, and the anisotropy coefficients (n) obtained from the Freundlich model are all greater than 1. These coefficients imply that the adsorption of Cs$^+$ from aqueous solutions by fly-ash-based zeolites is the favorable equilibrium adsorption region [33].

3.3. Desorption Performances

In the desorption test of Cs$^+$, a hydrochloric acid solution with a pH = 1 was used as the eluent, and the mass percentage of concentrated hydrochloric acid was 37%. Table 5 highlights the desorption capacity and desorption rate of Cs$^+$ on the T120 products at different desorption times under the condition of 25 °C. The initial concentration of Cs$^+$ is selected as 150 mg/L (to ensure sufficient adsorption and low concentrations, as the Cs$^+$ in real LILW is rather low). It can be seen from the adsorption isotherm curves in Figure 7 that the equilibrium adsorption capacity of chabazite/garronite-rich geopolymers at 150 mg/L Cs$^+$ solution is 31.51 mg/g. However, Table 5 shows that when the desorption time is 60 min, the desorption capacity of Cs$^+$ (32.06 mg/g) in the solution after desorption is greater than the equilibrium adsorption capacity of 31.51 mg/g. This result may be caused by a deviation during the dilution process of the test samples. Therefore, the pseudo-first-order and pseudo-second-order kinetic fittings were performed using the desorption data of the (5~45 min) time period. Table 6 outlines the fitting parameters of the pseudo-first-order and pseudo-second-order kinetic models [34,35]. Figures 9 and 10 outline the pseudo-first-order and pseudo-second-order kinetic fitting curves at the time period of 5~45 min, respectively.

Table 5. Desorption rate and amount of Cs^+ from T120 products at different desorption times.

Desorption Time (min)	5	10	15	30	45	60
Desorption rate (%)	83.38	90.63	80.52	87.73	88.25	101.74
Desorption capacity (mg/g)	26.28	28.56	25.38	27.65	27.81	32.06

Table 6. Desorption kinetic equation and fitting parameters of Cs^+ from T120 products.

Kinetic Equation	Desorption Kinetic Equation	Desorption Rate	Theoretical Equilibrium Desorption Capacity/mg·g^{-1}	R^2
Pseudo-first-order	$\lg(Q_e - Q_t) = 0.68 - 2.28 \times 10^{-3}\,t$	$5.25 \times 10^{-3}\,\text{min}^{-1}$	4.74	0.09
Pseudo-second-order	$t/Q^t = 1.71 \times 10^{-2} + 3.576 \times 10^{-2}\,t$	$7.48 \times 10^{-2}\,\text{mg}/(\text{g·min}^{-1})$	27.96	0.99

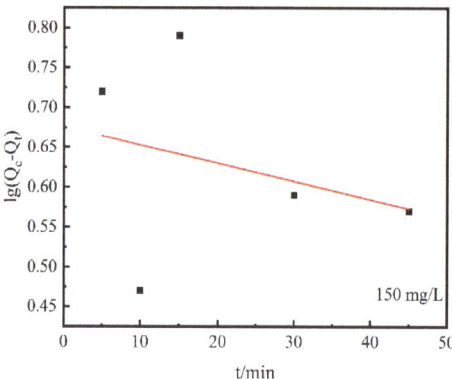

Figure 9. The fitting diagram of pseudo-first-order kinetic model to the desorption of Cs^+ from T120 products.

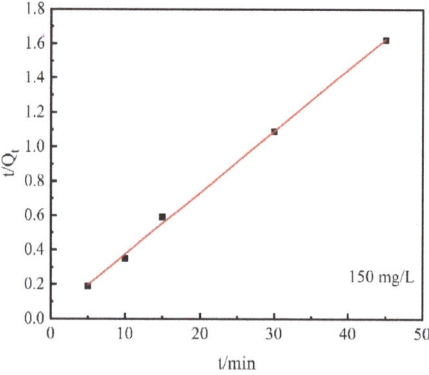

Figure 10. The fitting diagram of pseudo-second-order kinetic model to the desorption of Cs^+ from T120 products.

From the fitting parameters of Table 6, it can be seen that the desorption data of Cs^+ on the chabazite/garronite-rich geopolymers after saturated adsorption at 150 mg/L have a good correlation with the pseudo-second-order kinetic equation during the desorption period of 5~45 min. The correlation is 0.99, and the pseudo-second-order desorption rate is 7.48×10^{-2} mg/(g·min^{-1}). The theoretical equilibrium desorption capacity during the 5–45 min period is 27.96 mg/g, which is close to the desorption capacity of 27.81 mg/g at 45 min, and the relative error is 0.55%. Combined with the adsorption model mentioned above, it can be inferred that the adsorption and desorption processes of Cs^+ on chabazite/garronite-rich geopolymers are mainly chemical processes [36,37].

4. Conclusions

The products of fly-ash-based geopolymers and the adsorption performance of the products to the simulated radionuclide Cs^+ were controlled by $NaNO_3$ concentrations, NaOH concentrations, and curing temperatures. Specific findings from this work include the following:

(1) In the low $NaNO_3$ system (100 g/L), an increase in the NaOH concentration from 0.66 M to 2 M promoted the formation of Y-type zeolites and chabazite, while an increase in the NaOH concentration from 2 M to 8 M led to the transformation of zeolites into cancrinite. In the high $NaNO_3$ (300~500 g/L) system, the increase in NaOH concentrations above 2 M had no obvious effect on the product transformation, and the products were mainly cancrinite and chabazite.

(2) In the low $NaNO_3$ system (100 g/L) with NaOH concentrations of 2 M, Y-type zeolite was formed at 90 °C. With the increase in curing temperatures (90~150 °C), the Y-type zeolite was firstly transformed into garronite and chabazite, and then cancrinite at last. It can be concluded that $NaNO_3$ concentrations, NaOH concentrations, and curing temperatures all promote the crystallization of cancrinite and chabazite.

(3) At the $NaNO_3$ concentration of 100 g/L and 120 °C, the adsorption capacity of Cs^+ decreased with the increase in the initial concentration of Cs^+, and the adsorption rate of Cs^+ increased with the increase in the initial concentration of Cs^+ initial. The adsorption of Cs^+ in the low initial concentration range was more suitable for the Freundlich equation, while the Langmuir equation fit the adsorption process at the high initial concentration range. The adsorption of Cs^+ on the chabazite/garronite-rich geopolymer was dominated by physical and chemical adsorption at the same time.

(4) In the range of 5~45 min, the desorption kinetic process of Cs^+ on the chabazite/garronite-rich geopolymer was in good agreement with the pseudo-second-order equation. Indeed, the desorption of Cs^+ on the chabazite/garronite-rich geopolymer was a chemical desorption.

Author Contributions: Conceptualization, Z.Z. and Y.L.; Methodology, Z.Z.; Software, J.Y.; Formal analysis, Z.Z.; Investigation, M.C.; Resources, J.Y., M.C. and K.Y.; Data curation, M.C. and H.S.; Writing—original draft, Z.Z.; Writing—review & editing, J.Y., K.Y. and H.S.; Supervision, X.M. and Y.L. All authors have read and agreed to the published version of the manuscript.

Funding: This work was supported by the Scientific Research Fund of Southwest University of Science and Technology (No. 21zx7124) and the Natural Science Basic Research Program of Shaanxi (Program No. 2022JQ-897).

Data Availability Statement: Data are contained within the article.

Acknowledgments: The authors gratefully acknowledge the School of Civil Engineering and Architecture, School of Materials and Chemistry (Southwest University of Science and Technology) for providing their facilities to carry out this work.

Conflicts of Interest: Author Kui Yang was employed by the company Sino Shaanxi Nuclear Industry Group. The remaining authors declare that the research was conducted in the absence of any commercial or financial relationships that could be construed as a potential conflict of interest.

References

1. Kosowski, P.; Kosowska, K.; Janiga, D. Primary Energy Consumption Patterns in Selected European Countries from 1990 to 2021: A Cluster Analysis Approach. *Energies* **2023**, *16*, 6941. [CrossRef]
2. Fernandez-Jimenez, A.; Macphee, D.; Lachowski, E.; Palomo, A. Immobilization of cesium in alkaline activated fly ash matrix. *J. Nucl. Mater.* **2005**, *346*, 185–193. [CrossRef]
3. Fu, S.; He, P.; Wang, M.; Cui, J.; Wang, M.; Duan, X.; Yang, Z.; Jia, D.; Zhou, Y. Hydrothermal synthesis of pollucite from metakaolin-based geopolymer for hazardous wastes storage. *J. Clean. Prod.* **2019**, *248*, 119240. [CrossRef]
4. Tyupina, E.A.; Kozlov, P.P.; Krupskaya, V.V. Application of Cement-Based Materials as a Component of an Engineered Barrier System at Geological Disposal Facilities for Radioactive Waste—A Review. *Energies* **2023**, *16*, 605. [CrossRef]
5. He, P.; Wang, Q.; Fu, S.; Wang, M.; Zhao, S.; Liu, X.; Jiang, Y.; Jia, D.; Zhou, Y. Hydrothermal transformation of geopolymers to bulk zeolite structures for efficient hazardous elements adsorption. *Sci. Total. Environ.* **2021**, *767*, 144973. [CrossRef]
6. Wang, C.F.; Li, J.S.; Wang, L.J.; Sun, X.Y. Influence of NaOH concentrations on synthesis of pure-form zeolite A from fly ash using two-stage method. *J. Hazard. Mater.* **2008**, *155*, 58–64. [CrossRef]
7. Khalid, H.R.; Lee, N.K.; Park, S.M.; Abbas, N.; Lee, H.K. Synthesis of geopolymer-supported zeolites via robust one-step method and their adsorption potential. *J. Hazard. Mater.* **2018**, *353*, 522–533. [CrossRef]
8. Huseien, G.F.; Mirza, J.; Ismail, M.; Ghoshal, S.; Hussein, A.A. Geopolymer mortars as sustainable repair material: A comprehensive review. *Renew. Sustain. Energy Rev.* **2017**, *80*, 54–74. [CrossRef]
9. Zhao, X.; Liu, C.; Zuo, L.; Wang, L.; Zhu, Q.; Liu, Y.; Zhou, B. Synthesis and characterization of fly ash geopolymer paste for goaf backfill: Reuse of soda residue. *J. Clean. Prod.* **2020**, *260*, 121045. [CrossRef]
10. Yarusova, S.; Shichalin, O.; Belov, A.; Azon, S.; Buravlev, I.Y.; Golub, A.; Mayorov, V.Y.; Gerasimenko, A.; Papynov, E.; Ivanets, A.; et al. Synthesis of amorphous KAlSi3O8 for cesium radionuclide immobilization into solid matrices using spark plasma sintering technique. *Ceram. Int.* **2021**, *48*, 3808–3817. [CrossRef]
11. Shichalin, O.; Papynov, E.; Nepomnyushchaya, V.; Ivanets, A.; Belov, A.; Dran'kov, A.; Yarusova, S.; Buravlev, I.; Tarabanova, A.; Fedorets, A.; et al. Hydrothermal synthesis and spark plasma sintering of NaY zeolite as solid-state matrices for cesium-137 immobilization. *J. Eur. Ceram. Soc.* **2022**, *42*, 3004–3014. [CrossRef]
12. Penilla, R.P.; Bustos, A.G.; Elizalde, S.G. Immobilization of Cs, Cd, Pb and Cr by synthetic zeolites from Spanish low-calcium coal fly ash. *Fuel* **2006**, *85*, 823–832. [CrossRef]
13. Cardoso, A.M.; Paprocki, A.; Ferret, L.S.; Azevedo, C.M.; Pires, M. Synthesis of zeolite Na-P1 under mild conditions using Brazilian coal fly ash and its application in wastewater treatment. *Fuel* **2015**, *139*, 59–67. [CrossRef]
14. Wu, D.; Sui, Y.; Chen, X.; He, S.; Wang, X.; Kong, H. Changes of mineralogical–chemical composition, cation exchange capacity, and phosphate immobilization capacity during the hydrothermal conversion process of coal fly ash into zeolite. *Fuel* **2008**, *87*, 2194–2200. [CrossRef]
15. Wang, Y.; Han, F.; Mu, J. Solidification/stabilization mechanism of Pb (II), Cd (II), Mn (II) and Cr (III) in fly ash based geopolymers. *Constr. Build. Mater.* **2018**, *160*, 818–827. [CrossRef]
16. Wdowin, M.; Wiatros-Motyka, M.M.; Panek, R.; Stevens, L.A.; Franus, W.; Snape, C.E. Experimental study of mercury removal from exhaust gases. *Fuel* **2014**, *128*, 451–457. [CrossRef]
17. Lei, H.; Muhammad, Y.; Wang, K.; Yi, M.; He, C.; Wei, Y.; Fujita, T. Facile fabrication of metakaolin/slag-based zeolite microspheres (M/SZMs) geopolymer for the efficient remediation of Cs+ and Sr2+ from aqueous media. *J. Hazard. Mater.* **2021**, *406*, 124292. [CrossRef]
18. Lee, N.K.; Khalid, H.R.; Lee, H.K. Adsorption characteristics of cesium onto mesoporous geopolymers containing nano-crystalline zeolites. *Microporous Mesoporous Mater.* **2017**, *242*, 238–244. [CrossRef]
19. Hui, K.S.; Chao CY, H.; Kot, S.C. Removal of mixed heavy metal ions in wastewater by zeolite 4A and residual products from recycled coal fly ash. *J. Hazard. Mater.* **2005**, *127*, 89–101. [CrossRef]
20. Querol, X.; Moreno, N.; Umaña, J.C.; Alastuey, A.; Hernández, E.; López-Soler, A.; Plana, F. Synthesis of zeolites from coal fly ash: An overview. *Int. J. Coal Geol.* **2002**, *50*, 413–423. [CrossRef]
21. Kumar, M.M.; Jena, H. Direct single-step synthesis of phase pure zeolite Na-P1, hydroxy sodalite and analcime from coal fly ash and assessment of their Cs+ and Sr2+ removal efficiencies. *Microporous Mesoporous Mater.* **2022**, *333*, 111738. [CrossRef]
22. Palomo, Á.; Alonso, S.; Fernandez-Jiménez, A.; Sobrados, I.; Sanz, J. Alkaline Activation of Fly Ashes: NMR Study of the Reaction Products. *J. Am. Ceram. Soc.* **2004**, *87*, 1141–1145. [CrossRef]
23. Zheng, Z.; Ma, X.; Zhang, Z.; Li, Y. In-situ transition of amorphous gels to Na-P1 zeolite in geopolymer: Mechanical and adsorption properties. *Constr. Build. Mater.* **2019**, *202*, 851–860. [CrossRef]
24. Ren, X.; Liu, S.; Qu, R.; Xiao, L.; Hu, P.; Song, H.; Wu, W.; Zheng, C.; Wu, X.; Gao, X. Synthesis and characterization of single-phase submicron zeolite Y from coal fly ash and its potential application for acetone adsorption. *Microporous Mesoporous Mater.* **2019**, *295*, 109940. [CrossRef]
25. Zheng, Z.; Li, Y.; Sun, H.; Zhang, Z.; Ma, X. Coupling effect of NaOH and NaNO3 on the solidified fly ash-cement matrices containing Cs+: Reaction products, microstructure and leachability. *J. Nucl. Mater.* **2020**, *539*, 152252. [CrossRef]
26. Aono, H.; Takeuchi, Y.; Itagaki, Y.; Johan, E. Synthesis of chabazite and merlinoite for Cs+ adsorption and immobilization properties by heat-treatment. *Solid State Sci.* **2019**, *100*, 106094. [CrossRef]

27. Munthali, M.W.; Johan, E.; Aono, H.; Matsue, N. Cs+ and Sr2+ adsorption selectivity of zeolites in relation to radioactive decontamination. *J. Asian Ceram. Soc.* **2015**, *3*, 245–250. [CrossRef]
28. Perná, I.; Šupová, M.; Hanzlíček, T.; Špaldoňová, A. The synthesis and characterization of geopolymers based on metakaolin and high LOI straw ash. *Constr. Build. Mater.* **2019**, *228*, 116765. [CrossRef]
29. Chen, Z.; Li, J.S.; Zhan, B.J.; Sharma, U.; Poon, C.S. Compressive strength and microstructural properties of dry-mixed geopolymer pastes synthesized from GGBS and sewage sludge ash. *Constr. Build. Mater.* **2018**, *182*, 597–607. [CrossRef]
30. Liu, Y.; Yan, C.; Zhao, J.; Zhang, Z.; Wang, H.; Zhou, S.; Wu, L. Synthesis of zeolite P1 from fly ash under solvent-free conditions for ammonium removal from water. *J. Clean. Prod.* **2018**, *202*, 11–22. [CrossRef]
31. Guo, X.; Wang, J. Comparison of linearization methods for modeling the Langmuir adsorption isotherm. *J. Mol. Liq.* **2019**, *296*, 111850. [CrossRef]
32. Majd, M.M.; Kordzadeh-Kermani, V.; Ghalandari, V.; Askari, A.; Sillanpää, M. Adsorption isotherm models: A comprehensive and systematic review (2010−2020). *Sci. Total. Environ.* **2021**, *812*, 151334. [CrossRef] [PubMed]
33. Xiang, Y.; Hou, L.; Liu, J.; Li, J.; Lu, Z.; Niu, Y. Adsorption and enrichment of simulated 137Cs in geopolymer foams. *J. Environ. Chem. Eng.* **2021**, *9*, 105733. [CrossRef]
34. Allaoui, M.; Berradi, M.; Bensalah, J.; Es-Sahbany, H.; Dagdag, O.; Ibn Ahmed, S. Study of the adsorption of nickel ions on the sea shells of Mehdia: Kinetic and thermodynamic study and mathematical modelling of experimental data. *Mater. Today Proc.* **2021**, *45*, 7494–7500. [CrossRef]
35. Jin, H.; Zhang, Y.; Wang, Q.; Chang, Q.; Li, C. Rapid removal of methylene blue and nickel ions and adsorption/desorption mechanism based on geopolymer adsorbent. *Colloid Interface Sci. Commun.* **2021**, *45*, 100551. [CrossRef]
36. Deng, H.; Li, Y.; Huang, Y.; Ma, X.; Wu, L.; Cheng, T. An efficient composite ion exchanger of silica matrix impregnated with ammonium molybdophosphate for cesium uptake from aqueous solution. *Chem. Eng. J.* **2016**, *286*, 25–35. [CrossRef]
37. Deng, H.; Li, Y.; Wu, L.; Ma, X. The novel composite mechanism of ammonium molybdophosphate loaded on silica matrix and its ion exchange breakthrough curves for cesium. *J. Hazard. Mater.* **2017**, *324*, 348–356. [CrossRef]

Disclaimer/Publisher's Note: The statements, opinions and data contained in all publications are solely those of the individual author(s) and contributor(s) and not of MDPI and/or the editor(s). MDPI and/or the editor(s) disclaim responsibility for any injury to people or property resulting from any ideas, methods, instructions or products referred to in the content.

Article

The Sealing Performance of Cement Sheaths under Thermal Cycles for Low-Enthalpy Geothermal Wells

Anisa Noor Corina * and Al Moghadam

TNO Applied Geosciences, 3584 CB Utrecht, The Netherlands
* Correspondence: anisa.corina@tno.nl

Abstract: The repetitive process of shut-in and production in geothermal wells promotes thermal stress on the wellbore components, including annular cement. A cement sheath at a relatively shallow depth undergoes the most significant stress change due to the high differential temperature between the geothermal gradient and the production fluid's temperature. Understanding the impact of cyclical thermal stresses on cement is critical for assessing the barrier integrity at a shallow depth that serves as aquifer protection. A novel large-scale setup simulating a 1.5 m-long casing-cement-casing well section was built to study the changes in cement's sealing performance of low-enthalpy geothermal wells during production. Using this setup, a cement sheath can be cured similarly to the in situ conditions, and the annular temperature can be cycled under realistic operating conditions. The change in flow rate through the cement sheath before and after cycling is quantified through leak tests. UV dye is injected at the end of the experiment to identify the location and type of damage in the cement sheath. A hydromechanically coupled finite element model was used to estimate the stress evolution in cement during the tests. The model incorporated the impact of cement hydration and strength development during curing. The numerical results were used as a guide to ensure the test design closely mimicked in situ conditions. The results show the presence of a small microannulus immediately after curing due to hydration shrinkage. Thermal cycles reduced the permeability of the microannulus. The size of the micro-annulus was observed to be sensitive to the backpressure applied to the cement sheath, indicating the need for pressure to maintain an open microannulus. Thirty-nine thermal cycles between 80 and 20 °C did not change the permeability of the cement sheath significantly. Tensile cracks in the cement sheath were not continuous and may not be a significant pathway. The new setup allows for measuring cement's effectiveness in withstanding in situ stress conditions when exposed to thermal cycles such as geothermal and CCS wells.

Keywords: well integrity; well cementing; numerical modeling; geothermal

Citation: Corina, A.N.; Moghadam, A. The Sealing Performance of Cement Sheaths under Thermal Cycles for Low-Enthalpy Geothermal Wells. *Energies* **2024**, *17*, 239. https://doi.org/10.3390/en17010239

Academic Editors: Carlo Roselli, Yongjun Deng, Lin Chen, Kai Wang and Jie Wu

Received: 7 November 2023
Revised: 8 December 2023
Accepted: 25 December 2023
Published: 2 January 2024

Copyright: © 2024 by the authors. Licensee MDPI, Basel, Switzerland. This article is an open access article distributed under the terms and conditions of the Creative Commons Attribution (CC BY) license (https://creativecommons.org/licenses/by/4.0/).

1. Introduction

Geothermal energy plays a crucial role in the global energy transition towards a more sustainable and low-carbon future. The ambition of geothermal energy in The Netherlands is to contribute 55.5 + TWh to the heat demand by 2050 [1]. By the end of 2022, there were a total of 27 geothermal production installations available in The Netherlands with depths ranging between 1600–2800 m [2] and an operating temperature of <100 °C. During the life of the well, geothermal producer wells undergo thermal cycles of heating and cooling as a result of the repetitive production and shut-in processes. These cycles cause the wellbore components to alternately expand and contract, which induces cyclic stress. Of specific concern is the annular cement, which may sustain damage in the form of cracks or microannuli due to cyclic stress. Such damage could increase the risk of creating leak pathways, compromising the cement integrity.

In the producer wells, the heating effect is particularly significant for the shallow section. During the production phase, the cement sheath is heated up from the geothermal gradient to a temperature close to the reservoir brine's temperature. During shut-in, the temperature will return to the initial condition. It is also important to consider that the properties of the thermal expansion coefficient and thermal conductivity, differ between the components of casing, cement, and formation. The difference in the thermal conductivity results in a radial temperature profile, hence causing each component to expand and contract at a different rate.

Several experimental studies have evaluated the impact of thermal cycles on cement sheath properties. Therond et al. [3] performed a large-scale experiment investigating the change in the annular flow of a cement sheath that is subjected to cooling/heating cycles and pressure cycles in the context of water injection wells. The result shows that the cooling effect has a significant impact on the sealing of the cement sheath as it promotes the creation of microannulus at the inner and outer interface. The radial cracks from the pressure increase were found to have a negligible effect on the annular permeability. The results, however, are not relevant in the context of producer wells where the thermal load from heating is more dominant.

Goodwin and Crook [4] indirectly simulated the thermal loading of heating/cooling on cement sheath by cycling the inner casing pressure. The change in the annular permeability was measured in each cycle. Significant cement damage in the form of radial cracks was observed after the inner pressure was raised to 2.8 MPa. This pressure is equivalent to a temperature change of 100–200 °C for various casing sizes and types. However, the approach is less accurate to mimic thermal stress without involving heat transfer. Recent works by De Andrade et al. [5], Vrålstad et al. [6], and De Andrade et al. [7] used a scaled-down wellbore setup to directly simulate thermal cycles in various downhole scenarios (e.g., effects of different confining formations and casing eccentricity). The evolution of cement damage was monitored through a CT scan, and the volume of the damage was quantified. Their results suggest that the thermal heating–cooling cycle (110 °C to 40 °C) has a minor influence on the integrity of the cement sheath. A similar work by Lin et al. [8] and Kuanha et al. [9] used a large-scale setup that simulated a 0.76 m length of cement sheath between 5″ inner and 7″ outer casings. It was shown that the cement sheath was damaged from thermal cycling, likely due to plastic deformation, and the damage is more significant at higher temperature loading. These works, however, did not quantify the change in the annular permeability.

In this work, the evolution of the sealing properties of annular cement subjected to thermal loading is evaluated through laboratory investigation using a novel large-scale wellbore simulator setup. The large-scale setting was selected in order to accommodate crack growth, especially in the axial direction, and understand the impact of localized cracks on the annular permeability. We applied cycles of heating/cooling on the cement sheath, simulating the operating conditions of low-enthalpy geothermal production wells. The change in cement sheath's effective permeability is measured before and after thermal and pressure cycles. The experiment was replicated using a finite element model to gather more insight into cement stress development and to compare and contrast the numerical results with experimental observations.

2. Materials and Methods

2.1. Laboratory Experiments

A unique large-scale experimental setup was built to replicate a ~1.2 m long cement sheath placed between a 7–5/8″ inner casing and a 9–5/8″ outer casing (Figure 1). The inner and outer diameter of the cement sheath is 194 and 242 mm, respectively. A vessel (OD: 340 mm) is placed around the outer casing to provide confinement. The entire setup is placed inside a 2 × 2 m heating enclosure to provide a stable temperature. The fluid within the 7–5/8″ inner casing can be circulated with hot and cold water for thermal cycling and can be pressurized for pressure cycling. The sealing of the cement sheath is evaluated using a leak test, which is performed by recording the flow rate at both inflow and outflow at one or more differential pressures (ΔP) and different backpressures (BP) or outlet pressures. Using Darcy's flow equation, the equivalent effective permeability can be calculated following the equation:

$$k_{eff} = 1.01 \times 10^{15} \frac{Q\mu L}{\Delta P A} \quad (1)$$

where Q is the water flow rate (m^3/s), μ is the water viscosity at 20 °C (Pa/s), L is the length of the cement sheath (m), A is the cross-section area of the annulus (m^2), ΔP is the differential pressure (Pa), and k is the effective permeability (milliDarcy).

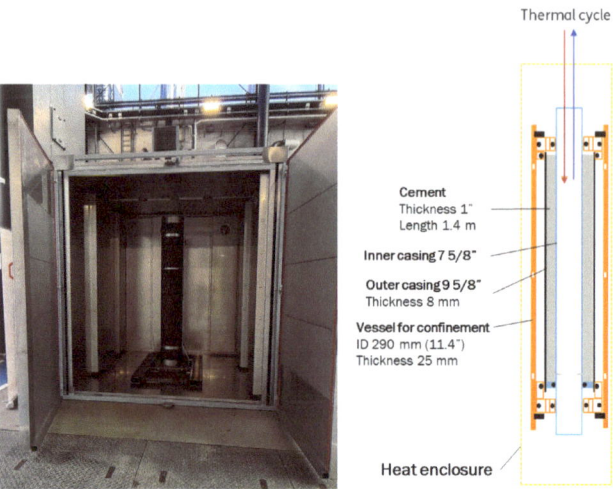

Figure 1. (**Left**) Realization of the large-scale setup and (**right**) the schematic drawing of the setup with a description of the components.

One experiment was conducted using class-G neat cement with 44% BWOC of water. Approx. Twenty liters of a cement slurry was mixed and injected into the annulus. The slurry was then cured for 11 days under an elevated annular pressure of 5 MPa, a confined pressure of 5 MPa, and a temperature of 30 °C. During the leak test, the confined pressure was kept constant at ~4.2 MPa. The chronological steps of testing are as follows:

1. An initial leak test was conducted to generate the initial effective permeability of the sample.
2. Thermal (heating–cooling) cycling. The hot water injected in the inner casing had a temperature of 80 °C, while the cold water had a temperature of 20 °C. This process of alternating between hot and cold water was repeated for a certain number of cycles. Following the thermal cycles, a leak test was performed.
3. Pressure cycling. The pressure cycle was performed by alternately pressurizing the inner casing to 30 MPa and releasing it back to atmospheric pressure. Afterward, a leak test was performed.
4. Rhodamine B dye is injected into the cement sheath, and the location of cement damage and the form of damage are observed in the post-mortem analysis.

The realization of the testing steps is summarized in Figure 2. Following curing and initial leak testing, two thermal cycles were conducted with a low backpressure (BP) of 0.5 MPa and a high backpressure of 4 MPa. The intention of cycling at two different backpressures was to generate observation of the influence of thermal cycles at different states of microannulus opening. At high backpressure, the microannulus is expected to open at both ends of the annular cement. On the contrary, the microannulus is expected to close or partially open at the outlet at low back pressure. After pressure cycling, two leak tests were performed: one with an inner casing pressure of 30 MPa and one without inner pressure.

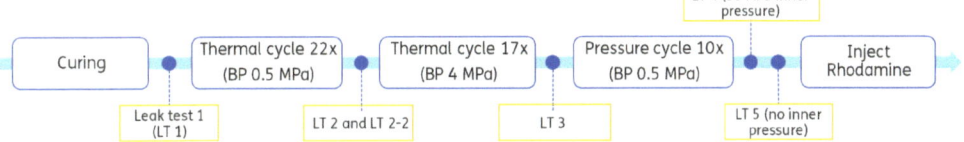

Figure 2. Realization of testing steps and reference of the leak test (LT) from LT 1 to LT 5.

2.2. Numerical Modeling

We developed a hydromechanical model to investigate the evolution of stress in the cement sheath in the test setup. The modeling is conducted using the Abaqus finite element package. The model mimics the same geometry as the test setup presented in Figure 1. Cement is modeled as a poroelastic material. The initial pressure of the cement was assumed to be equal to the curing pressure. The hydration rate of class G cement was used to estimate the water consumption, pressure drop, and stress change in cement during curing [10,11]. After curing, the model imposes a cyclical temperature or pressure boundary condition on the inside of the inner casing. The evolution of cement stress is monitored over time to understand the impact of pressure and temperature on the stresses in cement and whether damage can be expected. Details of the modeling methodology are described in Moghadam and Loizzo [11]. Figure 3 provides a schematic of the geometry that was considered in the modeling exercise.

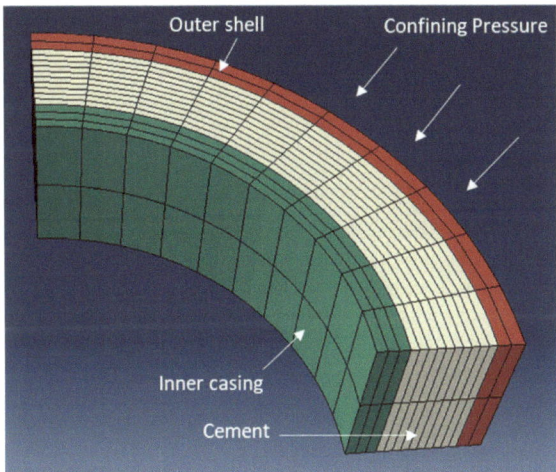

Figure 3. The geometry of the setup in the numerical model. The model uses a plane strain assumption in the z-direction (equivalent to a 2D geometry).

3. Results

3.1. Experimental Results

The plots of flow rate vs. ΔP from all leak tests are presented in Figure 4, grouped by the BP. The slurry had shrunk by 4.1% during curing. This value is similar to existing data for neat cement [12]. The flow rate measured just after curing (LT 1) was relatively high, as shown in Figure 4a,b, equivalent to an effective of 4 mD. This indicates the potential formation of a microannulus after curing.

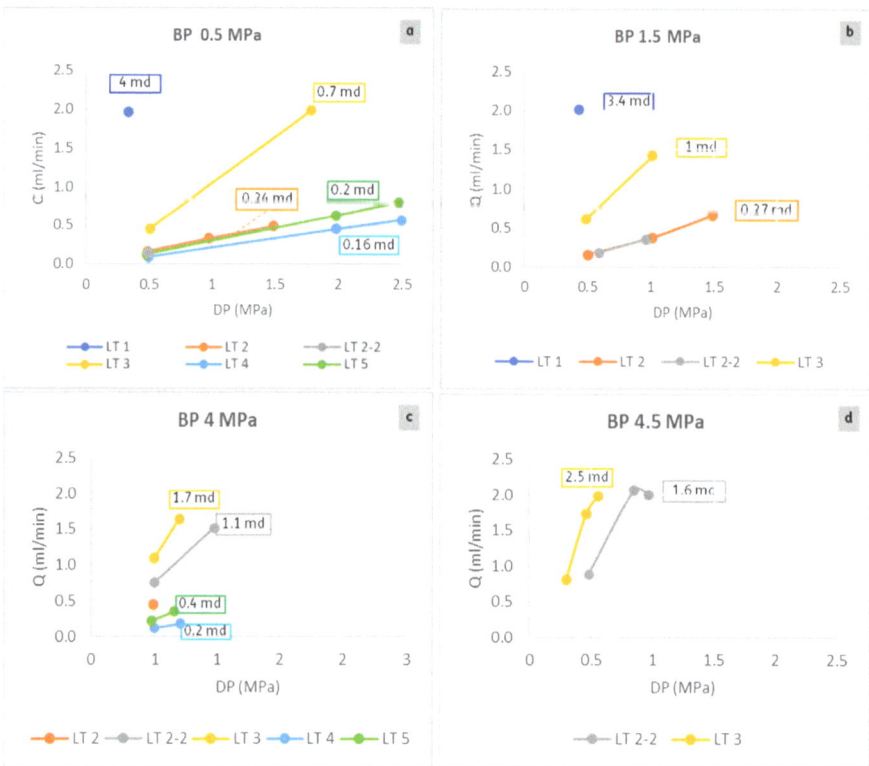

Figure 4. Plots of leak rate vs. ΔP from all leak tests (LT 1 until LT 5) grouped by the backpressure of 0.5 MPa (**a**), 1.5 MPa (**b**), 4 MPa (**c**), and 4.5 MPa (**d**). The equivalent k-eff is shown in the text box close to the line graph with the corresponding color coding.

During the first set of thermal cycles (22 cycles), the differential pressure was kept at 0.5 MPa. The evolution of both inflow and outflow rates after every cycle of heating–cooling is shown in Figure 5. If the sample is leak-tight, the inflow and outflow rate should be zero. Since a leak was already detected in LT 1, the flow rate in the first cycle is not zero. With increasing cycles, both inflow and outflow gradually reduce, indicating a reduction in permeability. After completing the thermal cycles, the flow through the sample (LT 2) was measured. The equivalent permeability at BP of 0.5 and 1.5 MPa is ~0.2 mD, which is significantly smaller than the initial permeability. When the leak test was repeated at a higher backpressure of 4 MPa (see Figure 4c), the flow increased, and permeability was equivalent to ~1 mD. Such a significant increase could be evidence of the re-opening of microannulus.

Afterward, the backpressure was kept at 4 MPa, and the second set of thermal cycles was performed. A differential pressure of 1 MPa was kept during the cycles. The flow rate after each cycle of heating–cooling is shown in Figure 6. Both inflow and outflow stayed constant at around 1.7 mL/min and −1.6 mL/min, respectively. This response is different from the thermal cycles at low backpressure. The flow through cement after this cycle (LT 3) is higher than the previous cycle (LT 2) but still less than the initial flow rate. The permeability ranges between 0.7–2.5 mD at various backpressures. The reason behind the overall decrease in permeability after the thermal cycles is unclear. This could be due to further cement hydration. However, the initial curing period at elevated temperatures should have been sufficient according to experimental measures of hydration degree [11].

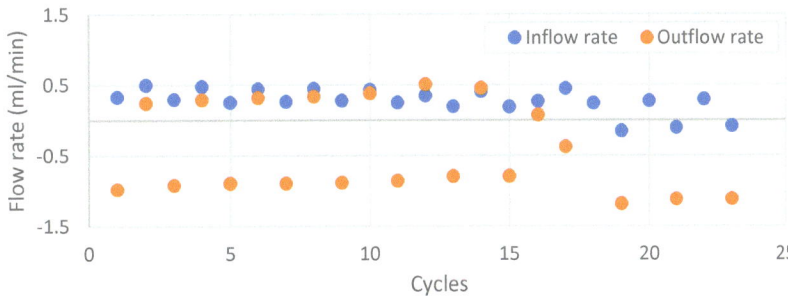

Figure 5. Inflow and outflow rates after each cycle of thermal heating and cooling performed at backpressure of 0.5 MPa and a differential pressure of 0.5 MPa. The thermal cycle was temporarily stopped after cycle 17. The positive sign means flow into the sample, and the negative sign means flow out from the sample.

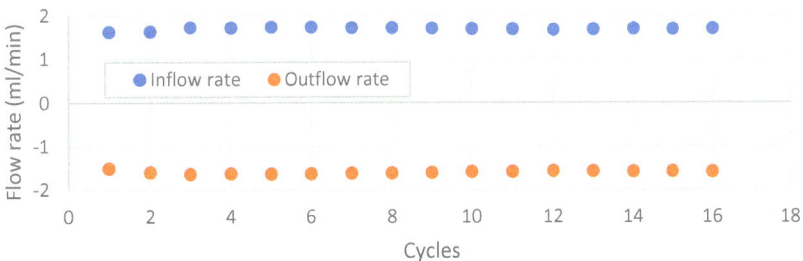

Figure 6. Inflow and outflow rates after each cycle of thermal heating–cooling performed at a back pressure of 4 MPa and differential pressure of 1 MPa. Lost data communication in cycle 17 (omitted from graph). The positive sign means flow into the sample, and the negative sign means flow out from the sample.

The next step was pressure cycling at a low backpressure of 0.5 MPa. After cycling, the inner casing pressure was kept at 30 MPa, and the measured flow (LT 4) was relatively low, with a permeability of ~0.2 mD. The flow measurement was then repeated at 0 MPa of inner casing pressure, and the effective permeability (LT 5) slightly increased to 0.2–0.4 mD. In a previous experiment (not described here), performing a leak test at high casing pressure caused the radial cracks formed during the pressure cycle to open. This led to an increase in flow. However, this was not observed based on results from LT 4 and 5. Overall, pressure cycling was found to reduce the cement's permeability. In Figure 4a, the flow after pressure cycles (LT 4 and 5) lies in the same range as those after the first thermal cycle (LT 2).

The post-mortem observations (Figure 7) suggest that the flow through the cement was mainly through the microannulus. The microannulus was mostly concentrated at the outer annulus. There was no clear evidence of radial cracks, which were expected to form during the pressure cycle. During the dye injection, no pressure was applied to the inner casing, which is likely causing the cracks to close and thus limiting the penetration of the dye in the crack. In addition, the dye might be restricted in penetrating micro- to nano-scale cement defects compared to water. The dye adsorption could also reduce the tracing effectivity.

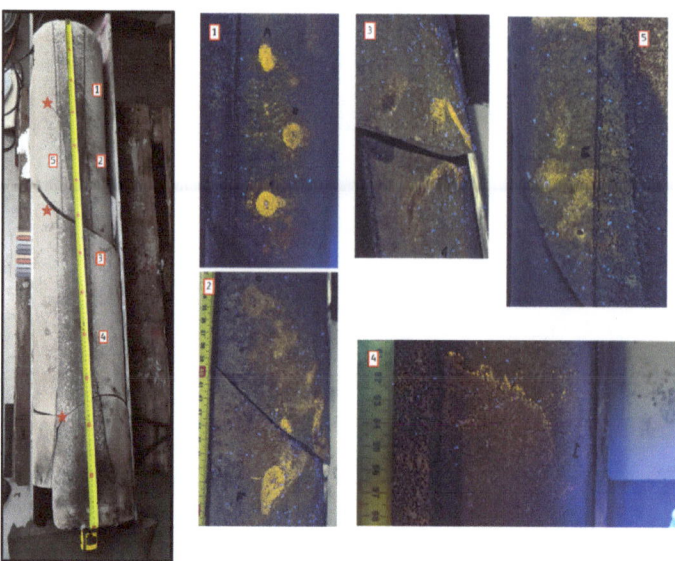

Figure 7. Visual observations of one side of the cement sheath using the UV light at different locations from top to bottom (indices 1–5). The dye traces (yellow) indicate the leakage pathway. The big cracks pointed out by the star sign are the artificial cracks created when removing the sample.

3.2. Numerical Modeling

We use the modeling results for additional insight into experiments. Modeling provides estimates of cement stress evolution, which controls potential cement failure. Therefore, numerical results can be used to further explain experimental observations and investigate potential failure mechanisms. Figure 8 shows the results of the temperature cycle modeling. The cement's inner and outer temperatures are plotted against time. The initial temperature of the system is 30 °C. First, the inner casing temperature is lowered to 20 °C and subsequently increased to 80 °C. The outer cement temperature changes with a delay and reaches a smaller magnitude. During the initial cooling period, an inner microannulus opens on the inner interface with a maximum aperture of 10 microns. As the system heats up, the cement expands, and the microannulus closes. When the system starts cooling down by changing the inner casing temperature to 20 °C (equivalent to cold water injection), the microannulus opens to nearly 30 microns. As the temperatures between the inner casing and cement equilibrate, the microannulus gradually closes (depending on the temperature evolution). Therefore, during the temperature cycles, a microannulus can temporarily open on the inner interface during the cooling phase. Hoop stress also changes with the temperature cycles. It becomes tensile during the cooling down period with a maximum of 5 MPa. This magnitude is above the range of cement's tensile strength (1 to 3 MPa) and may cause radial cracks [13].

Figure 9 demonstrates the magnitude of the hoop stress in the cement sheath during a pressure cycle from 0 to 30 MPa in the inner casing. The hoop stress (tangential direction) is slightly compressive after curing (negative indicates compressive stress, and positive values indicate tensile stress). During the pressure cycle, the hoop stress increases to nearly 7 MPa of tensile stress. This value surpasses the tensile strength of class G cement, typically in the range of 1 to 3 MPa [13]. Therefore, during the pressure increase in the experiments, we expect radial cracks to open along the cement sheath. Once the pressure drops, the stress drops back to near zero; therefore, there is a possibility of the radial cracks closing. Overall, the modeling shows that in the present system, pressure cycles open radial cracks and temperature cycles open microannuli. However, both are expected to close as the cycle is over.

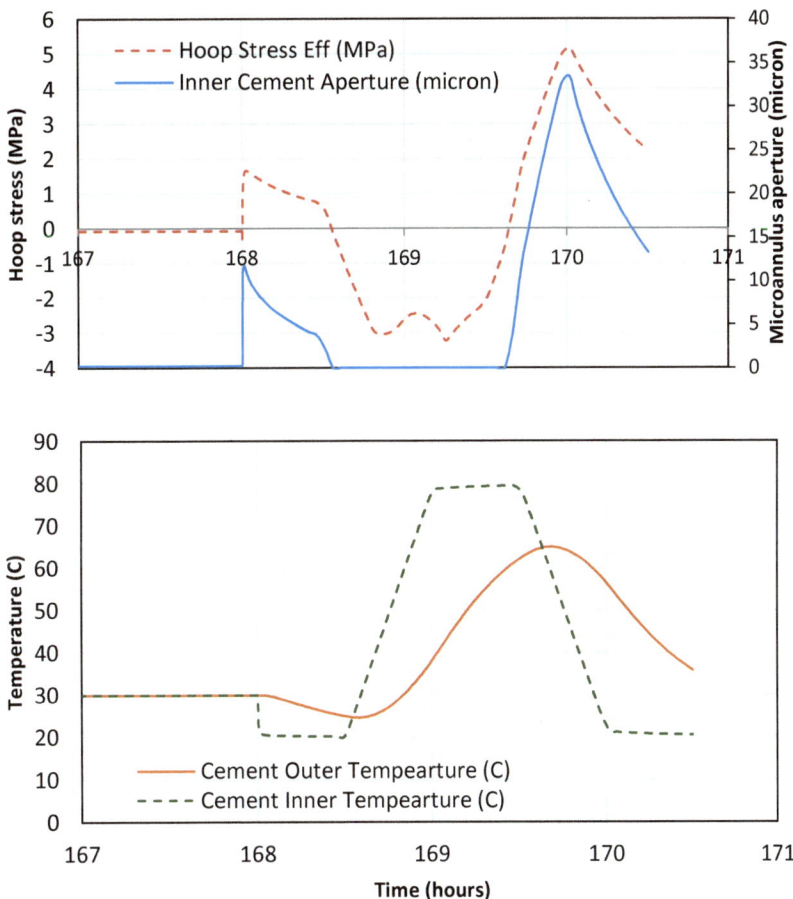

Figure 8. Cement temperature, hoop stress, and microannulus size during the temperature cycles.

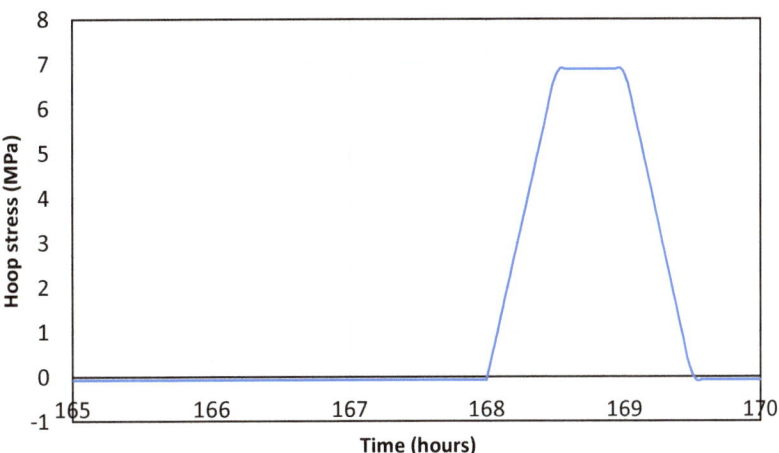

Figure 9. The magnitude of the hoop stress (tangential direction) during a pressure cycle between 0 and 30 MPa in the inner casing as performed in the experiments.

In order to assess the cyclical stress damage, the stress path at the cement (inner side) is plotted for both the pressure and temperature cycles in Figure 10. The initial size of the yield envelope for cement is also plotted in Figure 10, along with the critical state line [14]. Both the pressure and temperature cycle paths are largely within the elastic region. This indicates no shear failure under the lab conditions. However, cyclical damage may accumulate at levels below the shear capacity. The stress path for the temperature cycles varies over a wide region. Part of this region exceeds the yield envelope on the "dry side" (to the left of the critical state line). This typically leads to dilation and shear cracks and may lead to permeability. However, no significant permeability gain was observed in the lab tests. Based on the present model, it is not clear whether the cyclical damage accumulates or plateaus when a high number of cycles occur.

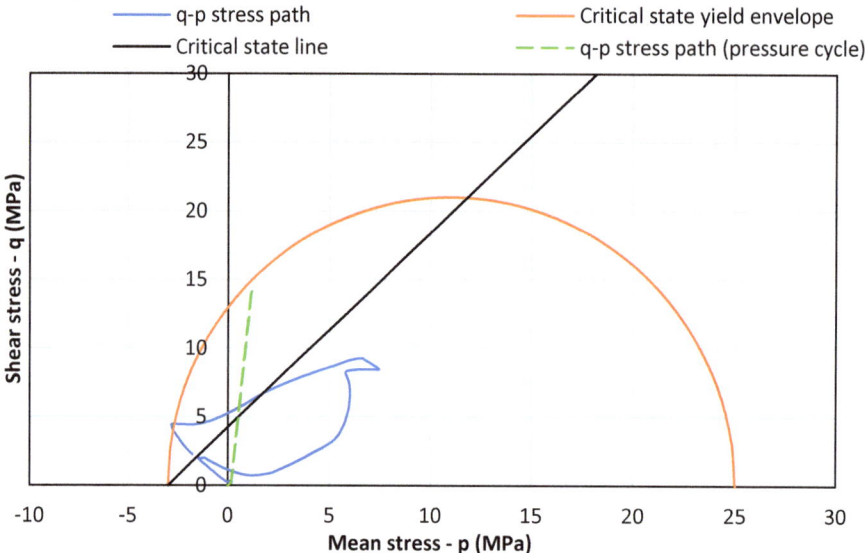

Figure 10. The stress path in the cement sheath during the pressure and temperature cycles, along with the initial size of the yield envelope from Soustelle et al. (2023) [14]. In this plot, negative indicates tensile stress.

4. Discussion

As cement cures, its pore pressure drops due to hydration reactions, which leads to a change in its stress conditions. In our experiments, the top and bottom of the cement sheath are exposed to water during curing. This leads to water flowing into the cement matrix during curing, which minimizes the pressure drop in cement at both ends of the sample and helps retain the radial stress. Therefore, we anticipate high (total) radial stress at either end of the sample compared to the middle of the cement sheath. The drop in the radial stress is expected to cause the creation of a microannulus as the total stress in the sample drops below the water pressure at the top of the cement. The stress drop has been observed to be higher on the outer interface, which indicates a higher probability of an outer microannulus [11].

In the current experiment, the leakage results from LT 1 show an initial microannulus with an equivalent permeability of 4 mD. This is likely due to the hydration shrinkage of cement and the subsequent pore pressure drop. The post-mortem observation confirms a microannulus at the outer interface. The current experiment also shows that the microannulus size or permeability is also a strong function of the back pressure. Higher backpressure provides the force to push apart the cement and casing interfaces at the microannulus. This indicates that the mere debonding of the cement interface microannulus is not sufficient for leakage to occur. Fluid pressure in the microannulus is also required to keep it open to flow.

Overall, thermal cycles were found to reduce the microannulus size and cement permeability. The mechanism for this is unclear. According to Wolterbeek and Hangx [15], the thermal expansion of cement is likely close to 1.3×10^{-5} 1/K. This means that the thermal expansion of cement is higher than that of casing. Therefore, as the temperature increases, the cement will expand more than steel. This differential expansion can close the microannulus and even compress the cement. Once the temperature drops, the cement may not return to its initial volume, and some inelastic strains may occur, which would explain the reduction in permeability observed here. We should note that overall, the permeability of the microannulus in this work is fairly low throughout the entire experiment.

The backpressure could provide another explanation. The temperature cycles at a backpressure of 0.5 MPa decrease the equivalent permeability of the system. However, as the backpressure is raised to 4 MPa and the second set of temperature cycling is performed, the flow rate recovers slightly but still less than the initial permeability. The thermal response of an open microannulus with higher fluid pressure is different from that of a closed microannulus. At low backpressure, the microannulus gradually closes with more thermal cycles, as indicated by the flow response in Figure 5. On the other hand, at high backpressure, the microannulus was kept open during the thermal cycles (Figure 6).

The numerical model considers the middle of the cement sheath, ignoring the impact of water pressure at either end of the sample (plane strain). During curing, the pore pressure of cement drops to zero. This reduces the outer interface stress from the initial slurry pressure of 5 to 0.4 MPa (compressive). In reality, the fluid pressure above the cement sheath (5 MPa) will be enough to open the interface. This explains why immediately after curing, a microannulus is already present (likely on the outer interface). During the temperature cycles, the microannulus closes when the cement is heated, but the cooling part of the cycle opens a microannulus on the *inner interface*. This opening is only temporary and closes as the temperature between the casing and cement equilibrates. The closing of the microannulus during heating and the temporary opening of the microannulus during cooling could not be captured in the experiment. Instead, the steady-state flow was measured after completing the temperature cycles (LT 2 and 3) and not during. The hoop stress values during the temperature cycles in the present experiments can exceed 5 MPa (tensile). This value is higher than the tensile strength of cement. Therefore, the likelihood of sustained damage due to this may be present but did not yield significant continuous damage in the leak tests.

The stress path during the thermal cycles shows a wide range of shear and mean stress. At higher temperatures, the cement compresses, reaching a mean pressure of 7 MPa and shear stress of 9 MPa (Figure 10). This is mostly within the elastic regime. However, during the cooling period, the mean pressure and shear stress reaches −2.5 MPa and 5 MPa, respectively. This point just crosses the yield envelope on the dry side, which points to a dilatative (brittle) failure of cement and should increase permeability. However, as the permeability in the cement sheath did not show a marked increase, we conclude that the damage is likely not well-connected after 39 cycles in this experiment. Kuanhai et al. [11] conducted cyclical experiments on cement under both pressure and temperature conditions and observed permeability generation. However, the range of temperature and pressure in their work was much more extreme, between 30 and 150 °C and 0 to 70 MPa, respectively. The cement sheath length was also half the size of the present setup (0.76 m).

During pressure cycles, the inner casing pressure is raised to 30 MPa. The numerical model indicates hoop stress of 7 MPa (tensile). This value is firmly above the tensile strength of cement and should cause radial tensile cracks. The experimental results show that the cement permeability after pressure cycles remains low, and it is close to the permeability after the first set of thermal cycles (LT 2). In the leak test with 30 MPa casing pressure (LT 4), the crack is deliberately kept open, and yet the permeability stays low. This indicates that the tensile cracks are likely not continuous and have a low overall permeability. When the leak test is repeated with 0 MPa casing pressure (LT 5), the permeability slightly increases. Reducing the casing pressure allows for a bigger microannulus opening. In addition, the pressure cycles further close the gap that originated from thermal cycles at high backpressure (LT 3). Since the pressure cycles are conducted at low backpressure of 0.5 MPa, the increase in casing pressure may squeeze the water out of the microannulus.

In the present experiments, the cement's permeability from microannulus flow only increased for the case of thermal cycles at higher backpressure, though still not as high as the initial permeability. This indicates that the impact of backpressure may be significant. The effect of backpressure in the field case is akin to the annular casing pressure, assuming the path of the microannuli extends to the wellhead. This means that maintaining a low annular casing pressure can reduce the size of the microannulus and the resulting leakage. In the case of sustained casing pressure, as the pressure increases, the microannulus opens further, which could exacerbate the leakage issue.

5. Conclusions

In this paper, we conducted a large-scale experiment representing a 1.2-m-long cement sheath between two casings to simulate thermal heating–cooling cycles and pressure cycles. The results show a complex picture in terms of cement's permeability evolution. The cement sheath shows an initial permeability of 4 mD immediately after curing, which indicates a small microannulus. Temperature cycles are found to reduce the size of the microannulus and, in turn, the permeability. In addition, the impact of the thermal cycles seems to be dependent on the back pressure. Higher backpressure overall shows a higher permeability by widening the gap between the casing and cement. The gap also stayed open from high back pressure during the thermal cycle. This simulation of backpressure is similar to the sustained casing pressure at the wellhead. High casing annular pressure at the surface could open the microannulus further and lead to higher leakage rates.

The combination of numerical modeling and experiments adds further insight to the analysis. The numerical results show a high likelihood of tensile cracks during the pressure tests. However, cement's permeability from the experiment does not increase, which points to a lack of connectivity and permeability in tensile cracks. This is evidence suggesting that well leakage is driven by microannuli rather than tensile cracks. The temperature cycles also show a stress path that should cause dilatative failure during cooling. However, under the mild conditions of the experiments and after 39 thermal cycles, no significant permeability was generated. This indicates that under low-temperature geothermal conditions, cyclical cement damage may be minimal.

Author Contributions: Conceptualization, A.N.C. and A.M.; methodology, A.N.C. and A.M.; software, A.N.C. and A.M.; validation, A.N.C. and A.M.; formal analysis, A.N.C. and A.M.; investigation, A.N.C. and A.M.; data curation, A.N.C. and A.M.; writing—original draft preparation, A.N.C. and A.M.; writing—review and editing, A.N.C. and A.M.; visualization, A.N.C. and A.M.; supervision, A.M.; project administration, A.N.C. and A.M. All authors have read and agreed to the published version of the manuscript.

Funding: This research was funded by the Geothermica project "Sustainable Geothermal Well Cements for Challenging Thermo-Mechanical Conditions (TEST-CEM)", number 2003184001. The project was subsidized through the Cofund GEOTHERMICA by the DoE (the USA), RVO NL (The Netherlands), and the Research Council of Norway.

Data Availability Statement: The data presented in this study are available on request from the corresponding author.

Acknowledgments: Contributions from our partners Equinor (Norway), SINTEF (Norway), EBN (The Netherlands), Imerys (France), and CURISTEC (France) are greatly acknowledged. We would like to thank Vincent Soustelle, Tatiana Pyatina, and Marcel Naumann for the fruitful scientific and technical discussion.

Conflicts of Interest: The authors declare no conflicts of interest.

References

1. Geothermie; DAGO; Warmtenetwerk; EBN. Master Plan Geothermal Energy in The Netherlands: A Broad Foundation for Sustainable Heat Supply. 2018. Available online: https://geothermie.nl/images/bestanden/Masterplan_Aardwarmte_in_Nederland_ENG.pdf (accessed on 6 November 2023).
2. NLOG. Natural Resources and Geothermal Energy in The Netherlands: Annual Review 2022. 2023. Available online: https://www.nlog.nl/sites/default/files/2023-09/annual_review_2022_-_natural_resources_and_geothermal_energy_in_the_netherlands.pdf (accessed on 6 November 2023).
3. Therond, E.; Bois, A.-P.; Whaley, K.; Murillo, R. Large-Scale Testing and Modeling for Cement Zonal Isolation in Water-Injection Wells. *SPE Drill. Complet.* **2017**, *32*, 290–300. [CrossRef]
4. Goodwin, K.J.; Crook, R.J. Cement Sheath Stress Failure. *SPE Drill. Eng.* **1992**, *7*, 291–296. [CrossRef]
5. De Andrade, J.; Sangesland, S.; Skorpa, R.; Todorovic, J.; Vrålstad, T. Experimental Laboratory Setup for Visualization and Quantification of Cement-Sheath Integrity. *SPE Drill. Complet.* **2016**, *31*, 317–326. [CrossRef]
6. Vrålstad, T.; Skorpa, R.; Opedal, N.; De Andrade, J. Effect of Thermal Cycling on Cement Sheath Integrity: Realistic Experimental Tests and Simulation of Resulting Leakages. In Proceedings of the SPE Thermal Well Integrity and Design Symposium, Banff, AB, Canada, 23–25 November 2015; p. D011S001R002.
7. De Andrade, J.; Torsæter, M.; Todorovic, J.; Opedal, N.; Stroisz, A.; Vrålstad, T. Influence of Casing Centralization on Cement Sheath Integrity During Thermal Cycling. In Proceedings of the IADC/SPE Drilling Conference and Exhibition, Fort Worth, TX, USA, 4–6 March 2014; p. SPE-168012-MS.
8. Lin, Y.; Deng, K.; Yi, H.; Zeng, D.; Tang, L.; Wei, Q. Integrity tests of cement sheath for shale gas wells under strong alternating thermal loads. *Nat. Gas Ind. B* **2020**, *7*, 671–679. [CrossRef]
9. Kuanhai, D.; Yue, Y.; Yi, H.; Zhonghui, L.; Yuanhua, L. Experimental study on the integrity of casing-cement sheath in shale gas wells under pressure and temperature cycle loading. *J. Pet. Sci. Eng.* **2020**, *195*, 107548. [CrossRef]
10. Moghadam, A.; Corina, A.N. Modelling Stress Evolution in Cement Plugs During Hydration. In Proceedings of the 56th U.S. Rock Mechanics/Geomechanics Symposium, Santa Fe, NM, USA, 26–29 June 2022; p. ARMA-2022-0966.
11. Moghadam, A.; Loizzo, M. Cement Integrity Assessment Using a Hydration-Coupled Thermo-Mechanical Model. In Proceedings of the SPE Offshore Europe Conference & Exhibition, Aberdeen, UK, 5–8 September 2023; p. D021S008R004.
12. Reddy, B.R.; Xu, Y.; Ravi, K.; Gray, D.; Pattillo, P.D. Cement-Shrinkage Measurement in Oilwell Cementing—A Comparative Study of Laboratory Methods and Procedures. *SPE Drill. Complet.* **2009**, *24*, 104–114. [CrossRef]
13. Teodoriu, C.; Amani, M.; Yuan, Z.; Schubert, J.; Kosinowski, C. Investigation of the mechanical properties of Class G cement and their effect on well integrity. *Int. J. Eng.* **2013**, *3*, 2305–8269.
14. Soustelle, V.; Moghadam, A.; Corina, A.N. Modified Cam-Clay Model Parameters for Well-Cement. In Proceedings of the SPE EuropEC—Europe Energy Conference featured at the 84th EAGE Annual Conference & Exhibition, Vienna, Austria, 5–8 June 2023; p. D022S003R005.
15. Wolterbeek, T.K.T.; Hangx, S.J.T. The thermal properties of set Portland cements—A literature review in the context of CO_2 injection well integrity. *Int. J. Greenh. Gas Control* **2023**, *126*, 103909. [CrossRef]

Disclaimer/Publisher's Note: The statements, opinions and data contained in all publications are solely those of the individual author(s) and contributor(s) and not of MDPI and/or the editor(s). MDPI and/or the editor(s) disclaim responsibility for any injury to people or property resulting from any ideas, methods, instructions or products referred to in the content.

Article

Study on the Mechanism of Gas Intrusion and Its Transportation in a Wellbore under Shut-in Conditions

Haifeng Zhu [1], Ming Xiang [1], Zhiqiang Lin [1], Jicheng Yang [2], Xuerui Wang [3,*], Xueqi Liu [2] and Zhiyuan Wang [2]

1. CNOOC International Limited, Beijing 100028, China; zhuhf@cnoocinternational.com (H.Z.); xiangming@cnoocinternational.com (M.X.); linzhq@cnoocinternational.com (Z.L.)
2. School of Petroleum Engineering, China University of Petroleum (East China), Qingdao 266555, China; jicheng0827@gmail.com (J.Y.); lxq6snow7@163.com (X.L.); wangzy1209@126.com (Z.W.)
3. College of Computer Science and Technology, China University of Petroleum (East China), Qingdao 266555, China
* Correspondence: 20190038@upc.edu.cn; Tel.: +86-0532-86983137

Abstract: This paper presents a comprehensive study based on multiphase-seepage and wellbore multiphase-flow theories. It establishes a model for calculating the rate of gas intrusion that considers various factors, including formation pore permeability, bottomhole pressure difference, rheology of the drilling fluid, and surface tension. Experiments were conducted to investigate the mechanism of gas intrusion under shut-in conditions, and the experimental results were employed to validate the reliability of the proposed method for calculating the gas intrusion rate. Furthermore, this research explores the transportation rates of single bubbles and bubble clusters in drilling fluid under shut-in conditions. Additionally, empirical expressions were derived for the drag coefficient for single bubbles and bubble clusters in the wellbore. These expressions can be used to calculate gas transportation rates for various equivalent radii of single bubbles and bubble clusters. The initial bubble size of intrusive gas, the transportation speed of intrusive gas in the wellbore, the rate of gas intrusion, and variations in the wellbore pressure after gas intrusion were analyzed. Additionally, a method was developed to calculate the rising velocity of bubble clusters in water based on experimental results. The study reveals that the average bubble size in the bubble cluster is significantly smaller than the size of single bubbles generated from the orifice. When the viscosity of the drilling fluid is low, the transportation velocity of the bubble cluster exhibits a positive correlation with the average bubble diameter. When the average bubble diameter exceeds 1 mm, the bubble velocity no longer varies with changes in the bubble-cluster diameter. The research results provide theoretical support for wellbore pressure prediction and pressure control under shutdown conditions.

Keywords: offshore well drilling; gas intrusion rate; bubble initial size; bubble transport velocity

Citation: Zhu, H.; Xiang, M.; Lin, Z.; Yang, J.; Wang, X.; Liu, X.; Wang, Z. Study on the Mechanism of Gas Intrusion and Its Transportation in a Wellbore under Shut-in Conditions. *Energies* **2024**, *17*, 242. https://doi.org/10.3390/en17010242

Academic Editors: Edo Boek and Dameng Liu

Received: 25 October 2023
Revised: 20 November 2023
Accepted: 13 December 2023
Published: 3 January 2024

Copyright: © 2024 by the authors. Licensee MDPI, Basel, Switzerland. This article is an open access article distributed under the terms and conditions of the Creative Commons Attribution (CC BY) license (https://creativecommons.org/licenses/by/4.0/).

1. Introduction

As oil and gas exploration extends into deep and ultra-deep formations, the geological structures encountered during the drilling of oil and gas wellbores become notably complex. Studies show that various factors can lead to the intrusion of formation fluids into these wellbores. During offshore drilling, if gas invasion and overflow are not detected in time, blowouts can rapidly occur. Influx fluids tend to have a high pressure; a gas kick is much more detrimental than a liquid kick due to gas expansion and, as a result, has greater variation in pressure [1]. To ensure the safety of drilling operations and facilitate the rapid progress of drilling, the a precise model for gas intrusion rates and bubble transportation must be developed. Such a model can provide the foundation for predicting wellbore pressure dynamics.

The mechanisms responsible for gas intrusion into the wellbore can be categorized into three classes: differential-pressure gas intrusion, diffusion gas intrusion, and gas intrusion driven by gravity displacement. The present study focuses on the most common mechanism, which is differential-pressure gas intrusion. The majority of gas intrusion models based on differential-pressure gas intrusion were developed using percolation theory. Numerous investigations have considered various factors influencing pressure-differential gas intrusion. For instance, Rommetveit [2] conducted field experiments to investigate the impact of diverse factors on gas intrusion. Stefan and Samuel [3] focused on the pore pressure and permeability of the formation in their prediction model.

To enhance the accuracy of multiphase flow models and simulate gas intrusion accurately, numerous investigations have focused on gas-bubble migration. Gas migration is a crucial process in gas-liquid two-phase flows, and it is also a theoretical and fundamental problem in wellbores during well drilling. Studies show that the mechanisms governing gas migration are closely related to factors such as drilling conditions, properties of the drilling fluid, and gas intrusion into the wellbore. Davies and Taylor [4] proposed a predictive expression for gas slip velocity in various flow types such as Taylor bubble flow, segmented plug flow, and churn flow within a round vertical tube. Harmathy [5] performed experiments and derived an empirical expression for estimating bubble rising velocity in stationary liquids. Wallis [6] studied the flow of gas bubbles in stationary liquids in a round tube and modified the empirical expression proposed by Harmathy for single bubbles. Mendelson, Fan, and Tamiyama [7] introduced different models for the rising velocity of gases in non-Newtonian fluids under various experimental conditions. In recent years, extensive investigations have been carried out focusing on bubble rising behavior, employing advanced experimental equipment and numerical simulation techniques. For instance, Tai Wang et al. [8] studied the fusion of two bubbles along the same axis, finding that bubbles do not fuse when the surface tension is high. Conversely, it was found that bubbles are prone to fragmentation when the surface tension is very low. Liu Yipeng et al. [9] combined the theory of flow transformation with experiments and developed a predictive expression for the formation position of Taylor bubbles in cryogenic pipelines. Krzan et al. [10] performed comprehensive experiments and demonstrated that the addition of a small quantity of ionic surfactant can effectively reduce the rising rate of bubbles in organic solutions. Furthermore, Azzopardi et al. [11] experimentally investigated the rising of large bubbles in highly viscous liquids and established a model for periodic oscillation of the free liquid surface induced by gas lift. Keshavarzi et al. [12] experimentally analyzed the interface deformation during the rise of individual bubbles and validated the theoretical models of VOF and CLSVOF. Cano-Lozano et al. [13] correlated the Rastello linear-rise model for bubbles with the Clift curve-rise model for bubbles, yielding a correlation equation applicable to the calculation of bubble rise velocities for a wide range of bubble diameters and liquid properties. Xiao Kang Yan [14] conducted numerous experiments using a high-speed video system and proposed an equation for predicting the drag coefficient. Khodayar [15] investigated the dynamics of rising bubbles on vertical walls under different wettability conditions, discovering that bubbles with contact angles less than $90°$ move faster than bubbles with contact angles greater than $90°$. Du Jingyu [16] found that bubble lift-off diameter is related to wall superheat, latent heat, liquid velocity, fluid properties, bulk-liquid subcooling, etc. Francesca [17] examined the effects of density and viscosity on the behavior of bubble populations in turbulent channels, highlighting the influence of viscosity on fracture and agglomeration. He Hongbin [18] used numerical simulations to investigate the effects of initial bubble diameter, horizontal spacing and arrangement on the kinematic state of three parallel bubbles.

Recently, gas intrusion in conjunction with engineering challenges has become a hot topic in research. However, the majority of studies have focused on gas intrusion during normal drilling, and only a few comprehensive investigations are related to wells. This gap is especially pronounced in studies on deep-water oil and gas fields located long distances (often more than 300 km) offshore. During these offshore operations, which may also involve lengthy periods of equipment maintenance due to severe weather conditions, gas can intrude into the wellbore and subsequently accumulate at the wellhead. Over time, this accumulation poses a significant risk to open-well operations [19]. In the case of pressure-differential gas intrusion under well shut-in conditions, the behavior of gas-liquid two-phase flow within the wellbore differs from that under other conditions in which gas intrusion occurs, possibly because the drilling fluid cannot flow back out from the wellhead. Studies show that the initial characteristics of the invading gas bubbles and their rising velocity within the wellbore are influenced by a variety of factors, including surface tension, rheological properties, density of the drilling fluid under well shut-in conditions, reservoir properties, and the pressure difference between the reservoir and the wellbore bottom. It is worth noting that the rise speed of intrusion gas bubbles within the wellbore is an important parameter for calculating the wellbore pressure and the total gas intrusion volume. A review of the literature indicates that although numerous investigations have been conducted in the field of wellbore drilling and gas intrusion, further research is required in how characteristics such as the influence of surface tension, rheological characteristics, density of drilling fluids, reservoir properties, and pressure difference between the reservoir and the wellbore bottom influence the size of intrusion gas bubbles and flow patterns of individual and cluster bubbles within the wellbore. With the aim of addressing this shortcoming, the present study investigates the gas intrusion mechanism when gas enters the wellbore due to pressure differences during well shutdown. This investigation considers various formation properties, characteristics of the drilling fluid, and pressure-difference conditions between the formation and the wellbore. The study focuses on the initial size of bubbles, the transportation speed of the intruding gas within the wellbore, and changes in gas intrusion rates. The findings of this article are expected to provide theoretical support for the prediction and control of wellbore pressure under shutdown conditions.

This research holds significant practical value in the field of oil and gas exploration. A thorough understanding and mastery of the mechanism of bubble formation can contribute to accurately predicting and effectively controlling gas intrusion within wellbores. This knowledge, in turn, enhances the efficiency of oil and gas exploration and production while concurrently reducing operational risks. The study of the initial size of bubbles further aids in assessing reservoir hydrocarbons. Monitoring and analyzing the initial size of bubbles yields more precise information about the reservoir, offering a scientific and reliable basis for oil and gas resource evaluation and decision-making in exploration. This expansion of knowledge not only contributes to optimizing production strategies in the course of oil and gas exploration, but also provides robust technical support for the rational development of oil and gas resources, thereby promoting the sustainable development of the oil and gas exploration sector.

2. Visualization of Gas Intrusion

2.1. Experimental System

A schematic of the experimental setup is shown in Figure 1. The experimental setup for simulating gas intrusion under well-shutdown conditions consists of a stratigraphic system, an experimental wellbore system, and a data-acquisition system.

The stratigraphic system consists of a gas source, a pressure-reducing valve, a pressure gauge, a microporous aerator (92 mm diameter, 80 mesh screen) and several valves. During the experiment, gas passes through the microporous aerator and is transported upward in the wellbore in the form of bubble clusters.

The main structure of the experimental wellbore system is constructed from Plexiglas and has an inner diameter of 150 mm, a wall thickness of 10 mm, a height of 1000 mm, and a pressure resistance of 5 MPa. The bottom center of the wellbore is connected to the stratigraphic system, and the side of the wellbore is equipped with a drain hole for discharging the experimental solution. Meanwhile, a scale is provided within the wellbore and on the walls to calibrate the size of air bubbles and the void ratio of the wellbore. The upper part of the wellbore is equipped with a manometer, a liquid-injection hole and an air-vent hole.

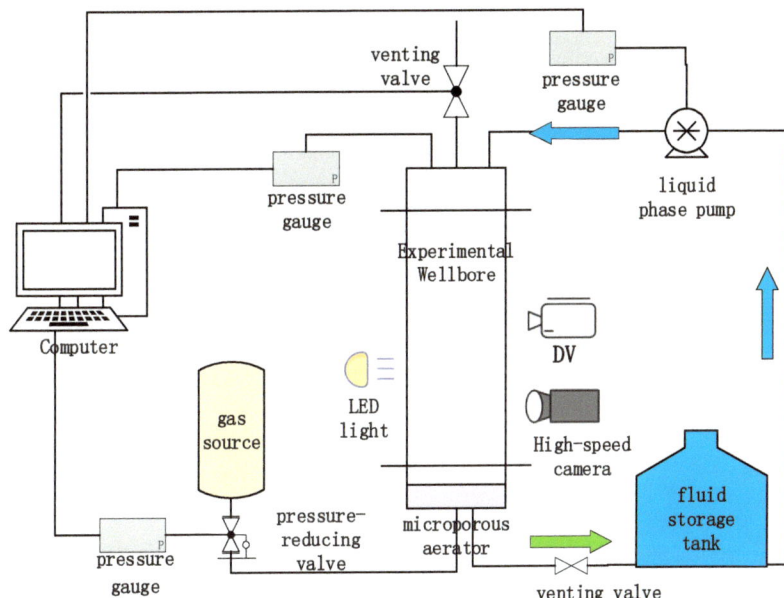

Figure 1. Schematic of gas intrusion test system.

The data-acquisition system includes a pressure-acquisition module and an image-acquisition module. The former module consists of a pressure gauge at the gas source and a manometer at the top of the wellbore, which measure the pressure and transmit the collected data to the computer (FPS01W-XD, Changzhou, China). The latter module consists of a high-speed camera (OLYMPUS I-Speed 3, Tokyo, Japan), a DV, and a computer. The high-speed camera has a maximum frame rate of 2000 fps and a pixel accuracy of 13 μm.

2.2. Experimental Materials

Xanthan gum (XC) is a commonly used viscosity enhancer for drilling fluids. In the present study, deionized water and aqueous solutions with various concentrations of xanthan gum were utilized to simulate drilling fluids in the experiments. The experimental process considered the influence of liquid-phase rheology on the gas intrusion speed and the size of the intruding bubbles, while the effects of gas dissolution on the intrusion were ignored. Nitrogen was used as the experimental gas phase. The rheological test results of the experimental solutions with various concentrations of xanthan gum are shown in Figure 2.

The solution rheology was fitted using a power law function with the specific parameters shown in Table 1.

Table 1. Properties of experimental materials.

Type	Consistency Factor K ($Pa \cdot s^{-n}$)	Fluidity Index n	Density ($kg \cdot m^{-3}$)	Surface Tension ($N \cdot m^{-1}$)
Water	0.01	1	1000	72.35
0.05% XC	0.07224	0.51458	1000	73.67
0.1% XC	0.12377	0.48544	1000	74.24
0.2% XC	0.34337	0.43297	1000	75.05
0.4% XC	1.28381	0.34793	1000	76.75

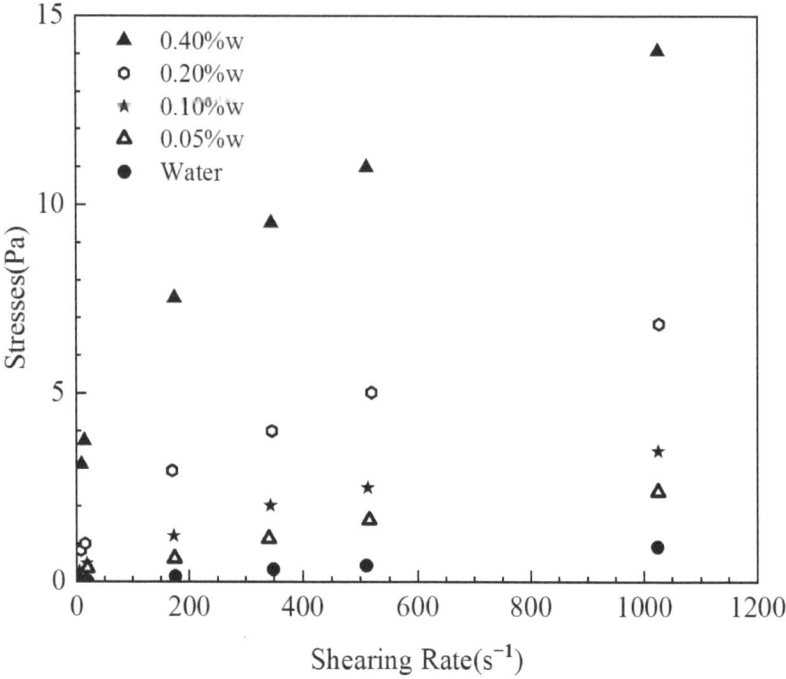

Figure 2. The rheology properties of experiment solutions.

2.3. Experimental Procedure

The experiments were divided into the simulation of gas intrusion into the wellbore under shut-in conditions and the simulation of intruded-gas transportation velocity in the wellbore under shut-in conditions. Prior to the experiment, instruments were calibrated, the connection joints between instruments and pipelines were sealed, and the gas tightness test was conducted across the experimental loop. To minimize experimental errors, each test was repeated five times and the average values were recorded. The main experimental procedures are as follows:

(1) Simulation of gas intrusion into the wellbore under shut-in conditions

This part of the test consists of six parts as follows (The schematic diagram of the experimental process is shown in Figure 3):

① Connect the experimental device and check its gas tightness;

② Inject the experimental solution into the liquid surface at a height of 70 cm, with the upper part of the air at atmospheric pressure;

③ Close the top exhaust valve of the wellbore, adjust the decompression valve in the system to the set pressure, open the ball valve to inject the formation gas into the wellbore, record variations in the gas pressure within the wellbore using a manometer and record the gas intrusion into the wellbore using a high-speed camera. Additionally, record the size and rising speed of bubbles, as well as the gas level in the gas intrusion using DV;

④ When the pressure in the upper part of the wellbore has stabilized, close the ball valves between the injection system and the wellbore and open the venting valve at the top of the wellbore to exhaust the accumulated gas;

⑤ Repeat steps ③ and ④ 5 times and record the data;

⑥ Adjust the pressure of the formation-simulation system, and repeat steps ③–⑤ to study gas intrusion under different pressure conditions. After the experiment is complete, open the drain valve at the bottom of the wellbore, drain the experimental solution, and inject deionized water to inflate and clean the wellbore 3–5 times. Replace the experimental solution and repeat steps ③–⑥ to simulate gas intrusion under various pressure differentials.

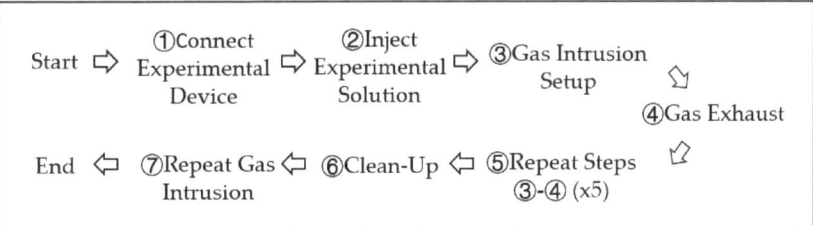

Figure 3. The experimental flow chart.

(2) Obtaining intrusive-gas transportation velocity within the wellbore under shut-in conditions

Bubble-cluster transport velocity experiments were conducted using various solutions under shut-in conditions, following the same experimental procedure described for experiment (1). Subsequently, the microporous aerator was replaced with a 0.5 mm-diameter nozzle, and experiments were conducted to measure the transport velocity of individual bubbles in solutions with various properties. The measurement of bubble speed involves the following steps:

① Time measurement: Set the shooting rate of the high-speed camera to 1000 fps;

② Coordinate conversion: There is a scale within the wellbore with marks at 1 mm intervals along the horizontal direction. In the vertical direction, the heights of the scale lines vary as follows: the ordinary scale line is 3 mm high, the secondary scale line, which occurs every 5 mm, is 4 mm high, and the main scale line, which occurs every 10 mm, is 5 mm high. Use this scale to convert pixel coordinates to mm-scale coordinates, thereby eliminating the "convex lens" effect on the bubble geometry within the circular wellbore;

③ Measurement of geometric features of bubbles: Utilize the Edit-Draw function in Image-Pro Plus (V8.0) software to trace the outline of the selected bubble. Then, use the "count size" function to determine parameters at the center of the bubble, including horizontal and vertical coordinates (X, Y), maximum radius (R_{max}), and minimum radius (R_{min}). Additionally, use the Measure function to directly measure the geometric features of the bubble;

④ Measurement of bubble rising speed: While observing the movement of bubbles in the camera, select a bubble in the ith frame, measure its geometric features $(X_1, Y_1, R_{1max},$ and $R_{1min})$, record its geometric features in the $i + n$th frame $(X_2, Y_2, R_{2max},$ and $R_{2min})$, and calculate the average rising speed of bubbles over the test period using the following expression:

$$u_z = \frac{Y_2 - Y_1}{\Delta t} = \frac{Y_2 - Y_1}{0.001n} \tag{1}$$

3. Results and Analysis

3.1. Simulation of Gas Intrusion into the Wellbore under Shutdown Conditions

3.1.1. Variations in the Size of the Intrusion Bubbles

The pore diameter of the microporous aerator is 0.178 mm, corresponding to the dimensionless bubble size $(2R_b/D_o)$. The experimental results are shown in Figure 4. When the viscosity of the drilling fluid is low, the initial size of intruding bubbles is affected mainly by the surface tension at the orifice when the bubbles are dislodged. Moreover, Figure 4 indicates that there is a positive correlation between the size of intruding bubbles and the viscosity of the liquid phase. The average sizes of bubbles in the 0.05% XC solution and in the aqueous solution are nearly identical. However, the size of bubbles in the 0.05% XC solution is significantly smaller than those observed in the 0.1% and 0.2%XC solutions.

In summary, the bubble size in the 0.05% XC solution is significantly smaller than those in the 0.1% and 0.2% XC solutions.

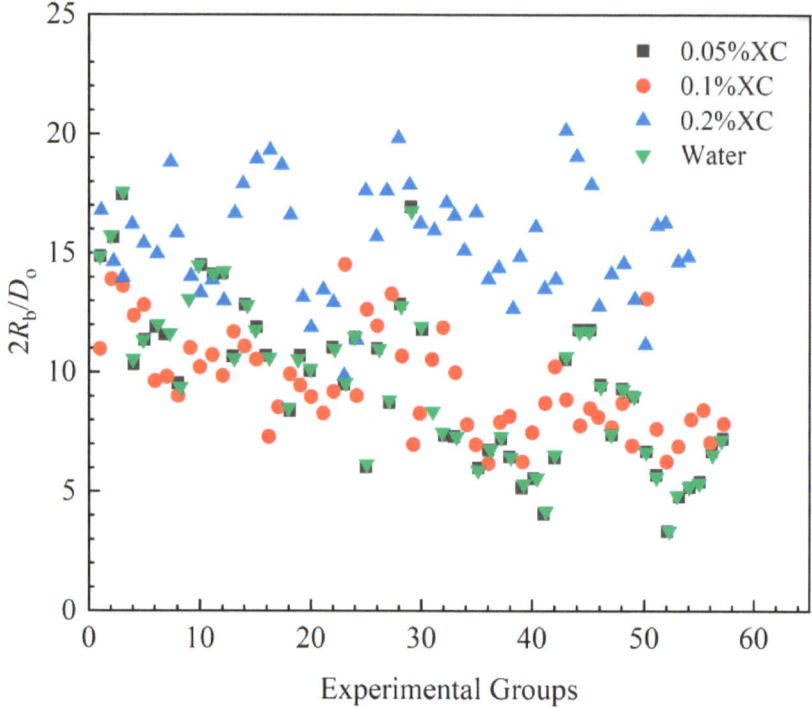

Figure 4. Distribution of dimensionless bubble diameter.

Figure 5 indicates that the average dimensionless diameter of gas bubbles is affected by the pressure difference. More specifically, the greater the pressure difference between the formation and the wellbore, the larger the average size of intruding gas bubbles. This phenomenon may be attributable to the substantial pressure difference between the formation and the wellbore caused by the gas intrusion into the wellbore. Consequent to this pressure difference, the gas seepage rate increases, intruding gas bubbles expand, and the average size of bubbles increases when they detach from the borehole.

3.1.2. Modeling Gas Intrusion Rate under Shut-in Well Conditions

When a well is shut down after an overflow and the formation pressure exceeds the bottomhole pressure, the formation fluid intrudes into the wellbore due to the pressure difference. The flow behavior can be described using Darcy's law. As the gas intrusion into the wellbore is an unsteady planar radial seepage process, the unsteady gas intrusion binomial flow equation is employed to describe the seepage process [20]:

$$p_e^2 - p_w^2 = \left(\frac{\mu_g Z}{\pi k h} \frac{p_a T}{Z_a T_a} \ln \frac{r_e}{r_w}\right) Q_g + \frac{\alpha \rho_a}{2\pi^2 h^2} \frac{p_a Z T}{Z_a T_a} \left(\frac{1}{r_w} - \frac{1}{r_e}\right) Q_g^2 \qquad (2)$$

where p_e is formation pressure, Pa; p_w is the bottomhole pressure, Pa; μ_g is the average gas viscosity, Pa·s; Z is the gas compression factor at the temperature and pressure of the formation; k is the formation permeability, m^2; h is the thickness of the open gas layer, m; p_a is the pressure of the standard atmosphere, Pa; T is the formation temperature, K; Z_a is the gas compression factor under the standard conditions; T_a is the temperature under the standard conditions, K; r_e is the effective gas intrusion radius, m; r_w is the bottomhole radius, m; Q_g is the gas intrusion flow rate under standard conditions, m³/s; α is the coefficient of inertial drag caused by turbulence; and ρ_a is the gas density under standard conditions, g/m³.

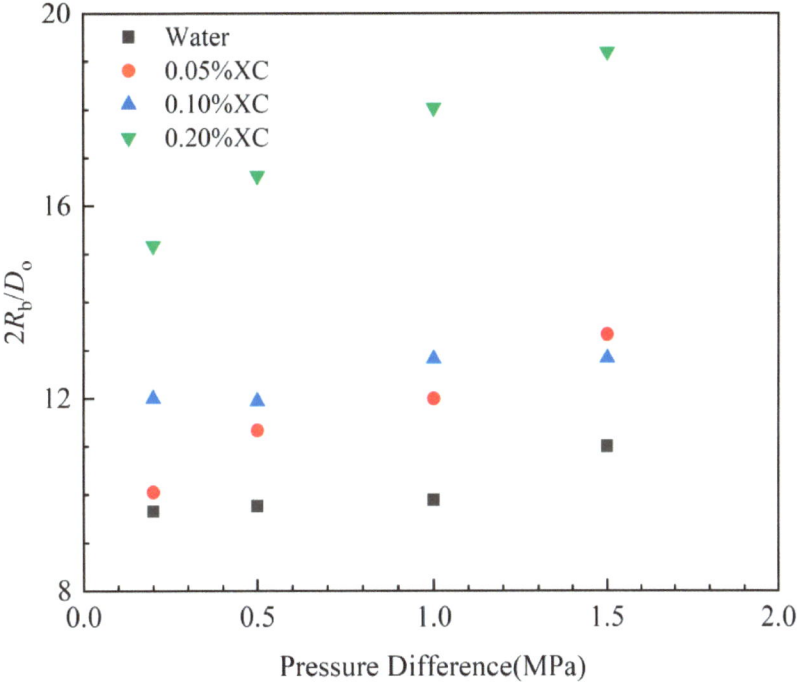

Figure 5. Plot of dimensionless bubble diameter against pressure difference.

With continuous gas intrusion into the wellbore, the wellbore pressure gradually increases and the pressure difference between the formation and the wellbore gradually decreases. As a result, the gas intrusion rate gradually decreases. The total volume of the intruded gas under standard conditions can be obtained using the following expression:

$$\Delta V_g = \int_0^{t_1} Q_g(t) dt \tag{3}$$

Based on Equation (2), the gas intrusion rate is related to the bottomhole pressure [20]:

$$A = \left(\frac{\mu_g Z}{\pi K h} \frac{p_a T}{Z_a T_a} \ln \frac{r_e}{r_w} \right) \tag{4}$$

$$B = \frac{\alpha \rho_a}{2\pi^2 h^2} \frac{p_a Z T}{Z_a T_a} \left(\frac{1}{r_w} - \frac{1}{r_e} \right) \tag{5}$$

$$Q_g = \frac{\sqrt{A^2 + 4B(p_e^2 - p_w^2(t))} - A}{2B} \tag{6}$$

The total gas intrusion volume under standard conditions can be obtained through the following expression:

$$\Delta V_{g,b} = \int_0^{t_1} \frac{\sqrt{A^2 + 4B(p_e^2 - p_w^2(t))} - A}{2B} dt \tag{7}$$

where t_1 is the total intrusion time from the onset of gas intrusion to its cessation, s.

Equation (7) indicates that the total gas intrusion is related to the wellbore pressure, which is affected by numerous factors such as gas rising speed. Therefore, calculating the total gas intrusion requires the establishment of a model for calculating the gas transport velocity. Ignoring the effects of gas expansion on the wellbore pressure, the wellbore pressure is related only to the amount of gas intrusion into the wellbore. In the experiments, the volume of the upper part of the wellbore was denoted as V_0, and the temperature was maintained at 290 K. Ignoring the temperature change caused by variations in the pressure, instantaneous wellbore pressure can be calculated using the following expression:

$$p_w(t) = \frac{P_a Z T_s}{V_0 Z_a T_a}\left(\frac{Z_a T_a}{Z_s T_s}V_0 + \Delta V_{g,b}\right) \tag{8}$$

where T_s is the experimental temperature, K; Z_s is the gas compression factor at 0.1 MPa and experimental temperature.

$$p_w(t) = \frac{P_a Z T_s}{Z_a T_a}\left(\frac{Z_a T_a}{Z_s T_s}V_0 + \int_0^{t_1}\frac{\sqrt{A^2 + 4B(p_e^2 - p_w^2(t))} - A}{2B}dt\right) \tag{9}$$

Equations (3)–(8) can be rewritten in the form below:

$$p_w(t) = \frac{Z}{Z_s}P_a + \frac{P_a T_s}{V_0 Z_a T_a}\sum_{i=1}^{N}\frac{Z\sqrt{A^2 + 4B\left(p_e^2 - p_w^2\left(\frac{i}{N}t_1\right)\right)} - A}{2B}\frac{i}{N}t_1 \tag{10}$$

Figure 6 illustrates the distribution of wellbore pressure over time, calculated using Equation (10). The highest wellbore pressures were obtained for water, 0.05% XC solution, 0.1% XC solution and 0.2% XC solution.

Figure 6 indicates that the model predictions align with the experimental results. The results show that as the negative pressure difference between the formation and the wellbore increases, the time from the initiation of gas intrusion to its cessation increases. Meanwhile, as the solution viscosity increases, the time required to reach equilibrium between wellbore pressure and formation pressure after gas intrusion decreases. The changes in the initial size of gas bubbles arising from invading gases in the wellbore are due to variations in solution viscosity: the greater the solution viscosity, the larger the initial size of gas bubbles from invading gases in the wellbore. As a result, less time is needed for invading gases to reach the upper section of the wellbore, thereby rapidly increasing the wellbore pressure.

3.2. Simulation of Intrusive-Gas Transportation Velocity in Wellbore under Shut-in Conditions
3.2.1. Single-Bubble Transport Experiment

This section describes an investigation of the transportation velocity of individual bubbles in experimental solutions. Variations in the transportation velocity of individual bubbles with the geometry of bubbles in different solutions are shown in Figure 7. When the viscosity of the solution is low, bubbles injected through the orifice are prone to rupture during their ascent, resulting in a smaller average equivalent diameter of bubbles in low-viscosity solutions. In 0.05% XC solution, the bubble speed changes slightly compared to that in water, and the change in bubble speed with increasing bubble equivalent radius is small. However, as solution viscosity increases, the range of bubble equivalent diameters expands. Concurrently, the size variations in the bubble speed with respect to changes in bubble equivalent radius change also increases.

When the equivalent radius of a bubble is small, it experiences resistance from the viscous solution, causing slower ascent. Consequently, the higher the solution viscosity, the slower the bubble rise. On the other hand, as the bubble equivalent radius increases, the bubble rise rate also increases. In this case, bubble transport speed is affected by both viscous forces and buoyancy. Furthermore, when the solution viscosity increases, the equivalent radius of individual bubbles generated at the orifice increases. Consequently, the rise rate of bubbles increases and the buoyant force that pushes bubbles upward becomes more dominant than the viscous force. As a result, the transport speed of bubbles may exceed the transport speed of gas in low-viscosity fluids. The buoyancy force pushes the bubble upward more than the viscous force resists bubble rise, resulting in a bubble upward velocity that may exceed the gas transport velocity in lower-viscosity fluids. In order to elucidate single-bubble transport velocity, the experimental results were analyzed as shown in Figure 8.

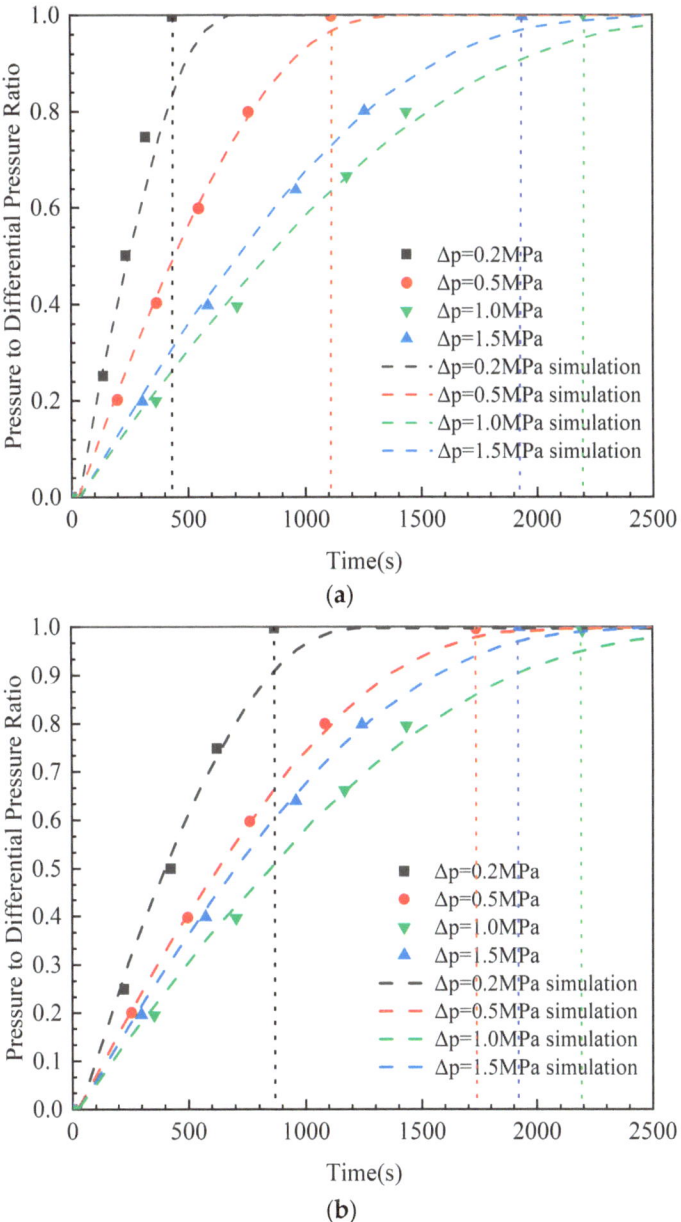

Figure 6. *Cont.*

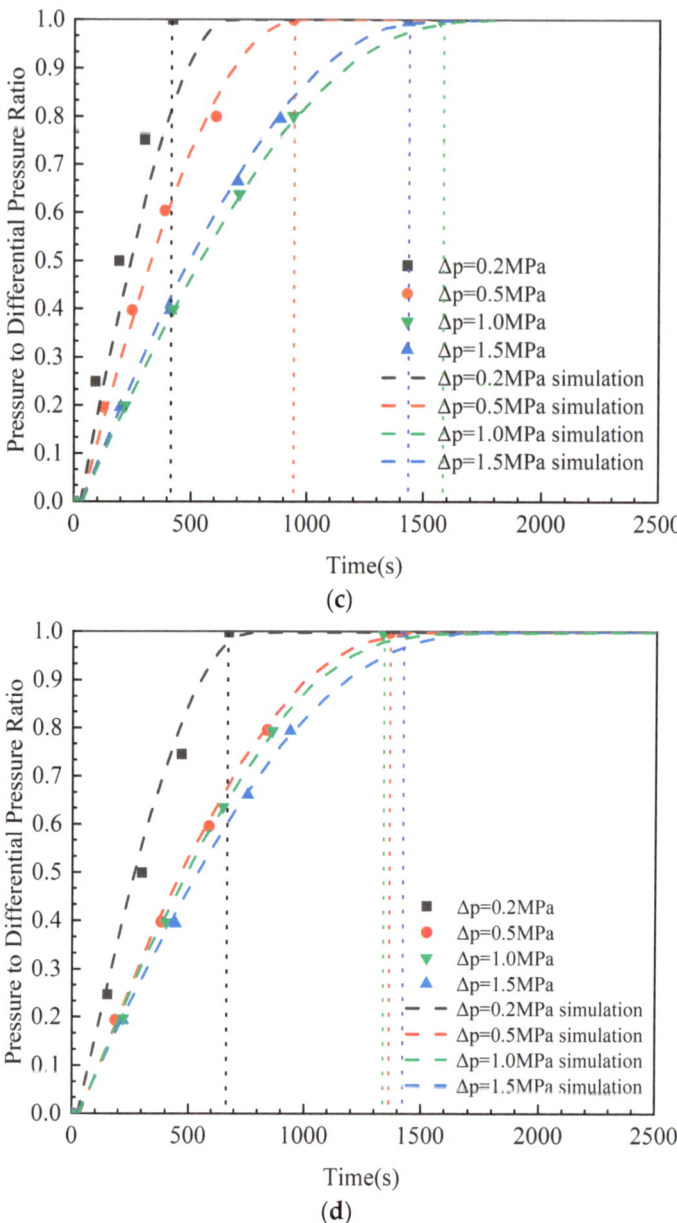

Figure 6. Variations in the wellbore pressure during gas intrusion for various concentrations of XC solution. (**a**) Variations of the wellbore pressure over time for water; (**b**) Variations in the wellbore pressure over time for a 0.05% XC solution; (**c**) Variations in the wellbore pressure over time for a 0.1% XC solution; (**d**) Variations in the wellbore pressure over time for a 0.2% XC solution.

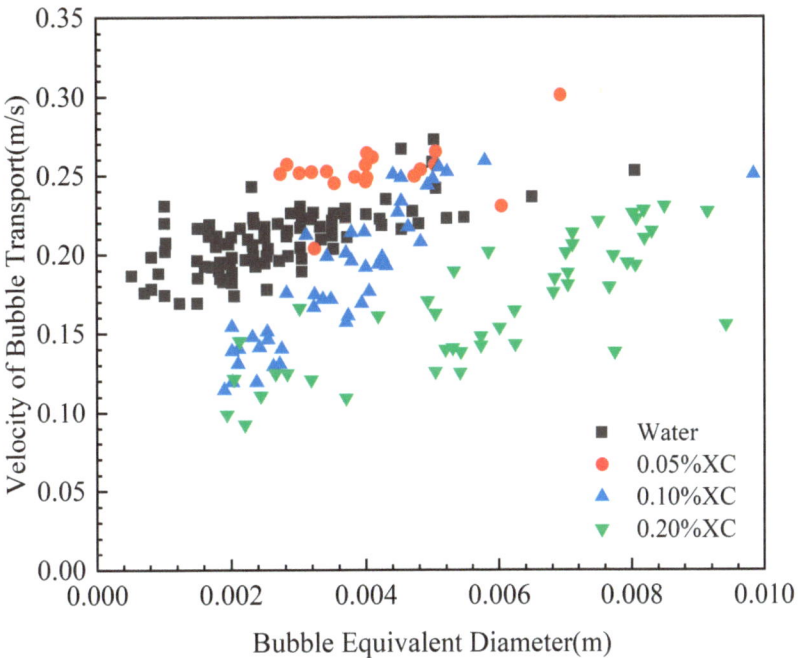

Figure 7. Velocity of a single bubble versus equivalent bubble radius.

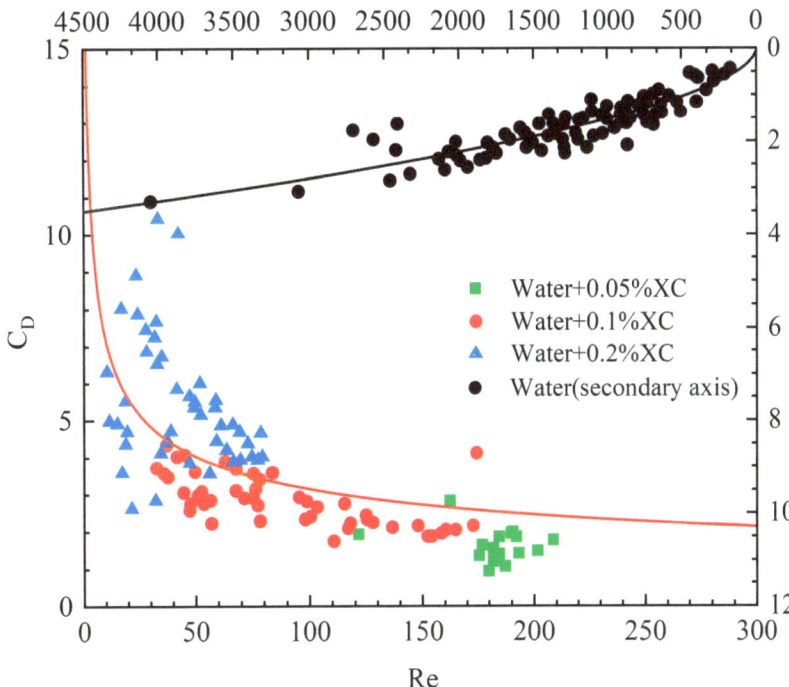

Figure 8. The drag coefficient for various Reynolds numbers.

As the Reynolds number for bubble transport in water is much larger than that in the XC solution [17], the experimental data in water are represented using the sub-coordinate axes (the left numerical axis is the vertical axis, and the upper numerical axis is the horizontal axis). The red solid line in the figure is the fitted model of the drag force coefficient method with the Reynolds number in the xanthan-gum solution, while the black solid line is the fitted model of the drag force coefficient method with the Reynolds number in the water, in which the intrinsic model for the XC solution follows a power law model, and its parameters are provided in Table 1, with the Reynolds number defined as

$$\text{Re} = \frac{2\rho u_z R_b}{K\dot{\gamma}^{n-1}} = \frac{\rho u_z R_b}{K\left(\frac{u_z}{R_b}\right)^{n-1}} = \frac{\rho u_z R_b^{2-n}}{K u_z^{n-1}} \qquad (11)$$

The drag force coefficient C_D is defined as [21]

$$C_D = \frac{8g(\rho_l - \rho_g)R_b}{3\rho_l u_z^2} \qquad (12)$$

An empirical relationship between the drag coefficient and Reynolds number for bubbles in water and XC solution is derived by fitting the experimental data:

$$C_{D,XC} = \frac{22.136 \pm 3.88}{\text{Re}^{0.432 \pm 0.046}} \qquad (13)$$

The empirical relationship for the drag coefficient in water is as follows:

$$C_{D,w} = (0.0292 \pm 0.004)\text{Re}^{0.568 \pm 0.029} \qquad (14)$$

3.2.2. Experiments on the Transport of Bubble Populations

When gas intrudes into the wellbore due to the pressure difference between the formation and the bottom of the well, the gas typically passes through the porous medium in the form of a bubble cluster. It is worth noting that the transport speed of a bubble cluster differs from that of an individual bubble. This section describes the experimentally investigation of the transport speed of bubble clusters in the wellbore under various gas intrusion conditions.

Compared with single bubbles injected through an orifice, intruding gas bubbles in porous media exhibit a smaller average size. The maximum equivalent bubble radius observed in the experiments for a bubble cluster was 1.81 mm, as shown in Table 2 and Figure 9. However, it should be noted that the gas rise velocity of a bubble cluster is typically greater than that of a single bubble of the same size.

In contrast to the positive correlation between the rise velocity of individual bubbles and their equivalent radius, the velocity of gas transport within a bubble cluster exhibits a positive correlation with the equivalent radius of bubbles when the viscosity of the liquid phase is low. However, when the viscosity is high, no correlation is detected between the gas transport velocity and the equivalent radius of bubbles. In order to investigate the transportation velocity of a bubble cluster in the wellbore, a dimensionless analysis of the bubble-cluster transportation velocity was carried out.

The following relationship between the bubble Reynolds number and the drag coefficient during bubble rise was established through fitting based on experimental results shown in Figure 10.

$$C_{D,XC} = (27.12 \pm 3.7)\text{Re}^{-0.955 \pm 0.045} \qquad (15)$$

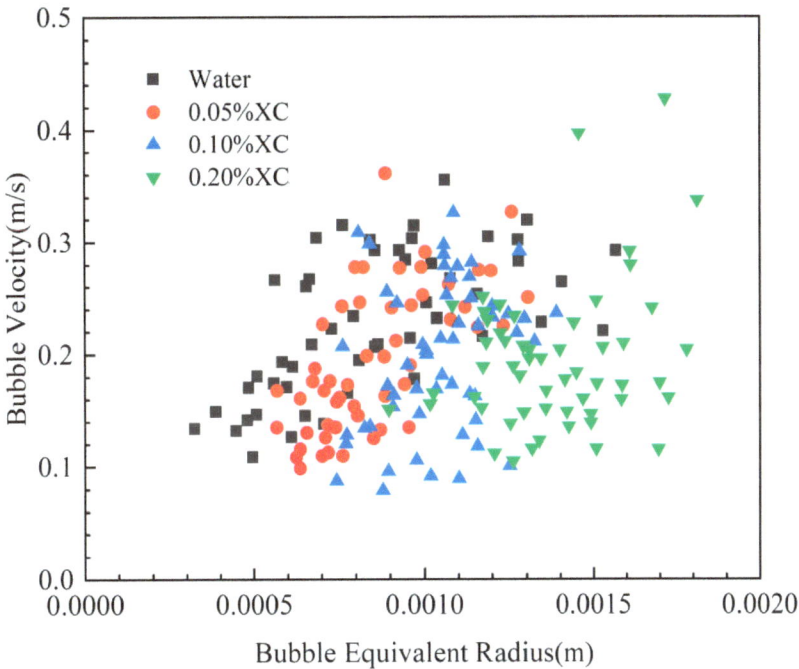

Figure 9. Velocity distribution of bubbles plotted against equivalent bubble radius.

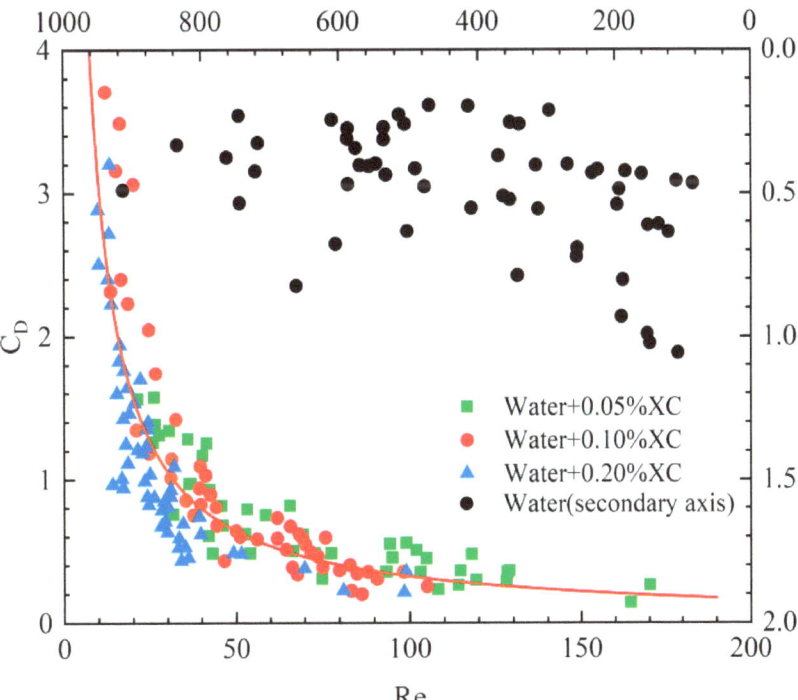

Figure 10. Plot of the drag coefficients for various Re numbers.

Table 2. Range of bubble velocity and equivalent radius.

Type	Range of Bubble Equivalent Radius (mm)		Bubble Velocity Range (mm/s)	
	Single Bubble	Bubble Group	Single Bubble	Bubble Group
Water	0.5–8	0.32–1.6	170–267	110–330
0.05% XC	2.7–6.9	0.56–1.29	213–303	100–360
0.1% XC	2–9.8	0.74–1.43	116–259	77.9–321
0.2% XC	1.9–9.4	0.89–1.81	95–231	106–428

Figure 11 reveals that the bubble-cluster transport velocity in water is related to its equivalent radius. The black dashed line represents the predicted migration speed curve when the equivalent radius is less than 1 mm, while the red dashed line represents the predicted migration speed curve when the equivalent radius is greater than 1 mm. The bubble-cluster transport velocity in water can be calculated using the following expressions:

$$u_z = 2.5\sqrt{gR_b} \quad R_b \leq 1 \text{ mm} \tag{16}$$

$$u_z = 0.26 \text{ m/s} \quad R_b \geq 1 \text{ mm} \tag{17}$$

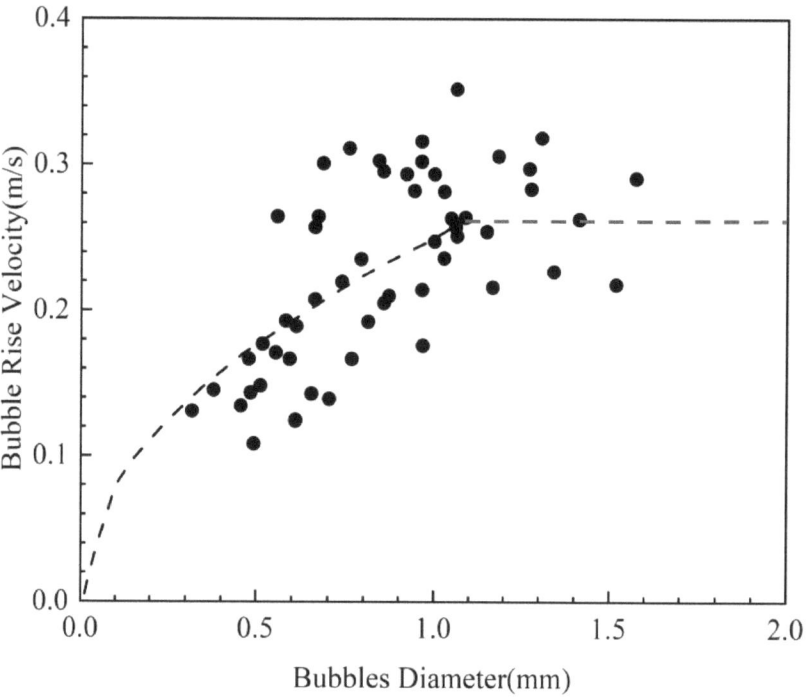

Figure 11. The velocity of a bubble cluster in water.

4. Conclusions

The present study incorporates the theories of multiphase seepage and multiphase flow in wellbores to investigate gas intrusion into a wellbore induced by differential pressure after well shutdown. In this context, a model was established to calculate the rate of gas intrusion, considering various parameters including the porosity of the formation, permeability of the formation, pressure difference between the formation and the wellbore, rheology of the drilling fluid, and surface tension. The experimental results were utilized to validate the prediction model for the initial size of gas bubbles and the calculation method for gas intrusion rates.

(1) Ignoring the change in wellbore pressure caused by the upward movement of gas, the lower the viscosity of the drilling fluid, the longer the time needed for the wellbore pressure to reach equilibrium with the formation pressure.

(2) The larger the pressure difference between the formation and the wellbore during gas intrusion, the longer the time needed for the wellbore pressure to reach equilibrium with the formation pressure.

(3) The initial amount of gas intruding into the wellbore and the rate of gas intrusion are crucial factors for calculating the transportation speed of the intruding gas within the wellbore and the changes in the wellbore pressure due to gas intrusion. Based on the experimental results, empirical expressions were established to predict the drag coefficient of single bubbles and bubble clusters flowing in the wellbore. These expressions can be used to calculate the gas transport velocity for different equivalent radii of single bubbles and the average equivalent radius of bubble clusters. Additionally, the calculation method for the rise velocity of bubble clusters in water was derived based on experimental results.

(4) In 0.05% XC solution, the change in bubble velocity compared to that of bubbles in water is negligible. Moreover, the change in bubble velocity with respect to bubble equivalent radius is also small.

(5) As the solution viscosity increases with the change from 0.1% to 0.2% XC solution, the range of bubble equivalent diameters expands. Simultaneously, the magnitude of change in bubble velocity with bubble equivalent radius also increases. The equivalent radius of bubbles within a bubble cluster is significantly smaller than the equivalent radius of single bubbles generated by the orifice.

(6) When the viscosity of the drilling fluid is low, the transportation speed of bubble clusters exhibits a positive correlation with the average bubble diameter. However, when the average bubble diameter exceeds 1 mm, the bubble-cluster transportation speed no longer changes with variations in bubble diameter. On the other hand, when the viscosity of the drilling fluid is high, there is no apparent relationship between the transportation velocity of bubble clusters and bubble equivalent radius. This observation is of great significance when considering the influence of drilling fluid on the bubble transport velocity in multiphase flow modeling.

Author Contributions: Writing—Original Draft, Methodology, H.Z.; Writing—Original Draft, Visualization, M.X.; Data Curation, Validation, Z.L.; Writing—Original Draft, Experiments, J.Y.; Writing—Review & Editing, Funding acquisition, X.W.; Data Curation, Validation, X.L.; Writing—Review & Editing, Funding acquisition, Z.W. All authors have read and agreed to the published version of the manuscript.

Funding: The authors acknowledge the support from the National Key R&D Program of China (2022YFC2806502), National Natural Science Foundation (52004315, U21B2069), Natural Science Foundation of Shandong Province (ZR2020QE113).

Data Availability Statement: Data are contained within the article.

Conflicts of Interest: Xuerui Wang was employed by CNOOC China Limited, Hainan Branch. The remaining authors declare that the research was conducted in the absence of any commercial or financial relationships that could be construed as a potential conflict of interest.

References

1. Yin, H.; Si, M.; Li, Q.; Zhang, J.; Dai, L. Kick Risk Forecasting and Evaluating during Drilling Based on Autoregressive Integrated Moving Average Model. *Energies* **2019**, *12*, 3540. [CrossRef]
2. Rommetveit, R.; Vefring, E.H. Comparison of results from an advanced gas kick simulator with surface and downhole data from scale gas kick experiments in an inclined well. In Proceedings of the SPE Annual Technical Conference and Exhibition, SPE 22558, Dallas, TX, USA, 6–9 October 1991.
3. Stefan, M.; Samuel, G.; Azar, J. Modeling of pressure buildup on a kicking well and its practical application. In Proceedings of the Permian Basin Oil and Gas Recovery Conference, Midland, TX, USA, 27–29 March 1996.

4. Davies, R.M.; Taylor, G. The mechanics of large bubbles rising through extended liquids and through liquids in tubes. *Dyn. Curved Front.* **1988**, *200*, 377–392.
5. Harmathy, T.Z. Velocity of large drops and bubbles in media of infinite or restricted extent. *Aiche J.* **1960**, *6*, 281–288. [CrossRef]
6. Wallis, G.B. *One Dimensional Two-Phase Flows*; McGraw-Hill: New York, NY, USA, 1969; pp. 97–112.
7. Fan, W.Y. *Study of Bubble Behaviors and Flow Field around Moving Bubble in Non-Newtonian Fluid*; Tianjin University: Tianjin, China, 2008.
8. Wang, T.; Li, H.X.; Li, Y. Numerical Investigation on Coaxial Coalescence of Two Gas Bubbles. *J. Xi'an Jiaotong Univ.* **2013**, *47*, 1–6.
9. Liu, Y.P.; Wang, P.Y.; Lin, S.N. Correlation of Position of Taylor Bubble Formation in Cryogenic Tube. *J. Shanghai Jiaotong Univ.* **2013**, *47*, 1509–1514+1519.
10. Ulaganathan, V.; Krzan, M.; Lotfi, M.; Dukhin, S.; Kovalchuk, V.; Javadi, A.; Gunes, D.; Gehin-Delval, C.; Malysa, K.; Miller, R. Influence of β-lactoglobulin and its surfactant mixtures on velocity of the rising bubbles. *Colloids Surf. A Physicochem. Eng. Asp.* **2014**, *460*, 361–368. [CrossRef]
11. Azzopardi, B.J.; Pioli, L.; Abdulkareem, L.A. The properties of large bubbles rising in very viscous liquids in vertical columns. *Int. J. Multiph. Flow* **2014**, *67*, 160–173. [CrossRef]
12. Keshavarzi, G.; Pawell, R.S.; Barber, T.J.; Yeoh, G.H. Transient analysis of a single rising bubble used for numerical validation for multiphase flow. *Chem. Eng. Sci.* **2014**, *112*, 25–34. [CrossRef]
13. Cano-Lozano, J.C.; Bolaños-Jiménez, R.; Gutiérrez-Montes, C.; Martínez-Bazán, C. The use of volume of fluid technique to analyze multiphase flows: Specific case of bubble rising in still liquids. *Appl. Math. Model.* **2015**, *39*, 3290–3305. [CrossRef]
14. Yan, X.; Zheng, K.; Jia, Y.; Miao, Z.; Wang, L.; Cao, Y.; Liu, J. Drag Coefficient Prediction of a Single Bubble Rising in Liquids. *Ind. Eng. Chem. Res.* **2018**, *57*, 5385–5393. [CrossRef]
15. Khodayar, J.; Davoudian, S.H. Surface Wettability Effect on the Rising of a Bubble Attached to a Vertical Wall. *Int. J. Multiph. Flow* **2018**, *109*, 178–190.
16. Du, J.; Zhao, C.; Bo, H.; Ren, X. The Modeling of Bubble Lift-Off Diameter in Vertical Subcooled Boiling Flow. *Energies* **2022**, *15*, 6857. [CrossRef]
17. Mangani, F.; Soligo, G. Influence of density and viscosity on deformation, breakage, and coalescence of bubbles in turbulence. *Phys. Rev. Fluids* **2022**, *7*, 34. [CrossRef]
18. He, H.; Liu, Z.; Ji, J.; Li, S. Analysis of Interaction and Flow Pattern of Multiple Bubbles in Shear-Thinning Viscoelastic Fluids. *Energies* **2023**, *16*, 5345. [CrossRef]
19. Guo, Y.L. Study on Gas Invasion Mechanism and Wellbore Pressure during Well Shut-In in Deepwater Drilling. Ph.D. Thesis, China University of Petroleum, Qingdao, China, 2018.
20. Zhang, J.; Du, D.; Hou, J. *Seepage Mechanics of Oil and Gas Reservoirs*, 2nd ed.; China University of Petroleum Press: Dongying, China, 2009. (In Chinese)
21. Sun, B.; Guo, Y.; Wang, Z.; Yang, X.; Gong, P.; Wang, J.; Wang, N. Experimental study on the drag coefficient of single bubbles rising in static non-Newtonian fluids in wellbore. *J. Nat. Gas Sci. Eng.* **2015**, *26*, 867–872. [CrossRef]

Disclaimer/Publisher's Note: The statements, opinions and data contained in all publications are solely those of the individual author(s) and contributor(s) and not of MDPI and/or the editor(s). MDPI and/or the editor(s) disclaim responsibility for any injury to people or property resulting from any ideas, methods, instructions or products referred to in the content.

Article

The Insignificant Improvement of Corrosion and Corrosion Fatigue Behavior in Geothermal Environment Applying Boehmit Coatings on High Alloyed Steels

Anja Pfennig [1,*], Wencke Mohring [2] and Marcus Wolf [2]

[1] HTW Berlin, University of Applied Sciences Berlin, Wilhelminenhofstraße 75A, 12459 Berlin, Germany
[2] BAM Federal Institute of Materials Testing and Research, Unter den Eichen 87, 12205 Berlin, Germany; wencke.mohring@50hertz.com (W.M.)
* Correspondence: anja.pfennig@htw-berlin.de; Tel.: +49-3050194231

Abstract: The efficacy of alumina-sol based coatings in a water-free atmosphere at high temperatures suggests a potential solution for enhancing the corrosion resistance of high-alloyed steels in Carbon Capture and Storage (CCS) environments. In this study, coupons of X20Cr13, designed for use as injection pipes with 13% Chromium and 0.20% Carbon (1.4021, AISI 420), were sol-gel coated with water and ethanol-based alumina. These coated coupons were then exposed to CO_2-saturated saline aquifer water, simulating conditions in the Northern German Basin, for 1000 h at ambient pressure and 60 °C. Corrosion fatigue experiments were also conducted using specimens of X5CrNiMoCuNb16-4 (1.4542, AISI 630), a suitable candidate for geothermal applications, to assess the impact of the ethanol-based coating on the number of cycles to failure at different stress amplitudes. Unfortunately, the coating exhibited early spallation, resulting in corrosion kinetics and corrosion fatigue data identical to those of uncoated specimens. Consequently, the initially promising Boehmit coating is deemed unsuitable for CCS applications and further research therefore not advisable.

Keywords: alumina coating; high alloyed steel; pitting; surface corrosion; CO_2; pipeline; corrosion; CCS; CO_2-storage

Citation: Pfennig, A.; Mohring, W.; Wolf, M. The Insignificant Improvement of Corrosion and Corrosion Fatigue Behavior in Geothermal Environment Applying Boehmit Coatings on High Alloyed Steels. *Appl. Sci.* **2024**, *14*, 1575. https://doi.org/10.3390/app14041575

Academic Editors: Kai Wang, Jie Wu, Yongjun Deng and Lin Chen

Received: 30 December 2023
Revised: 10 February 2024
Accepted: 13 February 2024
Published: 16 February 2024

Copyright: © 2024 by the authors. Licensee MDPI, Basel, Switzerland. This article is an open access article distributed under the terms and conditions of the Creative Commons Attribution (CC BY) license (https://creativecommons.org/licenses/by/4.0/).

1. Introduction

Addressing global warming, one of society's significant challenges, involves mitigating the emission of the greenhouse gas CO_2, primarily originating from power plants. A proposed solution is Carbon Capture and Storage (CCS), a process comprising three key stages: the capture and compression of CO_2 and emission gases directly from combustion processes at the source (such as coal power plants), the transmission through pipelines, and the injection into suitable geological formations [1,2]. During this process, CO_2 is pressurized to a supercritical (or liquid) state [3,4].

Hence, a key area of research focuses on understanding the corrosion behavior of steels in supercritical CO_2 [5] or CO_2-saturated saline aquifer waters [6]. In the process of injecting CO_2 into deep geological saline aquifer reservoirs, as observed in the Northern German Basin [7–9], the dissolved CO_2 in the aquifer water creates a highly corrosive environment that can lead to the failure of pipe steels [10,11].

The general corrosion process is explained as follows: An $FeCO_3$ corrosion layer (siderite) forms on the alloy surface due to the anodic dissolution of iron from the pipe steel [10,12]. In a CO_2-rich environment, simultaneous anodic and cathodic reactions result in the formation of corrosion products that precipitate on the steel surface. These products mainly consist of $FeCO_3$ (siderite), $Fe(HCO_3)_2$ (iron bicarbonate), and $FeOOH$ (goethite) [10,13].

Corrosion processes in Carbon Capture and Storage (CCS) environments are significantly influenced various environmental factors, including the composition of the surrounding media and alloy, temperature, CO_2 partial pressure, flow conditions, contaminations, dissolved salts and the formation of protective scales [10–22] Although high-alloyed steel generally exhibits good corrosion resistance, corrosion in the injection pipe can occur in CO_2-rich aquifer water, particularly if phase boundaries form during intermissions of the injection process when the aquifer water flows back into the pipe [13–15].

1.4021 is a widely used martensitic stainless steel in industry, featuring 13% Cr and 0.20% C. Another notable stainless steel, 1.4542 (AISI 630, X5CrNiCuNb16-4), is a precipitation hardening martensitic variant with approximately 3% small copper particles distributed in the matrix for enhanced precipitation hardening [23,24]. Research indicates a direct correlation between the mechanical and corrosion properties of various steel types and their surface conditions resulting from machining processes [25–29]. Generally, corrosion resistance improves with reduced surface depth on carbon steel [26], austenitic stainless steel [28] after shot peening [29], and ferritic stainless steel when Ra exceeds 0.5 µm [28]. Interestingly, the impact of relative humidity on internal pipeline corrosion is more significant than that of initial surface roughness; reducing humidity proves more effective than altering initial surface conditions [27].

Smooth surfaces and compressive surface stress have a positive correlation with improved fatigue behavior in air under corrosive environments. This relationship extends to enhanced corrosion resistance, leading to improved corrosion fatigue behavior as well [30–33].

To ensure the required longevity of pipes utilized in Carbon Capture and Storage (CCS) technology, employing protective coatings composed of heat-resistant ceramics is a viable technical solution [34,35]. Alumina coatings emerge as promising choices owing to their high inertness and stability at elevated temperatures [36,37]. Various methods, such as plasma spraying, chemical vapor deposition, and ceramic slurry application, are employed to produce thick alumina coatings, while the sol-gel technique allows for the application of thin alumina coatings [34,35,38,39]. Alumina has demonstrated resilience in high-temperature environments and can be effectively applied as a coating through the sol-gel method [34,36,40–43]. However, despite these benefits it is important to note that this type of coating has great impact in terms of rising costs, more complicated manufacturing and implication in the supply chain.

Research by Dressler et al. [44] showcased the efficacy of a sol-gel alumina layer, generated via a modified Yoldas sol [41], in providing effective protection for Inconel-718 even after 4000 h of exposure at 800 °C in static laboratory air. Additionally, a 400 nm thick transition alumina layer, applied using the sol-gel method on commercial power plant steel X20, exhibited excellent protection against oxidation in laboratory air up to 650 °C [35]. The infusion of manganese and chromium ions into the alumina layer, leading to the formation of mixed oxides, contributes to the enhanced protective properties of the coating, preventing oxidation. Furthermore, the avoidance of local defects, attributed to substrate preparation through polishing, is identified as a significant factor in improving the coating process [34,35].

Enhancing the corrosion fatigue behavior of coated steels involves achieving high resistance to cracking, low porosity in the coating [45], and inducing compressive residual stresses within the substrate surface [30,46].

Schmitt-Thomas et al. [47] observed that anodic coatings generally improved the corrosion fatigue behavior of X20Cr13, offering cathodic protection in case of coating breakdown. No corrosion occurred in the substrate when using coatings less noble than the steel, as anodic dissolution took place in the coatings [47,48]. Alumina coating provided cathodic protection for both X20Cr13 and X5CrNiCuNb16-4. Oliveira et al. [48] noted that the fatigue behavior of AISI4043 (36CrNiMo4), a medium carbon, low alloy steel, remains similar to that in air. Fatigue cracks initiated at alumina particles in the substrate steel rather than corrosion pits formed during testing. The corrosion fatigue strength of ceramic-sprayed medium carbon steel (S45C) was slightly higher than that of the substrate steel at medium stress levels, but the coating's sealing nature had little effect on improving corrosion fatigue strength at higher stress levels. Once coating cracks formed, corrosion fatigue strength became equivalent to that of the substrate steel. Voorwald et al. [46] indicated that tungsten carbide WC–17Co and WC–10Co–4Cr thermal spray coatings applied by high velocity oxygen fuel coating (HVOF) process result in higher fatigue strength. In contrast, the fatigue strength of AISI 4340 steel associated with chromium electroplating was significantly reduced.

The study introduces the prospect of reducing corrosion rates and pit initiation by applying an alumina coating via the sol-gel process to coupons of 1.4021 and fatigue specimens of 1.4542. However, experiments indicate that this approach is not favorable in a CCS environment. Nonetheless, the results provide valuable insights into corrosion management during carbon compression.

2. Materials and Methods

Immersive tests were conducted under ambient pressure using coupons measuring 50 mm × 20 mm × 4 mm, fabricated from high-alloyed martensitic stainless steel 1.4021 (X20Cr13, AISI 420). Corrosion fatigue tests were carried out using high-alloyed martensitic corrosion resistant stainless steel 1.4542 (X5CrNiCuNb16-4, AISI 630, PRE number 15.0–18.9 (PREN = %Cr + 3.3% Mo + 16% N) [49,50]). The chemical composition was analyzed using spark emission spectrometry (SPEKTROLAB M) and the Electron Probe Microanalyzer JXA8900-RLn (Table 1).

Table 1. Chemical composition of 1.4542 (X5CrNiCuNb16-4, AISI 630) and 1.4021 (X20Cr13, AISI 420) (in mass percent).

Elements	C	Si	Mn	P	S	Cr	Mo	Ni	Cu	Nb
				1.4542 (X5CrNiCuNb16-4, AISI 630)						
acc standard [a]	<0.07	≤0.70	≤1.50	≤0.04	≤0.015	15.0–17.0	≤0.60	3.00–5.00	3.00–5.00	0.20–0.45
analysed [b]	0.03	0.42	0.68	0.018	0.002	15.75	0.11	4.54	3.00	0.242
				1.4021 (X20Cr13, AISI 420)						
acc standard [a]	0.17–0.25	≤01.00	≤1.00	≤0.045	≤0.03	12.0–14.0				
analysed [b]	0.22	0.39	0.32	0.007	0.006	13.3		0.123		

[a] elements as specified according to DIN EN 10088-3 in %; [b] spark emission spectrometry ±1%.

The microstructure of alloy 1.4542, revealed through etching before exposure and fatigue testing, displays martensite with varying needle-shaped sizes and a minor percentage of delta phase precipitation.

For the preparation of water and ethanol-based sols and coatings, the procedures outlined in [34,35,43,44,51–53], and illustrated in [34] were followed. Initially, an aluminum nitrate (V) solution was heated to 87 °C, and aluminum tri-sec-butylate dissolved in secbutanole was added under vigorous stirring for up to 1 h. After cooling to room temperature, the sol was directly used for the dip-coating process. The preparation of aqueous modified Yoldas sols followed the method described in [51,52], resulting in sols with a solid content of 10 wt.% Al_2O_3 and a NO_3^-/Al ratio of 0.6, yielding an acidity of pH = 3.5. Additional information on this type of sol is provided in [51,53]. The sol was applied before exposure to the corrosive environment on polished (coupons) or fine machined (fatigue specimen) surfaces by pulling substrates ((coupons: water based sol: 170 mm/min, ethanol based sol: 40 mm/min and 170 mm/min) and (corrosion fatigue specimen: 15 mm/s and 30 mm/s)). Then the samples were dried at 120 °C for 2:30 min and heat treated in inert atmosphere at 500 °C for 30 min h to form a well adhesive alumina coating. The coating thickness on the flat coupons was measured via TEM and varied according to the pulling velocity (Table 2). Measurements of the coating thickness on round fatigue specimen were not successful due to unevenness and lateral detachment at the specimen center:

Table 2. Chemical composition of 1.4542 (X5CrNiCuNb16-4, AISI 630) and 1.4021 (X20Cr13, AISI 420) (in mass percent).

Sample Nr.	Sol	Pulling Velocity	Coating Thickness	Remarks
1.4542 (X5CrNiCuNb16-4, AISI 630) corrosion fatigue specimen				
6 TOT	water based	30 mm/s	not applicable < 0.5 µm	uneven, layered, lateral detachments at center
7 TOT	water based	15 mm/s	not applicable < 0.2 µm	uneven, layered, lateral detachments
12 TOT	water based	15 mm/s	not applicable < 0.5 µm	uneven
21 TOT	water based	30 mm/s	not applicable < 0.6 µm	uneven
1.4021 (X20Cr13, AISI 420) coupons				
20, 21, 22	water based	170 mm/min	ca. 1.4 µm	
23, 24	ethanol based	40 mm/min	ca. 0.8 µm	
25, 26	ethanol based	170 mm/min	ca. 1.9 µm	

For ambient pressure (X20Cr13) and fatigue experiments (X5CrNiCuNb16-4) technically clean CO_2 (99.999%) was used to saturated the aquifer water. The laboratory brine similar to the Stuttgart Aquifer [7] (Table 3) was synthesized in a strictly orderly way to avoid precipitation of salts and carbonates.

Table 3. Chemical composition of the synthetic aquifer electrolyte according to the Stuttgart Formation.

	According to Stuttgart Formation							
	NaCl	KCl	$CaCl_2 \times 2H_2O$	$MgCl_2 \times 6H_2O$	$Na_2SO_4 \times 10H_2O$	KOH	$NaHCO_3$	pH value
g/L	224.6	0.3902	6.452	10.62	12.074	0.3206	0.0475	8.2–9
	or							
	Ca^+	K^{2+}	Mg^{2+}	Na^{2+}	Cl^-	SO_4^{2-}	HCO_3^-	pH value
g/L	1.76	0.43	1.27	90.1	14.33	3.6	0.04	8.2–9

Exposure of the coupons (a set of two or three per parameter) was disposed according to Pfennig [13] in a chamber kiln at 60 °C and ambient pressure. Flow control of CO_2 (3 NL/h) was done by a capillary meter GDX600_man by QCAL Messtechnik GmbH, München [13].

Corrosion fatigue specimens were manufactured by means of precision turning without an additional surface finish in order to replicate the prefabricated technological conditions (Rz = 4, surface roughness 2.6 µm to 4.7 µm with mean arithmetic value of 3.65 (1.4542).

Coating of 1.4542 (X5CrNiCuNb16-4) fatigue specimens was done via dip-coating into ethanol-based alumina sols according to the procedure demonstrated above and schematically shown in [34]. The sol was applied before exposure to the corrosive environment on fine machined fatigue specimen that had reached endurance limits in previous experiments (run-outs) surfaces by pulling substrates (corrosion fatigue specimen: 15 mm/s and 30 mm/s)). Because of the length of the specimen the coating procedure was undertaken in two consecutive steps resulting a visual barrier on the surface. Then the samples were dried at 120 °C for 2:30 min and heat treated in inert atmosphere at 500 °C for 30 min h to form a well adhesive alumina coating and loaded into the corrosion chamber.

Corrosion fatigue test were conducted using hourglass specimens (critical cross-section is 12.5 mm in diameter) according to the standard DIN EN ISO 11782 1 and to the recommendations of the FKM Research Issue [54,55]. The resonant testing machine (sinusoidal dynamic test loads, R = −1; resonant frequency ~30 Hz) and corrosion fatigue testing have been explained has been in detail by Wolf and Pfennig [13,49]. The set up (Figure 1) was used to test 9 specimens between 280 MPa and 390 MPa.

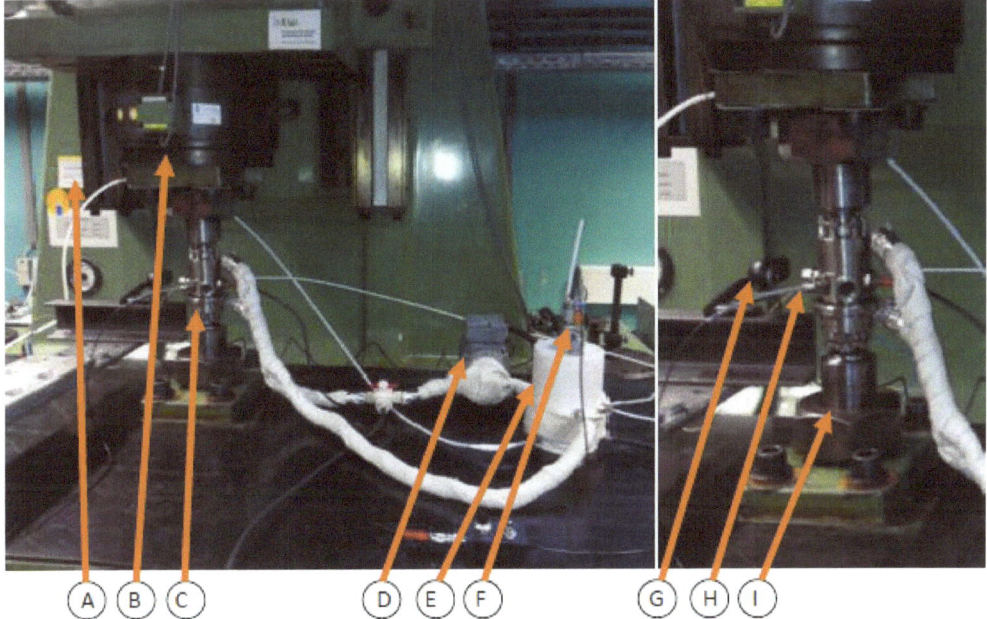

Figure 1. Schematic set-up of operating corrosion chamber for in-situ corrosion fatigue testing (vertical resonant testing machine (A), clamping socket and plate (B), corrosion chamber (C), gear pump (D), aquifer reservoir with heating (E), pH-measuring and mixing unit (F), camera (G), thermocouple (H) and union nut (I)).

X-ray diffraction analysis was performed using a URD-6 (Seifert-FPM) equipped with CoKα-radiation, featuring automatic slit adjustment, a step size of 0.03, and a 5-s count. Phase analysis was carried out using AUTOQUAN® by Seifert FPM. For gravimetric measurement, samples were descaled by exposure to 37% HCl for 24 h, and the mass gain was analyzed according to DIN 50 905 part 1–4. Surface corrosion characterization utilized SEM/EDX with a Leo Gemini 1530 VP at an acceleration voltage of 15 kV. Pitting corrosion analysis involved 3D imaging with the Microprof TTV double optical system by FRT.

Sections of non-descaled samples were embedded in cold resin (Epoxicure, Buehler), cut, and polished with SiC-Paper ranging from 180 µm to 1200 µm under water. The final polishing stages included diamond paste with grit sizes of 6 µm, 3 µm, and 1 µm. Measurement of layer thicknesses, residual pipe wall thicknesses, and microstructure analysis were conducted through light and electron microscopy, utilizing the semi-automatic analyzing program Analysis Docu ax-4 by Aquinto.

3. Results and Discussion

To assess the potential damage to injection pipe steels and their protective coatings, experimental conditions were selected to replicate a harsh carbon capture and storage (CCS) environment, specifically focusing on the specimen fully immersed into the brine. This represents injection pauses under decreased pressure in the injection pipe where the water level may rise, affecting the conditions [13,21].

3.1. Influence of Alumina Coating on Static Corrosion Behavior of 1.4021

Figure 2 shows the sol gel coated (water based as well as ethanol based, binder: polyvinylbutyral) coupons of X20Cr13 before and after exposure to CO_2 saturated saline aquifer water at 60 °C and ambient pressure. The macroscopic degradation of the coating is demonstrated by lateral detachment of the coating and dissolution of the alumina sol (Figure 3).

Regions of the coupon surfaces with degraded coating are highly susceptible towards CO_2 corrosion and show typical corrosion products and patterns of uncoated samples as demonstrated earlier [13,21,56]. Surface corrosion layers and pits reveal both, $FeCO_3$ and FeOOH as the main precipitation phases with no dependence on the coating. The lack of coating elements within the corrosion products indicate that the coating is detached before surface reaction take place at the newly exposed metal surface. Pit formation is driven by the formation of carbonic acid and existence of HCO_3 as well as a transient ferrous hydroxide phase $Fe(OH)_2$ reacting to corrosion products -mainly siderite- on the pits as detected on the surface elsewhere.

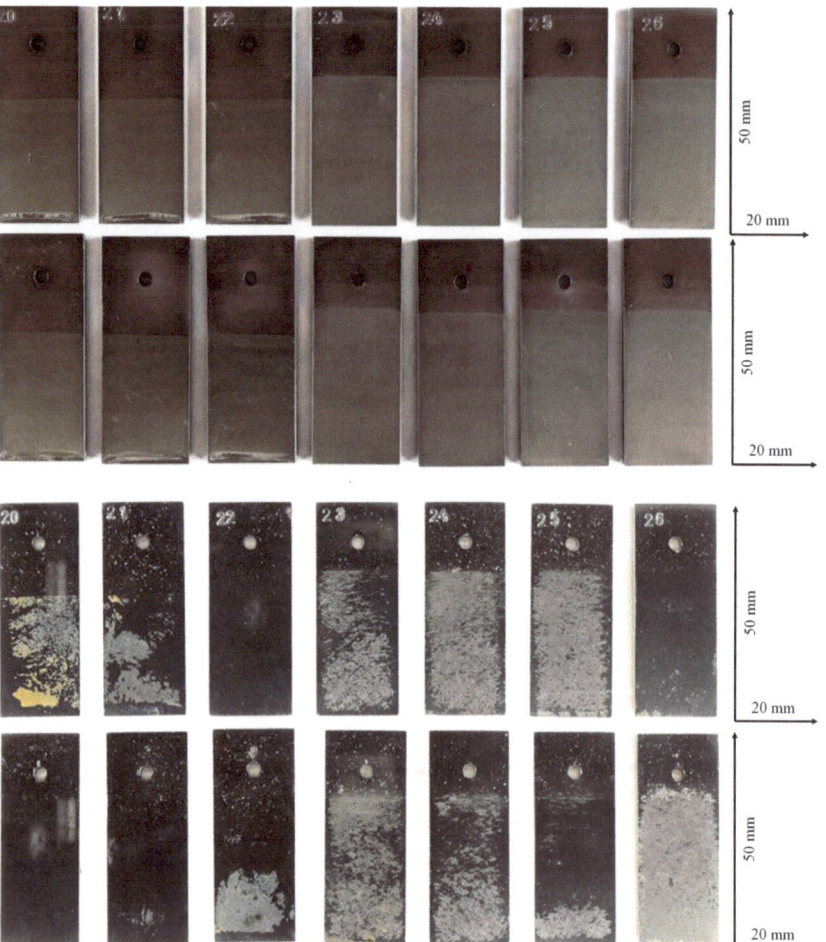

Figure 2. Sample surfaces of water and ethanol based alumina coated martensitic stainless steel X20Cr13 after 1000 h of exposure to CO_2 saturated saline aquifer water at 60 °C and ambient pressure. Top: coated, upper row: front side, lower row: back side. Bottom: corroded after exposure: upper row: front side, lower row: back side.

The corrosion kinetics of 1.4021 have been extensively studied [13,21] indicating that exposure at ambient pressure serves as a worst-case scenario simulation. The corrosion rates of coated coupons align with earlier findings for X20Cr13 at ambient pressure remained approximately 0.1 mm/year after 1000 h of exposure. Locally, regions fully coated showed very good corrosion resistance and low corrosion rates in the liquid phase (where coupons are fully immersed in the aquifer) are low, indicating that the CO_2 partial pressure is not sufficient to initiate the corrosive reactions described by Wei et al. and Pfennig et al. [10,13]. Figures 3 and 4 clearly show that ethanol based alumina sol does not coat the entire surface after exposure. Regions with locally destroyed coating reveal corrosion products according to the reactions and products stated earlier.

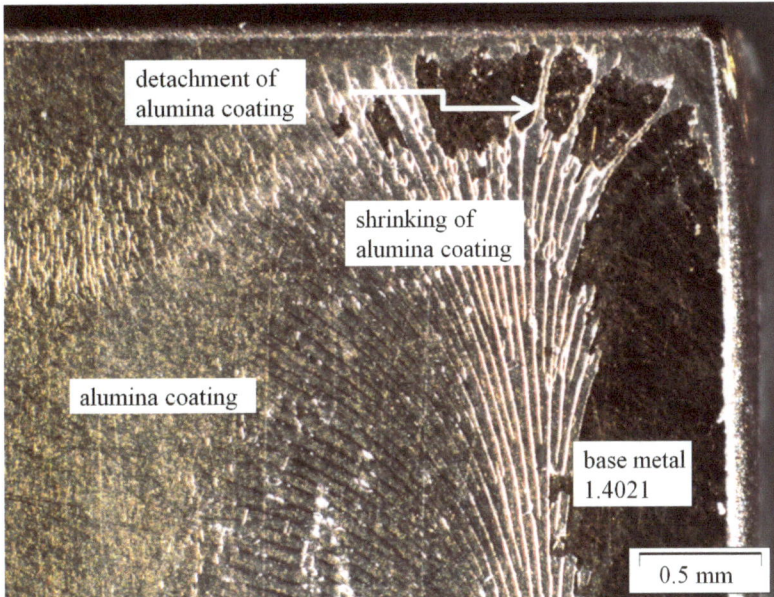

Figure 3. Shrinking and detachment of water and ethanol based alumina coating on martensitic stainless steel X20Cr13 after 1000 h of exposure to CO_2 saturated saline aquifer water at 60 °C and ambient pressure.

There are three possible reasons for early failure of the coating (note that all samples were wetted perfectly prior to exposure to CCS environment):

1. Due to the surface tension the coating is thinner towards the edges of the coupons than in the center. Therefore, stress causes micro cracking at the edges with the capillary forces driving the saline aquifer brine underneath the coating wetting the base metal surface.
2. The high porosity may also be cause for the instability of the coating. The brine may have contact to the base metal surface through pores that connect surface and environment. Early corrosion reactions cause the coating to detach then in lateral direction.
3. The reliability of the coating may also be reason for failure, because the method of coating directly influences the continuity and thickness of the coating. Dip coating leads to an increasing thickness of the coating towards the bottom of the samples because gravity forces the sol to flow before its gel status. Due to the "flowing" of the sol a bulge precipitates that has higher micro stress than the remaining coating. Therefor micro cracking of the coating is initialized within this bulge. In general, the reliability of the coating is directly dependent on the homogeneity of the coating and its consistent thickness avoiding stress gradients along the surface.

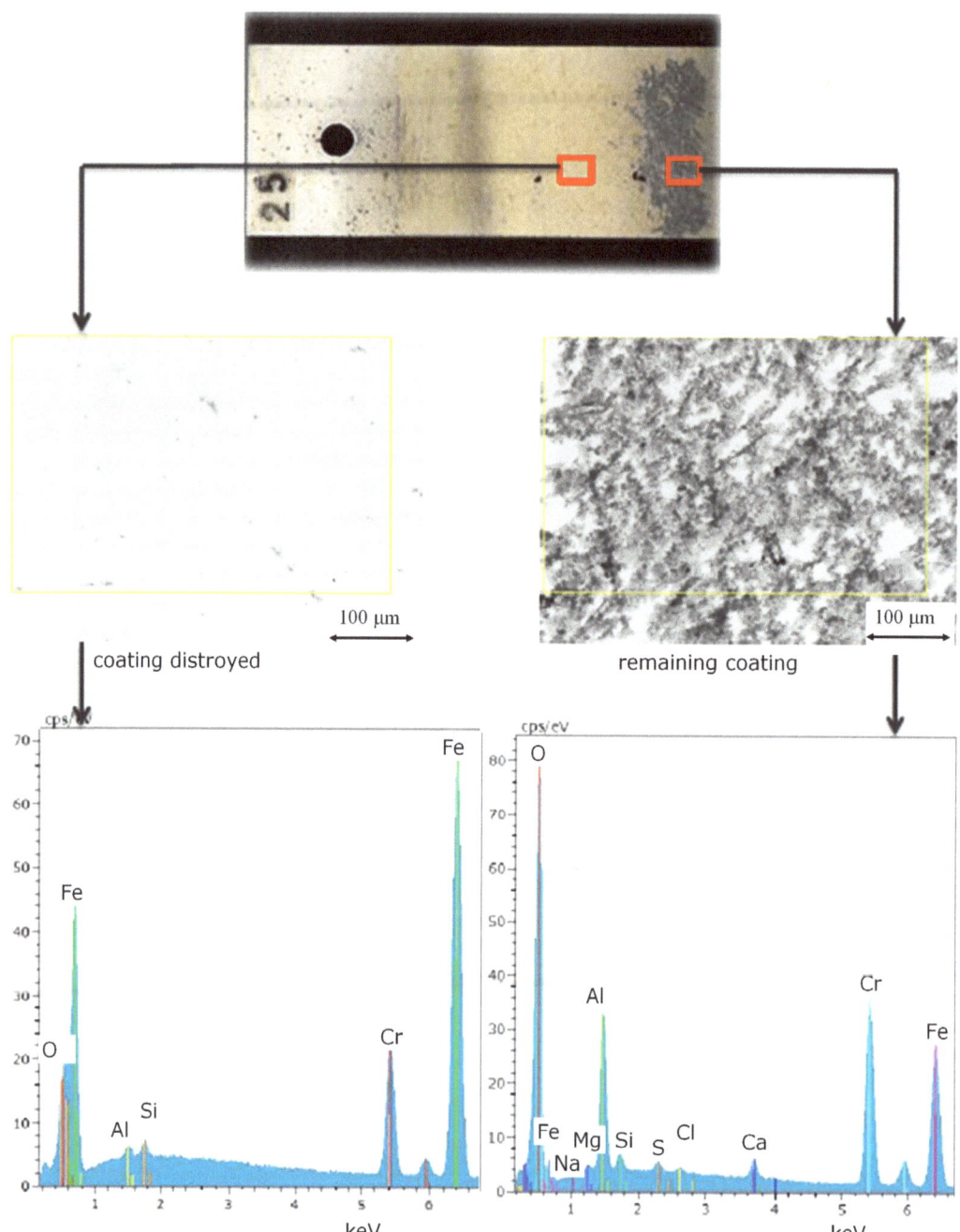

Figure 4. Element distribution and surface images of different local regions on coupons of ethanol based alumina coated martensitic stainless steel X20Cr13, 1.4021 after exposure to and CO_2 saturated saline aquifer water at 60 °C and ambient pressure.

As Schulze et al. [34] state chromium and manganese diffusion into alumina improves the adhesion and protection of the coating. Especially, the role of chromium diffusion is important for the adhesion of the coating. In case of low alloyed steels containing less than 9 wt.% chromium, the alumina layer spalled off after the heat treatment for 0.5 h at service temperature of the respective steel [34]. X20Cr13 with 13% steel might have enough chromium for sufficient diffusion into the coating, but the heat treatment temperatures during exposure to CCS environment were too low for diffusion processes to be effective.

3.2. Influence of Alumina Coating on the Corrosion Fatigue Behavior of 1.4542

Earlier the authors presented the influence of corrosive media on the mechanical behavior of stainless steel AISI 630 (X5CrNiCuNb16-4, 1.4542) during carbon capture and storage as well as in geothermal energy production [13,21]. Earlier results do not change with alumina coatings indicating that the coating has no influence on the corrosion fatigue behavior. This includes S-N curve with low coefficient of correlation (r^2 = 0.33) and a large scattering range (TN = 1:34.4), corrosion fatigue strength 60% below the endurance limit measured in air (620 MPa), reaching a maximum of 10×10^7 cycles at a stress amplitude of 150 MPa (Wöhler-exponent of k = 3.59), unusual corrosion pattern and failure related to statistical crack initiation due to either pitting or the formation of micro cracks depending on the stress amplitude [13].

The influence of ethanol based alumina coatings of the fatigue behavior is demonstrated in Figure 5. The coating is distributed homogeneously over the test area of the specimen. Four specimen were tested, one failed due to machine failure but three reached 10^7 cycles at stress amplitude of 200 MPa. However, these endurance limits were only reached because of the low stress amplitude and are not related to the protection of the coating. The alumina coating was almost completely detached leaving the bare surface in contact with the corrosive environment resulting in severe pitting (Figure 5, bottom). An important finding is that this type of coating (first developed for high temperature alloys (use temperature > 550 °C) and nickel-base superalloys) does not perform in saline-water based aquifer environments with process gasses such as CO_2. Also the considerable low drying and heating temperatures do not guarantee the adhesion of the coating.

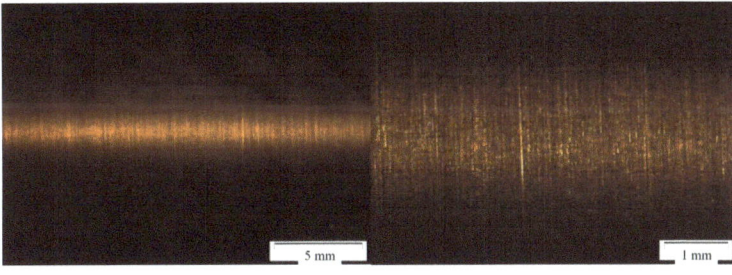

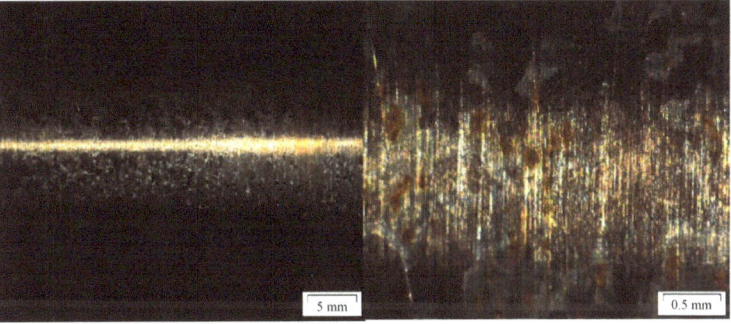

Figure 5. Fatigue specimen and surface image after coating with ethanol based alumina sol prior (**top**) and after (**bottom**) to exposure to CCS environment.

When a consistent protective coating is present on the metal surface, fatigue cracks are observed to initiate at alumina particles embedded in the substrate steel matrix during blasting, rather than at corrosion pits formed during testing, as highlighted by Oliveira et al. [48]. The researchers argue that the corrosion fatigue strength of a coated substrate is governed by the same mechanism that influences the fatigue behavior of the material in the air.

In the case of the detached ethanol-based alumina sol-gel coating on X5CrNiCuNb16-4, pits precipitate early, causing crack initiation comparable to the non-coated specimen within a corrosive environment (CO_2-saturated saline water), according to Oliveira et al. [48]. Voorwald et al. [46] suggest that shot peening prior to coating provides an excellent alternative to increase the fatigue strength of AISI 4340 steel. In AISI 630, X5CrNiCuNb16-4, the residual stresses in the coating and substrate may change from tensile near the coating surface to compressive inside the coating center. These compressive stresses may then enable the coating to remain attached to the metal surface during fatigue testing.

4. Conclusions

Coupons and fatigue specimens made of X20Cr13 and X5CrNiCuNb16-4 were coated with water-based and ethanol-based alumina sol and exposed to CO_2-saturated saline aquifer water for up to 1000 h, simulating conditions in the Northern German Basin, at ambient pressure and 60 °C in laboratory experiments. Corrosion fatigue experiments were conducted on coated X5CrNiCuNb16 specimens at 200 MPa stress amplitude, under ambient pressure, applying the same CCS conditions as for coupon tests.

Macroscopic surface images and the absence of coating elements within corrosion products indicate that the coating detaches before surface reactions occur. Surface corrosion layers and pits reveal carbonate corrosion products ($FeCO_3$ and FeOOH) as the main precipitation phases, independent of the coating. Coating failure was observed in both static and dynamic tests accounting for water-based as well as ethanol-based coatings, rendering the alumina coating non-functional.

Early failure of the coating is possibly attributed to micro cracking due to surface tension gradient driving the saline aquifer brine underneath the coating which then wets the base metal surface. High porosity may also cause the instability of the coating resulting in early detachment. The reliability of the coating is directly dependent on the homogeneity of the coating and its consistent thickness avoiding bulges and stress gradients along the surface.

Although alumina-sol based spin coatings have been proven to show very good corrosion resistance in dry process gas atmosphere at high temperature this type of coating does not apply for CCS or general geothermal environment with high temperature and high water solubility being the most critical factors in CO_2 gas mixtures.

Author Contributions: Conceptualization, A.P., W.M. and M.W.; methodology, A.P., W.M. and M.W.; software, M.W.; validation, A.P.; formal analysis, W.M. and M.W.; investigation, A.P., W.M. and M.W.; resources, A.P., W.M. and M.W.; data curation, A.P.; writing—original draft preparation, A.P.; writing—review and editing, A.P.; visualization, A.P.; supervision, A.P.; project administration, A.P.; funding acquisition, A.P. All authors have read and agreed to the published version of the manuscript.

Funding: This research received no external funding.

Data Availability Statement: Data are contained within the article.

Conflicts of Interest: The authors declare no conflict of interest.

References

1. Thomas, D.C. Carbon Dioxide Capture for Storage in Deep Geologic Formations—Results from CO_2 Capture Project, Volume 1: Capture and Separation of Carbon Dioxide from Combustion Sources. In *CO_2 Capture Project*; Elsevier Ltd.: London, UK, 2005; ISBN 0080445748.
2. Ruhl, A.S.; Goebel, A.; Kranzmann, A. Corrosion Behavior of Various Steels for Compression, Transport and Injection for Carbon Capture and Storage. *Energy Procedia* **2012**, *23*, 216–225. [CrossRef]

3. Gale, J.; Davison, J. Transmission of CO_2—Safety and economic considerations. *Energy* **2004**, *29*, 1319–1328. [CrossRef]
4. Eldevik, F.; Graver, B.; Torbergsen, L.E.; Saugerud, O.T. Development of a Guideline for Safe, Reliable and Cost Efficient Transmission of CO_2 in Pipelines. *Energy Procedia* **2009**, *1*, 1579–1585. [CrossRef]
5. Russick, E.M.; Poulter, G.A.; Adkins, C.L.; Sorensen, N. Corrosive effects of supercritical carbon dioxide and cosolvents on metals. *J. Supercrit. Fluids* **1996**, *9*, 43–50. [CrossRef]
6. Nešić, S. Key issues related to modelling of internal corrosion of oil and gas pipelines—A review. *Corros. Sci.* **2007**, *49*, 4308–4338. [CrossRef]
7. Förster, A.; Norden, B.; Zinck-Jørgensen, K.; Frykman, P.; Kulenkampff, J.; Spangenberg, E.; Erzinger, J.; Zimmer, M.; Kopp, J.; Borm, G.; et al. Baseline characterization of the CO2SINK geological storage site at Ketzin, Germany. *Environ. Geosci.* **2006**, *13*, 145–161. [CrossRef]
8. Förster, A.; Schöner, R.; Förster, H.-J.; Norden, B.; Blaschke, A.-W.; Luckert, J.; Beutler, G.; Gaupp, R.; Rhede, D. Reservoir characterization of a CO_2 storage aquifer: The Upper Triassic Stuttgart Formation in the Northeast German Basin. *Mar. Pet. Geol.* **2010**, *27*, 2156–2172. [CrossRef]
9. Kissinger, A.; Noack, V.; Knopf, S.; Scheer, D.; Konrad, W.; Class, H. Characterization of reservoir conditions for CO_2 storage using a dimensionless Gravitational Number applied to the North German Basin. *Sustain. Energy Technol. Assess.* **2014**, *7*, 209–220. [CrossRef]
10. Wei, L.; Pang, X.; Liu, C.; Gao, K. Formation mechanism and protective property of corrosion product scale on X70 steel under supercritical CO_2 environment. *Corros. Sci.* **2015**, *100*, 404–420. [CrossRef]
11. Carvalho, D.; Joia, C.; Mattos, O. Corrosion rate of iron and iron–chromium alloys in CO_2 medium. *Corros. Sci.* **2005**, *47*, 2974–2986. [CrossRef]
12. Cui, Z.D.; Wu, S.L.; Zhu, S.L.; Yang, X.J. Study on corrosion properties of pipelines in simulated produced water saturated with supercritical CO_2. *Appl. Surf. Sci.* **2006**, *252*, 2368–2374. [CrossRef]
13. Pfennig, A.; Wolf, M.; Kranzmann, A. Corrosion and Corrosion Fatigue of Steels in Downhole CCS Environment—A Summary. *Processes* **2021**, *9*, 594. [CrossRef]
14. Eslami, M.; Wang, X.; Choi, Y.-S. Electrochemical Study of Corrosion Resistant Alloys in Supercritical CO2 Environment. SSRN. 2023. Available online: https://ssrn.com/abstract=4571105 (accessed on 18 December 2023).
15. Bowman, S.; Agrawal, V.; Sharma, S. Evaluating the Impact of Redox Potential on the Corrosion of Q125, 316L, and C276 Steel in Low-Temperature Geothermal Systems. *Corros. Mater. Degrad.* **2023**, *4*, 573–593. [CrossRef]
16. Banaś, J.; Lelek-Borkowska, U.; Mazurkiewicz, B.; Solarski, W. Effect of CO_2 and H_2S on the composition and stability of passive film on iron alloys in geothermal water. *Electrochim. Acta* **2007**, *52*, 5704–5714. [CrossRef]
17. Choi, Y.-S.; Colahan, M.; Nešić, S. Effect of Flow on the Corrosion Behavior of Pipeline Steel in Supercritical CO_2 Environments with Impurities. *Corrosion* **2023**, *79*, 497–508. [CrossRef] [PubMed]
18. Liu, J.; Yao, D.; Chen, K.; Wang, C.; Sun, C.; Pan, H.; Meng, F.; Chen, B.; Wang, L. Effect of H_2O Content on the Corrosion Behavior of X52 Steel in Supercritical CO_2 Streams Containing O_2, H_2S, SO_2 and NO_2 Impurities. *Energies* **2023**, *16*, 6119. [CrossRef]
19. Choi, Y.-Y.; Nešić, S. Determining the corrosive potential of CO_2 transport pipeline in high pCO_2-water environments. *J. Green House Gas Control.* **2011**, *5*, 788–797. [CrossRef]
20. Han, J.; Zhang, J.; Carey, J.W. Effect of bicarbonate on corrosion of carbon steel in CO2 saturated brines. *Int. J. Greenh. Gas Control* **2011**, *5*, 1680–1683. [CrossRef]
21. Pfennig, A.; Kranzmann, A. Understanding the Anomalous Corrosion Behaviour of 17% Chromium Martensitic Stainless Steel in Laboratory CCS-Environment—A Descriptive Approach. *Clean Technol.* **2022**, *4*, 239–257. [CrossRef]
22. Mu, L.J.; Zhao, W.Z. Investigation on carbon dioxide corrosion behavior of HP13Cr110 stainless steel in simulated stratum water. *Corros. Sci.* **2010**, *52*, 82–89. [CrossRef]
23. Islam, A.W.; Sun, A.Y. Corrosion model of CO_2 injection based on non-isothermal wellbore hydraulics. *Int. J. Greenh. Gas Control* **2016**, *54*, 219–227. [CrossRef]
24. Wang, J.; Zou, H.; Li, C.; Zuo, R.; Qiu, S.; Shen, B. Relationship of microstructure transformation and hardening behavior of type 17-4 PH stainless steel. *J. Univ. Sci. Technol. Beijing Miner. Met. Mater.* **2006**, *13*, 235–239. [CrossRef]
25. Zhang, W.; Fang, K.; Hua, Y.; Wang, S.; Wang, X. Effect of machining-induced surface residual stress on initiation of stress corrosion cracking in 316 austenitic stainless steel. *Corros. Sci.* **2016**, *108*, 173–184. [CrossRef]
26. Evgeny, B.; Hughes, T.; Eskin, D. Effect of surface roughness on corrosion behaviour of low carbon steel in inhibited 4 M hydrochloric acid under laminar and turbulent flow conditions. *Corros. Sci.* **2016**, *103*, 196–205. [CrossRef]
27. Xu, M.; Zhang, Q.; Yang, X.X.; Wang, Y.; Liu, J.; Li, Z. Impact of surface roughness and humidity on X70 steel corrosion in su-percritical CO_2 mixture with SO_2, H_2O, and O_2. *J. Supercrit. Fluids* **2016**, *107*, 286–297. [CrossRef]
28. Lee, S.M.; Lee, W.G.; Kim, Y.H.; Jang, H. Surface roughness and the corrosion resistance of 21Cr ferritic stainless steel. *Corros. Sci.* **2012**, *63*, 404–409. [CrossRef]
29. Ahmed, A.A.; Mhaede, M.; Basha, M.; Wollmann, M.; Wagner, L. The effect of shot peening parameters and hydroxyapatite coating on surface properties and corrosion behavior of medical grade AISI 316L stainless steel. *Surf. Coat. Technol.* **2015**, *280*, 347–358. [CrossRef]
30. Kleemann, U.; Zenner, H. Structural component surface and fatigue strength—Investigations on the effect of the surface layer on the fatigue strength of structural steel components. *Mat. Wiss. U. Werkst.* **2006**, *37*, 349–373. [CrossRef]

31. Sanjurjo, P.; Rodríguez, C.; Pariente, I.; Belzunce, F.; Canteli, A. The influence of shot peening on the fatigue behaviour of duplex stainless steels. *Procedia Eng.* **2010**, *2*, 1539–1546. [CrossRef]
32. Abdulstaar, M.; Mhaede, M.; Wollmann, M.; Wagner, L. Investigating the effects of bulk and surface severe plastic deformation on the fatigue, corrosion behaviour and corrosion fatigue of AA5083. *Surf. Coat. Technol.* **2014**, *254*, 244–251. [CrossRef]
33. Wu, X.; Guan, H.; Han, E.H.; Ke, W.; Katada, Y. Influence of surface finish on fatigue cracking behavior of reactor pressure vessel steel in high temperature water. *Mater. Corros.* **2006**, *57*, 868–871. [CrossRef]
34. Schulz, W.; Nofz, M.; Feigl, M.; Dörfel, I.; Saliwan Neumann, R.; Kranzmann, A. Corrosion of uncoated and alumina coated steel X20CrMoV12-1 in H_2O-CO_2-O_2 and air at 600 °C. *Corros. Sci.* **2013**, *68*, 44–50. [CrossRef]
35. Schulz, W.; Feigl, M.; Dörfel, I.; Nofz, M.; Kranzmann, A. Influence of a sol–gel alumina coating on oxidation of X20CrMoV12-1 in air up to 650 °C. *Thin Solid Film.* **2013**, *539*, 29–34. [CrossRef]
36. Agüero, A.; Muelas, R.; Gutiérrez, M.; Van Vulpen, R.; Osgerby, S.; Banks, J.P. Cyclic oxidation and mechanical behaviour of slurry aluminide coatings for steam mturbine components. *Surf. Coat. Technol.* **2007**, *201*, 6253–6260. [CrossRef]
37. Hübert, T.; Schwarz, J.; Oertel, B. Sol-gel alumina coatings on stainless steel for wear protection. *J. Sol-Gel Sci. Technol.* **2006**, *38*, 179–184. [CrossRef]
38. Darut, G.; Ben-Ettouil, F.; Denoirjean, A.; Montavon, G.; Ageorges, H.; Fauchais, P. Dry Sliding Behavior of Sub-Micrometer-Sized Suspension Plasma Sprayed Ceramic Oxide Coatings. *J. Therm. Spray Technol.* **2010**, *19*, 275–285. [CrossRef]
39. Choy, K. Chemical vapour deposition of coatings. *Prog. Mater. Sci.* **2003**, *48*, 57–170. [CrossRef]
40. Fritsch, M.; Klemm, H.; Herrmann, M.; Schenk, B. Corrosion of selected ceramic materials in hot gas environment. *J. Eur. Ceram. Soc.* **2006**, *26*, 3557–3565. [CrossRef]
41. Yoldas, B.E. Alumina Sol Preparation from Alkoxides. *Am. Ceram. Soc. Bull.* **1975**, *54*, 289–290.
42. Vasconcelos, D.; Oréfice, R.; Vasconcelos, W. Processing, adhesion and electrical properties of silicon steel having non-oriented grains coated with silica and alumina sol–gel. *Mater. Sci. Eng. A* **2007**, *447*, 77–82. [CrossRef]
43. Dressler, M. Sol-Gel Preparation and Characterization of Corundum Based Ceramic Oxidation Protection Coatings. Ph.D. Thesis, TU Bergakademie Freiberg, Saxony, Germany, 2006.
44. Dressler, M.; Nofz, M.; Dörfel, I.; Saliwan-Neumann, R. Influence of sol–gel derived alumina coatings on oxide scale growth of nickel-base superalloy Inconel-718. *Surf. Coat. Technol.* **2008**, *202*, 6095–6102. [CrossRef]
45. Tokaji, K.; Ogawa Hwang, J.U.; Kobayashi, Y.; Harada, Y. Corrosion Fatigue Behavior of a Steel with Sprayed Coatings. *J. Therm. Spray Technol.* **1996**, *3*, 269–276. [CrossRef]
46. Voorwald, H.J.C.; Souza, R.C.; Pigatin, W.L.; Cioffi, M.O.H. Evaluation of WC-17Co and WC-10Co-4Cr thermal spray coatings by, HVOF on the fatigue and corrosion strength of AISI 4340 steel. *Surf. Coat. Technol.* **2005**, *190*, 155–164. [CrossRef]
47. Schmitt-Thomas, K.G.; Meisel, H.; Seoler, W. Einfluß von Beschichtungen auf das Schwingungsrißkorrosionsverhalten des Chromstahls X20CrI3. *Werkst. Und Korros.* **1986**, *37*, 36–44. [CrossRef]
48. Oliveira, F.; Hern′andez, L.; Berr′os, J.A.; Villalobos, C.; Pertuz, A.; Puchi Cabrera, E.S. Corrosion fatigue properties of a 4340 steel coated with Colmonoy 88 alloy, applied by HVOF thermal spray. *Surf. Coat. Technol.* **2001**, *140*, 128–135. [CrossRef]
49. Wolf, M.; Afanasiev, R.; Böllinghaus, T.; Pfennig, A. Investigation of Corrosion Fatigue of Duplex Steel X2CrNiMoN22-5 3 Exposed to a Geothermal Environment under Different Electrochemical Conditions and Load Types. *Energy Procedia* **2014**, *63*, 5773–5786. [CrossRef]
50. Gümpel, P.; Boskovic, L.; Straub, J.; Bogatzky, T.; Henkel, G.; Sorg, M.; Hörtnagl, A.; Bauer, A. *Rostfreie Stähle: Grundwissen, Konstruktions- und Verarbeitungshinweise*, 4th ed.; Expert Verlag: Renningen, Germany, 2008; ISBN 978-3-8169-2689-4.
51. Nofz, M.; Pauli, J.; Dressler, M.; Jügeg, C.; Altenburg, W. 27Al NMR Study of Al-Speciation in Aqueous Alumina-Sols. *J. Sol-Gel Sci. Technol.* **2006**, *38*, 25–35. [CrossRef]
52. Dressler, M.; Nofz, M.; Gemeinert, M. Rheology, UV-vis transparency and particle size of modified Yoldas sols. *J. Sol-Gel Sci. Technol.* **2006**, *38*, 261–269. [CrossRef]
53. Dressler, M.; Nofz, M.; Malz, F.; Pauli, J.; Jäger, C.; Reinsch, S.; Scholz, G. Aluminum speciation and thermal evolution of aluminas resulting from modified Yoldas sols. *J. Solid State Chem.* **2007**, *180*, 2409–2419. [CrossRef]
54. Buschermöhle, H. Vereinheitlichung von Proben für Schwingungsversuche. In *FKM Forschungsheft 217*; VDMA Services GmbH: Frankfurt, Germany, 1996.
55. DIN 50905-1:2009-09; GmbH: Korrosion der Metalle—Korrosionsuntersuchungen—Teil 1: Grundsätze. Beuth Verlag: Berlin, Germany, 2009.
56. Bäßler, R.; Sobetzki, J.; Klapper, H.S. Corrosion Resistance of High-Alloyed Materials in Artificial Geothermal Fluids. In Proceedings of the Vol. NACE Inter Nr. Corrosion 2013, Orlando, FL, USA, 17–21 March 2013; p. 2327.

Disclaimer/Publisher's Note: The statements, opinions and data contained in all publications are solely those of the individual author(s) and contributor(s) and not of MDPI and/or the editor(s). MDPI and/or the editor(s) disclaim responsibility for any injury to people or property resulting from any ideas, methods, instructions or products referred to in the content.

Article

Load Calculation and Strength Analysis of the Deepwater Landing Drill Pipe-Lowering Operation

Guolei He [1,2,3], Linqing Wang [1,2,3,*], Jiarui Wang [1,2,3], Kaixiang Shen [4,5], Hengfu Xiang [6], Jintang Wang [7], Haowen Chen [1,2,3], Benchong Xu [1,2,3], Rulei Qin [1,2,3] and Guole Yin [1,2,3]

1. Institute of Exploration Techniques, Chinese Academy of Geological Sciences, Langfang 065000, China; hguolei@mail.cgs.gov.cn (G.H.); wjiarui@mail.cgs.gov.cn (J.W.); chaowen@mail.cgs.gov.cn (H.C.); xbenchong@mail.cgs.gov.cn (B.X.); qrulei@mail.cgs.gov.cn (R.Q.); yguole@mail.cgs.gov.cn (G.Y.)
2. Technology Innovation Center for Directional Drilling Engineering, Ministry of Natural Resources, Langfang 065000, China
3. Innovation Base for Automatic and Intelligent Drilling Equipment, Geological Society of China, Langfang 065000, China
4. Guangzhou Marine Geological Survey, China Geological Survey, Guangzhou 511458, China; shenkaixiang@mail.cgs.gov.cn
5. National Engineering Research Center of Gas Hydrate Exploration and Development, Guangzhou 511458, China
6. College of Mechanical and Electronic Engineering, China University of Petroleum (East China), Qingdao 266580, China; hfxiang@upc.edu.cn
7. School of Petroleum Engineering, China University of Petroleum (East China), Qingdao 266580, China; wangjintang@upc.edu.cn
* Correspondence: wanglinqing@mail.cgs.gov.cn

Abstract: A landing string is directly exposed to seawater and subjected to significant stresses and complex deformations due to environmental loads such as wind, waves, and ocean currents during the phase in which the drill string carries the casing to the wellhead. Meanwhile, as the water depth increases, the weight of the drill string increases, leading to an increase in the tensile loads borne by the drill string, which can easily cause a risk of failure. Therefore, a quasi-static load calculation model for the deepwater insertion of the pipe column was established. Using the Ansys platform, simulations were conducted for average wind, wave, and ocean current conditions during different months throughout the year. The ultimate loads and stress distributions of the string were derived from theoretical analyses and numerical simulations for different operational sea states, and the suggested safe operating window and desired BOP trolley restraining reaction force for landing strings' lowering are given according to the existing industry standards. The research findings can help in identifying the potential risks and failure modes of the deepwater landing string under different working conditions.

Keywords: deepwater; landing string; ocean current load; lateral displacement; Mises stress; ultimate tensile load

1. Introduction

During the well-completion testing phase of natural gas hydrates, it is necessary to lower various instruments, pipelines, and cables below the testing string. As the entire string is directly exposed to seawater during the lowering process, the vortex-induced lift and drag forces generate periodic lateral oscillations, which create instability in the testing string and fixed pipelines when subjected to wave action. As a result, collisions and interference occur, causing damage to the fixed-coupled testing pipelines, instrument cables, and other components. These damages significantly impede the progress of well-completion testing. An urgent investigation is required to examine the lateral displacement and stress distribution of the gas hydrate testing column under extreme operating conditions at sea.

This investigation aims to effectively limit the displacement and buffer collision forces of the testing column caused by wave action, ultimately providing a theoretical foundation for calculating the load when inserting the hydrate completion into the testing column. An inadequate handling capacity to accommodate the sinking of the wellhead has been consistently observed during deepwater drilling operations in Brazil, West Africa, and Gulf of Mexico, leading to the abandonment of wellbores [1,2].

A great deal of research has been devoted to the analysis of the forces and loads to which the landing lines are subjected. Azar et al. introduced the notion of quasi-static environmental loads [3]. In 2005, Everage et al. presented mathematical models for predicting the dynamic axial loads imposed on the deepwater landing string by the response of the drilling vessel to the oceanic wave action. The model could be adapted to a given landing string's geometric and mechanical constraints [4]. Bradford et al. used five sets of deepwater landing string hangers to perform caving tests on a tubular column, including elastic load tests near yield strength and overload tests beyond yield strength, to determine wellhead stresses on a tubular column during deepwater drilling when lowering a heavy tubular string [5]. Zhang H. et al. proposed a tension load-based design method for the descent riser [6]. Gao D. et al. considered the characteristics of deep-water environmental loads and non-segregating water pipe operations, mainly focusing on axial static loads for design. As a result, they developed static strength design and dynamic verification methods for inserting drill pipes into deep-water non-segregating water pipes [7–9]. Guan Z. et al. developed a calculation model for quasi-static loads on drill pipes in a non-segregating water environment, based on the Euler–Bernoulli beam theory. They utilized the weighted residual method to solve and analyze the effects of marine environmental loads, drilling ship offset motion, and the gravity of the bottom casing column on the forces, deformation, and strength of the drill pipe, thus offering optimized structural solutions [10]. He Li et al. researched the impact of the dynamic tensile load caused by the vertical movement of a drillship on the landing string. They created a mathematical model that considers the Bottom Hole Assembly (BHA) and Guide Base Plate (GBP) effects, aspects frequently overlooked by other scholars [11]. Liu [12], Dutta [13], Karampour [14], and Zhao [15] addressed the impact of factors such as wind, waves, and constraints on the forces, deformations, and strength of the inserted pipe column during deepwater drilling. They presented numerical modeling, nonlinear dynamic analysis, and experimental assessments to understand the behavior and failure risks of the landing string in offshore environments. In addition, some scholars, taking into account the characteristics of deepwater environmental loads and the operational conditions of unsegregated water pipes, have developed a design approach primarily based on axial static loads. This approach focuses on the static strength design and dynamic verification of the deepwater unsegregated water pipe descent riser, leading to the acquisition of certain regular insights [16–20].

Although there are numerous reports on the landing string design from foreign sources, little attention has been focused on the analysis of the constraints on the lateral movements of the rope and the impact on the tension of the rope during the descent of support [21]. Consequently, it is imperative to undertake a thorough mechanical analysis to examine the load characteristics associated with landing strings.

The influence of marine environmental loads and gravity on the bottom casing column is comprehensively examined in this study. Utilizing theories on column mechanics and marine environmental load, the problem is solved using Ansys software 19.2. Through the analysis of various factors using engineering examples, the study investigates how these factors impact the forces, deformations, and strength safety of the inserted casing column during deep-water drilling. This research provides a theoretical foundation for the load calculations and strength design of the inserted casing column in deep-water drilling.

2. Landing Pipe Column-Lowering Condition and Environmental Load Analysis

Figure 1 illustrates the connection of the hydrate test string to the derrick of the drilling pipe at the upper end and the running tool and casing string at the lower end. During the running process, the test string lacks lateral compensation and comes in contact with seawater. It experiences various loads including wind force, gravity, buoyancy, ocean current force, wave force, frictional force, and internal and external pressure from the casing string. Consequently, the test string undergoes lateral bending deformation. As water depth increases, the exposed test string encounters a more complex load environment from both seawater and the entire running string. The inability to mitigate string deformation through tensioning leads to the generation of significant bending stresses, thereby endangering string safety and impacting running operations. To tackle the issues of load, deformation, and strength safety encountered by the test string in deep water, it is necessary to develop a mathematical model for calculating the load [22]. The following assumptions are made: (1) the string is a slender rod along its axis, disregarding joints and variations in cross-section; (2) the string material is linearly elastic, homogeneous, and isotropic; (3) the heave motion of the drilling vessel and the variation in the position of the running string along the vertical direction are disregarded, along with the lifting force of ocean currents; (4) the influence of the dynamic effects, such as the dynamic response of the landing string under wave-induced vibrations, are not considered.

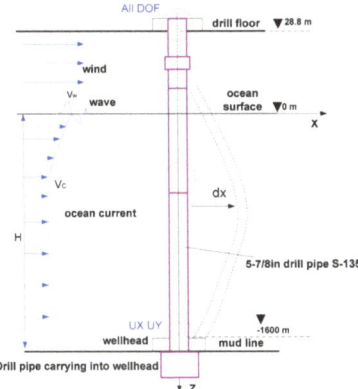

Figure 1. Schematic diagram of the landing pipe column subjected to wind and wave and ocean current loads.

The deformation control equation of the test string can be established according to the schematic diagram in Figure 1, which shows the string subjected to wind, wave, and current. This equation is derived from the Euler–Bernoulli beam theory and is given as follows:

$$EI\frac{d^4y}{dx^4} - T(x)\frac{d^2y}{dx^2} - W\frac{dy}{dx} = F(x) \quad (1)$$

In the equation, E represents the modulus of elasticity, Pa; I represents the moment of inertia of the cross-section, m^4; $T(x)$ represents the effective axial load on the cross-section of the column at position x, N; W represents the gravitational force on the unit length of the column submerged in seawater, N; $F(x)$ represents the wave-induced load on the unit length of the column at position x, in N.

$$T(x) = T_0 + \int_0^x [W - T_1]dx \quad (2)$$

Among them,

$$T_1 = \frac{\pi}{4}D^2\rho_w g - \frac{\pi}{4}d^2\rho_m g \quad (3)$$

$$W = \frac{\pi}{4}\left(D^2 - d^2\right)\rho_s g + \frac{\pi}{4}d^2\rho_m g - \frac{\pi}{4}D^2\rho_w g \qquad (4)$$

In the equation, T_0 represents the tensile load applied at the bottom of the column, N; T_1 represents the virtual tensile force induced by the pressure difference between the internal and external static liquid columns on the unit length of the column, N; D and d represent the outer and inner diameters of the column, respectively, m; ρ_s, ρ_m, and ρ_w represent the densities of the column material, drilling fluid, and seawater, respectively, kg/m^3.

Lateral forces acting on the column are primarily induced by the combined effects of waves and ocean currents. Neglecting the dynamic influences, the resultant force, known as the wave–current load, can be determined using the formula provided below:

$$F(x) = \frac{1}{2}C_D\rho_w D(v_w + v_c)|v_w + v_c| + \frac{\pi}{4}C_m\rho D^2 \frac{\partial v_c}{\partial t} \qquad (5)$$

In the equation, C_D represents the drag coefficient, v_w represents the velocity of ocean currents at a specific depth beneath the surface of the sea, m/s; v_c represents the horizontal velocity of waves, m/s; C_m represents the coefficient of inertial force of seawater. As per the "Riser Analysis Design Premises CNOOC (10001089936-PDC-000)", the hydrodynamic parameters for the stability analysis of the column are as follows [23]: drag coefficient is 1.2 for depths ranging from 0 to 150 m below the sea surface, and 0.7 for depths from 150 m to the seabed; the coefficient of inertial force of seawater is 2.0. The hydrodynamic load refers to the resistance imposed by the movement of water on a length of conduit:

$$f_c(x) = \frac{1}{2}C_D\rho_w D v_w^2(x) \qquad (6)$$

The wave load is caused by the relative motion between the wave water quality point and the landing column. Since the landing column is a small-diameter component with a diameter-to-wavelength ratio of much less than 0.2, Morison's equation and Stokes' fifth-order wave theory are used to calculate the wave load per unit length of the landing column [13], i.e.,

$$f_{wc}(x) = \frac{1}{2}C_D\rho_w D v_c|v_c| + \frac{\pi}{4}C_w\rho_w D^2 \frac{\partial v_c}{\partial t} \qquad (7)$$

Airy's linear micro-amplitude wave theory [24] was employed to compute the horizontal velocity and acceleration at the wave's water quality point, i.e.,

$$\begin{cases} v_c = \frac{\pi H_b ch(kx)}{T sh(kH)} \cos(ky - wt) \\ a = \frac{\partial v_c}{\partial t} = \frac{2\pi^2 H_b ch(kx)}{T^2 sh(kH)} \sin(ky - wt) \end{cases} \qquad (8)$$

In the equation, a represents the horizontal acceleration of water particles in waves, m/s^2; H_b represents the wave height, m; T represents the wave period, s; k represents the wave number within 2π length; w represents the angular frequency of the surface wave; t represents the moment when the wave takes effect, s; $ky - wt$ represents the phase angle of the surface wave.

Above the sea surface, hydrate test columns are also subjected to wind loads. Assuming the projected area of the casing that withstands wind pressure is A the total wind force can be expressed as follows:

$$F_w = KK_Z P_0 A \qquad (9)$$

In the equation, F_w represents the wind load, N; K represents the wind load shape factor, which is taken as 0.5; K_Z represents the coefficient of wind pressure height variation on the sea, which is taken as 1.0; A represents the wind-exposed area, m^2; P_0 represents the basic wind pressure, Pa, and $P_0 = 0.613 V_t^2$, where V_t represents the design wind speed, m/s.

3. Analysis of Axial Tensile Strength of Pipe Columns

According to the specifications outlined in "GB/T 19830-2011 Steel Pipes for Oil and Gas Wells Casing or Tubing [25]", the chosen drill string for insertion into the pipe column has a diameter of 5-7/8 inches and a steel grade of S-135. The drill string's outer diameter measures 149.2 mm, with a wall thickness of 15.25 mm and a density of 7850 kg/m^3. It has a tensile strength of 5974 kN, a yield strength of 930.83 MPa, an elastic modulus of 210 GPa, and a Poisson's ratio of 0.3. In contrast, the casing itself has a diameter of 9-5/8 inches, an outer diameter of 244.48 mm, and a wall thickness of 11.99 mm. The casing exhibits a tensile strength of 8450 kN, a yield strength of 758.42 MPa, an elastic modulus of 210 GPa, and a Poisson's ratio of 0.3. The drill pipe transports the casing into the wellbore. The length of the drill pipe measures 1628.8 m, with a corresponding depth of 1600 m in seawater. According to the recommended guidelines outlined in GB/T 24956-2010, which pertain to the design and operational limits of drilling columns within the oil and gas industry, the tensile strength of the pipe column is calculated and verified using an ultimate tensile load model to assess its safety [26,27]. The particular operational scenario being examined entails inserting the casing-carrying drill pipe into the wellbore, utilizing a 1628.8 m 5-7/8 inch drill pipe paired with a 400 m 9-5/8 inch casing. The maximum static tensile load on the pipe column is deemed a critical condition. By utilizing Equation (10), the buoyant weight of both the pipe column and casing column can be determined:

$$P = K_b \left(L_{dp} \cdot W_{dp} + L_c \cdot W_c \right) \tag{10}$$

In this equation, P denotes the buoyant weight of the drilling string located beneath the drill pipe, expressed in kilonewtons (kN); L_{dp} stands for the length of the drill pipe, measured in meters; W_{dp} represents the linear weight of the drill pipe in air, set at 0.69727 kN/m; L_c indicates the length of the casing, expressed in meters; W_c denotes the linear weight of the casing in air, set at 0.68551 kN/m; and K_b signifies the buoyancy factor, which is considered as 0.869.

According to the API standards [28], the specified theoretical tensile strength does not indicate the precise point where the material starts to show permanent deformation. Instead, it signifies the stress threshold at which a specific level of deformation has already taken place. This deformation comprises both elastic and partial plastic (permanent) deformation. If the drill pipe is subjected to a load exceeding the limit presented in the table, it may undergo a slight permanent elongation, leading to challenges in retaining its straightness. During the verification of design, the pipe column's maximum permissible tensile load is established at 90% of its theoretical tensile strength. This measure is taken to prevent the occurrence of the aforementioned situation:

$$P_a = 0.9 P_t \tag{11}$$

In the equation, P_a denotes the maximum acceptable design tensile load, measured in kilonewtons (kN); P_t stands for the theoretical tensile strength. The margin of pull (MOP) is defined as the disparity between the computed ultimate tensile load P and the maximum permissible tensile load P_a.

$$MOP = P_a - P \tag{12}$$

The ratio between the maximum allowable tensile load P_a and the calculated ultimate tensile load P is referred to as the safety factor SF.

$$SF = \frac{P_a}{P} \tag{13}$$

The calculation of the tensile load when running the 9-5/8 inch casing with a 5-7/8 inch drill pipe can be approached using two methods: the safety factor method, with a tension safety factor (SF) of 1.3, and the tension margin method, with an overpull (MOP) of 500 kN. During the operational phase, as shown in Figure 2, the curve of the tensile load illustrates how the drill pipe supports the casing at the wellhead. This figure provides an overview of three distinct pipe strength design approaches: ultimate tensile load, tension margin method, and safety factor method. The maximum tensile load for the conduit, as determined using the safety factor method, is calculated to be 2220.1 kN. This value is significantly lower than the maximum allowable design tensile load of 5376.6 kN.

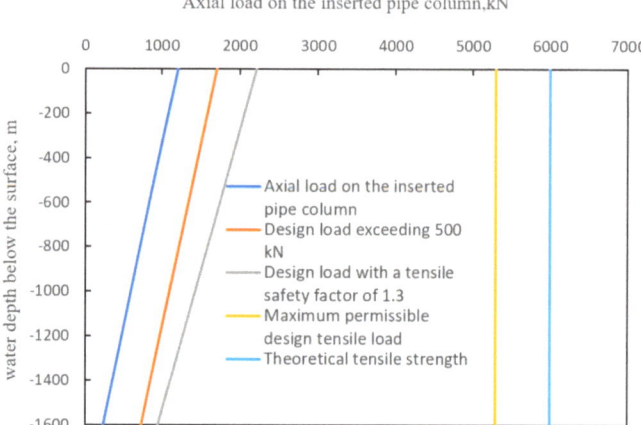

Figure 2. Tensile load curve of tubular column in drill pipe while casing is being carried into the wellhead stage.

4. Simulation Analysis of Landing String

The hydrate test column depicted in Figure 1 demonstrates the incorporation of the casing column into the wellbore through the drill pipe. At a specific location, precisely 28.8 m from the upper end, the column establishes contact with the drill floor, thereby serving as a fixed constraint. This particular position not only supports the weight of the column but also endures the lateral forces induced by wind and waves. It represents an area characterized by high levels of stress concentration. Above sea level, the column is influenced by wind forces, whereas from sea level downwards to the wellhead, it undergoes the combined impact of sea currents and waves. As a result, lateral displacement occurs, resulting in significant stress accumulation that could lead to column failure.

The numerical simulation will be conducted on the ANSYS software 19.2 platform using the Pipe59 element, which can withstand tensile, compressive, and bending forces. Additionally, this element allows for the simulation of uniaxial ocean waves and water currents.

A thorough investigation was conducted over the course of several years to calculate the impact of oceanic environmental loads on the forces, deformations, and stresses experienced by the riser. In this study, an average environmental load in the operational area was selected for analysis. Five specific load conditions, as presented in Table 1, were chosen for further examination.

Table 1. Distribution of wind and wave current loads for five operating conditions.

Operating Condition	Surface Flow Velocity (m/s)	Wind Speed (m/s)	Effective Wave Height	Wave Period (s)
1	0.08	4.3	1.4	5
2	0.16	5.3	1.6	5.2
3	0.2	4.4	1.3	4.7
4	0.25	7.4	1.7	5.4
5	0.345	9.4	2.2	5.9

Figure 3 presents the lateral displacement and Mises stress distribution of the riser under different operating conditions. Specifically, Figure 3a,b demonstrate the lateral displacement and Mises stress pattern under operating condition 4, whereas Figure 3c,d showcase the lateral displacement and Mises stress distribution under operating condition 5. The figures indicate that the surface velocity in operating condition 4 is lower than that in operating condition 5. Moreover, the lateral displacement and Mises stress in operating condition 4 are considerably smaller compared to operating condition 5. According to Formula 6, the lateral force caused by seawater is proportional to the square of the surface velocity. Hence, an increase in surface velocity leads to a rise in lateral displacement and Mises stress, thereby increasing the likelihood of failure for the riser.

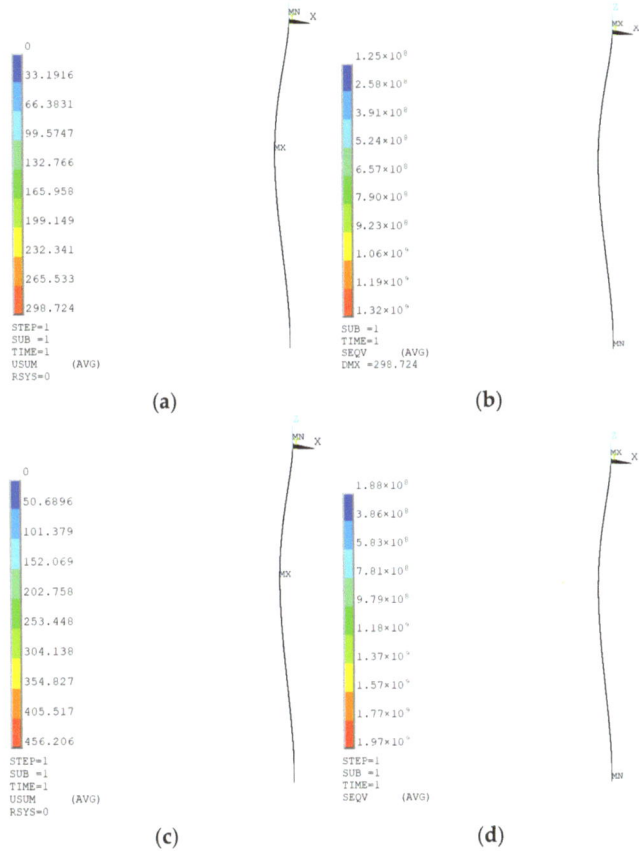

Figure 3. Distribution of lateral displacement and maximum Mises stress in drill pipe entry stage. (**a**) The lateral displacement under operating condition 4; (**b**) the Mises stress under operating condition 4; (**c**) the lateral displacement under operating condition 5; (**d**) the Mises stress under operating condition 5.

Figure 4 illustrates the correlation between the depth of seawater and the lateral displacement deformation of the drill string across five distinct operational scenarios. It can be observed from Figure 4 that the lateral displacement of the drill string increases with increasing water depth, reaching a maximum deformation at around a water depth of 645 m. This is because the drill string experiences lateral constraint at the seabed, and the velocity of the ocean current gradually decreases with increasing water depth, leading to a corresponding reduction in the ocean current load. This indicates that the lateral displacement input to the drill string is highly sensitive to ocean current velocity, with greater ocean current velocities resulting in larger lateral displacements of the drill string. However, wave loads have almost no effect on the lateral deformation and stress of the drill string. With higher ocean current velocities, the bending moment at the top of the drill string increases, as shown in Figure 5. In the presence of consistent ocean current velocities, wave loads exert a growing influence on the bending moment experienced at the upper end of the drill string. This underscores the substantial impact of wave loads on the bending moment transmitted to the upper section of the drill string.

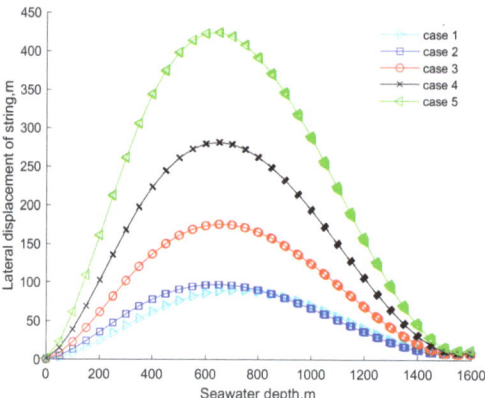

Figure 4. Relationship between seawater depth and lateral displacement of landing string.

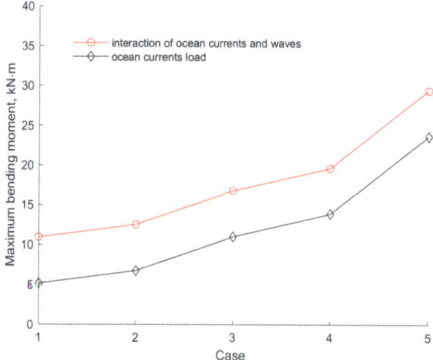

Figure 5. Effect of wave current loading on the bending moment at the top of the landing string.

Due to the fact that the top of the drill string is always fixedly suspended on the drilling rig floor, and the most likely collision point is in the moon pool area, it is proposed to use the BOP (Blowout Preventer) trolley in the mother ship's moon pool area as a device to carry the test drill string's stability protection system. The BOP trolley shown in Figure 6 is approximately 9 m long, 6 m wide, and designed with a safe load of 120 tons. The entire trolley can move freely along the bow-stern direction of the ship, with a travel distance of nearly 20 m. Additionally, there is an approximately 3 m long and 1 m wide opening designed in the middle of the trolley, which can accommodate most diameter drill strings. The trolley with a roller straightening mechanism is positioned at a depth of 15.5 m below sea level, as illustrated in Figure 7. The roller at the 15.5 m depth imposes lateral constraints on the drill string to mitigate the lateral displacement of the casing caused by seawater currents.

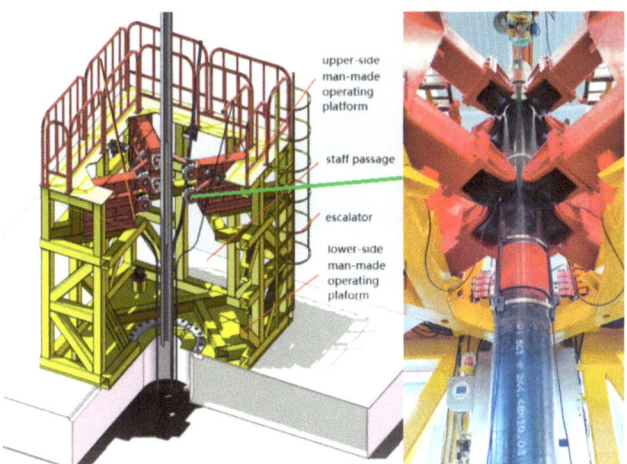

Figure 6. BOP (Blowout Preventer) trolley-constrained casing.

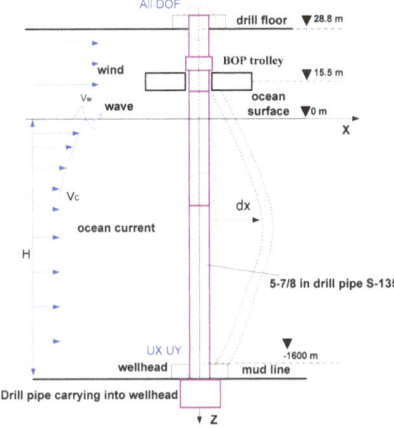

Figure 7. Schematic diagram outlines how wind, waves, and ocean currents impact the riser while constrained by the trolley.

Figure 8 depicts the lateral displacement and distribution of Mises stress in the riser when it is exposed to the lateral constraint of the trolley in working condition 4. Upon the application of the trolley constraint, the maximum lateral displacement of the riser is reduced from 298.772 m (Figure 3a) to 281.308 m, while the maximum Mises stress decreases from 1320 MPa (Figure 3b) to 1290 MPa. Additionally, the maximum stress point shifts from 28.8 m to 15.5 m. The trolley constraint significantly improves the lateral deformation and Mises stress of the riser.

Figure 9 illustrates the substantial impact of the trolley constraint on both the maximum lateral displacement and maximum Mises stress across five distinct working conditions. The trolley constraint successfully mitigates the lateral displacement and maximum Mises stress exerted on the casing. Specifically, the maximum lateral displacement decreases by approximately 5.5%, and the maximum Mises stress decreases by around 3%.

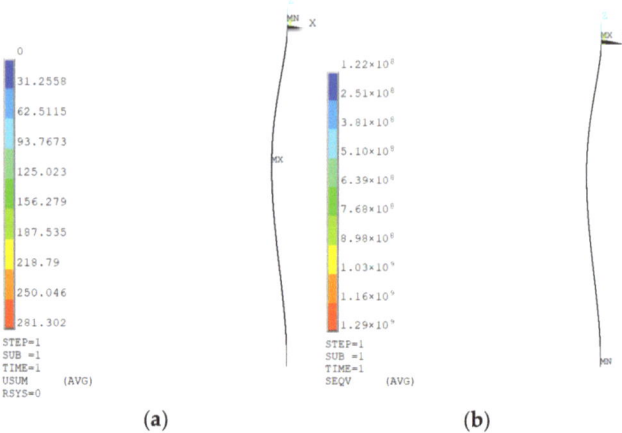

Figure 8. Distribution of lateral displacement and maximum Mises stress in drill pipe entry stage with trolley restraint in Case 4. (**a**) Lateral displacement of the landing string; (**b**) Mises stress of the landing string.

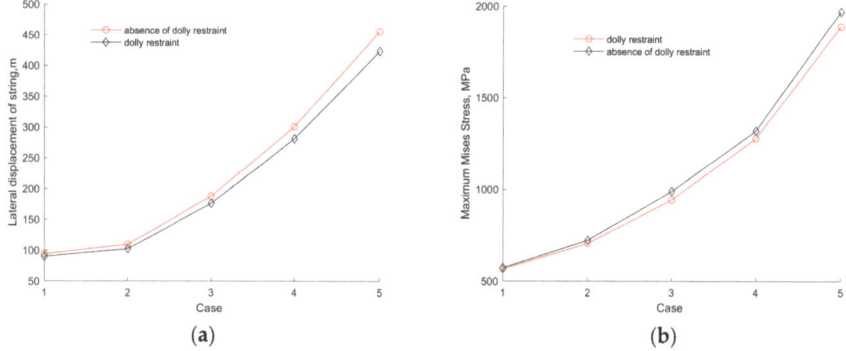

Figure 9. Effect of presence or absence of trolley restraint on maximum lateral displacement and maximum Mises stresses. (**a**) Maximum lateral displacement of the inserted column; (**b**) maximum Mises stress of the landing string.

5. Example Analysis

The first self-operated deepwater gas field in the South China Sea was discovered by an independent exploration of CNOOC (China National Offshore Oil Corporation), which is named LS17-2 block, where the water depth ranges from 1220 m to 1560 m. The deep water floater was determined based on the present offshore industrial technological capability of China and its adaptability to the South China Sea [29]. Statistical data regarding the monthly average wind, wave, and ocean current conditions within the operational area are presented in Table 2. A comprehensive analysis was conducted to assess the structural strength of the drilling riser under the combined effects of wind, waves, and ocean current. The maximum lateral displacement and Mises stress were determined under controlled and uncontrolled conditions, both with and without the use of a BOP trolley. Figure 9 depicts the maximum lateral displacement of the drilling riser under different conditions, while Figure 10 shows the maximum Mises stress. The results from both figures demonstrate that the implementation of a derrick constraint leads to a reduction in both the maximum lateral displacement and Mises stress of the drilling riser, effectively suppressing lateral offset.

Table 2. Statistics on the average environmental load over several years, calculated monthly, near the operational area.

Month	Water Depth (m)	Surface Layer	800 m	1600 m	Wind Speed (m/s)	Wind Direction (°)	Significant Wave Height (m)	Wave Period (s)	Wave Direction (°)
1	Velocity of flow (m/s)	0.254	0.036	0.008	7.8	54.3	1.8	6.4	58.4
	Flow direction (°)	217.2	218.2	237.1					
2	Velocity of flow (m/s)	0.1	0.029	0.006	6.3	71.5	1.5	5.2	67.2
	Flow direction (°)	201	180.3	86.5					
3	Velocity of flow (m/s)	0.08	0.024	0.005	4.8	106.5	1.4	5.2	84.3
	Flow direction (°)	83.5	130.1	92.3					
4	Velocity of flow (m/s)	0.2	0.026	0.006	5.4	135.7	1.3	4.6	90.4
	Flow direction (°)	64.4	175.9	79.8					
5	Velocity of flow (m/s)	0.22	0.035	0.01	5.9	156.1	1.2	4.8	125.7
	Flow direction (°)	55	164	336.4					
6	Velocity of flow (m/s)	0.19	0.021	0.008	4.8	185.7	1.1	4.2	139
	Flow direction (°)	34.9	175	286.2					
7	Velocity of flow (m/s)	0.23	0.036	0.009	4.8	180	1	4.2	160.5
	Flow direction (°)	33.3	185	300					
8	Velocity of flow (m/s)	0.055	0.018	0.007	3.9	195.2	1	4.5	156.9
	Flow direction (°)	26.3	211.2	307.4					
9	Velocity of flow (m/s)	0.017	0.01	0.003	2.4	125.5	1.1	4.4	123
	Flow direction (°)	65	170	297					
10	Velocity of flow (m/s)	0.16	0.018	0.007	4.3	58.9	1.6	5.3	78.7
	Flow direction (°)	222.6	175.5	325.2					
11	Velocity of flow (m/s)	0.282	0.024	0.01	7.5	47.2	1.9	5.6	60
	Flow direction (°)	218.8	203.6	35.3					
12	Velocity of flow (m/s)	0.345	0.026	0.014	9.1	45.4	2.1	5.9	59.4
	Flow direction (°)	221.5	233.4	231					

Based on the API Spec 16 standard recommended practice [17], the equivalent stress on the drillstring carrying the casing should not exceed 0.67 times the yield stress of the drillstring during casing run-in. Considering a casing yield stress of 930.83 MPa, the equivalent stress limit when running in the casing is determined to be 623.6561 MPa, as indicated by the red dashed line in Figure 11.

The simulation findings show that over the course of the year, the average stress levels experienced by the drill string casing entering the wellbore due to wind and waves fluctuate. The drill string casing experiences the minimum Mises stress in February (580 MPa without a trolley constraint and 552 MPa with a trolley constraint). Conversely, the casing experiences the maximum Mises stress in December (1970 MPa without a trolley constraint and 1890 MPa with a trolley constraint). During the October-to-January periods and from

April to July, the Mises stress surpasses the permissible yield strength of the casing. However, imposing a stricter trolley constraint significantly reduces the stress on the drill string casing. For safe casing-lowering operations, it is advisable to carry them out in February, March, August, and September. On the other hand, it is not recommended to perform casing-lowering operations between October and the subsequent January, as well as from April through July.

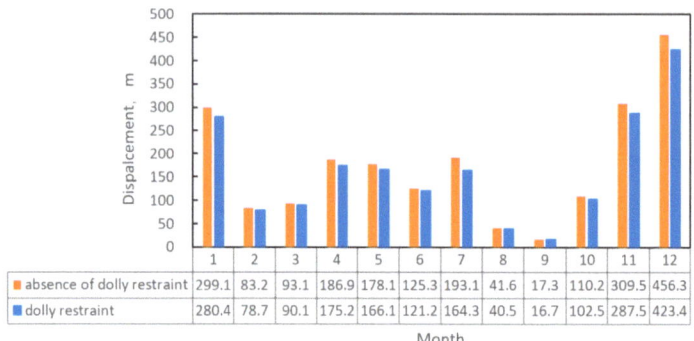

Figure 10. Comparison of maximum lateral displacements from January to December.

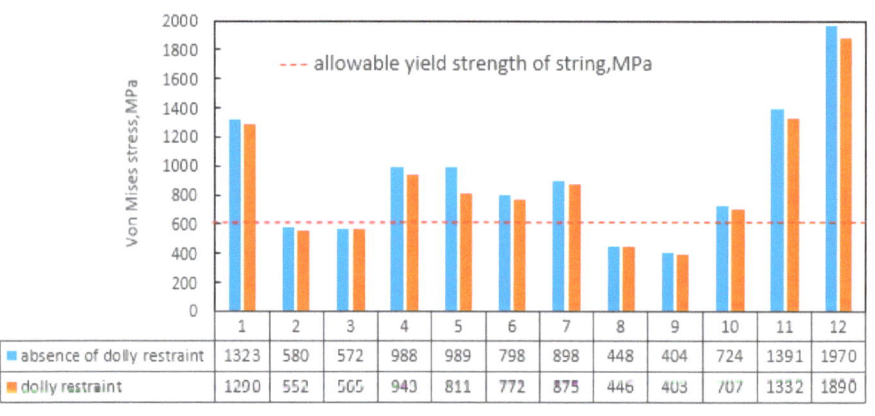

Figure 11. Comparison of maximum Mises stress from January to December.

In conclusion, the suggested safe operational timeframe for the drill pipe transporting casing to the wellhead stage all year round includes February, March, and August, with the surface velocity of flow for safe operations needing to remain below 0.18 m/s.

Figure 12 provides the required restraint reaction force for limiting the lateral displacement of the drill string during the year-round operation when the casing is carried into the wellhead through the BOP trolley. It is worth noting that the lowest constraint counterforce required for the drill pipe operation was in September (15.1 N), and the highest constraint counterforce required for the drill pipe operation was in December (1764.4 N).

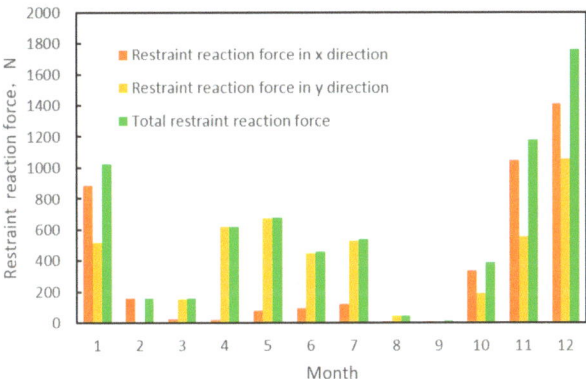

Figure 12. The restraining force exerted by the Blowout Preventer (BOP) trolley is intended to restrict the lateral movement of the drill string.

6. Example Validation

In this section, the results of pilot wells are incorporated as additional cases to validate the accuracy of the simulation results. According to the specifications outlined in "GB/T 19830-2011 Steel Pipes for Oil and Gas Wells Casing or Tubing [24]", the chosen casing of pilot wells has a diameter of 9-5/8 inches, an outer diameter of 244.48 mm, and a wall thickness of 11.99 mm. The length of the casing measures 760 m, with a corresponding depth of 749 m in seawater. Operations and construction were conducted in the South China Sea's western area in August 2020. The construction vessel employed was the Blue Whale I (Figure 13), utilizing a BOP trolley to restrict lateral displacement. The surface current velocity was 0.06 m/s, which was less than 0.18 m/s, the wind speed was 6.5 m/s, and the wave height was 1 m. During the casing run operation, there was no collision with the moon pool. This indicates that the simulation results are reliable.

Figure 13. Operations vessel "Blue Whale I" and BOP trolley.

7. Conclusions

This study focused on the effects of wind, waves, and ocean currents on the lateral displacement and Mises stress of the drill string during deepwater drill string-lowering operations. Based on the Euler–Bernoulli beam theory, a quasi-static load calculation model was developed, and a simulation analysis was performed using the ANSYS platform. Utilizing historical statistical data on the typical monthly wind and wave patterns within the operational zone spanning several years, the impacts of wind, waves, ocean currents, and the BOP trolley restraint on the lateral displacement, Mises stress, and top bending moment of the landing string were evaluated. In accordance with applicable standards, the recommended safe operational window and the results of the BOP trolley design's restraint reaction force are provided for the entire year of the landing string-lowering operation. The findings are as follows:

(1) The safety factor method calculates the landing string's ultimate tensile load as 2220.1 kN, significantly below the maximum allowable design tensile load of 5376.6 kN. Therefore, the ultimate tensile load satisfies the requirements for tensile design.

(2) The recommended safe operation windows are given according to the statistical data from the Lingshui 17-2 block. Under the constraint conditions of the BOP trolley, the lateral displacement reaches a minimum of 17 m in September while it reaches its maximum of 423 m in December. The maximum Mises stress on the landing string in February, March, August, and September remains below the allowable yield strength, rendering it suitable for landing string operations, and the suggested maximum safe working surface velocity is below 0.18 m/s.

While this study did not consider dynamic effects, such as the response of the landing string to vibrations induced by waves, it still holds engineering significance. Future research can delve deeper into exploring the impact of dynamic effects.

Author Contributions: Writing—original draft, methodology, funding acquisition, G.H. and L.W.; methodology, J.W. (Jiarui Wang) and K.S.; writing—original draft, investigation, H.X. and J.W. (Jintang Wang); resources, validation, H.C.; writing—original draft preparation, data curation, R.Q., B.X. and G.Y. All authors have read and agreed to the published version of the manuscript.

Funding: This work was supported by National Key Research and Development Program of China (2021YFC2800803; 2021YFC2800802), Marine Economy Development Foundation of Guangdong Province (GDNRC[2022]44) Technical Support for Stimulation and Testing of Gas Hydrate Reservoirs.

Data Availability Statement: Data will be made available on request.

Conflicts of Interest: The authors declare no conflict of interest. The authors declare that they have no known competing financial interests or personal relationships that could have appeared to influence the work reported in this paper.

References

1. Zhou, C.; Fu, Y.; Zhu, R. Design method and development trend of landing strings in deepwater drilling. *Pet. Drill. Tech.* **2014**, *42*, 1–7.
2. Lu, Q.P.; Yu, Y.J.; Wen-Wei, X.; Liang, J.Q.; Lu, J.A.; Xu, B.C.; Shi, H.X.; Yu, H.Y.; Qin, R.L.; Li, X.C.; et al. Design and feasibility analysis of a new completion monitoring technical scheme for natural gas hydrate production tests. *China Geol.* **2023**, *6*, 466–475. [CrossRef]
3. Azar, J.J.; Soltveitol, R.E. A Comprehensive Study of Marine Drilling Landing Strings. SPE 7200. 1978. Available online: https://www.researchgate.net/publication/35139419_A_comprehensive_study_of_marine_drilling_risers (accessed on 29 February 2024).
4. Everage, D.W.; Zheng, N.; Ellis, S.E. Evaluation of heave-induced dynamic loading on deepwater landing strings. *Spe Drill. Complet.* **2005**, *20*, 230–237. [CrossRef]
5. Bradford, D.W.; Payne, M.L.; Schultz, D.E.; Adams, B.A.; Vandervort, K.D. Defining the limits of tubular-handling equipment at extreme tension loadings. *SPE Drill. Complet.* **2009**, *24*, 72–88. [CrossRef]
6. Hui, Z.; Gao, D.; Tang, H. Landing string design and strength check in ultra-deepwater condition. *J. Nat. Gas Sci. Eng.* **2010**, *2*, 178–182. [CrossRef]
7. Gao, D.; Hui, Z. Mechanical analysis of tubes in deepwater drilling operation without landing string. *Sci. Technol. Rev.* **2012**, *30*, 37–42.
8. Gao, D.; Fang, J.; Wang, Y.B. Optimization analysis of the landing string top tension force in deep water drilling: Aiming at the minimum variance of lower flexible joint deflection angle. *J. Pet. Sci. Eng.* **2016**, *146*, 149–157.
9. Wang, Y.; Gao, D.; Fang, J. Static analysis of deep-water marine landing string subjected to both axial and lateral forces in its installation. *J. Nat. Gas Sci. Eng.* **2014**, *19*, 84–90. [CrossRef]
10. Guan, Z.; Li, J.; Han, C.; Zhang, B.; Zhao, X.; Teng, X.; Sun, B. Loads calculation and strength analysis of landing string during deepwater drilling. *J. China Univ. Pet.* **2018**, *42*, 71–78.
11. Li, H.; Gao, D.; Shi, W. A new mathematical model for evaluating the safe margin of deepwater landing operation. *Appl. Ocean Res.* **2020**, *104*, 102375. [CrossRef]
12. Liu, J.; Zhao, H.; Yang, S.X.; Liu, Q.; Wang, G. Nonlinear dynamic characteristic analysis of a landing string in deepwater riserless drilling. *Shock Vib.* **2018**, *2018*, 8191526. [CrossRef]
13. Dutta, S. Numerical Modeling of Large Deformation Behavior of Offshore Pipelines and Risers in Soft Clay Sea Beds. Ph.D. Thesis, Memorial University of Newfoundland, St. John's, NL, Canada, 2017.

14. Karampour, H.; Alrsai, M.; Khalilpasha, H.; Albermani, F. Experimental and numerical assessment on failure pressure of textured pipelines. *J. Off. Shore Mech. Arct. Eng.* **2022**, *144*, 031802. [CrossRef]
15. Zhao, S.; Yang, J.; Jiang, K.; Song, Y.; Chen, K. Loads calculation and strength calculation of landing string during deepwater drilling. *Energies* **2023**, *16*, 4854. [CrossRef]
16. Jellison, M.J.; Chan, A. Advanced Technologies and Practical Solutions for Challenging Drilling Applications. In *IADC/SPE Asia Pacific Drilling Technology Conference*; SPE-170566-MS; OnePetro: Calgary, AB, Canada, 2014.
17. Qin, K.; Di, Q.; Zhou, X. Nonlinear dynamic characteristics of the drill-string for deep-water and ultra-deepwater drilling. *J. Pet. Sci. Eng.* **2021**, *209*, 109905. [CrossRef]
18. Wang, X. Research on technology and application of tubing hanger landing string in deepwater well completion. *Ocean Eng. Equip. Technol.* **2020**, *7*, 352–355.
19. Tang, Y.; Wu, J.; Luo, Y.; Li, W. Development status of landing string control system and research difficulties for its localization. *China Offshore Oil Gas* **2022**, *34*, 138–145.
20. Zhang, M.H.; Yang, J.; Yang, Y.X.; Xu, D.S.; Zhou, Y.S. Study on load bearing characteristics of novel expandable deepwater drilling conductor based on la-boratory experiment and field test. *China Ocean. Eng.* **2023**, *37*, 16–28. [CrossRef]
21. Zheng, N.J.; Bakerjm, J.M.; Everage, S.D. Further Considerations of Heave-Induced Dynamic Loading on Deep Water Landing Strings. In *SPE/IADC Drilling Conference and Exhibition*; Curran Associates Inc.: Red Hook, NY, USA, 2005; p. SPE-92309.
22. *ISO 13625*; Petrolrum and Natural Gas Industries-Drilling and Proudction Equipment-Maring Drilling Riser Coupling. International Organization for Standardization: Geneva, Switzerland, 2002.
23. Wang, J.; Qin, R.; Feng, Q.; Chen, H.; Xu, B.; Lu, Q.; Liu, X. Design and modal analysis of wellhead suction module for ocean drilling. *Drill. Eng.* **2023**, *50*, 45–55.
24. Yu, J.; Wang, F.; Yu, Y.; Li, H.; Liu, X.; Sun, R. Study on fatigue spectrum analysis and reliability analysis of multilayer flexible riser. *J. Mar. Sci. Eng.* **2022**, *10*, 1561. [CrossRef]
25. *GB/T 19830-2011*; Petroleum and Natural Gas Industries-Steel Pipes for Use as Casing or Tubing for Wells. International Organization for Standardization: Geneva, Switzerland, 2011.
26. *GB/T 24956-2010*; Recommended Practice for Petroleum and Natural Gas Industries-Drill Stem Design and Operating Limits. International Organization for Standardization: Geneva, Switzerland, 2010.
27. American Petroleum Institute. *2RD: Design of Risers for Floating Production Systems and Tension-Leg Platforms*; American Petroleum Institute: Washington, DC, USA, 1998.
28. *API RP 16Q*; Recommended Practice for Design Selection Operation and Maintenance of Marine Drilling Riser System. American Petroleum Institute: Washington, DC, USA, 2001.
29. Zhu, H.; Li, D.; Wei, C.; Li, Q. Research on LS17-2 deep water gas field development engineering scenario in South China Sea. *China Offshore Oil Gas* **2018**, *30*, 170–177.

Disclaimer/Publisher's Note: The statements, opinions and data contained in all publications are solely those of the individual author(s) and contributor(s) and not of MDPI and/or the editor(s). MDPI and/or the editor(s) disclaim responsibility for any injury to people or property resulting from any ideas, methods, instructions or products referred to in the content.

Article

Study on the Mechanical and Permeability Characteristics of Gypsum Rock under the Condition of Crude Oil Immersion

Tingting Jiang, Xiurui Shang *, Dongzhou Xie, Dairong Yan *, Mei Li and Chunyang Zhang

School of Resources and Environment Engineering, Wuhan University of Technology, Wuhan 430070, China; ttjiang@whut.edu.cn (T.J.); xiedongzhou20@163.com (D.X.); mli@whut.edu.cn (M.L.); cyzhang@whut.edu.cn (C.Z.)
* Correspondence: sxr0505@whut.edu.cn (X.S.); dryan@whut.edu.cn (D.Y.)

Abstract: The utilization of gypsum mine goaf (GMG) for strategic oil reserves can realize the coordinated development of mining and oil storage. However, the variation in mechanical and permeability characteristics of gypsum rock under the action of crude oil erosion is not clear. At the same time, the deformation of gypsum rock caused by crude oil erosion will pose a threat to wellbore integrity. In this paper, a series of tests were carried out on gypsum rock before and after crude oil immersion to explore the effect of crude oil erosion on the mechanical and permeability characteristics of gypsum rock. The results show that crude oil soaking enhances the plastic deformation ability of gypsum rock. After soaking, the cohesion of gypsum rock increases by 14%, but the internal friction angle decreases by 7.2%. During the soaking process, crude oil invades the pores of gypsum rock, which can reduce the value of gypsum rock by 10^{-20} m^2. Crude oil immersion enhances the deformation resistance of gypsum rock surrounding rock and significantly reduces the permeability, which is conducive to the stability and sealing of gypsum rock goaf during oil storage. The research results are helpful in deepening the understanding of using GMG technology to construct crude oil storage and provide inspiration for the study of the influence of gypsum rock deformation on wellbore integrity under crude oil erosion.

Keywords: gypsum mine goaf; strategic oil storage; mechanical properties; permeability

Citation: Jiang, T.; Shang, X.; Xie, D.; Yan, D.; Li, M.; Zhang, C. Study on the Mechanical and Permeability Characteristics of Gypsum Rock under the Condition of Crude Oil Immersion. *Energies* **2024**, *17*, 1712. https://doi.org/10.3390/en17071712

Academic Editors: Manoj Khandelwal and Krzysztof Skrzypkowski

Received: 26 February 2024
Revised: 28 March 2024
Accepted: 30 March 2024
Published: 3 April 2024

Copyright: © 2024 by the authors. Licensee MDPI, Basel, Switzerland. This article is an open access article distributed under the terms and conditions of the Creative Commons Attribution (CC BY) license (https://creativecommons.org/licenses/by/4.0/).

1. Introduction

China is rich in gypsum mineral resources. By the end of 2017, 23 provinces or autonomous regions in China had gypsum mine output (Figure 1a). There are 169 mining areas with proven reserves, with total reserves of 98.5 billion tons [1]. The gypsum mine has a shallow buried depth and high backfilling cost. If gypsum mine goaf (GMG) is not effectively utilized and maintained, it may cause geological disasters such as land subsidence and surface subsidence. Making full use of the GMG and reducing the risk of geological disasters are key problems that need to be solved urgently in mining engineering.

It is an effective way to prevent large-scale geological disasters in GMG by using GMG for the oil strategic reserve (Figure 1c). Gypsum rock has the characteristics of low permeability and small porosity, which is a good sealing layer for many large oil and gas reservoirs [2]. For example, many super-large oil and gas fields in the world, such as the Gulf of Mexico, the Persian Gulf, and the Caspian Sea, are formed under gypsum layers [3–6]. At present, the deterioration mechanism of gypsum rock mechanics and permeability characteristics under crude oil immersion is not clear, and the stability and sealing of GMG under oil storage conditions are difficult to evaluate. Xiaxin gypsum mine has formed a number of gypsum mine goaves after mining, so it is necessary to find a way to reuse gypsum mine goaf. The Xiaxin gypsum mine is now studying the use of gypsum mine goaf for oil storage. Therefore, this paper takes the Xiaxin gypsum mine in Huangmei County, Hubei Province, as an example to study the variation law of mechanical properties and permeability characteristics of gypsum mine surrounding rock under the condition of crude oil immersion, which lays a theoretical foundation for the reconstruction of oil storage in GMG.

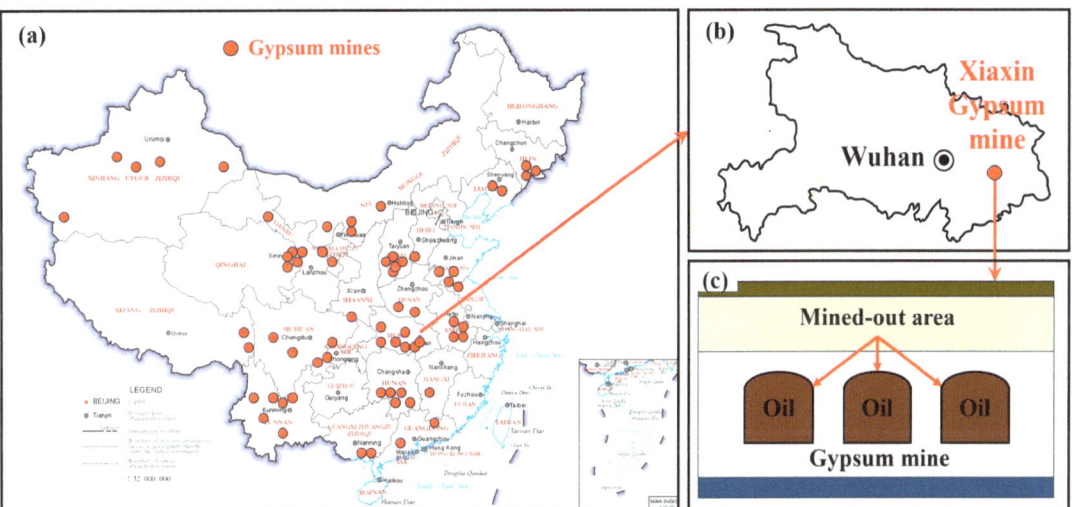

Figure 1. Xiaxin gypsum mine location and Oil storage diagram of GMG: (**a**) Xiaxin gypsum mine location in China; (**b**) Xiaxin gypsum mine location in Hubei Province; (**c**) the oil storage diagram of GMG.

Obtaining the mechanical properties of gypsum rock under various conditions is the prerequisite for evaluating the stability of GMG. Through a series of laboratory tests, the uniaxial compressive strength, elastic modulus, Poisson's ratio, and stress–strain curve characteristics of gypsum rock are analyzed [7–10]. The results show that the mechanical properties of gypsum rock are significantly different due to the gypsum rock's heterogeneity. The strength of gypsum rock is also related to factors such as water content and impurity content. Liu et al. [11] carried out mechanical tests on gypsum mudstone with different gypsum volume fractions. They found that the higher the volume fraction of anhydrite, the greater its compressive strength. The higher the water content, the greater the instantaneous and long-term creep deformation rate of gypsum rock; additionally, the water content has a significant effect on the long-term deformation of gypsum rock [12,13]. Auvray et al. [14] found that relative humidity is the main factor affecting the volume shrinkage of the gypsum mines by analyzing the stability of the Grozon gypsum mine in France. In order to further explore the erosion effect of water and other fluids on gypsum rock, the mechanical test of gypsum rock under water immersion conditions has been widely carried out [15–17]. The results show that the soaking effect of water not only significantly reduces the compressive strength and elastic modulus of gypsum rock but also changes the failure mode of gypsum rock, which has a negative impact on the stability of GMG. Wang et al. believed that the oil storage in GMG has good stability and tightness [18–22]. However, in the process of oil storage in the GMG, the deterioration of the mechanical strength of gypsum rock under the condition of crude oil immersion may lead to the instability of the goaf. In order to explore the change law of mechanical properties of gypsum rock during oil storage, the mechanical properties test of oil-immersed gypsum rock under different confining pressures was carried out [23–25]. The change rate of uniaxial compressive strength of gypsum rock under different oil immersion days is $-10\%\sim+33\%$. Mao found that the triaxial compressive strength of gypsum rock decreased by 26.4% after oil immersion. The above research describes the deterioration law of mechanical properties of gypsum rock under the action of crude oil erosion, but it still has the following shortcomings: (1) The erosion rate of crude oil to gypsum rock may change under the influence of temperature and pressure, but at present, no detailed research has been conducted. (2) The operation time of oil storage is very long, and the influence of long-term erosion on gypsum rock has not been studied. (3) Under high pressure, the infiltration of crude oil may change the permeability characteristics of gypsum rock, and no scholars have conducted in-depth research on this.

Therefore, the main purpose of this paper is to study the mechanics and permeability characteristics of gypsum rock under oil storage conditions. Firstly, the core of the Xiaxin gypsum mine in Huangmei County, Hubei Province, China, was obtained, and the gypsum rock immersion test scheme was formulated. Secondly, a series of rock mechanics tests of gypsum rock before and after crude oil immersion were carried out, and the influence of crude oil immersion on the mechanical properties of gypsum rock was analyzed. Third, the permeability and pore distribution characteristics of gypsum rock before and after crude oil immersion were analyzed. The research results can provide a theoretical basis for the use of GMG for crude oil reserves.

2. Preparation of Samples and Tests

2.1. Sample Preparation

Gypsum samples were taken from the Xiaxin gypsum mine (Figure 1b) in Huangmei County, Hubei Province, with sampling depths of 395.21–399.72 m and 404.36–409.48 m. Using the water-based grinding method, multiple sets of samples with a diameter of 50 mm and height of 100/25 mm were prepared for testing; some samples are shown in Figure 2. The prepared samples are the same batch of samples taken from the same place at the same time, and it is found that there is a high similarity among the samples. Therefore, the gypsum rock samples were randomly divided into an oil-soaked group and a non-soaked group to ensure that the two groups of samples could represent the surrounding rock of the gypsum rock goaf. A total of 59 samples were prepared: 32 soaked samples, including 27 samples with a height of 100 mm and 5 samples with a height of 25 mm, and 27 samples without crude oil soaking, including 22 samples with a height of 100 mm and 5 samples with a height of 25 mm. With the exception of the Brazilian splitting test, which required samples with a height of 25 mm, the other tests used samples with a height of 100 mm. The uniaxial compression test uses 4 samples without crude oil soaking and 4 after soaking, the triaxial compression test uses 15 samples without crude oil soaking and 15 after soaking, the Brazilian splitting test uses 5 samples without crude oil soaking and 5 samples after soaking, and the penetration test uses 3 samples without crude oil soaking and 3 samples after soaking. The size of the prepared samples was measured, and it was found that the diameter of the prepared sample was between 49.83 mm and 50.34 mm, with an error between −0.34% and 0.68%, and the height of the sample required for the Brazilian splitting test was between 24.91 mm and 25.08 mm, with an error between −0.36% and 0.32%. The height of the samples required for other tests is between 99.76 mm and 100.12 mm, and the error is between −0.24% and 0.12%, and it is found that the size difference between the prepared samples is slight. All samples meet the requirements of the rock mechanics test [26]. The composition and content of gypsum rock samples were quantitatively analyzed by X-ray diffractometer. The X-ray diffractometer uses a closed X-ray tube, the maximum tube current is 80 mA, the minimum step angle is 0.0001, the angle reproducibility is 0.0001, and the test temperature range is room temperature ~1600 °C. The sample can be placed on the carbon fiber translation axis, and the three-dimensional full-diameter scanning can be realized by rotating the ray source and moving the translation axis. The results are shown in Figure 2b. Anhydrite is the main component of the sample, with an average content of 96.32%. Montmorillonite and gypsum are secondary components, with an average content of 2.57% and 1.11%

 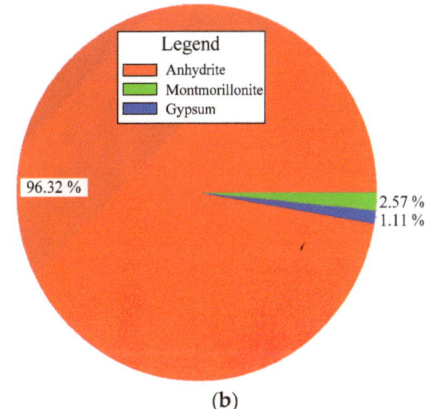

Figure 2. Gypsum samples and X-ray diffraction results: (**a**) gypsum samples; (**b**) the result of X-ray diffraction.

2.2. Crude Oil Soaking

The standard gypsum rock samples were soaked by high-temperature and high-pressure crude oil erosion experimental devices (Figure 3a). The maximum working pressure of the device is 50 MPa, the control accuracy is ±1%, the working temperature is from room temperature to 110 °C, and the control accuracy is ±1 °C. During the test, the gypsum rock sample was placed in the device, and the pressure and temperature were set to soak the crude oil at high temperature and high pressure. The target layer depth of the Xiaxin gypsum mine is about 400 m, and the formation temperature is about 32 °C. The density of crude oil is about 900 kg/m^3, and the pressure of the oil storage is about 3.6 MPa. Therefore, the temperature and pressure of the gypsum rock sample during immersion are set at 32 °C and 3.6 MPa. The crude oil used in the test was taken from Jianghan Oilfield, which has poor fluidity, high density, and a lot of solid precipitates at room temperature. In order to make the crude oil better intrude into the gypsum samples, the crude oil was diluted. The gypsum sample is placed in the autoclave of the high-temperature and high-pressure crude oil erosion experimental device. The cumulative immersion of crude oil erosion was about 2000 h.

2.3. Rock Mechanics Testing

In order to systematically study the changes in mechanical properties of gypsum rock during oil storage, the uniaxial compression, triaxial compression, and Brazilian splitting tests were carried out. Uniaxial and triaxial compression tests were carried out on the MTS rock mechanics test system (Figure 3b). The system can apply a maximum axial load of 4600 kN and a confining pressure of 140 MPa, and its frame stiffness is 11 GN/m. During the test, the sample was wrapped in a heat-shrinkable sleeve, the strain monitor was installed and placed in the confining pressure chamber, the confining pressure was set, the loading mode was selected, the stress loading was carried out, and the stress–strain curve was drawn. The Brazilian splitting test was carried out in the RMT150C rock mechanics test system (Figure 3c). The maximum vertical output is 1000 kN, the maximum horizontal output is 500.0 kN, and the maximum confining pressure that could be applied is 50.0 MPa. During the test, the sample was wrapped in a heat-shrinkable sleeve, and the axial and radial strain monitors were installed and placed in the confining pressure chamber. The loading mode was set, the stress was loaded, and the stress–strain curve was drawn. The number of tests in each group is not less than 3. The original in situ stress value of the sampling layer is about 12 MPa. The confining pressure in triaxial compression is taken as 6 MPa, 9 MPa, and 12 MPa, respectively.

Figure 3. Main instruments and equipment of the experiment: (**a**) high-temperature and high-pressure crude oil erosion test equipment; (**b**) MTS rock mechanics test system; (**c**) RMT150C rock mechanics test; (**d**) rock permeability test platform under multi-field coupling conditions; (**e**) magnetic resonance imaging systems.

2.4. Permeability Test

The permeability tests of gypsum rock were carried out before and after crude oil soaking. The test device is the experimental platform for the rock permeability test under multi-field coupling conditions (Figure 3d). The sampling rate is 100 MHz, the resolution is 16-bit, the positioning accuracy is less than 1 mm, the wavelength range is 1510~1590 nm, and the speed interval is 40 ns. Considering the low permeability of gypsum rock, the unsteady-state method is used to test the permeability. The test permeability medium is nitrogen (N_2), the confining pressure is 4 MPa, the temperature is 32 °C, and the gas pressure is 6 MPa.

The corresponding permeability can be calculated as:

$$k = -\frac{d}{dt}\left[\ln(\frac{\Delta p}{p_p})\right]\frac{\mu \delta L V_{up} V_{down}}{A(V_{up}+V_{down})} \quad (1)$$

where k is the permeability, m^2; A is the cross-sectional area of the sample, m^2; L is the length of the sample, m; μ is the dynamic viscosity of the test gas, Pa·s, where p_p is the initial pulse pressure, Pa; Δp is the pressure difference between upstream and downstream; V_{up} and V_{down} are the volume of upstream and downstream pressure chambers, respectively, m^3; δ is the gas compressive coefficient; t is the test time, s.

2.5. Pore Distribution Test

In order to evaluate the degree of crude oil erosion in gypsum samples, nuclear magnetic resonance (NMR) experiments were carried out on five gypsum samples soaked in crude oil. The instrument is the Macro MR12 NMR imaging system (Figure 3e). The magnetic field intensity is 0.3 ± 0.05 T, the main frequency is 12.8 MHz, and the probe coil diameter is 150 mm, which can effectively detect samples with a diameter of 150 mm. The sample is placed in the probe coil, the imaging parameters are set, and the imaging is run. The selected samples are randomly selected from the same batch of samples. Because the prepared samples are from the same area, after observation, it is considered that the similarity of each sample is high, and there is no significant difference. Therefore, the test results are considered to be repetitive and reliable. The principle of nuclear magnetic resonance is to use the hydrogen ions in crude oil under the action of the magnetic field to generate signals by collision on the pore wall surface. By capturing these signals, the porosity of the sample is measured. The sample is fully dried before the test, so there is no interference of free water. If the crude oil does not invade the micro cracks of the sample, the data cannot be measured. It can be seen from Figure 4 that the NMR T2 distribution of the samples is relatively clear and complete, and the regularity of the distribution is basically the same, indicating that the crude oil has intruded into the interior of the gypsum sample and filled tiny cracks in it. Meanwhile, the T2 zoning amplitude of the gypsum samples is small, generally about a dozen. According to the existing experimental results [27], the amplitude of granite with a porosity of about 2% is generally 300–400. This indicates that the porosity of the Xiaxin gypsum mine is very small, which is conducive to maintaining the airtight nature of crude oil storage. The results in Figure 4 also show that the proportions of the sum of the first and the second peak areas in the total area of different samples differ greatly. This indicates that the distribution uniformity of fractures in gypsum samples is poor, which may lead to great differences in rock mechanical parameters between different gypsum samples.

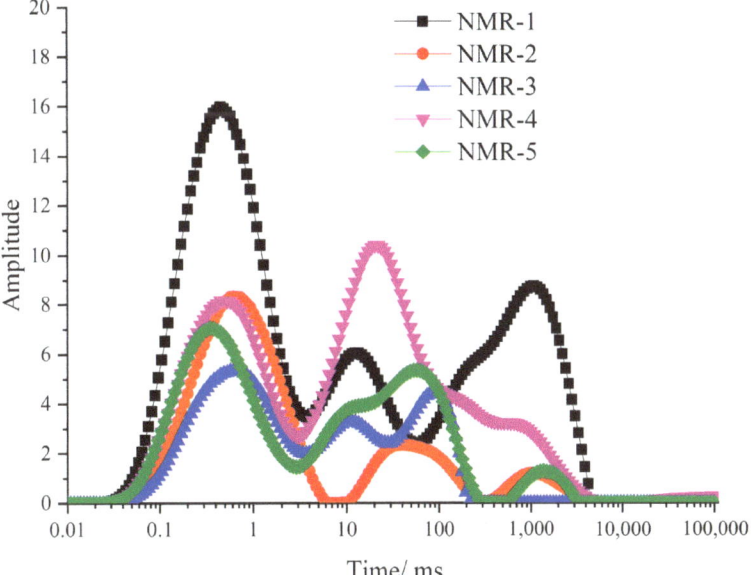

Figure 4. Results of NMR T2 zoning experiment of the five gypsum samples.

3. Results and Discussion

3.1. Uniaxial Compression Test

It can be seen from Figure 5 that the gypsum shows strong brittle failure characteristics during the uniaxial failure process. According to the experimental results, the stress–strain curve can be divided into the following four stages under uniaxial experimental conditions: (1) compaction stage, (2) elastic deformation stage, (3) unstable failure stage, and (4) post-peak stage.

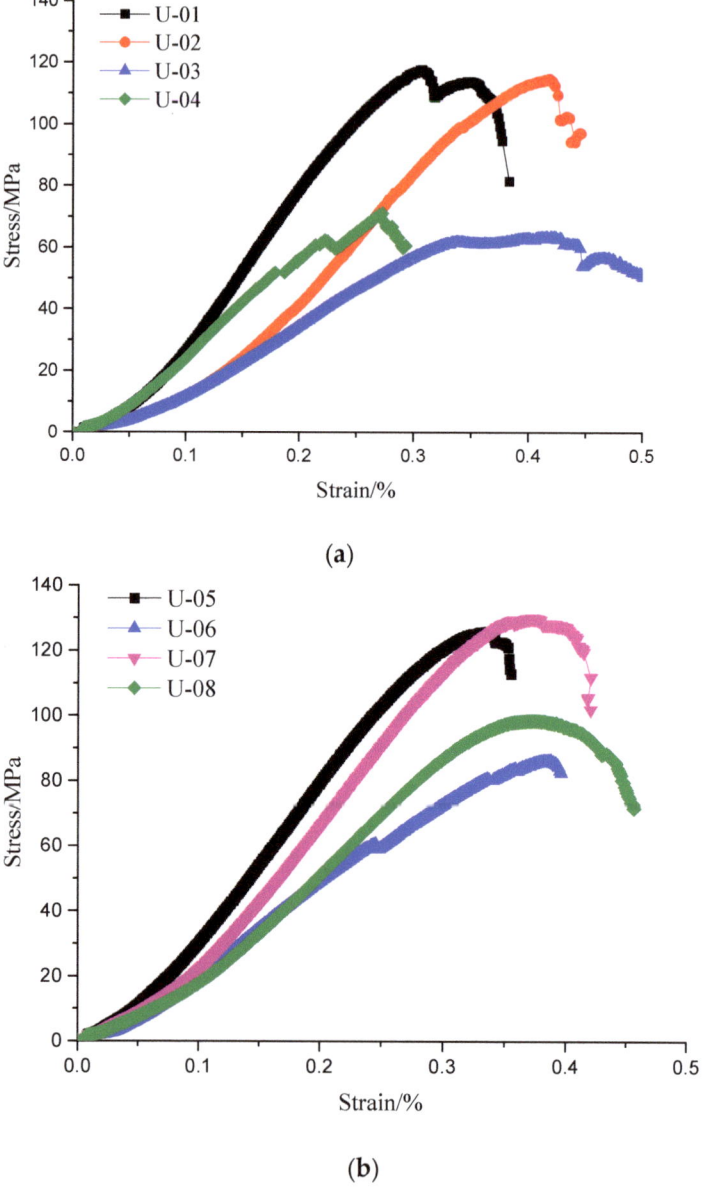

Figure 5. Axial stress–strain curve of gypsum samples: (**a**) original samples; (**b**) oil-soaked samples.

In order to explain these four stages and their characteristics more clearly, this section takes U-01 as an example to analyze the axial stress–strain curve of gypsum rock (Figure 6).

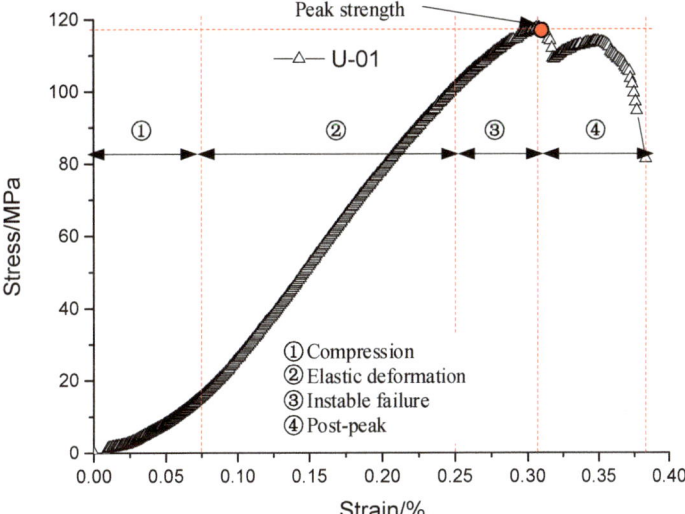

Figure 6. Axial stress–strain curve of U-01 sample.

Compaction stage: The slope of the stress–strain curve is small in the initial loading stage. Through the observation of the calculation results of samples, it can be found that the compaction stage occurs before the strain reaches 0.1%. By comparing the experimental results of soaked and unsoaked samples, it can be seen that the strain in the compaction stage of the soaked sample tends to be greater than the strain in the compaction stage of the sample without soaking. This is mainly because the cracks and micro-cracks in the soaked samples are filled with crude oil, which has partial bearing capacity when the sample is subjected to the external load, which slows down the closing speed of the cracks and micro-cracks. This indicates that the gypsum samples have been soaked sufficiently, and the oil has filled cracks and micro-cracks inside the samples.

Linear elastic stage: After the compaction stage, the stress–strain curve enters the linear elastic stage. The linear elastic stage of the gypsum sample is relatively short. The linear elastic strain of gypsum failure is 0.25%, which means the brittleness of gypsum is very weak. However, in the linear elastic stage of common brittle rock, such as granite, the linear elastic stage of granite can reach about 0.4%, which is significantly longer than that of the gypsum sample [26].

Plastic deformation stage: When the load increases to a certain value, it enters the unstable stage of asymptotic failure. Comparing the experimental results in Figure 5, it can be found that the duration of the plastic deformation stage of the crude oil-soaked samples is longer than that of the original samples. This may be because the crude oil in the pores and micro-cracks of the gypsum sample after soaking makes the surrounding rock around the pores and cracks bear more uniform stress, which improves the ductile deformation ability of the sample.

Post-peak stage: When the stress reaches the peak strength, the cracks develop gradually and form a macroscopic fracture surface until the sample loses its bearing capacity. By comparing the experimental results in Figure 5, it can be seen that the post-peak strength and post-peak duration of gypsum samples soaked with crude oil are significantly higher than those without soaking, showing typical ductile failure characteristics. This shows that crude oil immersion has an obvious effect on improving the post-peak strength and ductile deformation capacity of gypsum, which is beneficial to improving the stability of a GMG used for oil storage.

According to the above experimental results, the peak stress and elastic modulus of the gypsum sample without oil immersion and after oil immersion can be calculated. The average peak stress and average uniaxial compressive strength of unsoaked gypsum samples were 91.73 MPa and 30.59 GPa, respectively. In comparison, those of crude oil-soaked gypsum samples were 110.09 MPa and 41.14 GPa. After crude oil immersion, the uniaxial compressive strength and elastic modulus of gypsum rock increase by 20% and 34.49%, respectively. From the above experimental results, it can be seen that the uniaxial compressive strength of gypsum can be improved after crude oil soaking, which is conducive to the stability of GMG where crude oil is stored.

Figure 7 shows the photos of failure surfaces after uniaxial failure of crude oil-soaked and unsoaked gypsum samples. The sample numbers are U-01 and U-05, respectively. According to the comparison between Figure 7a,b, the main failure surface appears first in the gypsum samples without crude oil immersion and then slides along the main failure surface, finally resulting in the loss of bearing capacity of the whole sample. Therefore, there is a lot of gypsum powder generated by sliding friction on the failure surface. The bearing capacity of the samples after crude oil immersion is high, and the cracking occurs quickly after reaching the peak strength. Then, the slabs formed by cracking are unstable, so there is only a small amount of gypsum powder on the failure surface. This is consistent with the experimental results in Figure 6.

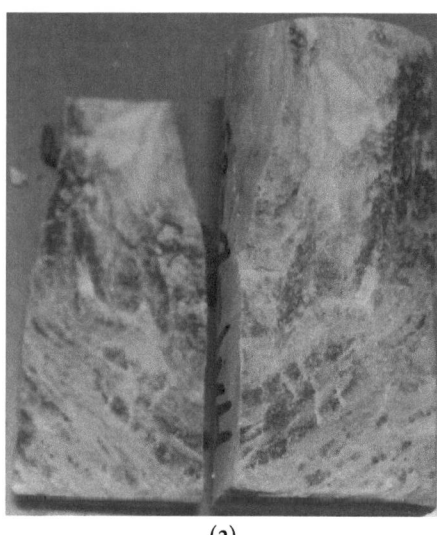

(a) (b)

Figure 7. Uniaxial failure surfaces of samples without and with oil soaking: (**a**) samples without crude oil soaking; (**b**) samples with crude oil soaking.

3.2. Triaxial Compression Tests

From Figure 8, with the increase in confining pressure, the brittleness characteristics of gypsum gradually weakened, the ductility gradually increased, and the strain softening phenomenon gradually weakened. The stress–strain curve can still be roughly divided into the following four stages.

Compaction stage: The compressive strain of the gypsum sample during the triaxial compression test is 0~0.1%, and that during the uniaxial compression test is 0~0.075%. The stress in the triaxial compression test compaction stage can reach about 14.73 MPa, and that in the uniaxial compression test is about 14.38 MPa. It was found that the stress–strain curve of the triaxial compression test sample only increased by 0.025% and 0.35 MPa compared with that of the uniaxial compression test sample. The duration of the compaction stage of soaked gypsum samples is shorter and less significant. This may be because the immersion makes the cracks and micro-cracks in these gypsum samples filled with crude oil, which improves the bearing capacity of the cracks and micro-cracks.

Linear elastic stage: After the compaction stage (about 0~0.1% strain), entering the linear elastic stage, the strains of the samples without oil immersion and after oil immersion are both 0.1~0.5% at this stage. This indicates that oil immersion has no significant effect on the linear elastic deformation of the gypsum samples under triaxial loading. With the increase in confining pressure, the slopes of the stress–strain relationship curve of the samples without and after crude oil immersion do not increase significantly, and the duration of the linear elastic stage is basically unchanged under the two conditions.

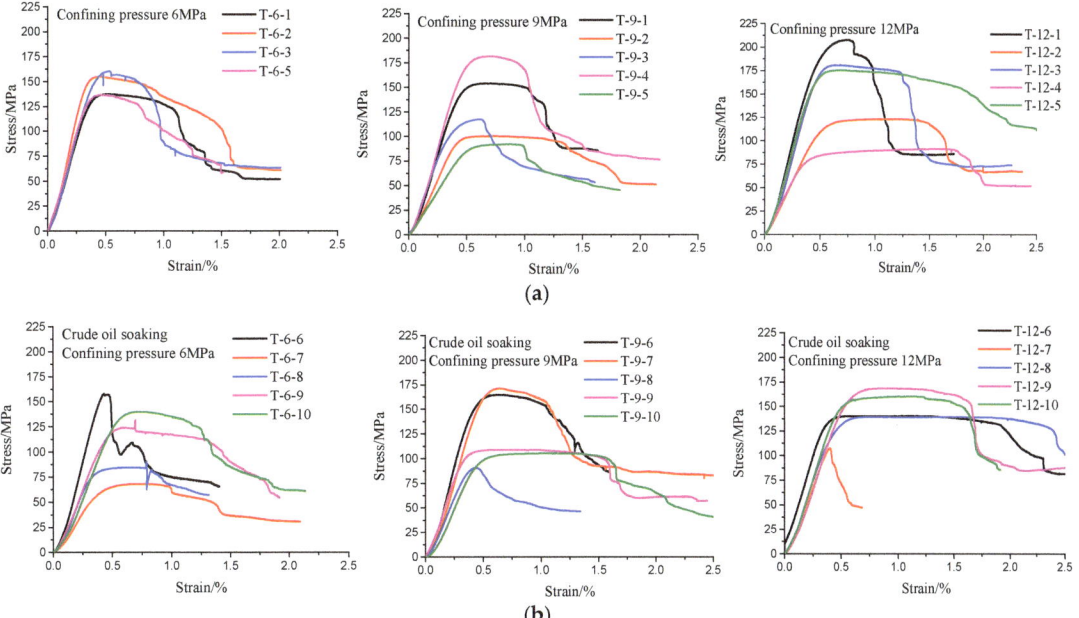

Figure 8. Axial stress–strain curves of gypsum samples under different confining pressures without and with crude oil soaking: (**a**) samples without crude oil soaking; (**b**) samples with crude oil soaking.

Plastic deformation stage: At low confining pressure (6 MPa), the yield stress point of elastic–plastic transformation of unsoaked samples is difficult to distinguish, which indicates that the expansion, penetration, and failure of gypsum microcracks are completed in a short period, and its brittleness is relatively strong. When the confining pressure increases, the stress–strain curve gradually flattens, the slope decreases, the plastic deformation stage gradually lengthens, the yield stress point of elastic–plastic transformation gradually becomes obvious, and the toughness of gypsum increases. When the confining pressure of the gypsum samples after crude oil immersion is low (6 MPa), the gypsum samples have different degrees of stress drop (Figure 8b). With the increase in confining pressure, the stress drop phenomenon gradually decreases (9 MPa) and finally disappears (12 MPa). Compared with the experimental results of the unsoaked gypsum samples, the plastic deformation ability of the soaked samples is enhanced, indicating that the crude oil soaking is beneficial in improving the plastic deformation ability of the gypsum. The large deformation of the rock mass will occur before the failure of the goaf.

Post-peak stage: By comparing the post-peak data of the stress–strain curves of soaked samples and non-soaked samples, it was found that the post-peak residual strength and residual deformation capacity of the stress–strain curves of soaked samples are greater than those of unsoaked samples. This indicates that crude oil soaking has an important influence on the residual strength and deformation capacity of gypsum samples.

The experimental results in Figure 8 also show that the peak stress of the gypsum samples without soaking and after soaking is not significantly affected by the increase in confining pressure. When gypsum is damaged, its plastic deformation capacity increases with the increase in confining pressure.

From Table 1, it can be seen that under the same confining pressure, the peak strength of differential stress of different samples varies greatly, which can be caused by the great difference in the development degree of internal cracks and micro-cracks in different samples. Due to the randomness of the development degree of internal cracks and micro-cracks in the samples, the peak strength of differential stress of the samples without crude oil immersion at the confining pressure of 9 MPa is the smallest. Compared with the experimental results of the unsoaked gypsum sample, the samples after soaking have better uniformity, but the peak values of differential stress are lower than those of the unsoaked gypsum sample under the same conditions. According to the experimental results in Table 1, it can be observed that the cohesion C of unsoaked gypsum is 17.5 MPa, and its internal friction angle φ is 45.6°. After crude oil immersion, the average cohesion and internal friction angle of the gypsum sample are 19.95 MPa and 42.30°, respectively.

Table 1. Experimental results of stress peak in triaxial compression tests of gypsum without immersion and after immersion.

Confining Pressure (MPa)	Original Sample Number	Peak Strengths (MPa)	Average (MPa)	Oi Soaking Sample Number	Peak Strengths (MPa)	Average (MPa)
6	T-6-1	137.52	147.52	T-6-6	157.96	117.74
	T-6-2	155.13		T-6-7	68.51	
	T-6-3	160.48		T-6-8	90.56	
	T-6-5	136.95		T-6-9	131.59	
				T-6-10	140.08	
9	T-9-1	154.02	129.01	T-9-6	164.88	128.38
	T-9-2	100.15		T-9-7	171.61	
	T-9-3	117.29		T-9-8	90.81	
	T-9-4	181.61		T-9-9	109.01	
	T-9-5	91.97		T-9-10	105.57	
12	T-12-1	207.58	155.65	T-12-6	140.27	143.22
	T-12-2	123.15		T-12-7	107.58	
	T-12-3	180.83		T-12-8	139.37	
	T-12-4	91.29		T-12-9	168.72	
	T-12-5	175.41		T-12-10	160.14	

Figure 9 shows that the failure mode of gypsum samples is relatively single, mainly shear failure. There is basically a macroscopic fracture surface in the samples. This is mainly because, with the increase in confining pressure when the axial stress reaches the peak strength, the fracture energy required for gypsum failure gradually increases, the elastic energy released is gradually insufficient to make gypsum further damage, the dynamic failure phenomenon weakening, and the fracture mode is mainly shear failure. With the increase in confining pressure, the number of fracture surfaces decreases gradually, the brittle fracture features gradually weaken, and the ductile fracture features become obvious. Therefore, the fracture mode is mainly single-shear failure.

Compared with the triaxial compression failure test results of gypsum samples without immersion, the samples after immersion with crude oil are more broken after failure dynamic damage degree of the gypsum samples after soaking is relatively strong, so many smaller fragments are formed after the sample failure.

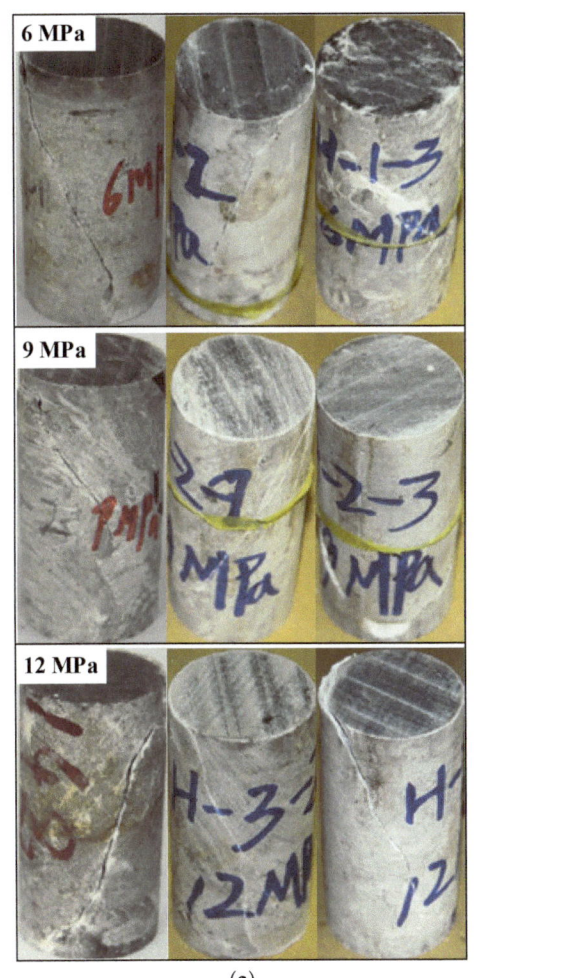
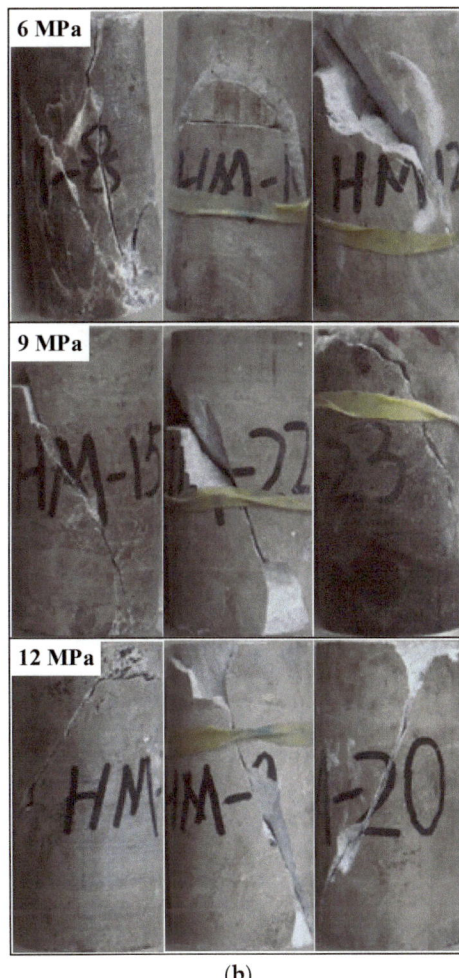

Figure 9. Gypsum samples after triaxial compression under different confining pressures: (**a**) original samples; (**b**) oil-soaked samples.

3.3. Brazilian Splitting Tests

It can be seen from Figure 10 that the vertical deformation gradually increases with the load increases, and the increasing rate presents a change rule of fast first and then slow. The load loses its bearing capacity immediately after the load reaches the peak during the loading process, and there is basically no residual strength. Compared with the gypsum samples without crude oil soaking, the load–deformation curve of the Brazilian splitting test of the gypsum samples soaked in crude oil is more uniform. This may be because the micro-cracks and cracks in the soaked samples are filled with crude oil. According to these results, the average tensile strength of the unsoaked gypsum sample is 7.00 MPa, and the average tensile strength of the soaked gypsum sample is 5.42 MPa.

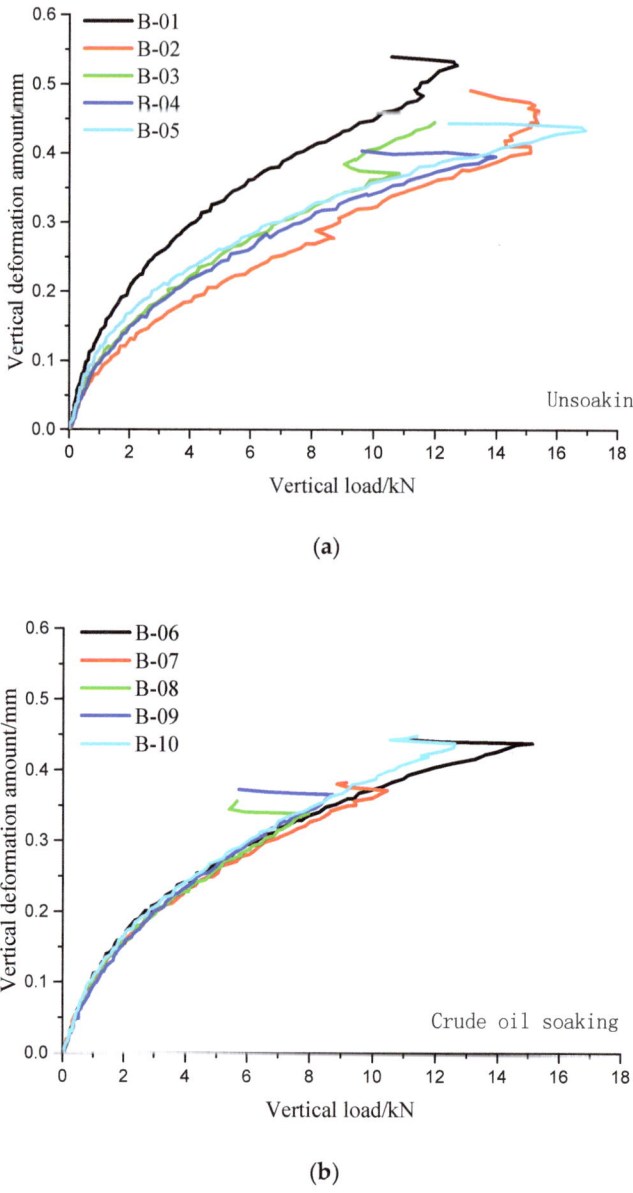

Figure 10. The load–deformation curves of gypsum samples during the Brazil splitting test under different conditions: (**a**) samples without soaking; (**b**) samples with soaking.

Figure 11 shows the experimental results of gypsum samples after Brazil splitting without soaking and after soaking. The gypsum samples have strong heterogeneity and uneven distribution of micro-cracks, so they cannot be spilt on the central axis during the Brazilian splitting test. Compared with the unsoaked samples, the symmetry of the two halves of the damaged gypsum sample after soaking is better, which indicates that the homogeneity of gypsum samples after soaking is better, indicating that the crude oil soaking is beneficial to improve the homogeneity of gypsum samples.

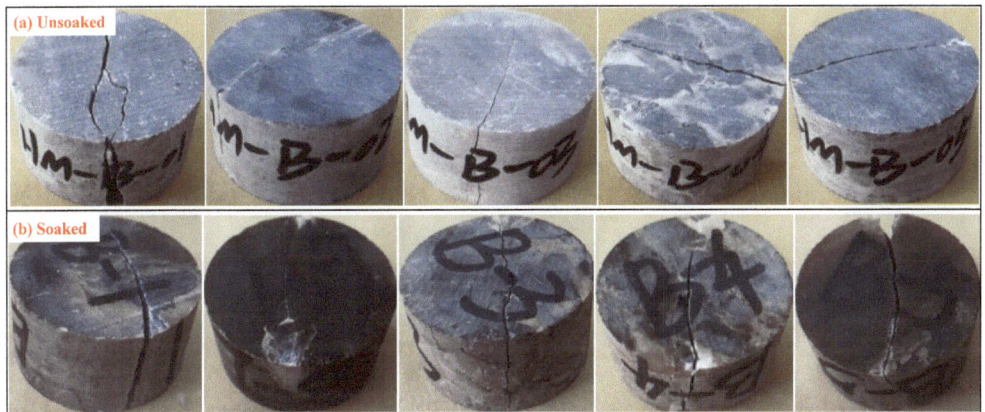

Figure 11. Photographs of gypsum samples after Brazil splitting experiment.

3.4. Permeability Tests

It can be seen from Figure 12 that the permeability of unsoaked gypsum samples decreases with the increase in confining pressure. This is mainly because the increase in confining pressure closes the micro-cracks in the sample. At the same time, the calculation results also show that the permeability of gypsum samples is extremely low and can reach the level of 10^{-19} m^2, indicating that the Xiaxin gypsum mine is relatively dense, which can provide effective storage of crude oil. The permeability of gypsum samples soaked with crude oil is significantly lower than that of unsoaked gypsum samples. This indicates that the permeability of the gypsum samples after crude oil immersion is lower than the minimum measurement accuracy of the experimental system, which is beneficial to the sealing of the storage.

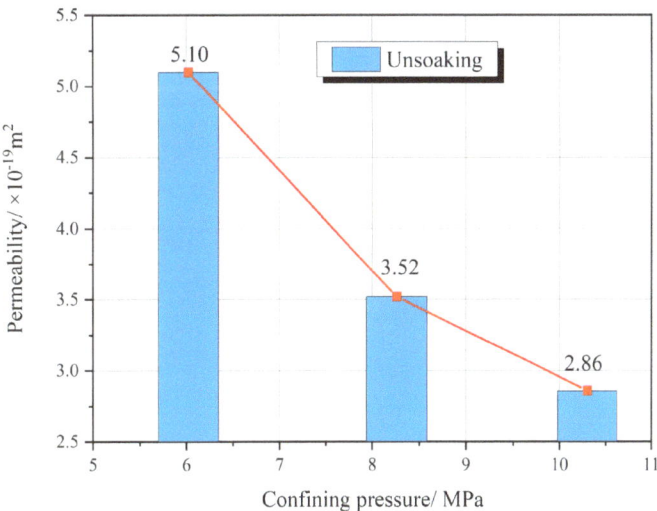

Figure 12. The relationship between permeability and confining pressure of unsoaked and soaked gypsum samples under different confining pressures.

4. Conclusions

In this paper, a series of mechanical and permeability tests of gypsum rock before and after crude oil immersion is carried out, including uniaxial compression, triaxial compression, Brazil splitting, and permeability tests. The conclusions are as follows:

(1) During the uniaxial compression experiment, the failure of the gypsum sample experienced a compaction stage, linear elastic stage, plastic deformation stage, and post-peak stage. Crude oil increases the post-peak strength and plastic deformation ability of the sample. The uniaxial compressive strength and elastic modulus of gypsum rock increased by 20% and 34.49%, respectively.

(2) During the triaxial compression experiment, the gypsum sample after crude oil immersion had better plastic deformation ability, and the sample after the failure test had stronger fragmentation. The cohesion of crude oil-soaked samples increased by about 14% compared with that of unsoaked samples, the internal friction angle decreased by about 7.2%, and the peak strength of differential stress under the same confining pressure had little difference.

(3) In the process of the Brazil splitting experiment, the load–deformation relationship curve of crude oil-soaked samples has better normalization. The average tensile strength of gypsum rock before and after crude oil immersion is 7.00 MPa and 5.42 MPa, respectively, which decreases by 22.6%. This is mainly because crude oil erosion intensifies the development of micro-cracks in the sample and significantly reduces the tensile strength of the sample.

(4) The permeability of gypsum rock can reach the level of 10^{-19} m^2, and the permeability of gypsum sample after immersion is even lower. This is mainly because the adhesion and blockage of pores by resins and asphaltenes in crude oil cause the permeability of gypsum samples to decrease.

(5) The results show that crude oil erosion is beneficial to oil storage in the goaf of the Xiaxin gypsum mine, but the properties of gypsum rock vary greatly in different areas. Subsequent studies may require comparative analysis of gypsum rocks from different regions to obtain results that are generally applicable nationwide or even worldwide.

Author Contributions: Conceptualization, T.J.; methodology, X.S.; software, D.X.; validation, D.Y.; formal analysis, M.L.; investigation, C.Z.; resources, T.J.; data curation, T.J.; writing—original draft preparation, X.S.; writing—review and editing, X.S.; visualization, D.X.; supervision, D.Y.; project administration, C.Z.; funding acquisition, D.X. All authors have read and agreed to the published version of the manuscript.

Funding: This research was funded by National Natural Science Foundation of China (No. 51804236); National Natural Science Foundation of Hubei Province (Study on hydraulic fracturing experiment and network fracture control technology of shale reservoir in western Hubei Province).

Data Availability Statement: The original contributions presented in the study are included in the article, further inquiries can be directed to the corresponding author.

Conflicts of Interest: The authors declare no conflict of interest.

Nomenclatures

A	Sample cross-sectional area [m^2]	t	Test time [s]
k	Permeability [m^2]	V_{up}	Volume of upstream pressure chambers [m^3]
L	Sample length [m]	V_{down}	Volume of downstream pressure chambers [m^3]
p_p	initial pulse pressure [Pa]	δ	Gas compressive coefficient
Δp	Pressure difference between upstream and downstream [Pa]	μ	Gas dynamic viscosity [Pa·s]

References

1. Ministry of Natural Resources. Report on reserves of China's major mineral resources published by Ministry of Natural Resources. *Geol. China* **2018**, *45*, 1315–1316.
2. Yin, S.; Zhou, W.; Shan, Y.; Ding, W.; Xie, R.; Guo, C. Assessment of the geostress field of deep-thick gypsum cap rocks: A case study of Paleogene Formation in the southwestern Tarim Basin, NW China. *J. Pet. Sci. Eng.* **2017**, *154*, 76–90. [CrossRef]
3. Jin, Z.; Zhou, Y.; Yun, J. Distribution of gypsum-salt cap rocks and near-term hydrocarbon exploration targets in the marine sequences of China. *Oil Gas Geol.* **2010**, *31*, 715–724.

4. Kamal, M.S.; Hussein, I.; Mahmoud, M.; Sultan, A.S.; Saad, M.A.S. Oilfield scale formation and chemical removal: A review. *J. Pet. Sci. Eng.* **2018**, *171*, 127–139. [CrossRef]
5. Liu, W.; Zhao, H.; Liu, Q.; Zhou, B.; Zhang, D.; Wang, J.; Lu, L.; Luo, H.; Meng, Q.; Wu, X. Significance of gypsum-salt rock series for marine hydrocarbon accumulation. *Pet. Res.* **2017**, *2*, 222–232. [CrossRef]
6. Yılmaz, I.; Sendır, H. Correlation of Schmidt hardness with unconfined compressive strength and Young's modulus in gypsum from Sivas (Turkey). *Eng. Geol.* **2002**, *66*, 211–219. [CrossRef]
7. Caselle, C.; Bonetto, S.; Colombero, C.; Comina, C. Mechanical properties of microcrystalline branching selenite gypsum samples and influence of constituting factors. *J. Rock Mech. Geotech. Eng.* **2019**, *11*, 228–241. [CrossRef]
8. Kong, L.; Ostadhassan, M.; Li, C.; Tamimi, N. Can 3-D Printed Gypsum Samples Replicate Natural Rocks? An Experimental Study. *Rock Mech. Rock Eng.* **2018**, *51*, 3061–3074. [CrossRef]
9. Meng, T.; Zhang, D.; Hu, Y.; Jianlin, X.; Sufang, S.; Xiaoming, L. Study of the deformation characteristics and fracture criterion of the mixed mode fracture toughness of gypsum interlayers from Yunying salt cavern under a confining pressure. *J. Nat. Gas Sci. Eng.* **2018**, *58*, 1–14. [CrossRef]
10. Sadeghiamirshahidi, M.; Vitton, S.J. Analysis of drying and saturating natural gypsum samples for mechanical testing. *J. Rock Mech. Geotech. Eng.* **2019**, *11*, 219–227. [CrossRef]
11. Liu, S.; Yang, S.; Shan, F.; Huang, T.; Yin, S.; Liu, X.X. Rock mechanics properties experiment and analysis of deep gypsum cap rocks. *Fault-Block Oil Gas Field* **2018**, *25*, 635–638.
12. Hoxha, D.; Giraud, A.; Homand, F. Modelling long-term behaviour of a natural gypsum rock. *Mech. Mater.* **2005**, *37*, 1223–1241. [CrossRef]
13. Hoxha, D.; Homand, F.; Auvray, C. Deformation of natural gypsum rock: Mechanisms and questions. *Eng. Geol.* **2006**, *86*, 1–17. [CrossRef]
14. Auvray, C.; Homand, F.; Hoxha, D. The influence of relative humidity on the rate of convergence in an underground gypsum mine. *Int. J. Rock Mech. Min. Sci.* **2008**, *45*, 1454–1468. [CrossRef]
15. Liang, W.; Yang, X.; Gao, H.; Zhang, C.; Zhao, Y.; Dusseault, M.B. Experimental study of mechanical properties of gypsum soaked in brine. *Int. J. Rock Mech. Min. Sci.* **2012**, *53*, 142–150. [CrossRef]
16. Meng, T.; Hu, Y.; Fang, R.; Fu, Q.; Yu, W. Weakening mechanisms of gypsum interlayers from Yunying salt cavern subjected to a coupled thermo-hydro-chemical environment. *J. Nat. Gas Sci. Eng.* **2016**, *30*, 77–89. [CrossRef]
17. Salih, N.; Mohammed, A. Characterization and modeling of long-term stressstrain behavior of water confined pre-saturated gypsum rock in Kurdistan Region, Iraq. *J. Rock Mech. Geotech. Eng.* **2017**, *9*, 741–748. [CrossRef]
18. Du, A. Study on Stability and Sealing of Surrounding Rock of Deep Anhydrite Cavern. Master's Thesis, China University of Geosciences, Beijing, China, 2022.
19. Liu, S.; Wang, H.; Du, A.; Zhang, B. Tightness Analysis of Anhydrite Mine-Out Used for Underground Crude Oil Storage Considering Seepage–Stress Coupling: A Case Study. *Energies* **2022**, *15*, 2929. [CrossRef]
20. Wang, H.-X.; Zhang, B.; Fu, D.; Ndeunjema, A. Stability and airtightness of a deep anhydrite cavern group used as an underground storage space: A case study. *Comput. Geotech.* **2018**, *96*, 12–24. [CrossRef]
21. Zhang, B.; Wang, H.; Wang, L.; Xu, N. Stability analysis of a group of underground anhydrite caverns used for crude oil storage considering rock tensile properties. *Bull. Eng. Geol. Environ.* **2019**, *78*, 6249–6265. [CrossRef]
22. Zhang, G.; Li, H.; Wang, M.; Li, X. Crack initiation of granite under uniaxial compression tests: A comparison study. *J. Rock Mech. Geotech. Eng.* **2020**, *12*, 656–666. [CrossRef]
23. Donggui, M. Study on Damage Evolution Law and Stability of Surrounding Rock in Underground Oil Storage Space of Abandoned Gypsum Mine. Master's Thesis, China University of Mining and Technology, Xuzhou, China, 2022.
24. Han, P.; Zhang, C.; Wang, X.; Wang, L. Study of mechanical characteristics and damage mechanism of sandstone under long-term immersion. *Eng. Geol.* **2023**, *315*, 107020. [CrossRef]
25. Wang, H.; Zhang, B.; Wang, L.; Yu, X.; Shi, L.; Fu, D. Experimental investigation on the long-term interactions of anhydrite rock, crude oil, and water in a mine-out space for crude-oil storage. *Eng. Geol.* **2020**, *265*, 105414. [CrossRef]
26. Professional Standard Compilation Group of People's Republic of China. *Specifications for Rock Tests in Water Conservancy and Hydroelectric Engineering*; China Standards Press: Beijing, China, 2007. (In Chinese)
27. Sun, Z.G.; Jiang, D.Y.; Xie, K.N.; Wang, K.; Li, L.; Jiang, X. Thermal damage study of Beishan granite based on low field magnetic resonance. *J. China Coal Soc.* **2020**, *45*, 1081–1088.

Disclaimer/Publisher's Note: The statements, opinions and data contained in all publications are solely those of the individual author(s) and contributor(s) and not of MDPI and/or the editor(s). MDPI and/or the editor(s) disclaim responsibility for any injury to people or property resulting from any ideas, methods, instructions or products referred to in the content.

Review

Anisotropic Mechanical Behaviors of Shale Rock and Their Relation to Hydraulic Fracturing in a Shale Reservoir: A Review

Peng-Fei Yin [1], Sheng-Qi Yang [2,3,*] and Pathegama Gamage Ranjith [4]

1. State Key Laboratory of Intelligent Construction and Healthy Operation and Maintenance of Deep Underground Engineering, China University of Mining and Technology, Xuzhou 221116, China; yinpengfei@cumt.edu.cn
2. School of Mechanics and Civil Engineering, China University of Mining and Technology, Xuzhou 221116, China
3. Key Laboratory of Rock Mechanics and Geohazards of Zhejiang Province, School of Civil Engineering, Shaoxing University, Shaoxing 312000, China
4. Deep Earth Energy Research Laboratory, Department of Civil Engineering, Monash University, Melbourne, VIC 3800, Australia; ranjith.pg@monash.edu
* Correspondence: yangsqi@hotmail.com; Tel.: +86-516-83995856; Fax: +86-516-83995678

Abstract: Shale gas is an important supplement to the supply of natural gas resources and plays an important role on the world's energy stage. The efficient implementation of hydraulic fracturing is the key issue in the exploration and exploitation of shale gas. The existence of bedding structure results in a distinct anisotropy of shale rock formation. The anisotropic behaviors of shale rock have important impacts on wellbore stability, hydraulic fracture propagation, and the formation of complex fracture networks. This paper briefly reviews previous work on the anisotropic mechanical properties of shale rock and their relation to hydraulic fracturing in shale reservoirs. In this paper, the research status of work addressing the lithological characteristics of shale rock is summarized first, particularly work considering the mineral constituent, which determines its physical and mechanical behavior in essence. Then the anisotropic physical and mechanical properties of shale specimens, including ultrasonic anisotropy, mechanical behavior under uniaxial and triaxial compression tests, and tensile property under the Brazilian test, are summarized, and the state of the literature on fracture toughness anisotropy is discussed. The concerns of anisotropic mechanical behavior under laboratory tests are emphasized in this paper, particularly the evaluation of shale brittleness based on mechanical characteristics, which is discussed in detail. Finally, further concerns such as the effects of bedding plane on hydraulic fracturing failure strength, crack propagation, and failure pattern are also drawn out. This review study will provide a better understanding of current research findings on the anisotropic mechanical properties of shale rock, which can provide insight into the shale anisotropy related to the fracture propagation of hydraulic fracturing in shale reservoirs.

Keywords: shale; hydraulic fracturing; anisotropic mechanical behavior; fracture toughness; brittleness evaluation; fracability evaluation

Citation: Yin, P.-F.; Yang, S.-Q.; Ranjith, P.G. Anisotropic Mechanical Behaviors of Shale Rock and Their Relation to Hydraulic Fracturing in a Shale Reservoir: A Review. *Energies* 2024, 17, 1761. https://doi.org/10.3390/en17071761

Academic Editor: Hossein Hamidi

Received: 11 March 2024
Revised: 31 March 2024
Accepted: 1 April 2024
Published: 7 April 2024

Copyright: © 2024 by the authors. Licensee MDPI, Basel, Switzerland. This article is an open access article distributed under the terms and conditions of the Creative Commons Attribution (CC BY) license (https:// creativecommons.org/licenses/by/ 4.0/).

1. Introduction

As a typical sedimentary rock, shale is widely distributed in the earth's crust. Shale gas has now become a strategically and globally significant unconventional resource [1,2]. Hydraulic fracturing is the core technology of shale gas development and is generally employed in the exploitation of shale gas [3–5]. The economic and efficient way for hydraulic fracturing to be performed is to create large-scale complex fracture networks, which is known as volume fracturing [6,7]. In addition, the effective implementation of hydraulic fracturing is closely related to the study of formation characteristics, including the in situ stress coefficient, strength and deformation properties of the rock, rock anisotropy and heterogeneity, distribution of natural fractures, and brittleness of the rock. Therefore, the mechanical properties of shale are an important part of shale gas reservoir evaluation [8–10]. The mechanical properties of the rock matrix and bedding development, as well as the hardness and brittleness of the rock, can lead the borehole wall of shale gas to collapse, causing leakage, and other instability problems [11,12]. Understanding the anisotropic behaviors of shale rock has important impacts for shale energy exploration, wellbore stability, the interpretation of micro-seismic monitoring, etc. [13]. Shale has many fine pores and micro-cracks due to its fine mineral particles and structure. The developmental characteristics of these micro-pores and micro-cracks will have certain impacts on the mechanical characteristics of shale at the macro level. The analysis of the internal microstructure of shale specimens can provide some auxiliary reference value for the subsequent test analysis. Therefore, it is of practical significance to investigate the anisotropic mechanical properties of shale, for the purpose of better application to the hydraulic fracturing of the shale reservoir.

As a highly differentiated stratified rock, shale has strong anisotropy, and is generally considered a transversely isotropic material [14–17]. The anisotropy of shale rock can be mainly attributed to certain factors, such as mineral orientation [18–20], microcracks [21], stress state [22,23], and kerogen content [24]. Overall, it is important to conduct comprehensive research on the shale's characteristics while considering its heterogeneity and anisotropy [25]. Scholars have summarized the effects of various minerals on the mechanical characteristics of shale [26,27]. These studies show that clay and a TOC content of 30–40 wt% are critical for micro-structural deformation [28], ensuring the various flow properties [29] and elastic properties [30], and revealing that mineralogy controls the mechanical response of the rock matrix. Therefore, to evaluate a shale reservoir as to whether it is a good candidate for fracture stimulation, rock mineralogy analysis is necessary, and a result indicating high levels of quartz and a low clay content is quite positive [31].

Shale anisotropy plays a significant role in engineering activities, and weak beddings have important impacts for the initiation, propagation, and formation of the hydraulic fracture networks of shale [32]. In several different studies, a distinct anisotropy of shale rock has been proven to have had a distinct effect on mechanical behaviors under different stress conditions [25,27,32–37], as well as predominate in the initiation and propagation of hydraulic fractures [38–40]. Many previous experimental investigations have studied the mechanical behaviors of different clayey rocks. In the prior experimental investigations, the shale was processed in different bedding inclinations. Many laboratory-based mechanical tests on shale specimens from outcrops and reservoirs have been conducted, including compressive tests [32,41–43], tensile tests [44], three-point bending tests [45], direct shear tests [36,46], etc. These studies provided adequate data and grounds for comprehending the anisotropic mechanical properties of the shale specimens.

Brittleness plays a critical role in hydraulic fracturing design [47]. However, brittleness cannot be measured directly from seismic data, or well-log data, either [48]. Using the mineralogy and mechanical parameters, various brittleness indices have been developed to evaluate the fracability of shale formations. A brittleness index based on mechanical parameters may provide a more accurate result, as these parameters can reflect the influence of stress on the brittleness [2], since the brittleness changes as the shale-formation depth increases [49]. An index based on mineralogy has been widely employed in the oil and gas industry because the input parameters are easy to obtain; however, due to the high cost of well exploration, the required data are difficult to obtain [50]. The reliability of the brittleness evaluation of shale should be improved. However, many of the brittleness indices function mostly according to the assumptions of isotropy theories [51], and it is well known that the anisotropy of shale rock rises to a high degree, thus, the anisotropic properties should be considered in the brittleness evaluation.

Moreover, the previous tests on shale mechanics [52–54] show that the anisotropy of the fracture toughness of shale is high, and these results show scatter between different shales. During the shale fracturing, the beddings are usually weaker than the rock matrix, and this can easily lead to crack propagation, i.e., when the fracture meets a bedding plane it can either extend along it or penetrate across it [36]. It is established that when a hydraulic fracture extends in the shale rock, it will be positively influenced by the bedding planes. The complex interaction between hydraulic fracture and bedding planes could be affected by several factors, i.e., the bedding strength, inclination, fluid injection rate, and so on. Compared to other discontinuities, bedding planes with large scales can be found more frequently in shale, particularly in continental stratum [55]. In general, the bedding planes have distinct effects on several aspects, including weakening the shear and tensile strength, disturbing fracture propagation, and complicating the interaction between the existing and induced fractures [56–60]. These results revealed that the stimulated fracture networks and the well extraction process are significantly affected by these discontinuities. Therefore, it is necessary to study the relationships between the bedding plane and the hydraulic fracture characteristics [61].

This paper briefly reviews previous work on the anisotropic mechanical properties of shale rock and their relation to hydraulic fracturing in shale reservoirs. The lithological characteristics of shale rock, particularly the mineral constituent, determine its physical and mechanical behavior in essence. In this paper, the status of the research addressing the lithological characteristics of shale rock is summarized first, and then the anisotropic physical and mechanical properties of shale specimens, including ultrasonic anisotropy, mechanical behavior under compression tests, and tensile property under the Brazilian test, as well as fracture toughness anisotropy in the literature, are summarized. The topics associated with anisotropic mechanical behavior under laboratory tests are addressed in this paper; in particular, the evaluation of shale brittleness based on its mechanical characteristics is discussed in detail. Finally, further concerns, such as the effects of the bedding on hydraulic fracturing failure strength, crack propagation, and failure pattern, are also drawn out. This review study will provide a better understanding of the current research findings on the anisotropic mechanical properties of shale rock, which might provide insight into the aspects of shale anisotropy related to hydraulic fracture propagation in shale reservoirs.

2. The Lithological Characteristics of Shale Rock

Rock mineralogy is essential for the evaluation of shale reservoirs [62,63], and can greatly determine the brittleness [47]. The anisotropy is induced by the packing density, which reveals that, in situ, the anisotropy of shale is from the deposition of the clay particles rather than from the intrinsic anisotropy of the mineral.

The mineral components of shale vary with different depositional and burial environments and aspects of its mineralogy [64]. The basic mineral components of shale rock are carbonates, quartz, clays, and feldspar. In many research works, the shales can be named according to the mineral enrichment, such as clay-rich shale [65–67], organic-rich shale [68–70], and carbonate-rich shale [71,72]. The relevant mineralogy plays an essential role in hydraulic fracturing implementation. Different proportions of quartz, carbonates, and clays could result in different mechanical properties in shale rocks. Shales with abundant quartz and a low clay content usually are of high brittleness, and shales that have a high clay content and low quartz content are low in brittleness [47,73].

As for the clay-rich shale, the common intrinsic clay minerals in petroleum reservoirs are kaolinite, chlorite, illite, smectite, and mixed-layer clays [74]. For instance, Pierre Shale [65], is a clay-rich rock, with its mineralogy dominated by mixed-layer illite–smectite, discrete illite, and quartz. Its microfabric shows a distinct amount of fine-grained rigid minerals floating in a clay matrix. This kind of shale is layered in places, with some high aspect-ratio rigid minerals, organic matter, and clays aligned sub-parallel to the bedding. Clay-rich shales have better reservoir properties, and they show higher porosity, pore volume, and specific surface. Clay-poor shales have higher concentrations of brittle minerals, resulting in limited shale gas storage, which makes the shale inadequate for shale gas exploration [66]. Organic-rich shales are usually considered to be good for petroleum and gas extraction [75]. The micro-pores are often developed at the nano-level in the organic-rich shale, leading to a prominent capillary effect and a high spontaneous imbibition effect [69]. As for carbonate-rich shale, such as Mancos Shale [71], as shown in the XRF maps shown in Figure 1, the light band from the petrographic data indicates the quartz-calcite-rich laminae. It can be seen that Si is distributed in the matrix, while Ca, Al, Fe, K, and Mg are distributed between the optically lighter and darker layers. It also can be found that Al, K, Fe, and Mg are intensively distributed in the more fine-grained (optically darker) layers, and Ca is intensively distributed in the more coarse-grained (optically lighter) layers. The micro-texture of the quartz-calcite-rich laminae consists of isolated organic material and coarser-grained quartz, dolomite, and feldspar minerals, cemented by calcite.

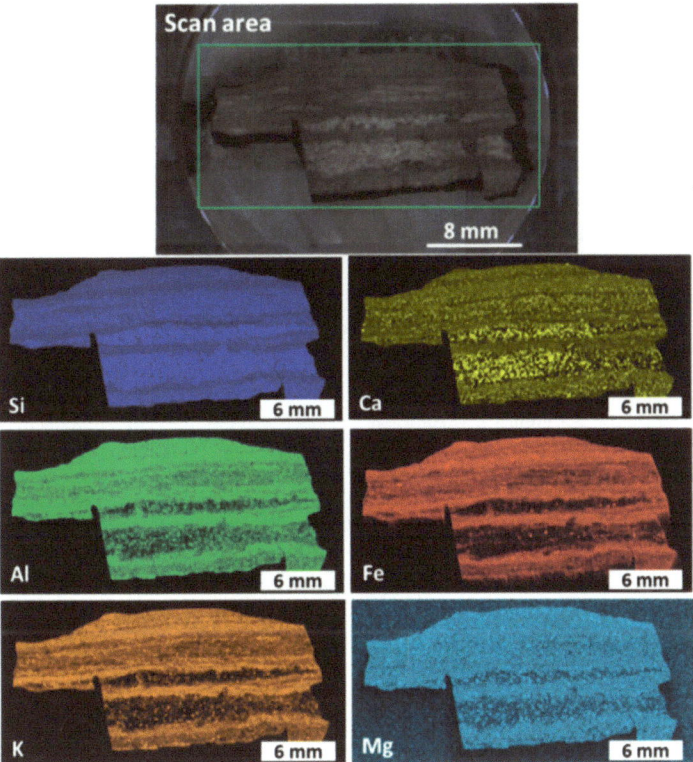

Figure 1. Micro-XRF maps for selected elements of Mancos Shale [71].

High quartz contents and low clay contents contribute to high levels of brittleness and help hydraulic fracture propagation. Li et al. [76] conducted research on black siliceous shale specimens from the southeastern edge of the Sichuan Basin of the Upper Yangtze Plate. The X-ray diffraction (XRD) results show that the brittle minerals of the shale specimen are 60–80 wt%, and that this gradually increases as the burial depth increases (Figure 2a). Figure 2b also shows that the clay content is low, and that the shale in the upper has clay-rich minerals, while the shale in the lower has brittleness-rich minerals like quartz [50].

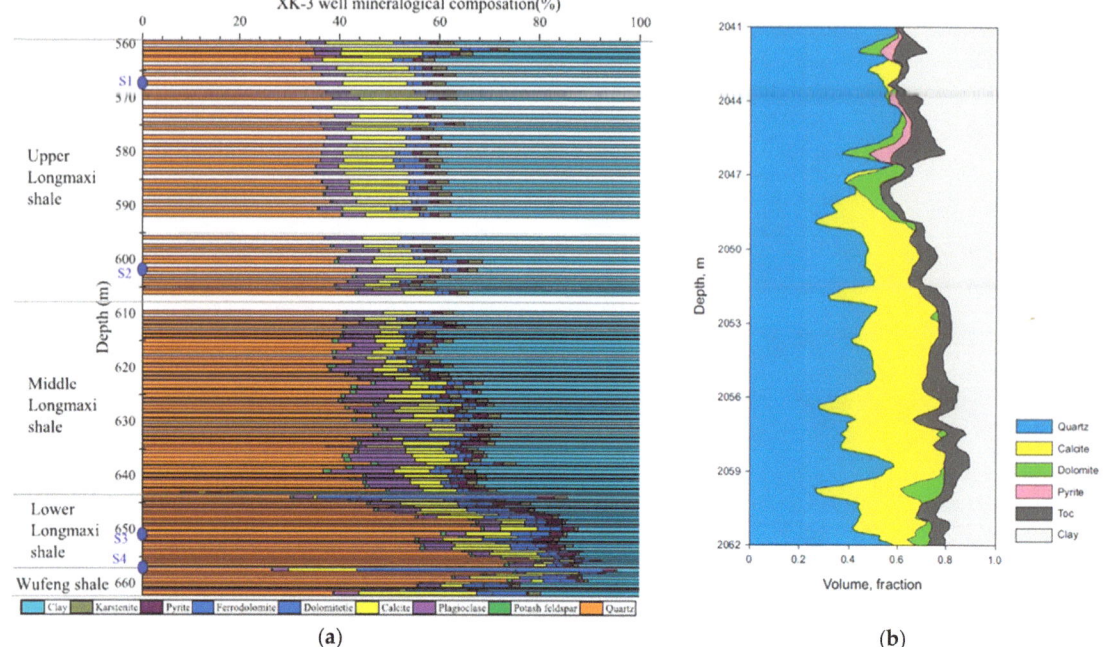

Figure 2. Mineral components of shale at different formation depths for the (**a**) shale of the Wufeng–Longmaxi Formation in well XK-3 [76]; (**b**) the mineralogy of target depths shows the volumetric fractions of constituents [50].

3. The Anisotropic Mechanical Behavior of Shale Specimens under Different Loading Conditions

The anisotropy should be carefully considered when attempting to predict the mechanical behaviors of shale [77]. Many research efforts have been conducted to study the mechanical behaviors of different rock materials containing high clay contents. As for those experiments, the shale specimen was processed in various loading directions relative to the bedding planes. In those studies, it was found that the anisotropic inclination of a shale specimen under a loading test is generally defined by two aspects: one is the angle of the loading line to the bedding plane [77–79], and the other one is the angle of the loading line relative to the normal direction of the bedding plane [11,25,80]. These two angles are complementary. To better compare and describe the test results of shale specimens in different studies, the anisotropic inclinations must be unified. As depicted in Figure 3, the angle θ of the loading line to the normal direction of the bedding plane is employed in this paper [2,11,81]. As illustrated in Figure 3a, the cylinder specimens with different bedding inclinations were loaded under the compression test, while, as seen in Figure 3b, the disc specimens were loaded under the Brazilian split test. Therefore, the anisotropic inclination θ is also the angle of the bedding plane relative to the horizontal line. The test data from different studies described in the following text are all standardized to this anisotropic inclination θ.

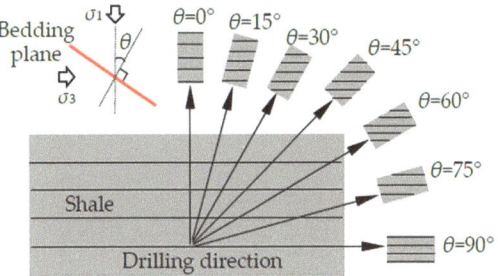 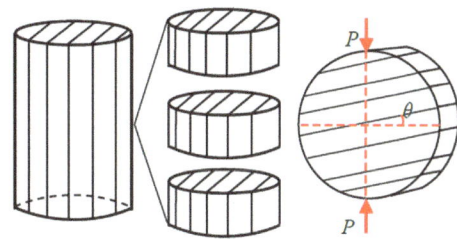

(a) Shale cylinder specimen with different drilling directions (b) Shale disc specimen with different loading inclination

Figure 3. Preparation of shale specimens in different bedding inclinations.

3.1. Ultrasonic Anisotropy of Shale Specimens

Ultrasonic techniques are maneuverable and nondestructive, and therefore useful for applications in laboratory conditions, and are promising for geophysical applications, including the interpretation of the dynamic mechanical behaviors of shale [77]. The P-wave travels faster than the S-wave. The anisotropy affecting wave velocity in the shale subsurface is considered to be a central difficulty, one resulting in significant issues for understanding dynamic elasticity in geophysical interpretation. Elastic constants of geologic materials, such as shale reservoirs, can be determined by measuring wave velocities [82]. Lo et al. [83] conducted a series of tests on the wave velocities associated with Chicopee shale by using ultrasonic techniques. Their experimental results show that the mineral orientation and the micro pores or cracks are the main factors resulting in elastic anisotropy. Additionally, the elastic anisotropy decreases with increasing confining pressure. Allan et al. [84] investigated the anisotropy of wave velocity in an organic-rich shale by proposing a multiscale methodology as well as various analyses based on XRD tests, BSE-SEM tests, and X-ray microtomography, which were applied to determine and understand the sources of wave velocity anisotropy.

Based on some ultrasonic test data from shale specimens [77,85,86], Figure 4 depicts the relationships between ultrasonic velocities and bedding angles. It can be clearly seen that the ultrasonic velocities are obviously affected by the bedding planes, and that the velocities of P- and S-waves parallel to the bedding plane are greater than those of the waves orthogonal to the bedding plane. The energy decreases due to the wave scattering on different beddings, leading to increases in the time duration of the wave traveling through the specimens. The above conclusions are similar to the results from Dewhurst et al. [87] and Zhubayev et al. [88]. The experimental results from Mokhtari et al. [44] show that the P-wave velocity of a shale specimen with a bedding inclination of 0° is 30% lower than a specimen with a bedding inclination of 90°.

Based on the propagation and polarization directions relative to the bedding plane, six ultrasonic wave velocities can be measured in shale specimens [89]. Figure 5 illustrates the six velocities, as follows [90]: one compressional wave (V_{pv}) and two shear waves ($V_{sv1} = V_{sv2}$), which propagate normally to the bedding plane in a vertical plug (as shown in Figure 5a, $\theta = 0°$); and one compressional wave (V_{ph}) and two shear waves ($V_{sh} > V_{sv}$), which propagate parallel to the bedding plane in a horizontal plug (as shown in Figure 5b, $\theta = 90°$). These six ultrasonic velocity parameters could then be employed to calculate the elastic parameters, which are used for describing the anisotropic elastic properties of a vertically transverse isotropic (VTI) material like shale rock [91,92].

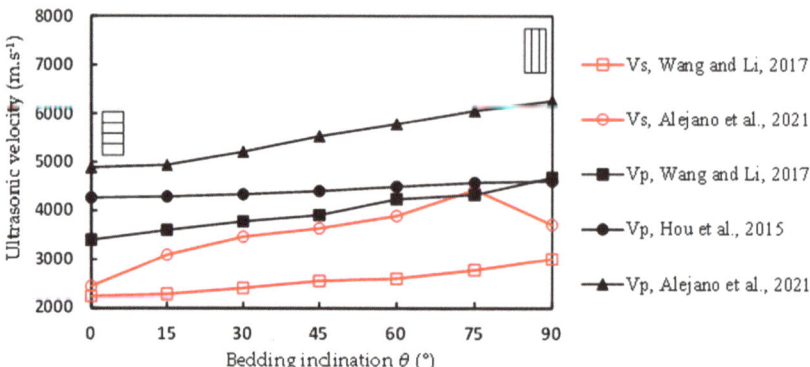

Figure 4. Ultrasonic velocity variation with respect to bedding inclination in shale specimens [77,85,86].

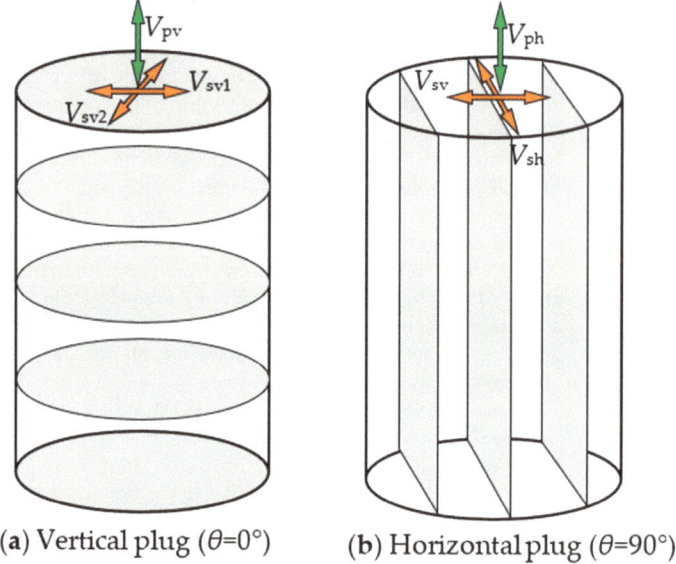

Figure 5. Schematic representation of wave velocity measurements commonly used in ultrasonic testing: (**a**) the wave travels normal to the bedding plane ($\theta = 0°$); and (**b**) the wave travels parallel to the bedding plane ($\theta = 90°$). (Adapted from Iferobia and Ahmad [90].)

In the study conducted by Yin [93], the wave's attenuation by bedding planes in the transversely isotropic rock is also described. As depicted in Figure 6, when the ultrasonic wave travels normal to the bedding, the transmission and reflection occur synchronously at the bedding plane (Figure 6a), and the wave's reflection results in energy loss and reduces the wave's velocity; with more bedding planes orthogonal to the propagation direction, greater energy losses occur, leading to a lower velocity of the ultrasonic wave. At a bedding inclination of 0° (Figure 6a), the bedding planes are orthogonal to the propagation direction of the P-wave, resulting in strong reflection at the bedding planes and a lower wave-velocity. With the increases of bedding inclination, the beddings are slowly brought parallel to the propagation direction, and the energy attenuation becomes less, leading to an increase in wave velocity. When at a bedding inclination of 90° (Figure 6b), the bedding planes are parallel to the propagation direction of the P-wave, and there is almost no reflection and the energy attenuation is minimum; thus, the wave velocity is at its maximum.

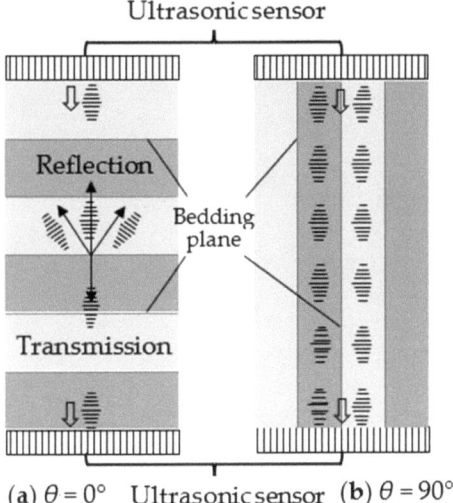

Figure 6. A schematic diagram of longitudinal wave propagation in transversely isotropic rock. (Adapted from Yin and Yang [93]).

3.2. Uniaxial and Triaxial Compression Test

Shale specimens with various bedding inclinations are frequently studied to investigate their anisotropic behaviors. Many researchers have conducted experiments on various shale specimens under compression conditions [36,41,42,46,78,94–104]. The above studies reported the mechanical properties of anisotropic rocks; in particular, in Ramamurthy's research [105], the anisotropy of rocks is generally classified into three modes: "U" type, undulatory type, and "shoulder" type. These are are shown in Figure 7.

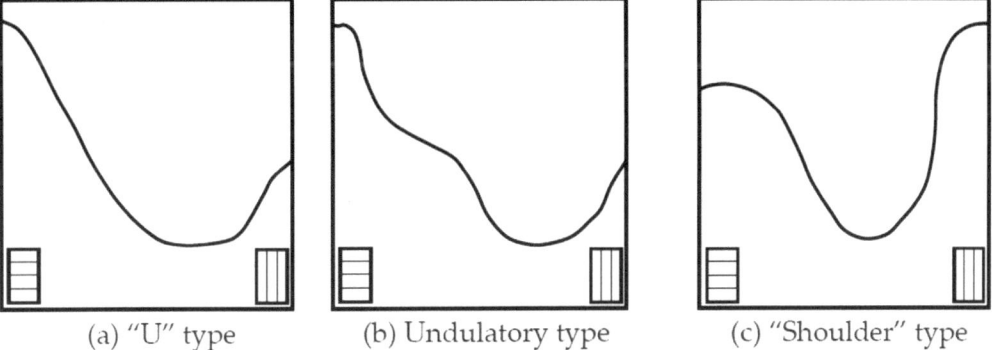

Figure 7. Three types of strength variation relative to the bedding inclination of the anisotropic rock.

Strength under triaxial compression conditions is the key characteristic of rock material. Some tests are conducted under uniaxial compression [2,11,25,36,77–80,85,86,95,106,107], and some tests are conducted under triaxial compression conditions [2,36,78,85,108]. The various shale specimens taken from different sampling sites in these studies display different magnitudes of strength, but their values all show a similar law of variation concerning the bedding inclination. Based on the test data from the above-mentioned studies, the uniaxial compressive strength (UCS) and triaxial compressive strength (TCS) variations with respect to bedding inclination are displayed in Figures 8 and 9, respectively. From Figures 8 and 9, it can be concluded that the peak strength instances all display a "U" type or a similar "shoulder" type variation with increased bedding inclination under different confining pressures. Particularly, the maximum peak strength is observed at $\theta = 0°$ or $90°$, and the minimum values are observed at $45° < \theta < 75°$.

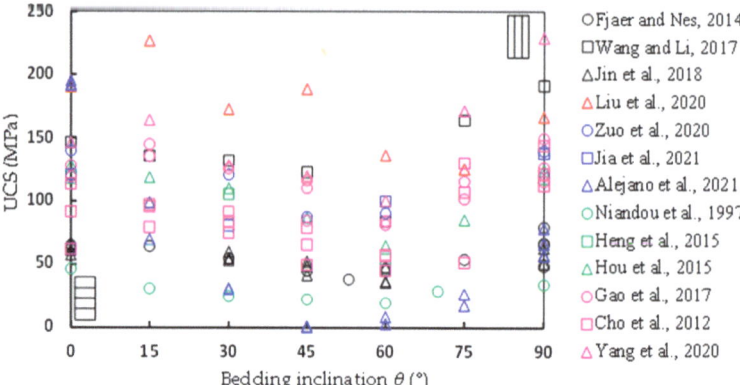

Figure 8. Uniaxial compressive strength (UCS) variation with respect to bedding inclination in shale specimens [2,11,25,36,77–80,85,86,95,106,107].

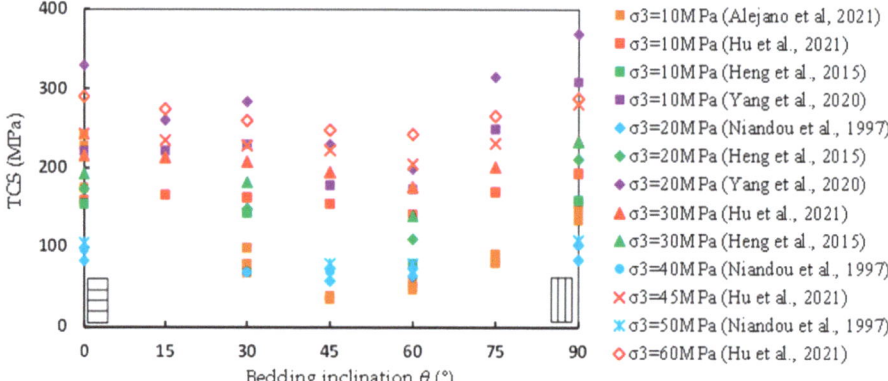

Figure 9. Triaxial compressive strength (TCS) variation with respect to the bedding inclination of shale specimens [2,36,78,85,108].

To better predict the strength of anisotropic rocks, many strength criteria have been proposed. The earliest attempt was the theory of a single weakness plane [109], and this was followed by various subsequent studies [110–117]; in one, an empirical strength criterion was proposed for predicting the non-linear strength behavior of intact anisotropic rocks [111,112]:

$$\frac{(\sigma_1 - \sigma_3)}{\sigma_3} = B_j \left(\frac{\sigma_{cj}}{\sigma_3}\right)^{\alpha_j} \qquad (1)$$

where σ_1 and σ_3 are the major and minor principal stresses, respectively, and σ_{cj} is the UCS at the particular θ. The values for α_j and B_j are defined herein to consider the anisotropy of the strength, and the functions are as follows:

$$\frac{\alpha_j}{\alpha_{90}} = \left(\frac{\sigma_{cj}}{\sigma_{c90}}\right)^{1-\alpha_{90}} \qquad (2)$$

$$\frac{B_j}{B_{90}} = \left(\frac{\alpha_{90}}{\alpha_j}\right)^{0.5} \qquad (3)$$

where σ_{c90} is the UCS at $\theta = 90°$, and α_{90} and B_{90} are the values of α_j and B_j at $\theta = 90°$. The values for α_j and B_j at each anisotropy orientation can be calculated by substituting the triaxial compression test data into Equation (1). Additionally, another empirical criterion under triaxial conditions can be expressed as follows [118]:

$$\frac{\sigma_1}{\sigma_{ci}} = \frac{\sigma_3}{\sigma_{ci}} + \left[\frac{1 + A(\sigma_3/\sigma_{ci})}{1 + B(\sigma_3/\sigma_{ci})}\right] - r \qquad (4)$$

where σ_{ci} is the UCS of intact rock; A and B are constant parameters; and r is the strength reduction factor, with a value equal to 0 for intact rock and 1 for highly jointed rock masses. To better apply the above failure criterion to transversely isotropic rocks, Saeidi et al. [114] modified the failure criterion (Equation (4)) to the following:

$$\sigma_1 = \sigma_3 + \sigma_{c\theta}\left[\frac{1 + A(\sigma_3/\sigma_{c\theta})}{\alpha + B(\sigma_3/\sigma_{c\theta})}\right] \qquad (5)$$

where $\sigma_{c\theta}$ is the UCS of anisotropic rock at θ, and α is the strength reduction parameter when taking into account the anisotropy, which is considered here in order to extend the generalization of Equation (4) to anisotropic rocks.

In addition, several mechanical models based on a damage theory have been proposed for anisotropic rocks by a series of researchers [119–123]. These models show a good capability for describing the main mechanical behaviors of anisotropic rocks, but still need to be verified by more experimental data for different types of anisotropic rocks. To evaluate the effect of the confining pressure on anisotropy, the strength anisotropy degree R_a, represented in terms of $\sigma_{pmax}/\sigma_{pmin}$, can be calculated in each confining pressure [113,124]. The strength anisotropy degree R_a indicates the maximum degree of strength deviation under different bedding inclinations. Figure 10 shows the variations of R_a with increases in confining pressure, the data for which are calculated based on the studies in Figures 8 and 9. It can be seen that the strength anisotropy degree R_a of a shale specimen decreases as the confining pressure increases. In other words, the confining pressure can reduce the strength anisotropy of rock material. The reason is that the confining pressure is hydrostatic pressure, and when it is applied to the specimen, the beddings are compacted, and thus the bedding strength is enhanced, and the bedding effect is weakened.

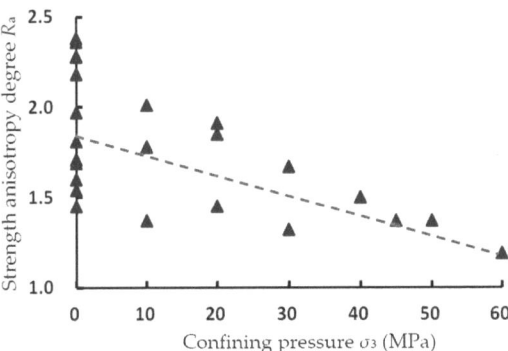

Figure 10. Strength anisotropy degree variation with respect to the bedding inclination of shale specimens.

In addition, Figure 11 shows the elastic modulus with bedding inclination increases under uniaxial compression of shale specimens, based on data drawn from some previous studies [11,25,36,77,80,86,106,107]. These studies all reported an increasing variation with the bedding inclination increases overall, even though there were some local fluctuations. The explanation for this can be developed as follows: (1) the directional alignment of the clay minerals and micro-cracks in the bedding plane is the inherent reason for the strong elastic anisotropy of the shale specimen [125], and the content of the clay and the development of micro-cracks are different in the specimens under different bedding inclinations; this results in the anisotropy of the elastic modulus [86]; (2) as typical sedimentary rock, the compaction degree of the bedding in shale rock is low in the diagenetic process. Therefore, when the principal stress is parallel to the bedding ($\theta = 90°$), the bedding is under a tensile state and the rock matrix is under compression; thus, the elastic modulus and deformation modulus are large and, this is in accord with the rock matrix in this case. When the principal stress is perpendicular to the bedding ($\theta = 0°$), the bedding is in a state of compaction, and the axial strain is larger, resulting in a smaller elastic modulus and deformation modulus.

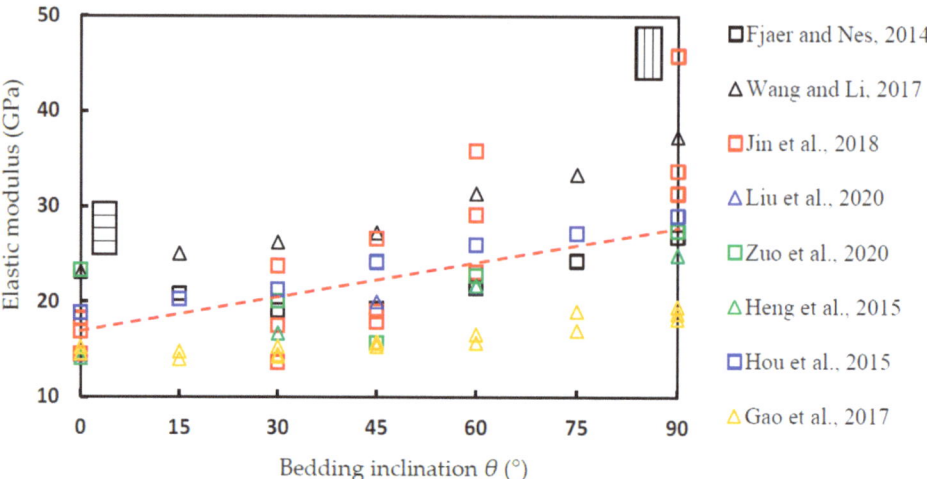

Figure 11. Elastic modulus with respect to inclination angle under uniaxial compression of shale specimens [11,25,36,77,80,86,106,107].

As for the failure behavior of shale rock under the conventional triaxial compression test, the fracture propagation during the failure process is generally affected by the bedding inclination and the confining pressure. The fracture angle strongly depends on the bedding inclination and the confining pressure. In the summary of Niandou et al. [78], for a bedding inclination of $0° < \theta < 30°$, tensile fracture along the loading direction occurs under low confining pressure and shear fracture across the bedding plane occurs under high confining pressure. For a bedding inclination of $30° < \theta < 75°$, the failure generally occurs by the bedding sliding; however, shear fracture in the matrix can occur and may cross the bedding plane when under a high confining pressure. For a bedding inclination of $75° < \theta < 90°$, when under low confining pressure, the failure occurs by the bedding plane splitting, and failure occurs by a shearing across the bedding plane when under a high confining pressure. In the study of Yang et al. [2], the researchers conducted conventional triaxial compression tests on Longmaxi Formation shale specimens under different bedding inclinations. According to the test results, they classified the shale specimens into four failure modes, as illustrated in Figure 12 and Table 1. Four failure modes are reported: tensile fracture through the bedding plane (T-T), tensile fracture along the bedding plane (T-A), shearing through the bedding plane (S-T), and shearing along the bedding plane (S-A). T-T generally occurs under uniaxial compression (or very low confining pressure) with low and medium bedding inclination (0°–45°). S-A mainly takes place in specimens with medium or high bedding inclinations (45°–75°). With increases in the confining pressure, S-T occurs in specimens with low or medium bedding inclinations (0°–45°), instead of T-T, and even in some high bedding inclinations, the S-T dominates the failure. As for the bedding inclination of 90°, T-A occurs when under uniaxial compression, as the bedding is activated by tension along the loading direction. With the increases in confining increases, S-T takes place in the rock matrix. It can be concluded that the bedding planes at medium and high inclinations have a distinct effect on the failure fracture, resulting in shear slipping along the bedding plane; with the increases in confining pressure, high confining pressure dominates the failure mode and shear fracture in the matrix (across the bedding plane) generally occurs, which is similar to the mode of isotropic rock material. These findings are also in accord with the studies of Liu et al. [80] and Zhao et al. [126].

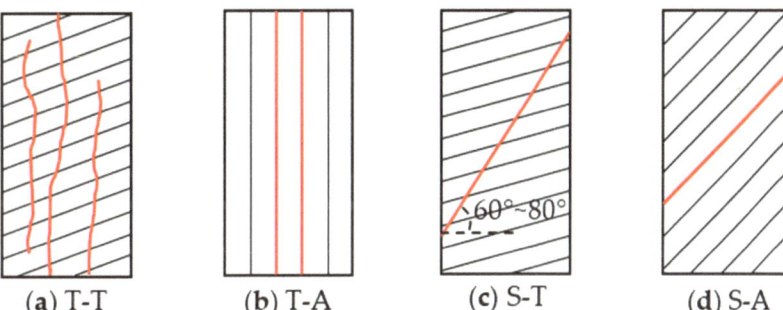

Figure 12. Classification of failure modes of shale specimens [2].

Table 1. Failure mode summary for shale specimens, from the study of Yang et al. [2].

	$\theta = 0°$	$\theta = 15°$	$\theta = 30°$	$\theta = 45°$	$\theta = 60°$	$\theta = 75°$	$\theta = 90°$
$\sigma_3 = 0$ MPa	T-T	T-T	T-T	T-T	S-A	T-T	T-A
$\sigma_3 = 5$ MPa	S-T	S-T	S-T	S-A	S-A	S-A, S-T	S-T
$\sigma_3 = 10$ MPa	S-T	S-T	S-T	S-A, S-T	S-A	S-A, S-T	S-T
$\sigma_3 = 15$ MPa	S-T	S-T	S-T	S-A, S-T	S-A	S-A	S-T
$\sigma_3 = 20$ MPa	S-T	S-T	S-T	S-A, S-T	S-A	S-A	S-T

3.3. Tensile Properties under Brazilian Split Test

The tensile strength of shale plays an important role in the initiation and propagation of hydraulic fractures. As a method classically used to evaluate the tensile strength, the Brazilian test with different bedding inclinations is widely employed for determining structural anisotropy. The analytical solution developed to measure the tensile strength of anisotropic rocks was developed by Amadei et al. [127] according to the theoretical relation between the stress and strain proposed by Lekhnitskii et al. [128]. The effects of bedding inclination and thickness on tensile strength and the fracture pattern of anisotropic rocks under diametrical loading conditions have been investigated by conducting various Brazilian tests [15,129–134].

Vervoort et al. [133] summarized the failure anisotropy of nine different anisotropic rocks under Brazilian test conditions into four trends, as follows (as shown in Figure 13).

Trend 1: Constant value over the entire range of anisotropy angles (Figure 13a).

Trend 2: Constant value between 0° and 45°, followed by a linear decrease (Figure 13b).

Trend 3: Decrease in the strength over the entire interval, but a rather systematic decrease, approximating a linear variation (Figure 13c).

Trend 4: Decrease from very low anisotropy angles (between 0° and 30°~40°), followed by a leveling off (Figure 13d).

The above four variation trends describe different types of anisotropic rocks with varying bedding planes. As for the shale rock specimens from different studies, the BTS variation with respect to bedding inclination is similar. Figure 14 summarizes some BTS data of shale specimens from different studies. From their test results, depicted in Figure 14, it can be seen that the BTS variation of shale specimens is more like the above trend 2 and trend 4. The studies of Jin et al. [25], Yang et al. [81], and Cho et al. [95] show that the BTS value is constant between 0° and 60° and then followed by a linear decrease, which is in accord with trend 2. The other studies, such as Yang et al. [135], Hou et al. [33,136], and Du et al. [137], show a trend similar to trend 4; their BTS decreases first to a low value (the anisotropy angle is around 60°) and then keeps on, unchanged. The BTS from the study of He and Afolagboye [138] shows the variation of trend 3, in which BTS decreases over the entire interval. From the above summary, it can be seen that, although the BTS variation of shale specimens from different sampling sites shows a slight difference, in general, their BTS values do not demonstrate much change from 0° to 30°, and then decrease to a low level around 45°–60°, and finally keep within a stable range around 75°–90°.

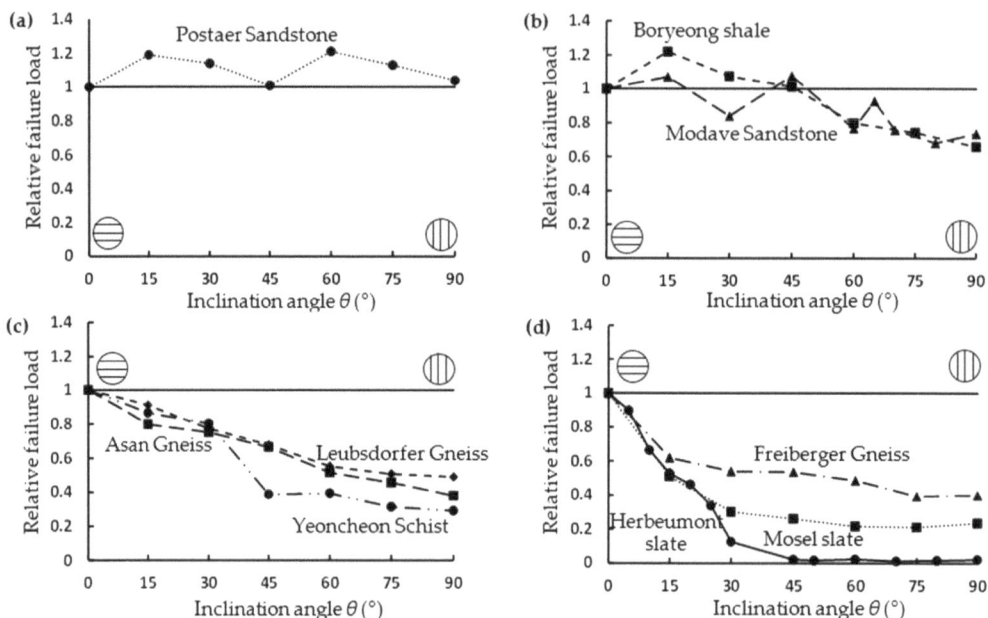

Figure 13. Variation of average failure load, relative to the failure load for loading perpendicular to weak planes: (**a**) trend 1; (**b**) trend 2; (**c**) trend 3; and (**d**) trend 4 [133].

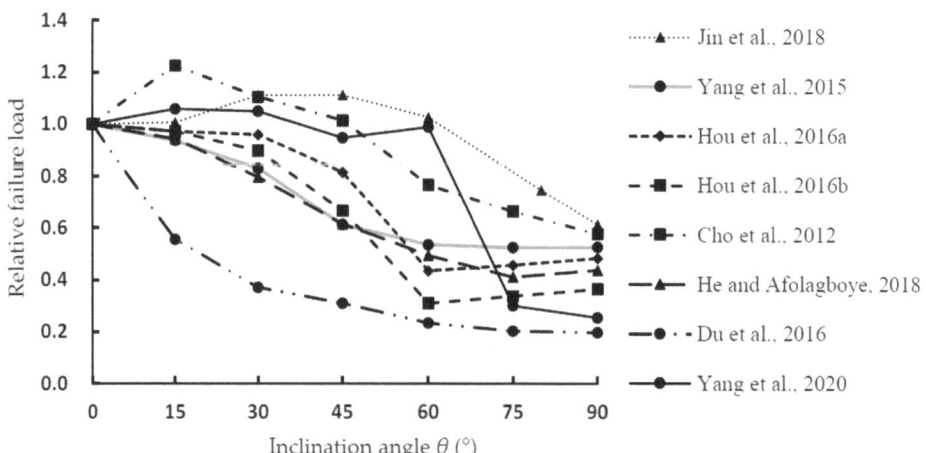

Figure 14. Variation of average failure load, relative to the failure load for loading perpendicular to weak planes, for shale specimens [25,33,81,95,135–138].

To further explain the failure behavior of the shale disc specimen with respect to the bedding plane, the failure pattern in the Brazilian test should be discussed. In the study of Hou et al. [136], shale disc specimens are loaded under Brazilian tests with different bedding inclinations. They classify the Brazilian split failure process of shale specimens into three stages: compaction, elastic deformation, and destruction stages. Their test results show that the anisotropic characteristics of tensile strength and failure pattern are highly distinct at the peak point of stress. The bedding plane has a strong effect at the bedding inclination of 60° and a weak effect at 0°. They also reported the transformation angles of the failure mechanisms, which are the bedding inclinations of 60° and 45°. At 60°, a tension fracture along the bedding turns to cross the bedding plane, and a cross-bedding plane fracture turns to shear slip at 45°. These findings are similar to those in the studies of Cho et al. [95] on Boryeong shale, Vervoort et al. [133] on Freiberger Gneiss, Du et al. [137] on Shaanxi shale, and Yin and Yang [104] on layered sandstone. Cho et al. [95] investigated the BTS of Boryeong shale under different loading angles. The study's results show that the maximum tensile strength occurs near 15°, and the minimum value occurs when the bedding line is parallel to the axial direction (90°). Additionally, the shale begins to fracture along the bedding when $\theta > 30°$. When $0° \leq \theta \leq 15°$, the crack propagates along the loading direction.

From the above studies and discussion on the anisotropic characteristics of BTS and failure patterns, it can be concluded that, with the bedding closely aligned to the loading direction, the bedding's effect on the BTS turns out to be significant. In other words, there seems to be a critical loading angle; the BTS could not be affected before the critical angle, and decreases rapidly when it exceeds it. In the research of Yin and Yang [104], many disc specimens of layered sandstone were tested and a critical bedding inclination θ of 71.6° was derived by theoretical analysis. In their studies, to better explain the stress state at the center, the normal and shear stress coefficients k_N and k_S are defined. As shown in Figure 15, when $0° < \theta < 71.6°$, the compressive stress and shear stress are applied on the bedding plane, and the disc specimen can evince tension in the rock matrix and shearing at the bedding; and when $71.6° < \theta < 90°$, the bedding plane is under tensile and shear states simultaneously, and when the shear stress is at a high level, the disc specimen can be shear or tensile along the bedding plane. The bedding may be under compression, shearing, or tension states with respect to the variation of bedding inclination. When the bedding inclination is low, the bedding is dominated by compression, does not easy fail. When at a high bedding inclination, the bedding is under shearing and tension simultaneously, and easily experiences shear slip and tensile split.

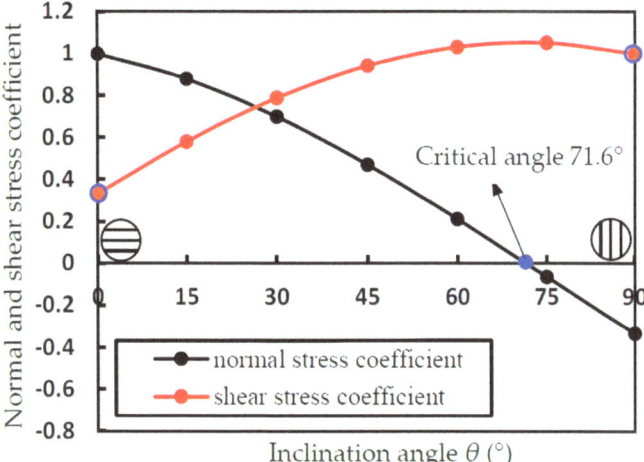

Figure 15. The variation of the stress state at the center with respect to bedding inclination [104].

Figure 16 presents some failure examples for shale disk specimens under the Brazilian test, with different bedding inclinations from different studies shown. From these fractured disk specimens with different loading directions, it can be seen that the fracture pattern of the anisotropic rock is much different from those of isotropic rock. Although the fractures generally propagate along the loading direction, the bedding directions have distinct effects on the fracture propagation to a certain extent. In some specimens with high bedding inclinations of 60° and 75°, such as the Mancos and Boryeong Shale with 60° (Figure 16a from Simpson et al. [139] and Figure 16b from Cho et al. [95]), Chongqing shale with 75° (Figure 16c from Hou et al. [136] and Figure 16d from Wang et al. [32]), the fractures extend along or divert to the bedding planes. The principal reason for this pattern is that the bedding plane is very closely aligned to the loading direction when under a high bedding inclination, and the shear stress or tensile stress is dominant in the bedding plane and this easily leads to shear or tensile fracture along it. This is in accord with the study and explanation of Yin and Yang [104] in Figure 15.

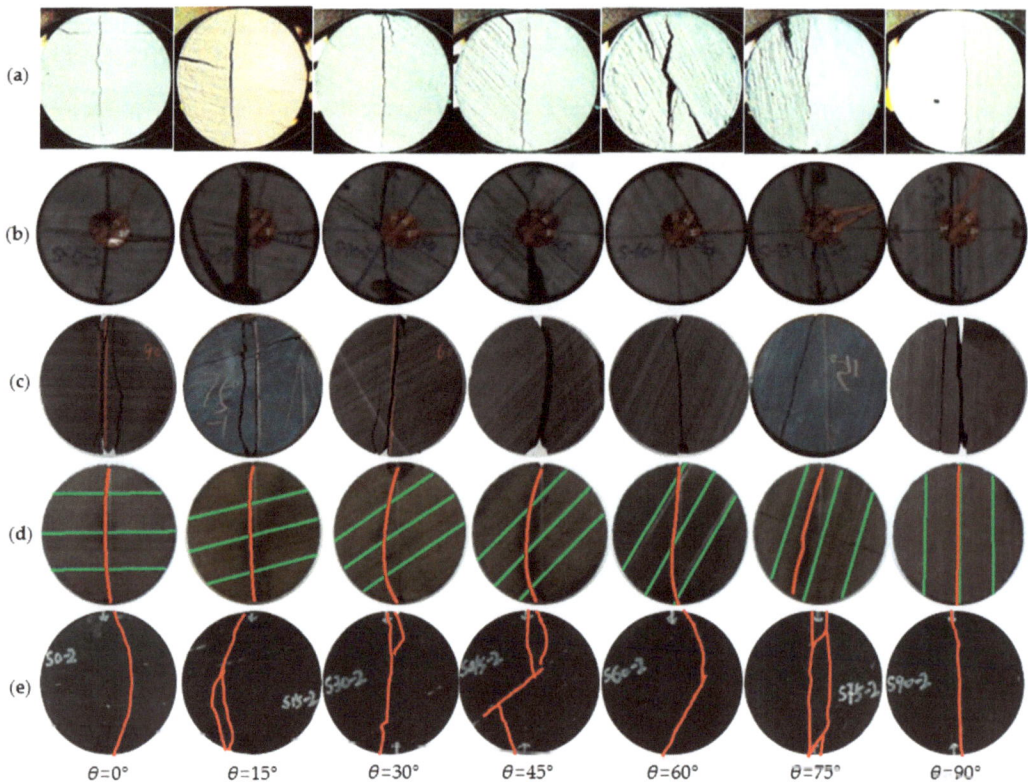

Figure 16. Examples of the failed disk specimen under the Brazilian test: (**a**) Mancos shale from Simpson et al. [139], (**b**) Boryeong shale from Cho et al. [95], (**c**) Longmaxi Formation, Chongqing shale from Hou et al. [136], (**d**) Longmaxi Formation, Chongqing shale from Wang et al. [32], and (**e**) Changsha shale from Yang et al. [81].

By conducting a series of Brazilian tests on layered sandstone, Tavallali and Vervoort [131] considered the specimens after failure and classified the observed fractures into three different types (see Figure 17): (1) layer activation (LA), in this pattern, the fractures extend along the bedding planes; (2) central fractures (CF), in this pattern, fractures are roughly located in the central part of the specimen and parallel to the loading direction; (3) non-central fractures, in these cases, the fractures are outside the central part. As shown in Figure 17, the mode of LA mostly occurs at a high loading direction of 60°–90°, and the mode of CF mostly occurs at a low loading direction of 0°–30°, while the combination of LA and CF appears at a loading direction of 45°–75°. The failure modes at $\theta = 0°$ and

90° look similar; these are all central fractures along the loading direction. However, their fracture mechanism is different: the central fracture in specimen $\theta = 0°$ extends across the beddings, which shows the rock matrix fractured by tension; and the central fracture in specimen $\theta = 90°$ extends at the beddings, which shows the bedding fractured by tension and is more in accord with the layer activation. This is also can be explained by Figure 15: the BTS of $\theta = 0°$ is rather similar to the BTS of the rock matrix, and the BTS of $\theta = 90°$ corresponds least to the BTS of the bedding plane [131].

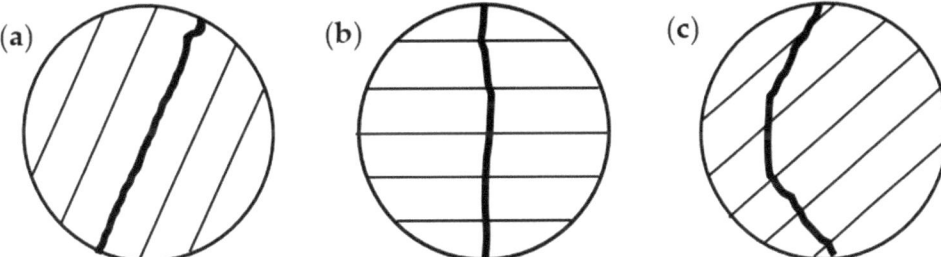

Figure 17. Schematic patterns of different fracture types in the Brazilian test [131]: (**a**) Layer activation (LA), (**b**) Central fracture (CF), and (**c**) Non-central fracture.

3.4. Fracture Toughness Anisotropy of Shale Specimens

The fracture toughness of mode I (K_{Ic}), is an important parameter that controls the hydraulic fracture propagation [140,141]. To determine the K_{Ic} of brittle rock material, many experimental methods and configurations have been designed; some common methods and their configurations are listed in Table 2. As described in Table 2, the configurations of CCCD, SCB, and SENB are specimens with straight-through notches, and the others are specimens with chevron notches. To make sure that fracture behavior can be characterized appropriately by the linear elastic fracture mechanics (LEFM), large specimens are selected, such as deep beam of NBD and CNDB.

Table 2. Summary of some common test methods and configurations for K_{Ic} determination of a rock specimen.

Methods and Specimen Configuration	Graphic Illustration	References
Centrally cracked circular disc (CCCD)		Awaji and Sato [142], Atkinson et al. [143]
Semicircular bend specimens (SCB)		Chong and Kuruppu [144], Lim et al. [145], Dai et al. [146], Funatsu et al. [147], Ren et al. [148]
Single-edged notched beam specimens (SENB)		ASTM [149]
Straight edge cracked round bar bending specimens (SECRBB)		Bush [150]
Single edge-notched deep beam specimens (SENDB)		Luo et al. [37,151]

Table 2. *Cont.*

Methods and Specimen Configuration	Graphic Illustration	References
Chevron-notched beam specimens (CNB)		Wu [152]
Chevron-notched deep beam specimens (CNDB)		Ren et al. [153]
Cracked chevron-notched Brazilian disc specimens (CCNBD)		Sheity et al. [154], Fowell [155], Dai et al. [156]
Chevron-notched semicircular bend specimens (CNSCB)		Kuruppu [157], Dai et al. [158]
Short-rod specimens (SR)		Barker [159], Ouchterlony [160]
Traditional Brazilian disk and flatted disk specimens without pre-existing flaws		Wang and Xing [161]

To determine the fracture toughness anisotropy of the shale specimens, three main crack orientations can be designed [162], as shown in Figure 18: arrester, divider, and short-transverse. By conducting loading tests on these three kinds of specimens using the different methods listed in Table 2, the fracture toughness tested in different crack orientations shows distinct differences. The fracture toughness of specimens with cracks oriented parallel to the bedding plane (short-transverse, Figure 18c) is smaller than in the other two crack orientations [54,151,153,162,163]. This result is easy to understand, and the reason for it is that the specimens with a pre-existing crack parallel to the bedding planes are easy to initiate, due to the weak tensile strength of the bedding plane [164].

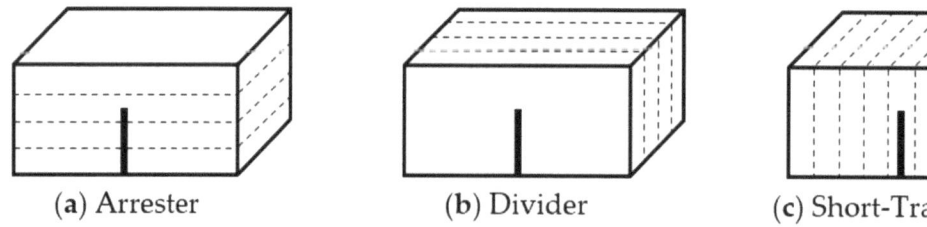

(a) Arrester **(b)** Divider **(c)** Short-Transverse

Figure 18. Sketch of the specimens with three main pre-existing crack orientations: (**a**) Arrester, (**b**) Divider, and (**c**) Short-Transverse.

From Figure 18, we can see that in the case of Arrester (Figure 18a), the crack extends through the isotropic plane (bedding plane). In the case of Divider (Figure 18b), the crack also propagates in the rock matrix but does not need to extend through the bedding plane. Thus, comparing the cases of Arrester and Divider, the crack propagation is slightly harder in Arrester, as the crack needs to expend more energy to penetrate through the isotropic plane. This is clearly illustrated in Table 3, which summarizes experimental results from various studies conducted using different testing methods. From Table 3, it can be found that almost all the fracture toughness values of $K_{Ic, A}$ are greater than $K_{Ic, D}$, while, meanwhile, their ratio is nearly always around 1, which indicates that the crack propagation in Arrester and Divider is dominated by the rock matrix. On the other hand, it can be clearly seen that in the case of Short-Transverse (Figure 18c) the crack initiation and propagation are mainly needed to open the beddings, and this expends less energy than used in the other two cases due to the weak tensile strength of bedding plane. Thus, as displayed in Table 3, the value $K_{Ic, ST}$ is much lower than the $K_{Ic, A}$ and $K_{Ic, D}$.

Table 3. Summary of the shale rock K_{Ic} values obtained from different studies.

Shale Material	References	Bedding Orientation	Method	K_{Ic} (MPa·m$^{1/2}$)	Fracture Toughness Ratios		
					$K_{Ic, A}/K_{Ic, ST}$	$K_{Ic, D}/K_{Ic, ST}$	$K_{Ic, A}/K_{Ic, D}$
Longmaxi shale	Luo et al. [37]	Arrester Short-Transverse	NDB	1.661 0.851	1.952	--	--
	Heng et al. [45]	Arrester Divider Short-Transverse	SECRBB	1.146 0.957 0.566	2.025	1.691	1.198
	Wang et al. [165]	Arrester Short-Transverse	CCNDB	0.9226 0.7028	1.313	--	--
	Wang et al. [165]	Arrester Short-Transverse	SCB	0.8297 0.6549	1.267	--	--
	Dou et al. [166]	Arrester Short-Transverse	SENB	1.366 0.927	1.476	--	--
	Ren et al. [153]	Arrester Divider Short-Transverse	CNDB	1.161 1.116 0.781	1.487	1.429	1.041
Mancos shale	Chandler et al. [54]	Arrester Divider Short-Transverse (low) Short-Transverse (high)	SR	0.44 0.44 0.12 0.31	3.667 (ST, low) 1.419 (ST, high)	3.667 (ST, low) 1.419 (ST, high)	1.000
	Chandler et al. [163]	Divider Short-Transverse	Double-torsion specimen	0.48 0.37	--	1.297	--
	Li et al. [167]	Arrester Divider Short-Transverse	SENB	0.912 1.200 0.917	0.995	1.309	0.760
	Lee et al. [53]	Arrester Divider	SCB (25.4 mm in diameter)	0.944 0.470	--	--	2.0099
		Arrester Divider	SCB (38.1 mm in diameter)	0.578 0.479	--	--	1.207
Nash Point shale	Inskip et al. [168]	Arrester Divider Short-Transverse	SCB	0.74 0.71 0.24	3.083	2.958	1.042
		Divider Short-Transverse	SR	0.73 0.30	--	2.433	--
Anvil Points shale (80 mL/kg kerogen content)	Schmidt [162]	Arrester Divider Short-Transverse	SENB	0.977 1.076 0.750	1.303	1.435	0.908
Anvil Points shale (160 mL/kg kerogen content)		Arrester Divider Short-Transverse	SENB	0.604 0.674 0.370	1.632	1.822	0.896

As for other anisotropic directions of crack propagation, Lei et al. [169] have conducted investigations regarding the fracture behavior of shale samples with different bedding strengths. Their beam specimens are processed with different bedding inclinations according to the straight-through notch and tested under three-point bending loading. The micro-crack propagation process was observed by the SEM equipment with a loading system. The results revealed that the outcrop shale of the Longmaxi Formation from Pengshui County, in the southeast of Chongqing, China, shows strong anisotropy and that this has a great effect on hydraulic fracture propagation. In Figure 19a, it can be observed that the micro-cracks initiate at the notch tip, and then propagate along the loading direction. Figure 19b shows the simulated failure patterns of three-point bending specimens under different bedding properties. The tensile strength of the smooth joint decreases from mode I to mode IV. The results revealed that the fracture characteristic of shale is affected by bedding inclination and bedding strength. It is also can be determined that the cracks exhibit strong tortuosity under a low tensile strength of the bedding plane.

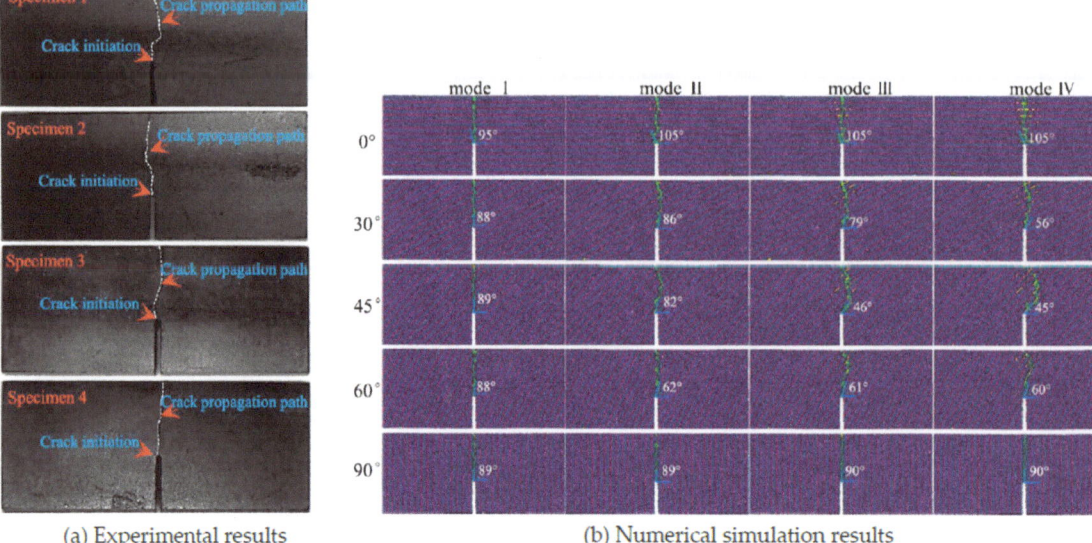

Figure 19. Failure patterns under different bedding properties [169].

On the other hand, as shown in Figure 20a, the study results of Lei et al. [169] reveal that, when under the same bedding strength, the fracture toughness decreases with the increases of bedding inclination, with a maximum at 0° and a minimum at 90°. Particularly, the fracture toughness decreases gently between 0° and 45°, and then decreases rapidly after 45°. It indicates that when the bedding inclination approaches the loading direction, the bedding's effect on fracture toughness turns to be more obvious. It can be clearly seen from Figure 19b that the crack propagation eventually occurs at the bedding planes more obviously when the bedding inclination increases after 45°. In particular, as displayed in Figure 19b, mode IV has the lowest bedding strength, resulting in the most obvious crack deflection towards the bedding plane. Luo et al. [37] and Shi et al. [170] have investigated fracture toughness anisotropy under different bedding inclinations. The shale specimens used by Luo et al. [37] and Shi et al. [170] are all sampled from the Longmaxi Formation. The former study processed NDB shale specimens, with a length-to-width ratio (L/W) of 2.0 and an edge crack with inclined angle $\alpha = 0°$ and bedding inclination $\theta = 0°, 30°, 60°,$ and 90°. The latter study employed the CCNBD specimens with chevron notches with angles of 0°, 30°, 45°, 60°, and 90°. Figure 20b depicts the K_{Ic} variations concerning the shale bedding inclination from the above two studies. The results show that the bedding inclination has a distinct effect on the fracture toughness, and the tested values all decrease with an increase of bedding inclination.

In the above studies, the test specimens with bedding inclinations of 0° and 90° are configured in accord with the cases of Arrester (shown in Figure 18a) and Short-Transverse (shown in Figure 18c), respectively. It can be determined from Figure 20 that the maximum value is at 0° and the minimum at 90°, which is in accord with the findings from Table 3. The crack's approach angle relative to the bedding plane and bedding strength are the key influence factors in hydraulic fracture propagation. The high approach angle and low bedding strength contribute to the fracture propagation along bedding planes. The Short-Transverse specimen has the lowest fracture toughness, due to the weaker bedding plane, and has a looser grain arrangement and is less resistant to crack propagation [171]. The microcracks could extend and coalesce, resulting in the failure of grain clusters [172].

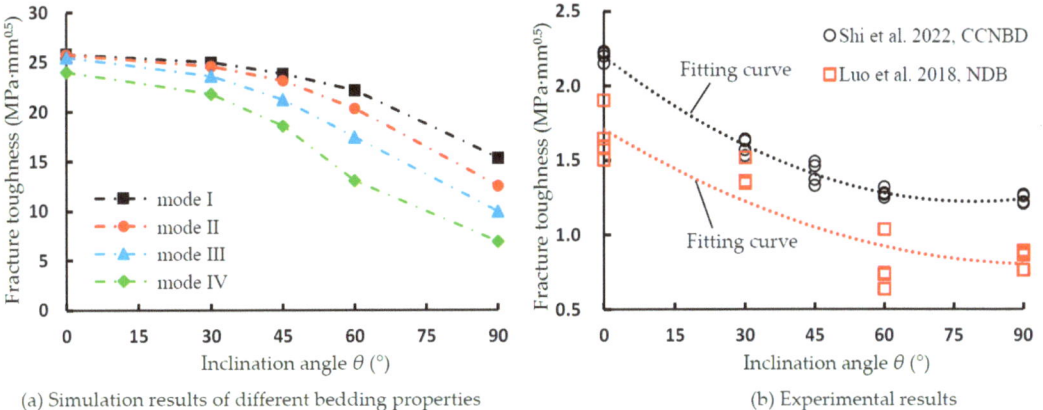

(a) Simulation results of different bedding properties (b) Experimental results

Figure 20. Variation of K_{Ic} with respect to the bedding inclination of the shale specimen [37,170].

4. The Relationship of Shale Mechanical Properties to Hydraulic Fracturing Evaluation

4.1. Quantitative Evaluation of Shale Brittleness Based on Mechanical Properties

Brittleness is commonly applied to characterize the failure behavior of the rock material [173]. For hydraulic fracturing, the high brittleness of shale may contribute to the deeper and more complex fracture propagation. Therefore, the evaluation of brittleness is important in engineering. Based on various previous studies, brittleness can be defined as the lack of ductility [174,175], as a failure at or only slightly exceeding the yield stress [176], as the cohesion destruction [177], as a failure with small or no plastic deformation [178], and as a self-sustaining failure process [179].

As for hydraulic fracturing in shale, the brittleness index (BI) is usually employed to quantify the brittleness degree of rock, which is useful for evaluating the ability of fracture propagation (fracability) and fracture network generation. To accurately measure the brittleness of shale rock, many factors should be considered. To date, many BIs have been proposed to quantify its extent [73,179–187]. Table 4 lists some common BIs, mainly based on mineral composition, and strength parameters.

Among the test methods in Table 4, mineral composition is easy to obtain and can be precisely determined by conducting laboratory analyses such as XRD testing. Methods based on strength parameters, stress–strain characteristics, and energy balance analysis can be determined by stress–strain curves. As the stress–strain curves can be obtained easily by conducting the triaxial compression test, this measurement is commonly used to determine strength parameters. The brittleness is a mechanical response to a specific stress state. For a material, the composition is constant, but the brittleness can vary if different stress states are applied. Therefore, the specific stress state must be considered when its brittleness is evaluated. Thus, BI definitions based on stress conditions, such as strength parameters, stress–strain characteristics, and energy balance analysis, are more significant [2].

As for the brittleness indices $BI_5 \sim BI_9$, which are based on strength parameters, parameters such as UCS and BTS are easily obtained by a laboratory test. All of these BIs display a positive relationship [179] with confining pressure. On the other hand, rocks that show different stress–strain curve shapes may be of the same BI values ($BI_{11} \sim BI_{13}$). Furthermore, the combination of strength and strain performances (BI_{10} and BI_{13}) is more precise in predicting rock brittleness [173]. Additionally, some BIs based on stress–strain ($BI_{11} \sim BI_{13}$) and energy balance (BI_{17}) only consider the effect of the pre-peak state and ignore the stress drop rate; however, some rock materials show a distinct ductility when high confining pressures are applied [185,195].

Table 4. Summary of some universal BI definitions.

Test Method	Formulae	References	Remarks
Mineral composition	$BI_1 = W_Q/W_{Q+C+Cl}$	Jarvie et al. [47]	W_x = weight fraction of component x; Q = quartz; C = carbonate; Cl = clay; TOC = total organic carbon; Lm = limestone; QFM = quartz+feldspar+mica
	$BI_2 = (W_Q + W_c)/W_{total}$	Rickman et al. [73]	
	$BI_3 = (W_Q + W_{Dol})/W_{Q+Dol+Lm+Cl+TOC}$	Wang and Gale [188]	
	$BI_4 = W_{QFM+C}/W_{total}$	Jin et al. [189]	
Strength parameters	$BI_5 = \sigma_c/\sigma_t$ $BI_6 = (\sigma_c - \sigma_t)/(\sigma_c + \sigma_t)$ $BI_7 = \sin(\varphi)$	Hucka and Das [180]	σ_c = uniaxial compressive strength; σ_t = Brazilian tensile strength; φ = internal friction angle; ρ = density
	$BI_8 = \sigma_c\sigma_t/2$	Altindag [181]	
	$BI_9 = 0.198\sigma_c - 2.174\sigma_c + 0.913\rho - 3.807$	Yagiz [184]	
Stress–strain characteristics	$BI_{10} = (\sigma_f - \sigma_r)/\sigma_f$	Bishop [190]	ε_p = sustained plastic strain at failure; ε_e = total elastic strain; ε_f = total strain at failure; ε_r = residual strain; σ_f = stress at failure; σ_r = residual strength; E = elastic modulus; M = post-peak modulus
	$BI_{11} = \varepsilon_e/\varepsilon_f$	Hucka and Das [180]	
	$BI_{12} = \varepsilon_p \times 100\%$	Andreev [191]	
	$BI_{13} = (\varepsilon_f - \varepsilon_r)/\varepsilon_f$	Andreev [191]	
	$BI_{14} = E/v$	Luan et al. [192]	
	$BI_{15} = E/M$ $BI_{16} = (M - E)/M$	Tarasov and Potvin [179]	
Energy balance analysis	$BI_{17} = U_{et}/(U_{et} + U_p)$	Hucka and Das [180]	U_{et} = total elastic energy; U_P = plastic energy; U_r = rupture energy; U_{ec} = consumed elastic energy; U_a = additional energy;
	$BI_{18} = U_r/U_{ec}$ $BI_{19} = U_a/U_{ec}$	Tarasov and Potvin [179]	
	$BI_{20} = U_{et}/(U_r + U_p)$ $BI_{21} = (U_{et} + U_p)/(U_r + U_p)$ $BI_{22} = U_{et}/U_r$	Munoz et al. [193]	
	$BI_{23} = (U_r + U_p)/(U_{ec} + U_p)$ $BI_{24} = U_a/(U_{ec} + U_p)$	Ai et al. [194]	

The drilling of horizontal wells for hydraulic fracturing in shale usually needs to cross bedding planes in different orientations; thus, the shale anisotropy should not be ignored in a brittleness evaluation [50]. According to the various studies on anisotropic mechanical properties of shale specimens from Sections 3.2 and 3.3, the UCS shows a "U" type variation and BTS shows a negative relationship with bedding inclination. Thus, as illustrated in Figure 21, from the equations of the BI_5 and BI_6 in Table 4, which are calculated by UCS and BTS, it can be concluded that the variation of the two BIs concerning bedding inclination will display a similar "U" type. Before the inflection point, the UCS and BTS decrease, and the UCS decreases faster than the BTS; therefore, the BI_5 and BI_6 decrease gently. When the bedding inclination increases to a high level, such as 90°, the BTS decreases to a very low level and UCS is at a high level; thus, the BI_5 and BI_6 increase to high magnitudes rapidly. It should be noted that the BI_5 and BI_6 variations before the inflection point closely depend on the margins of UCS and BTS reductions; as shown in Figures 8 and 14, the UCS and BTS reduction from various shale specimens are different. Therefore, in Figure 21 the BI_5 and BI_6 variation before the inflection point may show a decrease or a gentle increase, or, alternatively, it can be stable.

In the study on shale brittleness evaluation from Yang et al. [2], the researchers obtained data from the triaxial compression test. They proposed two new brittleness indices, BI_1^* and BI_2^*, to evaluate the shale brittleness variation with respect to bedding inclinations. By comparing the two new brittleness indices with the BI from Kivi et al. [49], which is referred to herein as BI_{25}, the BI variation concerning bedding inclinations is depicted in Figure 22. The new brittleness indices BI_1^* and BI_2^* from Yang et al. [2] and BI_{25} from Kivi et al. [49] are expressed as follows:

$$BI_1^* = \frac{\sigma_f - \sigma_r}{\sigma_f}\lg|M| \quad (6)$$

$$BI_2^* = U_{ei}/(U_{et} + U_p) \quad (7)$$

$$BI_{25} = 0.5(U_{ec}/U_r + U_{ec}/(U_{et} + U_p)) \quad (8)$$

where σ_f is the strength and σ_r is the residual strength, respectively, and M is the modulus after the peak. U_{ei} is the ideal elastic energy, as defined in Yang's study. [2]. The other parameters are noted in Table 4.

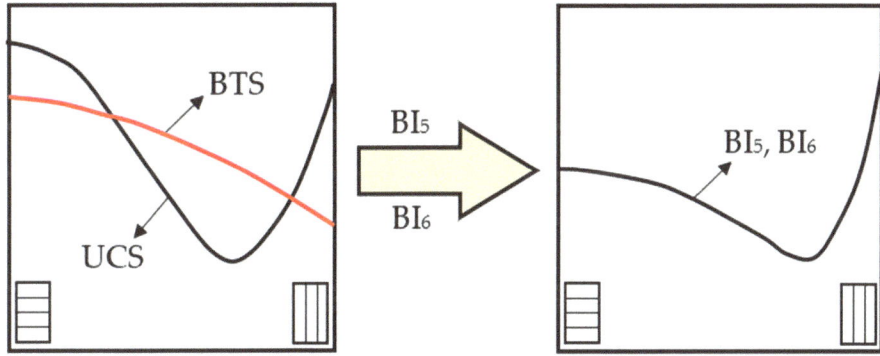

Figure 21. Shale brittleness variation with respect to bedding inclinations, using BI_5 and BI_6.

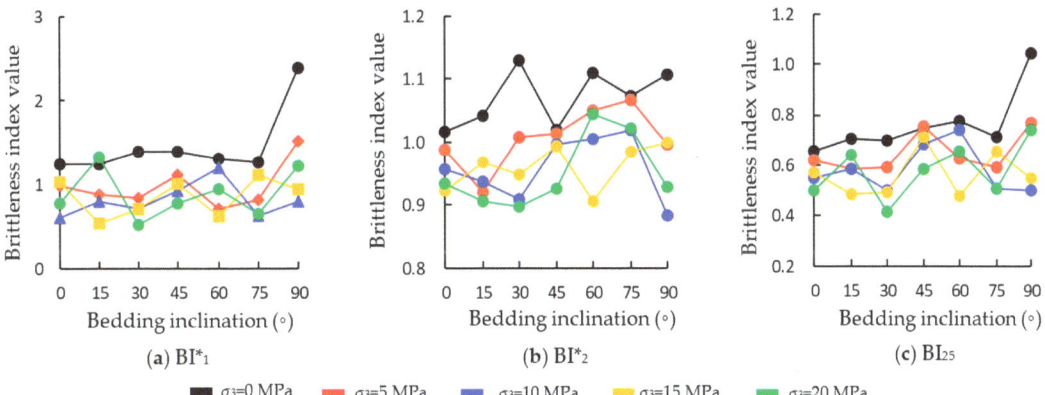

Figure 22. Shale brittleness variation with respect to bedding inclinations, using different BIs [2].

From Figure 22a,c, we can see that the BI_1^* and BI_{25} display very similar variations, roughly increasing with respect to the bedding inclination; in particular, they increase rapidly when the bedding inclination is at 90°. This indicates that the shale specimen shows higher brittleness when the bedding inclination is at a high level. This experimental finding is in accordance with the conclusion stated in Figure 21. In the study of Qian et al. [50], a novel BI is proposed for analyzing the anisotropic characteristics of brittleness [196,197]. Their theoretical research suggests that Young's modulus in the vertical is lower than that in the horizontal, and Poisson's ratio in the vertical can be either higher or lower than that in the horizontal. Thus, the BIs, which are determined by parameters of Young's modulus and Poisson's ratio, can vary significantly when drilling in shale formations with strong anisotropy.

The brittleness, under uniaxial compressive conditions, which results in tensile failure is higher than the brittleness when confining pressure is applied. Therefore, it can be concluded that the shale specimens that failed by tension show higher brittleness than those that failed by shearing. The reason may be attributed to the shearing slip along the fracture plane having led to a more plastic deformation. Specimens with fractures extending along the bedding show higher brittleness than those samples with fractures that extend across the bedding. [2]. Therefore, according to the conclusions from Figure 12 and Table 1, the order of brittleness can be as follows: T–A > T–T > S–A > S–T.

4.2. The Anisotropic Effect of Bedding Plane on Hydraulic Fracture Initiation and Propagation of Shale

The weak bedding has a key effect on hydraulic fracturing characteristics [61]. Many researchers have attempted to reveal the relationship between the breakdown pressure and the bedding inclination. A general conclusion has been reported that the breakdown pressure is considered to be decreased with increases of bedding inclination [40,198]. By comparing different experiments results, it is found that the breakdown pressure and BTS show similar variations, in which the maximum values and minimum values occur at the bedding inclinations of 0° and 90°, respectively [40]. However, the breakdown pressure also can be increased first and then decreased relative to the bedding inclination [61]. Figure 23 depicts the relationship of breakdown pressure and bedding inclination. As can be seen in Figure 23, the breakdown pressure first increases and then decreases, as the bedding inclination increases. The maximum and minimum breakdown pressures occur at 45° and 90°, respectively. This variation agrees well with the tendency determined in the research of Chong et al. [199], which synchronously considered the effect of confining pressure and under a condition of σ_v = 20 MPa, σ_c = 10 MPa, and Q_{inj} = 6 mL/min. In the research of Lin et al. [40] and Zhang et al. [198], the breakdown pressure under triaxial hydraulic fracturing (σ_v = 25 MPa, σ_c = 20 MPa, Q_{inj} = 12 mL/min) shows a decreasing tendency as the bedding inclination increases. The above two different conclusions reveal that the anisotropy of shale breakdown strength depends on the stress conditions, injection rates, and fluid viscosity [198,200]. However, the breakdown pressure seems to fluctuate within a bedding inclination range from 30° to 60°, and the breakdown pressure, mostly at the bedding inclination of 90°, is minimum. The reason for this variation may be attributed to the effects of tensile strength and the critical stress intensity factor (critical SIF). A smaller bedding inclination contributes to more stress components on the bedding planes, enhancing bedding compression, and thus, higher breakdown pressure is needed to initiate a fracture. Additionally, according to some relative theoretical analysis [201,202], it has been revealed that when $\theta < 45°$, the critical SIF increases as the bedding inclination increases, and when $\theta > 45°$, the critical SIF turns to become smaller as the bedding inclination increases. Higher fluid pressure is needed for the fracture propagation when under higher critical SIF, and then breakdown pressure turns to be at its maximum value at $\theta = 45°$.

There are significant interactions between a hydraulic fracture (HF) and joints, bedding planes (BP), and faults [38]. The development of BP is one of the key factors affecting hydraulic fracturing in shale formation [39]. The angle of the HF as it approaches the BP is generally considered to be an important parameter in determining the interaction between them [203–206]. At a low angle (<30°), the HF propagates along the BP, and crosses the BP at a high angle (>60°), while at moderate angles (30°~60°) the fracture arrests [38,39]. These weak BP would be activated during fracturing if the friction coefficient or cohesion is at a low level, which contributes to creating a complex fracture network [203,205–209]. Wang et al. [210] conducted experiments on shale specimens to investigate the fracturing mechanism and the effect of BP. The shale was marine black carbonaceous with obvious bedding structures, which lead to strong anisotropy. They summarized the fracturing failure patterns after hydraulic fracturing. As shown in Figure 24, the BP induce the HF propagating orientation to change. The HF is of a complex pattern controlled by bedding inclination and stress condition. From the test result, the failure patterns can be summarized as follows: (a) central planar fracture, which mostly occurs when $\theta = 0°$ and 90°; (b) deflected fracture, which mostly occurs when $\theta \leq 45°$; and (c) layer-activated fracture, which mostly occurs when $\theta > 45°$. It should be noted that the central planar fracture in the specimen of $\theta = 90°$ is also a layer-activated fracture, which is induced by the vertical bedding planes.

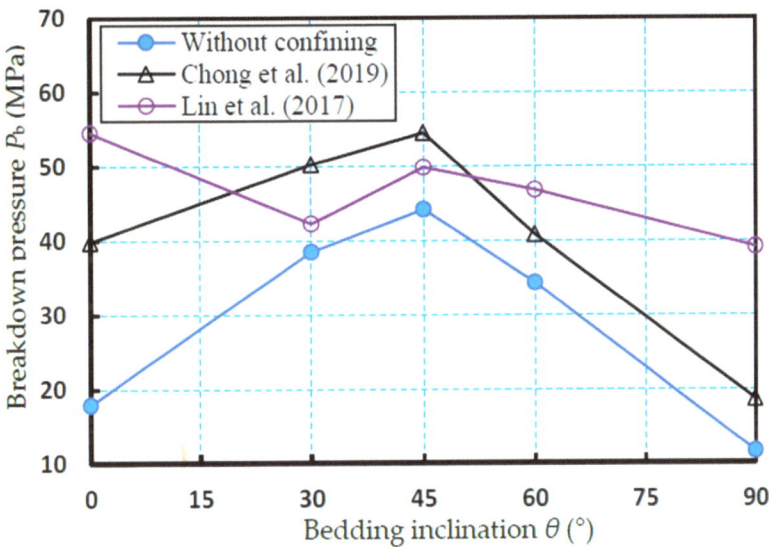

Figure 23. Breakdown pressure variation of shale under different bedding inclinations [40,61].

Figure 24. Anisotropic failure patterns of Longmaxi shale in hydraulic fracturing [210].

In general, two propagation modes can be classified when an HF encounters BP, namely crossing and opening. For further classification, as shown in Figure 25, an HF encountering BP can cause four cases: penetration, diversion, offset, and termination [206,211,212]. When under a true triaxial stress condition, the offset mode is observed during the intersection with BP [213]. Penetration (Figure 25a) refers to an HF that crosses the BP but does not change its propagation path. Diversion (Figure 25b) refers to a vertical HF that is deflected into the BP and is divided into two branches. Offset (Figure 25c) refers to an HF re-initiating and leaving a step-over at the BP. To explain these propagation modes, one possible reason is that the shear stress of the interface easily results in HFs crossing the barrier or extending along the BP [203]. Additionally, the strength of the BP may not be the same at different distances, leading to differences in the degree to which the bedding opens [214], and in that case some HFs may cross the BP and some terminate at the BP. The HF propagation is slightly affected by the BP and the natural fracture (NF) when under large vertical and horizontal stress anisotropy conditions [212].

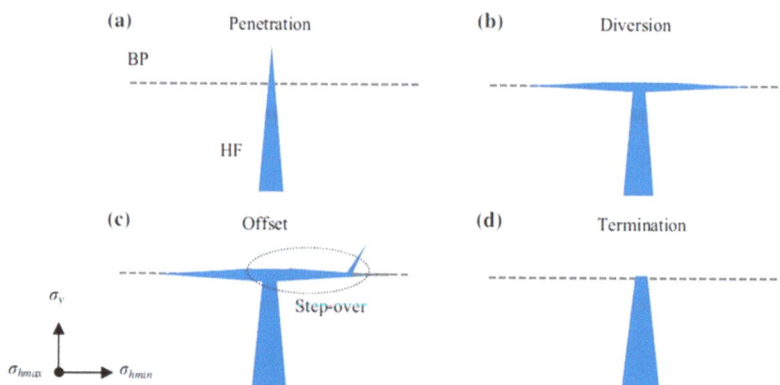

Figure 25. Four basic types of hydraulic fracture intersections with bedding (adapted from Thiercelin et al. [211] and Zou et al. [212]).

For the HF interacting with multi-BP in laminated shale formation, complex fracture networks are of great practical value. The activation of BP is the optimum condition for forming a complex fracture network in the vertical plane [214]. A possible fracture mode is summarized and shown in Figure 26. The depicted four fracture modes are very similar to the fracture types generalized in the study of Yin et al. [215], in which the complex fracture network is formed by primary cross-bedding fractures and secondary interbedding fractures. From the laboratory test results (depicted in Figure 27), it can be determined that interbedding fractures and cross-bedding fractures are generated.

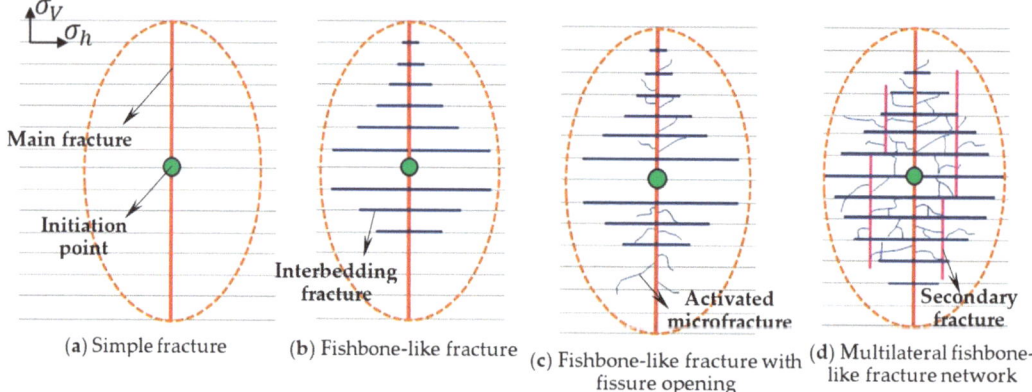

Figure 26. Some basic modes of fracture propagation in a vertical plane, as determined by experiments [214].

The basic HF patterns could be summarized as follows [214] (Figure 28): (1) Mode I (Figure 28a), an HF initiates and propagates perpendicular to the BP; (2) Mode II (Figure 28b), an HF initiates and propagates parallel to the BP; (3) Mode III (Figure 28c), an HF initiates and propagates normal to the BP and a complex fracture network is induced by the weak BP; (4) Mode IV (Figure 28d), an HF initiates and propagates parallel to the BP, and diverts into another propagation path when it meets a cemented fracture; and (5) Mode V (Figure 28e), an HF initiates and propagates from a few NFs, and is diverted into another propagation path by BP. To form the above patterns, the BP strength is the key: In the cases of Mode I and Mode III, the cement strength of the BP should be high, but in Mode III some weak points should be presented. In the cases of Mode II and Mode IV, the cement strength of the BP should be low, but some strong cement BP should be presented in Mode IV. In the case of Mode V, weak NFs occur near the initiation point and the cement of the BP is low. The above summaries are also consistent with the simulation results of Yin et al. [215], which reveal the important effect of the mechanical properties of BP on the propagation of HF. Additionally, the BP failure was closely related to its cohesion and friction coefficient. If the BP had a low friction coefficient or low cohesion, shear failure would obviously increase. If the cementing strength of the BP at different distances is not the same, the degree of the bedding opening would differ [206,216].

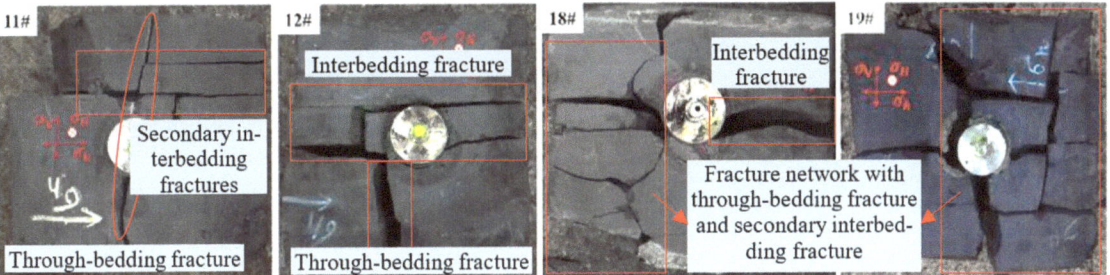

Figure 27. Hydraulic fracture network formed by interbedding fractures and cross-bedding fractures in experiments [214].

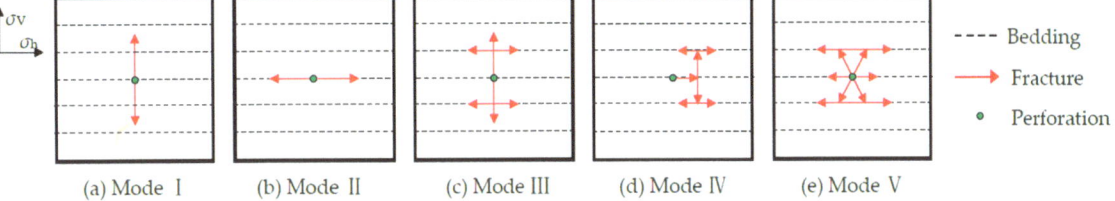

Figure 28. Basic model of hydraulic fracture propagation [214].

5. Conclusions

The efficient implementation of hydraulic fracturing is the key issue in the exploration of shale gas. The existence of bedding structure results in distinct anisotropy of the shale rock formation. The anisotropy has an important influence on wellbore stability and hydraulic fracturing implements. This paper first briefly reviews the previous research works on the lithological characteristics and anisotropic mechanical behavior of shale specimens, and then a brief discussion on the relationship of shale anisotropic mechanical properties to hydraulic fracturing evaluation is conducted. The concerns of anisotropic mechanical behaviors under laboratory tests are addressed; in particular, the evaluation of shale brittleness based on mechanical characteristics is discussed in detail. Further concerns, such as the bedding's effect on the hydraulic fracturing failure strength, crack propagation, and fracture pattern are also drawn. Some main conclusions are drawn, as follows:

(1) The rock mineralogy is critical for the intrinsic formation of anisotropy in shale rock. The proportions of quartz–carbonates–clays could result in very different rock mechanical properties. Clay-rich shale mostly contains highly developed weak planes. Shale samples containing high levels of quartz and low levels of clay have a relatively high Young's modulus and low Poisson's ratio, contributing to a high brittleness and helping hydraulic fracture propagation.

(2) The anisotropy of the wave velocity is thought to result in some geophysical interpretative problems for shale. The effects of mineral grain inclination and micro-cracks are the main factors attributed to an anisotropy. The wave velocities parallel to the bedding are higher than those normal to the bedding, as the ultrasonic wave energy is attenuated by the bedding's scattering. The attenuation anisotropy is affected by the stress, the bedding inclination, and the aspect ratios of micro-cracks.

(3) The mechanical properties of shale specimens under compression show distinct anisotropy. Most determinations of the compression strengths of shale specimens display a "U" type or a similar "shoulder" type variation trend with increasing bedding angle. Particularly, the maximum peak strength is observed at a θ of 0° or 90°, and the minimum values are observed at 45° < θ < 75°. The elastic modulus shows an increasing variation with the bedding inclination increases overall; this is in accord with the variation of ultrasonic velocities. The confining stress reduces the anisotropy of rock material, as the confining pressure is hydrostatic pressure, and when it is applied to the specimen, the beddings are compacted, resulting in the reduction of the primary defects effect. The fracture propagation during the failure process is generally affected by the bedding inclination and confining pressure. For bedding inclination of 0° < θ < 30°, tensile fracture along the loading direction occurs under low confining pressure, and shear fracture across the bedding plane occurs under high confining pressure; for a bedding inclination of 30° < θ < 75°, the fracture generally occurs by the sliding of the bedding, although shear fracture in the matrix can occur, and may cross the bedding; for 75° < θ < 90°, the failure occurs by bedding splitting when under low confining stress, and by bedding shearing under high confining stress.

(4) The tensile strength has a distinct effect on fracture initiation and propagation. The BTS variation in shale specimens from different sampling sites shows a slight difference; in general, their BTS values do not have much change from bedding inclination of 0° to 30°, and then decrease to a low level around 45°–60°, and finally keep within a stable range of around 75°-90°. The center of the disc specimen is in a compressive state when under low bedding inclination and experiencing shearing and tension when under a high bedding inclination. The layer activation occurs under a high bedding inclination, and the central fracture occurs under a low bedding inclination.

(5) The value $K_{Ic,\,ST}$ is much lower than the $K_{Ic,\,A}$ and $K_{Ic,\,D}$; additionally, $K_{Ic,\,A}$ is slightly greater than $K_{Ic,\,D}$. The bedding inclination has a distinct effect on the fracture toughness, the values for which all decrease as the bedding inclination increases. The high approach angle and low bedding strength contribute to the fracture propagation along the bedding planes.

(6) In shale hydraulic fracturing, the brittleness index (BI) is usually employed to quantify the brittleness of the shale rock formation, for the purpose of evaluating the possibilities of fracture propagation (fracability) and fracture network generation. The BI definitions considering mechanical response are more significant. The horizontal well for hydraulic fracturing usually crosses the bedding plane in different orientations; thus, shale anisotropy should not be ignored in brittleness evaluation. The BIs, which are determined by parameters of Young's modulus and Poisson's ratio, can vary significantly when drilling in shale formations with strong anisotropy.

(7) The bedding planes (BP) have an important effect on the fracturing characteristics. The breakdown pressure shows a similar variation with the BTS of shale specimens, in which the maximum and minimum values occur at the bedding inclinations of 0° and 90°, respectively. A smaller bedding inclination contributes to more stress components on the bedding planes, enhancing compression on the beddings and resulting in higher breakdown pressure. The bedding shear stress has an important effect on whether the HFs cross the rock formation or extend along the BP, as well as the fact that the strength of the BP may not be the same at different distances, leading to differences in the bedding opening, in which case some HFs may cross the BP and some terminate at the BP. Moreover, the activation of the BP is the optimum condition for forming fracture networks, as opposed to a simple fracture.

Author Contributions: Conceptualization, P.-F.Y. and S.-Q.Y.; methodology, P.-F.Y., S.-Q.Y., and P.G.R.; investigation, P.-F.Y. and S.-Q.Y.; writing—original draft preparation, P.-F.Y.; writing—review and editing, S.-Q.Y. and P.G.R.; supervision, S.-Q.Y. and P.G.R.; funding acquisition, P.-F.Y. and S.-Q.Y. All authors have read and agreed to the published version of the manuscript.

Funding: This research was supported by the National Natural Science Foundation of China (NO. 42202300, NO. 42077231), the Basic Research Program of Jiangsu Province (Natural Science Foundation) for Youth Foundation (NO. BK20221150), and the Fundamental Research Funds for the Central Universities-Special Funds for the State Key Laboratory (NO. Z21003).

Conflicts of Interest: The authors declare no conflicts of interests.

References

1. Wang, Y.; Zhang, D.; Hu, Y.Z. Laboratory investigation of the effect of injection rate on hydraulic fracturing performance in artificial transversely laminated rock using 3D laser scanning. *Geotech. Geol. Eng.* **2019**, *37*, 2121–2133. [CrossRef]
2. Yang, S.Q.; Yin, P.F.; Ranjith, P.G. Experimental study on mechanical behavior and brittleness characteristics of Longmaxi formation shale in Changning, Sichuan basin, China. *Rock Mech. Rock Eng.* **2020**, *53*, 2461–2483. [CrossRef]
3. Saldungaray, P.; Palisch, T.T. Hydraulic fracture optimization in unconventional reservoirs. In Proceedings of the Society of Petroleum Engineers—SPE Middle East Unconventional Gas Conference and Exhibition, UGAS—Unlocking Unconventional Gas: New Energy in the Middle East, Abu Dhabi, United Arab Emirates, 23–25 January 2012; SPE-151128-MS.
4. Nguyen, V.P.; Lian, H.; Rabczuk, T.; Bordas, S. Modelling hydraulic fractures in porous media using flow cohesive interface elements. *Eng. Geol.* **2017**, *225*, 68–82. [CrossRef]
5. Taghichian, A.; Hashemalhoseini, H.; Zaman, M.; Yang, Z.Y. Geomechanical optimization of hydraulic fracturing in unconventional reservoirs: A semi-analytical approach. *Int. J. Fract.* **2018**, *213*, 107–138. [CrossRef]
6. Jain, R. Natural resource development for science, technology, and environmental policy issues: The case of hydraulic fracturing. *Clean Technol. Environ. Policy* **2015**, *17*, 3–8. [CrossRef]
7. Guo, C.H.; Xu, J.C.; Wei, M.Z.; Jiang, R.Z. Experimental Study and Numerical Simulation of Hydraulic Fracturing Tight Sandstone Reservoirs. *Fuel* **2015**, *159*, 334–344. [CrossRef]
8. Ibanez, W.D.; Kronenberg, A.K. Experimental deformation of shale: Mechanical properties and microstructural indicators of mechanisms. *Int. J. Rock Mech. Min. Sci. Geomech. Abstr.* **1993**, *30*, 723–734. [CrossRef]
9. Horsrud, P.; Sqonstebo, E.F.; Boe, R. Mechanical and petrophysical properties of North Sea shales. *Int. J. Rock Mech. Min. Sci.* **1998**, *35*, 1009–1020. [CrossRef]
10. Al-Bazali, T.; Zhang, J.G.; Chenevert, M.E.; Sharm, M.M. Factors controlling the compressive strength and acoustic properties of shale when interacting with water-based fluids. *Int. J. Rock Mech. Min. Sci.* **2008**, *45*, 729–738. [CrossRef]
11. Fjaer, E.; Nes, O.M. The Impact of Heterogeneity on the Anisotropic Strength of an Outcrop Shale. *Rock Mech. Rock Eng.* **2014**, *47*, 1603–1611. [CrossRef]
12. Holt, R.M.; Larsen, I.; Fjaer, E.; Stenebraten, J.F. Comparing mechanical and ultrasonic behaviour of a brittle and a ductile shale: Relevance to prediction of borehole stability and verification of shale barriers. *J. Pet. Sci. Eng.* **2020**, *187*, 106746. [CrossRef]
13. Gao, Q.; Tao, J.L.; Hu, J.Y.; Yu, X. Laboratory study on the mechanical behaviors of an anisotropic shale rock. *J. Rock Mech. Geotech. Eng.* **2015**, *7*, 213–219. [CrossRef]
14. Amadei, B. Importance of anisotropy when estimating and measuring in situ stresses in rock. *Int. J. Rock Mech. Min. Sci.* **1996**, *33*, 293–325. [CrossRef]
15. Chen, C.; Pan, E.; Amadei, B. Determination of deformability and tensile strength of anisotropic rock using Brazilian tests. *Int. J. Rock Mech. Min. Sci.* **1998**, *35*, 43–61. [CrossRef]
16. Gale, J.F.W.; Reed, R.M.; Holder, J. Natural fractures in the Barnett Shale and their importance for hydraulic fracture treatments. *AAPG Bull.* **2007**, *91*, 603–622. [CrossRef]
17. Gale, J.F.W.; Laubach, S.E.; Olson, J.E.; Eichhuble, P.; Fall, A. Natural Fractures in shale: A review and new observations. *AAPG Bull.* **2014**, *98*, 2165–2216. [CrossRef]
18. Jones, L.E.A.; Wang, H.F. Ultrasonic velocities in Cretaceous shales from the Williston Basin. *Geophysics* **1981**, *46*, 288–297. [CrossRef]
19. Johnston, J.E.; Christensen, N.I. Seismic anisotropy of shales. *J. Geophys. Res. Solid Earth* **1995**, *100*, 5991–6003. [CrossRef]
20. Hornby, B.E. Experimental laboratory determination of the dynamic elastic properties of wet, drained shales. *J. Geophys. Res. Solid Earth* **1998**, *103*, 29945–29964. [CrossRef]
21. Hornby, B.E.; Schwartz, L.M.; Hudson, J.A. Anisotropic effective-medium modeling of the elastic properties of shales. *Geophysics* **1994**, *59*, 1570–1583. [CrossRef]
22. Holt, R.M.; Fjær, E.; Raaen, A.M.; Ringstad, C. Influence of stress state and stress history on acoustic wave propagation in sedimentary rocks. In *Shear Waves in Marine Sediments*; Springer: Dordrecht, The Netherlands, 1991; pp. 167–174.
23. Sayers, C.M. Stress-dependent seismic anisotropy of shales. *Geophysics* **1999**, *64*, 93–98. [CrossRef]
24. Vernik, L.; Landis, C. Elastic anisotropy of source rocks: Implications for hydrocarbon generation and primary migration. *AAPG Bullet.* **1996**, *80*, 531–544.
25. Jin, Z.F.; Li, W.X.; Jin, C.R.; Hambleton, J.; Cusatis, G. Elastic, strength, and fracture properties of Marcellus shale. *Int. J. Rock Mech. Min. Sci.* **2018**, *109*, 124–137. [CrossRef]
26. Loucks, R.G.; Reed, R.M.; Ruppel, S.C.; Hammes, U. Spectrum of pore types and networks in mudrocks and a descriptive classification for matrix-related mudrock pores. *AAPG Bullet.* **2012**, *96*, 1071–1098. [CrossRef]

27. Sone, H.; Zoback, M.D. Mechanical properties of shale-gas reservoir rocks—Part 1: Static and dynamic elastic properties and anisotropy. *Geophysics* **2013**, *78*, 381–392. [CrossRef]
28. Revil, A.; Grauls, D.; Brevart, O. Mechanical compaction of sand/clay mixtures. *J. Geophys. Res.* **2002**, *107*, 2293. [CrossRef]
29. Crawford, B.R.; Faulkner, D.R.; Rutter, E.H. Strength, porosity, and permeability development during hydrostatic and shear loading of synthetic quartz-clay fault gouge. *J. Geophys. Res. Solid Earth* **2008**, *113*, B03207. [CrossRef]
30. Kohli, A.H.; Zoback, M.D. Frictional properties of shale reservoir rocks. *J. Geophys. Res. Solid Earth* **2013**, *118*, 5109–5125. [CrossRef]
31. Bai, B.J.; Elgmati, M.; Zhang, H.; Wei, M.Z. Rock characterization of Fayetteville shale gas plays. *Fuel* **2013**, *105*, 645–652. [CrossRef]
32. Wang, J.X.; Xie, L.Z.; Xie, H.P.; Ren, L.; He, B.; Li, C.B.; Yang, Z.P.; Gao, C. Effect of layer orientation on acoustic emission characteristics of anisotropic shale in Brazilian tests. *J. Nat. Gas Sci. Eng.* **2016**, *36*, 1120–1129. [CrossRef]
33. Hou, P.; Gao, F.; Yang, Y.G.; Zhang, Z.Z.; Zhang, X.X. Effect of bedding orientation on failure of black shale under Brazilian tests and energy analysis. *Chin. J. Geotech. Eng.* **2016**, *38*, 930–937.
34. Li, C.B.; Gao, C.; Xie, H.P.; Li, N. Experimental investigation of anisotropic fatigue characteristics of shale under uniaxial cyclic loading. *Int. J. Rock Mech. Min. Sci.* **2020**, *130*, 104314. [CrossRef]
35. Li, C.B.; Wang, J.; Xie, H.P. Anisotropic creep characteristics and mechanism of shale under elevated deviatoric stress. *J. Pet. Sci. Eng.* **2020**, *185*, 106670. [CrossRef]
36. Heng, S.; Yang, C.H.; Zhang, B.P.; Guo, Y.T.; Wang, L.; Wei, Y.L. Experimental research on anisotropic properties of shale. *Chin. J. Rock Soil Mech.* **2015**, *36*, 609–616.
37. Luo, Y.; Xie, H.P.; Ren, L.; Zhang, R.; Li, C.B.; Gao, C. Linear elastic fracture mechanics characterization of an anisotropic shale. *Sci. Rep.* **2018**, *8*, 8505. [CrossRef] [PubMed]
38. Zou, Y.S.; Zhang, S.C.; Zhou, T.; Zhou, X.; Guo, T.K. Experimental investigation into hydraulic fracture network propagation in gas shales using CT scanning technology. *Rock Mech. Rock Eng.* **2016**, *49*, 33–45.
39. Guo, T.K.; Zhang, S.C.; Qu, Z.Q.; Zhou, T.; Xiao, Y.S.; Gao, J. Experimental study of hydraulic fracturing for shale by stimulated reservoir volume. *Fuel* **2014**, *128*, 373–380. [CrossRef]
40. Lin, C.; He, J.M.; Li, X.; Wan, X.L.; Zheng, B. An experimental investigation into the effects of the anisotropy of shale on hydraulic fracture propagation. *Rock Mech. Rock Eng.* **2017**, *50*, 543–554. [CrossRef]
41. Masri, M.; Sibai, M.; Shao, J.F.; Mainguy, M. Experimental investigation of the effect of temperature on the mechanical behavior of Tournemire shale. *Int. J. Rock Mech. Min. Sci.* **2014**, *70*, 185–191. [CrossRef]
42. Rybacki, E.; Reinicke, A.; Meier, T.; Makasi, M.; Dresen, G. What controls the mechanical properties of shale rocks?—Part I: Strength and young's modulus. *J. Pet. Sci. Eng.* **2015**, *135*, 702–722. [CrossRef]
43. Rybacki, E.; Meier, T.; Dresen, G. What controls the mechanical properties of shale rocks?—Part II: Brittleness. *J. Pet. Sci. Eng.* **2016**, *144*, 39–58. [CrossRef]
44. Mokhtari, M.; Alqahtani, A.A.; Tutuncu, A.N. Failure Behavior of Anisotropic Shales. In Proceedings of the 47th US Rock Mechanics/Geomechanics Symposium, San Francisco, CA, USA, 23–26 June 2013.
45. Heng, S.; Yang, C.H.; Guo, Y.T.; Wang, C.Y.; Wang, L. Influence of bedding planes on hydraulic fracture propagation in shale formations. *Chin. J. Rock Mech. Eng.* **2015**, *34*, 228–237.
46. Heng, S.; Guo, Y.T.; Yang, C.H.; Daemen, J.J.K.; Li, Z. Experimental and theoretical study of the anisotropic properties of shale. *Int. J. Rock Mech. Min. Sci.* **2015**, *74*, 58–68. [CrossRef]
47. Jarvie, D.M.; Hill, R.J.; Ruble, T.E.; Pollastro, R.M. Unconventional shale-gas systems: The Mississippian Barnett shale of north-central Texas as one model for thermogenic shale-gas assessment. *AAPG Bull.* **2007**, *91*, 475–499. [CrossRef]
48. Li, Q.H.; Chen, M.; Jin, Y.; Wang, F.P.; Hou, B.; Zhang, B. Indoor evaluation method for shale brittleness and improvement. *Chin. J. Rock Mech. Eng.* **2012**, *31*, 1680–1685.
49. Kivi, I.R.; Ameri, M.; Molladavoodi, H. Shale brittleness evaluation based on energy balance analysis of stress-strain curves. *J. Pet. Sci. Eng.* **2018**, *167*, 1–19. [CrossRef]
50. Qian, K.R.; Liu, T.; Liu, J.Z.; Liu, X.W.; He, Z.L.; Jiang, D.J. Construction of a novel brittleness index equation and analysis of anisotropic brittleness characteristics for unconventional shale formations. *Pet. Sci.* **2020**, *17*, 70–85. [CrossRef]
51. Huang, X.R.; Huang, J.P.; Li, Z.C.; Yang, Q.Y.; Sun, Q.X.; Cui, W. Brittleness index and seismic rock physics model for anisotropic tight-oil sandstone reservoirs. *Appl. Geophys.* **2015**, *12*, 11–22. [CrossRef]
52. Schmidt, R.A.; Huddle, C.W. *Fracture Mechanics of Oil Shale: Some Preliminary Results*; Report No. SAND-76-0727; Sandia Labs: Albuquerque, NM, USA, 1997; 29p. [CrossRef]
53. Lee, H.P.; Olson, J.E.; Holder, J.; Gale, J.F.W.; Myers, R.D. The interaction of propagating opening mode fractures with preexisting discontinuities in shale. *J. Geophys. Res. Solid Earth* **2015**, *120*, 169–181. [CrossRef]
54. Chandler, M.R.; Meredith, P.G.; Brantut, N.; Crawford, B.R. Fracture toughness anisotropy in shale. *J. Geophys. Res. Solid Earth* **2016**, *121*, 1706–1729. [CrossRef]
55. Yin, S.; Lv, D.W.; Jin, L.; Ding, W.L. Experimental analysis and application of the effect of stress on continental shale reservoir brittleness. *J. Geophys. Eng.* **2018**, *15*, 478–494. [CrossRef]
56. Zhao, Y.; He, P.F.; Zhang, Y.F.; Wang, C.L. A new criterion for a toughness-dominated hydraulic fracture crossing a natural frictional interface. *Rock Mech. Rock Eng.* **2019**, *52*, 2617–2629. [CrossRef]

57. Zhao, Y.; Zhang, Y.F.; He, P.F. A composite criterion to predict subsequent intersection behavior between a hydraulic fracture and a natural fracture. *Eng. Fract. Mech.* **2019**, *209*, 61–78. [CrossRef]
58. Kou, M.M.; Liu, X.R.; Tang, S.D.; Wang, Y.T. 3-D X-ray computed tomography on failure characteristics of rock-like materials under coupled hydro-mechanical loading. *Theor. Appl. Fract. Mech.* **2019**, *104*, 102396. [CrossRef]
59. Kou, M.M.; Liu, X.R.; Wang, Y.T. Study on rock fracture behavior under hydromechanical loading by 3-D digital reconstruction. *Struct. Eng. Mech.* **2020**, *74*, 283–296.
60. Zhang, Y.F.; Zhao, Y.; Yang, H.Q.; Wang, C.L. A semianalytical solution for a Griffith crack nonuniformly pressurized by internal fluid. *Rock Mech. Rock Eng.* **2020**, *53*, 2439–2460. [CrossRef]
61. Zhao, Y.; Zhang, Y.F.; Wang, C.L.; Liu, Q. Hydraulic fracturing characteristics and evaluation of fracturing effectiveness under different anisotropic angles and injection rates: An experimental investigation in absence of confining pressure. *J. Nat. Gas Sci. Eng.* **2022**, *97*, 104343. [CrossRef]
62. Yaalon, D.H. Mineral composition of average shale. *Clay Min. Bull.* **1962**, *5*, 31–36. [CrossRef]
63. Sliwinski, J.; Harrington, J.; Power, M.; Hughes, P.; Yeung, B. A high definition mineralogical examination of potential gas shales. Abstract Volume. In Proceedings of the AAPG—Annual Convention and Exhibition, New Orleans, LA, USA, 11–14 April 2010; p. 239.
64. Lyu, Q.; Shi, J.D.; Ranjith, P.G. Effects of testing method, lithology and fluid-rock interactions on shale permeability: A review of laboratory measurements. *J. Nat. Gas Sci. Eng.* **2020**, *78*, 1033023. [CrossRef]
65. Minaeian, V.; Dewhurst, D.N.; Rasouli, V. Deformational behaviour of a clay-rich shale with variable water saturation under true triaxial stress conditions. *Geomech. Energy Environ.* **2017**, *11*, 1–13. [CrossRef]
66. Wu, C.J.; Tuo, J.C.; Zhang, L.F.; Zhang, M.F.; Li, J.; Liu, Y.; Qian, Y. Pore characteristics differences between clay-rich and clay-poor shales of the Lower Cambrian Niutitang Formation in the Northern Guizhou area, and insights into shale gas storage mechanisms. *Int. J. Coal Geol.* **2017**, *178*, 13–25. [CrossRef]
67. Zhang, J.J.; Kamenov, A.; Zhu, D.; Hill, A.D. Development of new testing procedures to measure propped fracture conductivity considering water damage in clay-rich shale reservoirs: An example of the Barnett Shale. *J. Pet. Sci. Eng.* **2015**, *135*, 352–359. [CrossRef]
68. Amalokwu, K.; Spikes, K.; Wolf, K. A simple effective medium approach for the bulk electrical and elastic properties of organic-rich shales. *J. Appl. Geophys.* **2019**, *169*, 98–108. [CrossRef]
69. Wang, Y.P.; Liu, X.J.; Liang, L.X.; Xiong, J. Experimental study on the damage of organic-rich shale during water-shale interaction. *J. Nat. Gas Sci. Eng.* **2020**, *74*, 103103. [CrossRef]
70. Wei, M.M.; Zhang, L.; Xiong, Y.Q.; Peng, P.A. Main factors influencing the development of nanopores in over-mature, organic-rich shales. *Int. J. Coal Geol.* **2019**, *212*, 103233. [CrossRef]
71. Ilgen, A.G.; Aman, M.; Espinoza, D.N.; Rodriguez, M.A.; Griego, J.M.; Dewers, T.A.; Feldman, J.D.; Stewart, T.A.; Choens, R.C.; Wilson, J. Shale-brine-CO$_2$ interactions and the long-term stability of carbonate-rich shale caprock. *Int. J. Greenh. Gas Control* **2018**, *78*, 244–253. [CrossRef]
72. Teklu, T.W.; Abass, H.H.; Hanashmooni, R.; Carratu, J.C.; Ermila, M. Experimental investigation of acid imbibition on matrix and fractured carbonate rich shales. *J. Nat. Gas Sci. Eng.* **2017**, *45*, 706–725. [CrossRef]
73. Rickman, R.; Mullen, M.J.; Petre, J.E.; Grieser, W.V.; Kundert, D. A practical use of shale petrophysics for stimulation design optimization: All shale plays are not clones of the Barnett shale. In Proceedings of the SPE Annual Technical Conference and Exhibition, Society of Petroleum Engineers, Denver, CO, USA, 21–24 September 2008.
74. Civan, F. *Reservoir Formation Damage*; Gulf Publishing Company: Houston, TX, USA, 2007.
75. Loucks, R.G.; Ruppel, S.C. Mississippian Barnett Shale: Lithofacies and depositional setting of a deep-water shale-gas succession in the Fort Worth Basin, Texas. *AAPG Bull.* **2007**, *91*, 579–601. [CrossRef]
76. Li, G.F.; Jin, Z.J.; Li, X.; Liu, K.Q.; Yang, W.C.; Qiao, M.T.; Zhou, T.T.; Sun, X.K. Experimental study on mechanical properties and fracture characteristics of shale layered samples with different mineral components under cyclic loading. *Mar. Pet. Geol.* **2023**, *150*, 106114. [CrossRef]
77. Wang, Y.; Li, C.H. Investigation of the p- and s-wave velocity anisotropy of a Longmaxi formation shale by real-time ultrasonic and mechanical experiments under uniaxial deformation. *J. Pet. Sci. Eng.* **2017**, *158*, 253–267. [CrossRef]
78. Niandou, H.; Shao, J.F.; Henry, J.P.; Fourmaintraux, D. Laboratory investigation of the mechanical behaviour of Tournemire shale. *Int. J. Rock Mech. Min. Sci.* **1997**, *34*, 3–16. [CrossRef]
79. Jia, Y.Z.; Tang, J.R.; Lu, Y.Y.; Lu, Z.H. Laboratory geomechanical and petrophysical characterization of Longmaxi shale properties in Lower Silurian Formation, China. *Mar. Pet. Geol.* **2020**, *124*, 104800. [CrossRef]
80. Liu, Y.; Ma, T.; Wu, H.; Chen, P. Investigation on mechanical behaviors of shale cap rock for geological energy storage by linking macroscopic to mesoscopic failures. *J. Energy Storage* **2020**, *29*, 101326. [CrossRef]
81. Yang, S.Q.; Yin, P.F.; Li, B.; Yang, D.S. Behavior of transversely isotropic shale observed in triaxial tests and Brazilian disc tests. *Int. J. Rock Mech. Min. Sci.* **2020**, *133*, 104435. [CrossRef]
82. Brahma, J.; Sircar, A. Estimation of the effect of anisotropy on Young's moduli and Poisson's ratios of sedimentary rocks using core samples in western and central part of Tripura, India. *Int. J. Geosci.* **2014**, *05*, 184–195. [CrossRef]
83. Lo, T.W.; Coyner, K.B.; Toksoz, M.N. Experimental determination of elastic anisotropy of Berea sandstone, Chicopee shale, and Chelmsford granite. *Geophysics* **1986**, *51*, 164–171. [CrossRef]

84. Allan, A.M.; Kanitpanyacharoen, W.; Vanorio, T. A multiscale methodology for the analysis of velocity anisotropy in organic-rich shale. *Geophysics* **2015**, *80*, 73–88. [CrossRef]
85. Alejano, L.R.; González-Fernández, M.A.; Estévez-Ventosa, X.; Song, F.; Delgado-Martín, J.; Muoz-Ibáez, A.; Gonzalez-Molano, N.; Alvarellos, J. Anisotropic deformability and strength of slate from NW-Spain. *Int. J. Rock Mech. Min. Sci.* **2021**, *148*, 104923. [CrossRef]
86. Hou, Z.K.; Yang, C.H.; Guo, Y.T.; Zhang, B.P.; Wei, Y.L.; Heng, S.; Wang, L. Experimental study on anisotropic properties of Longmaxi formation shale under uniaxial compression. *Chin. J. Rock Soil Mech.* **2015**, *36*, 2541–2550.
87. Dewhurst, D.N.; Siggins, A.F.; Sarout, J.; Raven, M.D.; Nordgård-Bolås, H.M. Geomechanical and ultrasonic characterization of a Norwegian Sea shale. *Geophysics* **2011**, *76*, 101–111. [CrossRef]
88. Zhubayev, A.; Houben, M.E.; Smeulders, D.M.J.; Barnhoorn, A. Ultrasonic velocity and attenuation anisotropy of shales, Whitby, United Kingdom. *Geophysics* **2015**, *81*, 45–56. [CrossRef]
89. Jin, G.; Ali, S.S.; Abdullah, A.; Dhamen, A. Mechanical anisotropy of unconventional shale—Build the correct relationship between static and dynamic properties. In Proceedings of the 2016 Abu Dhabi International Petroleum Exhibition & Conference, Society of Petroleum Engineers, Abu Dhabi, United Arab Emirates, 7–10 November 2016; pp. 1–9.
90. Iferobia, C.C.; Ahmad, M. A review on the experimental techniques and applications in the geomechanical evaluation of shale gas reservoirs. *J. Nat. Gas Sci. Eng.* **2020**, *74*, 103090. [CrossRef]
91. Rasouli, V. Geomechanics of gas shales. In *Fundamentals of Gas Shale Reservoirs*; Rezaee, R., Ed.; John Wiley & Sons, Inc.: Hoboken, NJ, USA, 2015; pp. 169–190.
92. Panfiloff, A. *Experimental Evaluation of Dynamic Elastic Properties and Anisotropy in Shales*; Colorado School of Mines: Golden, CO, USA, 2016.
93. Yin, P.F.; Yang, S.Q. Experimental study on strength and failure behavior of transversely isotropic rock-like material under uniaxial compression. *Geomech. Geophys. Geo-Energy Geo-Resour.* **2020**, *6*, 44. [CrossRef]
94. Kuila, U.; Dewhurst, D.N.; Siggins, A.F.; Raven, M.D. Stress anisotropy and velocity anisotropy in low porosity shale. *Tectonophysics* **2011**, *50*, 34–44. [CrossRef]
95. Cho, J.W.; Kim, H.; Jeon, S.; Min, K.B. Deformation and strength anisotropy of Asan gneiss, Boryeong shale, and Yeoncheon schist. *Int. J. Rock Mech. Min. Sci.* **2012**, *50*, 158–169. [CrossRef]
96. Kim, H.; Cho, J.W.; Song, I.; Min, K.B. Anisotropy of elastic moduli, p-wave velocities, and thermal conductivities of Asan gneiss, Boryeong shale, and Yeoncheon schist in Korea. *Eng. Geol.* **2012**, *147–148*, 68–77. [CrossRef]
97. Yang, S.Q.; Yin, P.F.; Huang, Y.H. Experiment and discrete element modelling on strength, deformation and failure behaviour of shale under Brazilian compression. *Rock Mech. Rock Eng.* **2019**, *52*, 4339–4359. [CrossRef]
98. Josh, M.; Esteban, L.; Piane, C.D.; Sarout, J.; Dewhurst, D.N.; Clennell, M.B. Laboratory characterisation of shale properties. *J. Pet. Sci. Eng.* **2012**, *88–89*, 107–124. [CrossRef]
99. Wu, S.; Ge, H.K.; Wang, X.Q.; Meng, F.B. Shale failure processes and spatial distribution of fractures obtained by AE monitoring. *J. Nat. Gas Sci. Eng.* **2017**, *41*, 82–92. [CrossRef]
100. Li, X.L.; Lei, X.; Li, Q.; Li, X. Experimental investigation of Sinian shale rock under triaxial stress monitored by ultrasonic transmission and acoustic emission. *J. Nat. Gas Sci. Eng.* **2017**, *43*, 110–123. [CrossRef]
101. Nasseri, M.H.; Rao, K.S.; Ramamurthy, T. Failure mechanism in schistose rocks. *Int. J. Rock Mech. Min. Sci.* **1997**, *34*, 219. [CrossRef]
102. Nasseri, M.H.; Rao, K.S.; Ramamurthy, T. Anisotropic strength and deformational behavior of Himalayan schists. *Int. J. Rock Mech. Min. Sci.* **2003**, *40*, 3–23. [CrossRef]
103. Hakala, M.; Kuula, H.; Hudson, J.A. Estimating the transversely isotropic elastic intact rock properties for in situ stress measurement data reduction. A case study of the Olkiluoto mica gneiss, Finland. *Int. J. Rock Mech. Min. Sci.* **2007**, *44*, 14–46. [CrossRef]
104. Yin, P.F.; Yang, S.Q. Experimental investigation of the strength and failure behavior of layered sandstone under uniaxial compression and Brazilian testing. *Acta Geophys.* **2018**, *66*, 585–605. [CrossRef]
105. Ramamurthy, T. Strength, modulus responses of anisotropic rocks. In *Compressive Rock Engineering*; Hudson, J.A., Ed.; Pergamon: Oxford, UK, 1993; Volume 1, pp. 313–329.
106. Gao, C.; Xie, L.Z.; Xie, H.P.; He, B.; Jin, W.C.; Li, F.; Yang, Z.P.; Sun, Y.Z. Estimation of the equivalent elastic modulus in shale formation: Theoretical model and experiment. *J. Pet. Sci. Eng.* **2017**, *151*, 468–479. [CrossRef]
107. Zuo, J.P.; Lu, J.F.; Ghandriz, R.; Wang, J.T.; Li, Y.H.; Zhang, X.Y.; Li, J.; Li, H.T. Mesoscale fracture behavior of Longmaxi outcrop shale with different bedding angles: Experimental and numerical investigations. *J. Rock Mech. Geotech. Eng.* **2020**, *12*, 297–309. [CrossRef]
108. Hu, J.J.; Gao, C.; Xie, H.P.; Wang, J.; Li, M.H.; Li, C.B. Anisotropic characteristics of the energy index during the shale failure process under triaxial compression. *J. Nat. Gas Sci. Eng.* **2021**, *95*, 104219. [CrossRef]
109. Jaeger, J.C. Shear failure of transversely isotropic rock. *Geol. Mag.* **1960**, *97*, 65–72. [CrossRef]
110. Hoek, E.; Brown, E.T. *Underground Excavations in Rock*; Institution of Mining and Metallurgy: London, UK, 1980.
111. Rao, K.S.; Rao, G.V.; Ramamurthy, T. A strength criterion for anisotropic rocks. *Indian Geotech. J.* **1986**, *16*, 317–333.
112. Ramamurthy, T.; Rao, G.V.; Singh, J. A strength criterion for anisotropic rocks. In Proceedings of the Fifth Australia–New Zealand Conference on Geomechanics, Sydney, Australia, 22–26 August 1988; Volume 1, pp. 253–257.

113. Tien, Y.M.; Kuo, M.C. A failure criterion for transversely isotropic rocks. *Int. J. Rock Mech. Min. Sci.* **2001**, *38*, 399–412. [CrossRef]
114. Saeidi, O.; Vaneghi, R.G.; Rasouli, V.; Gholami, R. A modified empirical criterion for strength of transversely anisotropic rocks with metamorphic origin. *Bull. Eng. Geol. Environ.* **2013**, *72*, 257–269. [CrossRef]
115. Saeidi, O.; Vaneghi, R.G.; Rasouli, V.; Gholami, R.; Torabi, S.R. A modified failure criterion for transversely isotropic rocks. *Geosci. Front.* **2014**, *5*, 215–225. [CrossRef]
116. Singh, M.; Samadhiya, N.K.; Kumar, A. A nonlinear criterion for triaxial strength of inherently anisotropic rocks. *Rock Mech. Rock Eng.* **2015**, *48*, 1387–1405. [CrossRef]
117. Shi, X.C.; Yang, X.; Meng, X.F. An anisotropic strength model for layered rocks considering planes of weakness. *Rock Mech. Rock Eng.* **2016**, *49*, 3783–3792. [CrossRef]
118. Rafiai, H. New empirical polyaxial criterion for rock strength. *Int. J. Rock Mech. Min. Sci.* **2011**, *48*, 922–931. [CrossRef]
119. Chen, L.; Shao, J.F.; Huang, H.W. Coupled elastoplastic damage modeling of anisotropic rocks. *Comput. Geotech.* **2010**, *37*, 187–194. [CrossRef]
120. Chen, L.; Shao, J.F.; Zhu, Q.Z.; Duveau, G. Induced anisotropic damage and plasticity in initially anisotropic sedimentary rocks. *Int. J. Rock Mech. Min. Sci.* **2012**, *51*, 13–23. [CrossRef]
121. Yao, C.; Jiang, Q.H.; Shao, J.F.; Zhou, C.B. A discrete approach for modeling damage and failure in anisotropic cohesive brittle materials. *Eng. Fract. Mech.* **2016**, *155*, 102–118. [CrossRef]
122. Qi, M.; Giraud, A.; Colliat, J.B.; Shao, J.F. A numerical damage model for initially anisotropic materials. *Int. J. Solids Struct.* **2016**, *100–101*, 245–256. [CrossRef]
123. Qi, M.; Shao, J.F.; Giraud, A.; Zhu, Q.Z.; Colliat, J.B. Damage and plastic friction in initially anisotropic quasi brittle materials. *Int. J. Plast.* **2016**, *82*, 260–282. [CrossRef]
124. Tien, Y.M.; Kuo, M.C.; Juang, C.H. An experimental investigation of the failure mechanism of simulated transversely isotropic rocks. *Int. J. Rock Mech. Min. Sci.* **2006**, *43*, 1163–1181. [CrossRef]
125. Deng, J.X.; Shi, G.; Liu, R.X.; Yu, J. Analysis of the velocity anisotropy and its affection factors in shale and mudstone. *Chin. J. Geophys.* **2004**, *47*, 862–868.
126. Zhao, C.X.; Liu, J.F.; Xu, D.; Zhang, L.Q.; Lyu, C.; Ren, Y. Investigation on Mechanical Properties, AE Characteristics, and Failure Modes of Longmaxi Formation Shale in Changning, Sichuan Basin, China. *Rock Mech. Rock Eng.* **2023**, *56*, 1239–1272. [CrossRef]
127. Amadei, B.; Rogers, J.D.; Goodman, R.E. Elastic Constants and Tensile Strength of Anisotropic Rocks. In Proceedings of the 5th ISRM Congress, Melbourne, Australia, 10–15 April 1983; ISRM-5CONGRESS-1983-030.
128. Lekhnitskii, S.G. *Anisotropic Plates*; Gordon and Breach: New York, NY, USA, 1968.
129. Tavallali, A.; Vervoort, A. Failure of transversely isotropic rock material: Effect of layer orientation and material properties. In Proceedings of the 6th International Symposium on Ground Support in Mining and Civil Engineering Construction, Cape Town, South Africa, 30 March–3 April 2008; pp. 317–328.
130. Tavallali, A.; Vervoort, A. Failure of layered sandstone under Brazilian test conditions: Effect of micro-scale parameters on macro-scale behaviour. *Rock Mech. Rock Eng.* **2010**, *43*, 641–653. [CrossRef]
131. Tavallali, A.; Vervoort, A. Effect of layer orientation on the failure of layered sandstone under Brazilian test conditions. *Int. J. Rock Mech. Min. Sci.* **2010**, *47*, 313–322. [CrossRef]
132. Tavallali, A.; Vervoort, A. Behaviour of layered sandstone under Brazilian test conditions: Layer orientation and shape effects. *J. Rock Mech. Geotech. Eng.* **2013**, *5*, 366–377. [CrossRef]
133. Vervoort, A.; Min, K.B.; Konietzky, H.; Cho, J.W.; Debecker, B.; Din, Q.D.; Frühwirt, T.; Tavallali, A. Failure of transversely isotropic rock under Brazilian test conditions. *Int. J. Rock Mech. Min. Sci.* **2014**, *70*, 343–352. [CrossRef]
134. Tan, X.; Konietzky, H.; Frühwirt, T.; Dan, D.Q. Brazilian tests on transversely isotropic rocks: Laboratory testing and numerical simulations. *Rock Mech. Rock Eng.* **2015**, *48*, 1341–1351. [CrossRef]
135. Yang, Z.P.; He, B.; Xie, L.Z.; Li, C.B.; Wang, J. Strength and failure modes of shale based on Brazilian test. *Chin. J. Rock Soil Mech.* **2015**, *36*, 3447–3464.
136. Hou, P.; Gao, F.; Yang, Y.G.; Zhang, Z.Z.; Gao, Y.N.; Zhang, X.X.; Zhang, J. Effect of bedding plane direction on acoustic emission characteristics of shale in Brazilian tests. *Chin. J. Rock Soil Mech.* **2016**, *37*, 1603–1612.
137. Du, M.P.; Pan, P.Z.; Ji, W.W.; Zhang, Z.H.; Gao, Y.H. Time-space laws of failure process of carbonaceous shale in Brazilian split test. *Chin. J. Rock Soil Mech.* **2016**, *37*, 3437–3446.
138. He, J.; Afolagboye, L.O. Influence of layer orientation and interlayer bonding force on the mechanical behavior of shale under Brazilian test conditions. *Acta Mech. Sin.* **2018**, *34*, 349–358. [CrossRef]
139. Simpson, N.D.J.; Stroisz, A.; Bauer, A.; Vervoort, A.; Holt, R.M. Failure mechanics of anisotropic shale during Brazilian tests. In Proceedings of the 48th US Rock Mechanics/Geomechanics Symposium, Minneapolis, MN, USA, 1–4 June 2014; ARMA 14-7399.
140. Chandler, M.; Meredith, P.; Crawford, B. Experimental determination of the fracture toughness and brittleness of the Mancos shale. *Utah Br. J. Psychiatry* **2013**, *184*, 110–117.
141. Chen, M.; Zhang, G.Q. Laboratory measurement and interpretation of the fracture toughness of formation rocks at great depth. *J. Pet. Sci. Eng.* **2004**, *41*, 221–231. [CrossRef]
142. Awaji, H.; Sato, S. Combined mode fracture toughness measurement by the disk test. *J. Eng. Mater. Technol.* **1978**, *100*, 175–182. [CrossRef]

143. Atkinson, C.; Smelser, R.E.; Sanchez, J. Combined mode fracture via the cracked Brazilian disk test. *Int. J. Fract.* **1982**, *18*, 279–291. [CrossRef]
144. Chong, K.P.; Kuruppu, M.D. New specimen for fracture toughness determination for rock and other materials. *Int. J. Fract.* **1984**, *26*, 59–62. [CrossRef]
145. Lim, I.L.; Johnston, I.W.; Choi, S.K.; Boland, J.N. Fracture testing of a soft rock with semi circular specimens under three point bending. Part 2—Mixed-mode. *Int. J. Rock Mech. Min. Sci.* **1994**, *31*, 199–212. [CrossRef]
146. Dai, F.; Chen, R.; Xia, K. A semi-circular bend technique for determining dynamic fracture toughness. *Exp. Mech.* **2010**, *50*, 783–791. [CrossRef]
147. Funatsu, T.; Shimizu, N.; Kuruppu, M.; Matsui, K. Evaluation of mode I fracture toughness assisted by the numerical determination of K-resistance. *Rock Mech. Rock Eng.* **2014**, *48*, 143–157. [CrossRef]
148. Ren, L.; Xie, L.Z.; Xie, H.P.; Ai, T.; He, B. Mixed-mode fracture behavior and related surface topography feature of a typical sandstone. *Rock Mech. Rock Eng.* **2016**, *49*, 3137–3153. [CrossRef]
149. *ASTM E399–12e3*; Standard Test Method for Linear-Elastic Plane-Strain Fracture Toughness KIc of Metallic Materials. ASTM International: West Conshohocken, PA, USA, 2009.
150. Bush, A.J. Experimentally determined stress-intensity factors for single-edge-crack round bars loaded in bending. *Exp. Mech.* **1976**, *16*, 249–257. [CrossRef]
151. Luo, Y.; Ren, L.; Xie, L.Z.; Ai, T.; He, B. Fracture behavior investigation of a typical sandstone under mixed-mode I/II loading using the notched deep beam bending method. *Rock Mech. Rock Eng.* **2017**, *50*, 1987–2005. [CrossRef]
152. Wu, S.X. Fracture toughness determination of bearing steel using chevron-notch three point bend specimen. *Eng. Fract. Mech.* **1984**, *19*, 221–232. [CrossRef]
153. Ren, L.; Xie, H.P.; Sun, X.; Zhang, R.; Li, C.B.; Xie, J.; Zhang, Z.T. Characterization of Anisotropic Fracture Properties of Silurian Longmaxi Shale. *Rock Mech. Rock Eng.* **2021**, *54*, 665–678. [CrossRef]
154. Sheity, D.K.; Rosenfield, A.R.; Duckworth, W.H. Fracture toughness of ceramics measured by a chevron-notch diametral-compression test. *J. Am. Ceram. Soc.* **1985**, *68*, C325–C327. [CrossRef]
155. Fowell, R.J. Suggested method for determining mode I fracture toughness using cracked chevron notched Brazilian disc (CCNBD) specimens. *Int. J. Rock Mech. Min. Sci.* **1995**, *32*, 57–64. [CrossRef]
156. Dai, F.; Wei, M.D.; Xu, N.W.; Ma, Y.; Yang, D.S. Numerical assessment of the progressive rock fracture mechanism of cracked chevron notched Brazilian disc specimens. *Rock Mech. Rock Eng.* **2014**, *48*, 463–479. [CrossRef]
157. Kuruppu, M.D. Fracture toughness measurement using chevron notched semi-circular bend specimen. *Int. J. Fract.* **1997**, *86*, L33–L38.
158. Dai, F.; Xia, K.; Zheng, H.; Wang, Y.X. Determination of dynamic rock Mode-I fracture parameters using cracked chevron notched semi-circular bend specimen. *Eng. Fract. Mech.* **2011**, *78*, 2633–2644. [CrossRef]
159. Barker, L.M. A simplified method for measuring plane strain fracture toughness. *Eng. Fract. Mech.* **1977**, *9*, 361–369. [CrossRef]
160. Ouchterlony, F. A Presentation of the ISRM Suggested Methods for Determining Fracture Toughness of Rock Material. In Proceedings of the 6th ISRM Congress, Montreal, QC, Canada, 30 August–3 September 1987.
161. Wang, Q.Z.; Xing, L. Determination of fracture toughness KIC by using the flattened Brazilian disk specimen for rocks. *Eng. Fract. Mech.* **1999**, *64*, 193–201. [CrossRef]
162. Schmidt, R.A. Fracture Mechanics of Oil Shale—Unconfined Fracture Toughness, Stress Corrosion Cracking, and Tension Test Results. In Proceedings of the 18th U.S. Symposium on Rock Mechanics (USRMS), Golden, CO, USA, 22–24 June 1977.
163. Chandler, M.R.; Fauchille, A.L.; Kim, H.K.; Ma, L.; Mecklenburgh, J.; Rizzo, R.; Mostafavi, M.; Marussi, S.; Atwood, R.; May, S.; et al. Correlative optical and X-ray imaging of strain evolution during double-torsion fracture toughness measurements in shale. *J. Geophys. Res. Solid Earth* **2018**, *123*, 10517–10533. [CrossRef]
164. Xiong, J.; Liu, K.Y.; Liang, L.X.; Liu, X.J.; Zhang, C.Y. Investigation of influence factors of the fracture toughness of shale: A case study of the Longmaxi formation shale in Sichuan basin, China. *Geotech. Geol. Eng.* **2019**, *37*, 2927–2934. [CrossRef]
165. Wang, H.J.; Zhao, F.; Huang, Z.Q.; Yao, Y.M.; Yuan, G.X. Experimental study of mode-I fracture toughness for layered shale based on two ISRM-suggested methods. *Rock Mech. Rock Eng.* **2017**, *50*, 1933–1939. [CrossRef]
166. Dou, F.K.; Wang, J.G.; Zhang, X.X.; Wang, H.M. Effect of joint parameters on fracturing behavior of shale in notched three-point-bending test based on discrete element model. *Eng. Fract. Mech.* **2019**, *205*, 40–56. [CrossRef]
167. Li, W.X.; Jin, Z.F.; Cusatis, G. Size effect analysis for the characterization of marcellus shale quasi-brittle fracture properties. *Rock Mech. Rock Eng.* **2019**, *52*, 1–18. [CrossRef]
168. Inskip, F.N.D.; Meredith, P.G.; Chandler, M.R.; Gudmundsson, A. Fracture properties of Nash Point shale as a function of orientation to bedding. *J. Geophys. Res. Solid Earth* **2018**, *123*, 8428–8444. [CrossRef]
169. Lei, B.; Zuo, J.P.; Liu, H.Y.; Wang, J.T.; Xu, F.; Li, H.T. Experimental and numerical investigation on shale fracture behavior with different bedding properties. *Eng. Fract. Mech.* **2021**, *247*, 107639. [CrossRef]
170. Shi, X.S.; Zhao, Y.X.; Danesh, N.N.; Zhang, X.; Tang, T.W. Role of bedding plane in the relationship between Mode-I fracture toughness and tensile strength of shale. *Bull. Eng. Geol. Environ.* **2022**, *81*, 81. [CrossRef]
171. Zhang, Y.; Li, T.Y.; Xie, L.Z.; Yang, Z.P.; Li, R.Y. Shale lamina thickness study based on micro-scale image processing of thin sections. *J. Nat. Gas Sci. Eng.* **2017**, *46*, 817–829. [CrossRef]

172. Zhou, Q.; Xie, H.P.; Zhu, Z.M.; He, R.; Lu, H.J.; Fan, Z.D.; Nie, X.F.; Ren, L. Fracture Toughness Anisotropy in Shale Under Deep in Situ Stress Conditions. *Rock Mech. Rock Eng.* **2023**, *56*, 7535–7555. [CrossRef]
173. Zhang, D.C.; Ranjith, P.G.; Perera, M.S.A. The brittleness indices used in rock mechanics and their application in shale hydraulic fracturing: A review. *J. Pet. Sci. Eng.* **2016**, *143*, 158–170. [CrossRef]
174. Morley, A. *Strength of Materials: With 260 Diagrams and Numerous Examples*; Longmans, Green and Company: New York, NY, USA, 1944.
175. Hetényi, M. *Handbook of Experimental Stress Analysis*; Wiley: New York, NY, USA, 1950.
176. Obert, L.; Duvall, W.I. *Rock Mechanics and the Design of Structures in Rock*; Wiley: New York, NY, USA, 1967.
177. Ramsey, J. *Folding and Fracturing of Rock*; McGraw-Hill: New York, NY, USA, 1968.
178. Howell, J.V. *Glossary of Geology and Related Sciences*; American Geological Institute: Washington, DC, USA, 1960.
179. Tarasov, B.; Potvin, Y. Universal criteria for rock brittleness estimation under triaxial compression. *Int. J. Rock Mech. Min. Sci.* **2013**, *59*, 57–69. [CrossRef]
180. Hucka, V.; Das, B. Brittleness determination of rocks by different methods. *Int. J. Rock Mech. Min. Sci.* **1974**, *17*, 389–392. [CrossRef]
181. Altindag, R. The evaluation of rock brittleness concept on rotary blast hole drills. *J. South. Afr. Inst. Min. Metall.* **2002**, *102*, 61–66.
182. Hajiabdolmajid, V.; Kaiser, P.; Martin, C.D. Mobilised strength components in brittle failure of rock. *Géotechnique* **2003**, *53*, 327–336. [CrossRef]
183. Nygård, R.; Gutierrez, M.; Bratli, R.K.; Høeg, K. Brittle–ductile transition, shear failure and leakage in shales and mudrocks. *Mar. Pet. Geol.* **2006**, *23*, 201–212. [CrossRef]
184. Yagiz, S. Assessment of brittleness using rock strength and density with punch penetration test. *Tunn. Undergr. Space Technol.* **2009**, *24*, 66–74. [CrossRef]
185. Holt, R.M.; Fjaer, E.; Nes, O.M.; Alassi, H.T. A shaly look at brittleness. In Proceedings of the 45th US Rock Mechanics/Geomechanics Symposium, San Francisco, CA, USA, 26–29 June 2011. 10p.
186. Jin, X.; Shah, S.N.; Roegiers, J.C.; Zhang, B. Fracability evaluation in shale reservoirs—An integrated petrophysics and geomechanics approach. In Proceedings of the SPE Hydraulic Fracturing Technology Conference, The Woodlands, TX, USA, 4–6 February 2014.
187. Jin, X.C.; Shah, S.N.; Truax, J.A.; Roegiers, J.C. A Practical Petrophysical Approach for Brittleness Prediction from Porosity and Sonic Logging in Shale Reservoirs. In Proceedings of the 2014 SPE Annual Technical Conference and Exhibition, Amsterdam, The Netherlands, 27–29 October 2014.
188. Wang, F.P.; Gale, J.F. Screening criteria for shale-gas systems. *Gulf Coast Assoc. Geol. Soc. Trans.* **2009**, *59*, 779–793.
189. Jin, X.C.; Shah, S.N.; Roegiers, J.C.; Zhang, B. An integrated petrophysics and geomechanics approach for fracability evaluation in shale reservoirs. *SPE J.* **2015**, *20*, 518–526. [CrossRef]
190. Bishop, A.W. Progressive failure with special reference to the mechanism causing it. *Proc. Geotech. Conf.* **1967**, *2*, 142–150.
191. Andreev, G.E. *Brittle Failure of Rock Materials: Test Results and Constitutive Models*; A.A. Balkema: Rotterdam, The Netherlands, 1995; p. 446.
192. Luan, X.Y.; Di, B.R.; Wei, J.X.; Li, X.Y.; Qian, K.; Xie, J.Y.; Ding, P.B. Laboratory Measurements of brittleness anisotropy in synthetic shale with different cementation. In Proceedings of the 2014 SEG Annual Meeting, Denver, CO, USA, 26–31 October 2014; pp. 3005–3009.
193. Munoz, H.; Taheri, A.; Chanda, E.K. Fracture energy-based brittleness index development and brittleness quantification by pre-peak strength parameters in rock uniaxial compression. *Rock Mech. Rock Eng.* **2016**, *49*, 4587–4606. [CrossRef]
194. Ai, C.; Zhang, J.; Li, Y.W.; Zeng, J.; Yang, X.L.; Wang, J.G. Estimation criteria for rock brittleness based on energy analysis during the rupturing process. *Rock Mech. Rock Eng.* **2016**, *49*, 4681–5698. [CrossRef]
195. Yang, Y.; Sone, H.; Hows, A.; Zoback, M.D. Comparison of Brittleness Indices in Organic-rich Shale Formations. In Proceedings of the 47th US Rock Mechanics/Geomechanics Symposium, San Francisco, CA, USA, 23–26 June 2013.
196. Qian, K.R.; Zhang, F.; Li, X.Y. A rock physics model for estimating elastic properties of organic shales. In Proceedings of the 76th EAGE Conference and Exhibition, Amsterdam, The Netherlands, 16–19 June 2014.
197. Qian, K.R.; Zhang, F.; Chen, S.Q.; Li, X.Y.; Zhang, H. A rock physics model for analysis of anisotropic parameters in a shale reservoir in Southwest China. *J. Geophys. Eng.* **2016**, *13*, 19–34. [CrossRef]
198. Zhang, Y.X.; He, J.M.; Li, X.; Lin, C. Experimental study on the supercritical CO_2 fracturing of shale considering anisotropic effects. *J. Pet. Sci. Eng.* **2019**, *173*, 932–940. [CrossRef]
199. Chong, Z.H.; Yao, Q.L.; Li, X.H. Experimental investigation of fracture propagation behavior induced by hydraulic fracturing in anisotropic shale cores. *Energies* **2019**, *12*, 976. [CrossRef]
200. Hadei, M.R.; Veiskarami, A.; Sherizadeh, T.; Sunkpal, M. A laboratory investigation of the effect of bedding plane inclination angle on hydro-fracturing breakdown pressure in stratified rocks. In Proceedings of the 53rd U.S. Rock Mechanics/Geomechanics Symposium, New York, NY, USA, 23–26 June 2019.
201. Asadpoure, A.; Mohammadi, S. Developing New Enrichment Functions for Crack Simulation in Orthotropic Media by the Extended Finite Element Method. *Int. J. Numer. Methods Eng.* **2007**, *69*, 2150–2172. [CrossRef]
202. Wang, X.L.; Shi, F.; Liu, H.; Wu, H.A. Numerical Simulation of Hydraulic Fracturing in Orthotropic Formation Based on the Extended Finite Element Method. *J. Nat. Gas Sci. Eng.* **2016**, *33*, 56–69. [CrossRef]
203. Warpinski, N.R.; Teufel, L.W. Influence of geologic discontinuities on hydraulic fracture propagation. *J. Pet. Technol.* **1987**, *39*, 209–220. [CrossRef]

204. Zhou, J.; Chen, M.; Jin, Y.; Zhang, G.Q. Analysis of fracture propagation behavior and fracture geometry using a triaxial fracturing system in naturally fractured reservoirs. *Int. J. Rock Mech. Min. Sci.* **2008**, *45*, 1143–1152. [CrossRef]
205. Gu, H.R.; Weng, X.W. Criterion for fractures crossing frictional interfaces at non-orthogonal angles. In *Rock Mechanics Symposium: Proceedings of the 44th US Rock Mechanics Symposium and 5th US–Canada Rock Mechanics Symposium, Salt Lake City, UT, USA, 27–30 June 2010*, American Rock Mechanics Association: Alexandria, VA, USA, 2010.
206. Gu, H.R.; Weng, X.W.; Lund, J.B.; Mack, G.M.; Ganguly, U.; Suarez-Rivera, R. Hydraulic fracture crossing natural fracture at nonorthogonal angles: A criterion and its validation applications. *SPE Prod. Oper.* **2012**, *27*, 20–26. [CrossRef]
207. Blanton, T.L. Propagation of hydraulically and dynamically induced fractures in naturally fractured reservoirs. In Proceedings of the SPE/DOE Unconventional Gas Technology Symposium, Louisville, KY, USA, 18–21 May 1986.
208. Renshaw, C.E.; Pollard, D.D. An experimentally verified criterion for propagation across unbonded frictional interfaces in brittle, linear elastic materials. *Int. J. Rock Mech. Min. Sci.* **1995**, *32*, 237–249. [CrossRef]
209. Sarmadivaleh, M.; Rasouli, V. Modified Reinshaw & Pollard criteria for a non-orthogonal cohesive natural interface intersected by an induced fracture. *Rock Mech. Rock Eng.* **2013**, *47*, 2107–2115.
210. Wang, J.; Xie, H.P.; Li, C.B. Anisotropic failure behaviour and breakdown pressure interpretation of hydraulic fracturing experiments on shale. *Int. J. Rock Mech. Min. Sci.* **2021**, *142*, 104748. [CrossRef]
211. Thiercelin, M.; Roegiers, J.C.; Boone, T.J.; Ingraffea, A.R. An investigation of the material parameters that govern the behavior of fractures approaching rock interfaces. In Proceedings of the 6th International Congress of Rock Mechanics, Montreal, Canada, 30 August–3 September 1987.
212. Zou, Y.S.; Ma, X.F.; Zhang, S.C.; Zhou, T.; Li, H. Numerical investigation into the influence of bedding plane on hydraulic fracture network propagation in shale formations. *Rock Mech. Rock Eng.* **2016**, *49*, 3597–3614.
213. Huang, B.X.; Liu, J.W. Experimental investigation of the effect of bedding planes on hydraulic fracturing under true triaxial stress. *Rock Mech. Rock Eng.* **2017**, *50*, 2627–2643. [CrossRef]
214. Tan, P.; Jin, Y.; Han, K.; Hou, B.; Chen, M.; Guo, X.F.; Gao, J. Analysis of hydraulic fracture initiation and vertical propagation behavior in laminated shale formation. *Fuel* **2017**, *206*, 482–493. [CrossRef]
215. Yin, P.F.; Yang, S.Q.; Gao, F.; Tian, W.L.; Zeng, W. Numerical investigation on hydraulic fracture propagation and multi-perforation fracturing for horizontal well in Longmaxi shale reservoir. *Theor. Appl. Fract. Mech.* **2023**, *125*, 103921. [CrossRef]
216. Cheng, W.; Jin, Y.; Chen, M.A.; Xu, T.; Zhang, Y.K.; Diao, C. A criterion for identifying hydraulic fractures crossing natural fractures in 3D space. *Pet. Explor. Dev.* **2014**, *41*, 371–376. [CrossRef]

Disclaimer/Publisher's Note: The statements, opinions and data contained in all publications are solely those of the individual author(s) and contributor(s) and not of MDPI and/or the editor(s). MDPI and/or the editor(s) disclaim responsibility for any injury to people or property resulting from any ideas, methods, instructions or products referred to in the content.

Article

The Mechanism Study of Fracture Porosity in High-Water-Cut Reservoirs

Ning Zhang [1,2], Daiyin Yin [1,2,*], Guangsheng Cao [1,2] and Tong Li [1,2]

[1] Key Laboratory of Enhanced Oil Recovery, Northeast Petroleum University, Ministry of Education, Daqing 163318, China; nzh1515@163.com (N.Z.); nepucgswz@163.com (G.C.); sinei1234@126.com (T.L.)
[2] College of Petroleum Engineering, Northeast Petroleum University, Daqing 163318, China
* Correspondence: nepuydy@163.com

Abstract: Many onshore oil fields currently adopt water flooding as a means to supplement reservoir energy. However, due to reservoir heterogeneity, significant differences in permeability exist not only between different reservoirs but also within the same reservoir across different planar orientations. After prolonged fluid flushing in the near-wellbore zone of injection wells, the resulting increased flow resistance between layers exacerbates inefficient and ineffective circulation. A considerable amount of remaining oil is left unexploited in untouched areas, significantly impacting the overall recovery. To investigate the multiscale plugging mechanisms of fracture-dominated pore channels in high-water-cut oil reservoirs and achieve efficient management of fractured large channels, this study explores the formation of the fracture-flushing zone-low saturation oil zone. A physical experimental model with fractures and high-intensity flushing is established to analyze changes in pore structure, mineral composition, residual oil distribution, and other characteristics at different positions near the fractures. The research aims to clarify the mechanism behind the formation of large channels with fracture structures. The results indicate that under high-intensity water flushing, cementing materials are washed away by the flowing water, clay particles are carried to the surface with the injected fluid, and permeability significantly increases, forming high-permeability zones with fracture structures. In the rock interior away from the fracture end, channels, corners, and clustered oil content noticeably decrease, while the content of film-like oil substantially increases, and clay minerals are not significantly washed away. Under strong flushing conditions, the number of residual clay particles near the fracture end is mainly influenced by flow velocity and flushing time; thus, the greater the flushing intensity, the faster the water flow, and the longer the flushing time, the fewer residual clay particles near the fracture end.

Keywords: high-water-cut oil reservoirs; fractures; large channels; inefficient and ineffective circulation

Citation: Zhang, N.; Yin, D.; Cao, G.; Li, T. The Mechanism Study of Fracture Porosity in High-Water-Cut Reservoirs. *Energies* 2024, 17, 1886. https://doi.org/10.3390/en17081886

Academic Editor: Hossein Hamidi

Received: 11 March 2024
Revised: 31 March 2024
Accepted: 9 April 2024
Published: 16 April 2024

Copyright: © 2024 by the authors. Licensee MDPI, Basel, Switzerland. This article is an open access article distributed under the terms and conditions of the Creative Commons Attribution (CC BY) license (https://creativecommons.org/licenses/by/4.0/).

1. Introduction

After prolonged development of water-flooded oilfields, injecting water near the wellbore area can easily lead to the formation of relatively complex high-permeability zones. Particularly, following extended water flooding, differences in the structure and saturation levels of fractures and surrounding pore channels (dominant fluid flow pathways within the reservoir) can establish dominant flow pathways, significantly impacting the development efficiency of water flooding [1–8]. Through research, scholars have primarily focused on the formation, distribution status, and identification methods of large channels within actual formations. However, there has been a lack of further investigation into the mechanisms behind their formation, and a clear classification of large channels primarily dominated by fractures has not been established [9–15].

The current research on large channels can be traced back to the 1950s. In the United States, well-to-well tracer technology was employed to study the heterogeneity of reservoirs, which used tracer curves to invert corresponding logging features of the reservoir.

Curve characteristics were utilized to calculate the variations in permeability, thereby reflecting the thickness and depth positions of high-permeability layers in the formation. In summary, early research on large channels primarily relied on logging methods to explore and calculate reservoirs, drawing conclusions based on exploration results [10,16,17]. Brigham et al. [18] utilized well-to-well tracer technology to describe the reservoir's heterogeneity, obtaining logging features of small layers through tracers and predicting changes in permeability to identify high-permeability zones within the layers. In 1966, Thomas M. Garland [19] proposed selecting injection wells during profile control to avoid damage to the formation. In 1973, Vatkins and Y.P. Richard [20] suggested two more precise methods for identifying large channels: using water-absorption profile data combined with production logging data and combining porosity gamma-ray logging data with injection volume data. In 1987, K.S. Chan [21] established a material balance equation for particle transport, incorporating different sedimentation and release rates, forming a complete model for particle transport. This laid the theoretical foundation for the identification mechanism of dominant flow pathways.

Furthermore, some experts and scholars have not only conducted research on the identification of dominant flow pathways but have also expanded their research focus to the study of inter-well connectivity. The study of inter-well connectivity is based on the concept of system analysis, which utilizes production dynamic data and combining geological statistical foundations to evaluate inter-well connectivity [22]. In 2005, Yousef AA [23] established a novel volumetric model for inferring inter-well connectivity that considers compressibility and conduction effects. This method has three advantages: it can be applied to wells that have been shut in or have been producing for an extended period; it is suitable for regions producing primarily residual oil; and it can be combined with bottomhole pressure data to enhance the understanding of inter-well connectivity.

In terms of quantifying dominant flow pathways, researchers have also conducted studies. In 1979, Felsenthal [24], building on planar radial flow, treated injection wells and production wells as sources and sinks of fluid flow, respectively, to establish a simple model for calculating high-permeability zones. In 1983, Maghsood Abbaszadeh-Dehghani [10] used inter-well tracer methods to invert bottom-layer parameters for identifying dominant flow pathways. In 2009, C.S. Kabir [25] proposed an improved Hall curve method to identify dominant flow pathways and quantitatively calculate parameters such as pore radius, permeability, effective thickness, and more. In 2010, Shawket Ghedan et al. [26], through a study on a fan-shaped model of heterogeneous carbonate reservoirs, determined the impact of varying reservoir heterogeneity parameters on the formation of dominant channels, including parameters such as horizontal permeability and thickness. The study also explored the effect of gravity on leakage. In 2012, Saeid Sadeghnejad [27] used flow theory to quantify connectivity between injection and production wells in such systems, considering hydraulic conductivity and breakthrough time behavior. The study considered a three-dimensional overlapping sand body model and validated the permeation method based on the Burgn offshore reservoir dataset. Compared to computer-calculated results, this method demonstrated good application effectiveness. In 2016, Ding S.W. [28] proposed a model based on reservoir information for each grid, utilizing automatic history matching and fuzzy methods to determine the depth and parameters of different regions in the reservoir.

At present, the identification methods of large pores can be divided into four categories: inter-well tracer monitoring method, well test method, logging data interpretation method, and reservoir engineering synthesis method [29–31]. Inter-well tracer monitoring method is to inject tracer from the well, then monitor whether the produced liquid contains tracer, and draw a curve according to the tracer production concentration and time and use the result curve given by the tracer to analyze the underground reservoir parameters more carefully. The well test method reflects the development of large pores through the changes in pressure and production parameters, and can predict whether there are large pores in the formation through the changes. Log interpretation data method is used to determine whether there is a large hole by logging methods, without complicated calculation or numerical simulation. Based on relevant logging data and logging curve waterlogging characteristics, formation development can be predicted. Reservoir engineering synthesis method is to combine mathematical model, laboratory experiment, and field data, analyze these data comprehensively to judge whether there are large pores in the formation, and calculate the characteristic parameters of large pores [32]. In conclusion, the inter-well tracer monitoring method has the advantage of relatively accurately determining the direction of large channels. However, this method involves complex on-site processes, high costs, and substantial workload. The logging interpretation data method can determine the development of large channels at different periods, and it is relatively simple to apply. Nevertheless, this method is influenced by factors such as geological conditions, construction techniques, and logging instruments, resulting in ambiguity and uncertainty in the interpretation results. Reservoir engineering methods involve multiple factors, exhibit ambiguity, and most existing data focus on large channel identification methods related to reservoirs, lacking mechanistic analysis of laboratory-based physical models. Further analysis of the characteristics of large channels is necessary.

Therefore, in order to understand the development effect of fractured large channels after water drive development and analyze the formation mechanism of fracture-erosion zone-low saturation oil zone in injection wells of high-water-cut reservoirs, this paper establishes a high-intensity water drive erosion experimental model with fractures. By means of scanning electron microscope, X-ray diffractometer, and constant velocity mercury injection instrument, the variation characteristics of mineral composition, pore structure, and microscopic pore characteristics at different locations under strong scour conditions were analyzed so as to clarify the formation mechanism of fracture-scour zone and low-saturation zone in high-water reservoir, and provide theoretical basis for subsequent efficient treatment.

2. Materials and Methods

2.1. Materials

The experimental materials include the following: Berea sandstone cores (with a permeability of 500×10^{-3} μm^2), crude oil (from the Daqing Oilfield South Block, Daqing, China), kerosene, xylene, distilled water, sodium chloride, potassium chloride, calcium chloride, sodium carbonate, sodium bicarbonate, and sodium sulfate.

Prior to the experiment, Berea sandstone cores with a permeability of 500×10^{-3} μm^2 were selected. Using a core-cutting machine, the cores were cut into two blocks with dimensions of 15 cm in length and 4.5 cm in height. The dry weight of the cores was measured using a balance. Subsequently, the cores were vacuum-saturated with formation water using a vacuum pump, followed by oil saturation simulation using a cubic core holder. Finally, the cores were removed, and the rock specifications were adjusted using the core-cutting machine.

2.2. Experimental Apparatus and Methods

2.2.1. High Intensity Water Drive Core Scour Facility

An experimental model of fractured high-intensity water flooding was established to analyze the change in characteristics of pore structure, mineral composition, and remaining oil distribution at different locations near fractures. The experimental process is shown in Figure 1.

The experiment equipment included a fractured pore channel simulation device, a cuboid core holder, a vacuum pump, a core-cutting machine, ASPE-730 Automatic Pore Structure Tester (constant-rate mercury injection apparatus), the instruments purchased from Beijing Bida Xingye Technology Development Co., Ltd. (Beijing, China) X-ray diffractometer, fluorescence microscope, scanning electron microscope, gas permeameter, balance, ISCO pump, among others.

In the high-strength scour core experiment with cracks, the core was split from the middle by the core splitting device, the cut core was fitted into the core holder, and it was sealed well, and the overlying rock pressure was set at 10 MPa (simulating the actual overlying rock pressure). The 50 PV simulated formation water was displaced at a high flow rate of 3 mL/min.

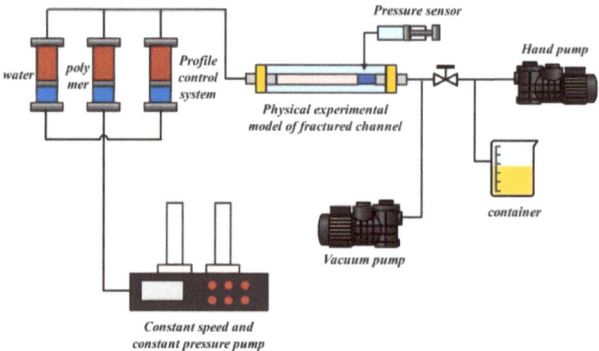

Figure 1. Flow chart of high strength erosion core experiment.

2.2.2. Pore Structure Types of Fractured Channels

The pore structure size and type of the core were observed by scanning electron microscopy. Before the scanning electron microscope test, the core after high intensity washing was cut by a cutting machine, and the core pieces of 5~10 mm were polished and sprayed with gold. The core sample was placed on a conductive adhesive and placed inside a scanning electron microscope instrument for observation.

2.2.3. Core Mineral Composition Analysis

The mineral composition and relative content of the rock were analyzed using X-ray diffraction (XRD) to determine the lithology of different regions within the fractured pore channels. After cutting the core into sections at different positions, some crushed samples were powdered and added to a glass slide circular groove. Another glass slide was used to compact the powder, ensuring a smooth surface. The XRD instrument was then activated and adjusted to analyze the mineral composition in the rock samples.

2.2.4. Microscopic Pore Size Distribution Characteristics of Fractured Channels

The crude oil inside the core was washed with xylene, and the mineral composition and pore structure of the core were observed by mercury injection at constant rate after treatment. In the constant velocity mercury injection experiment, the mercury injection was carried out at a rate of 0.00005 mL/s. The change in pore shape at the leading edge of mercury would cause the change in meniscus shape and lead to the change in capillary pressure in the system. In the experiment, the changes in pressure and the volume of mercury liquid were recorded constantly, and the data were used to analyze the core at different locations.

2.2.5. Residual Oil Observation Experiment

The core was taken out after the high intensity water drive core scour experiment, and the changes in remaining oil in the core at different positions were observed by fluorescence microscope.

A microscopic approach was employed to study the distribution of residual oil. The rock cores, subjected to high-intensity flushing, were cut into slices of approximately 5 mm at different positions. These slices were frozen, sealed, and stored for 24 h. Thin sections of 0.05 mm were prepared by polishing the slices, and the residual oil inside the rock cores was observed using a fluorescence microscope.

2.3. Numerical Simulations

2.3.1. Geometric Model

However, due to the limited experimental conditions, it was impossible to accurately describe the factors affecting the formation of fracture-scour zone and low-saturation zone in high-water reservoir. Using finite element simulation methods, fluid flow characteristics near the fracture tip were simulated, and the fluid flow and clay particle transport characteristics within the fracture-flushing zone and the flushing zone-low saturation zone were investigated. The simulation of the porous media and mesh division is illustrated in Figure 2.

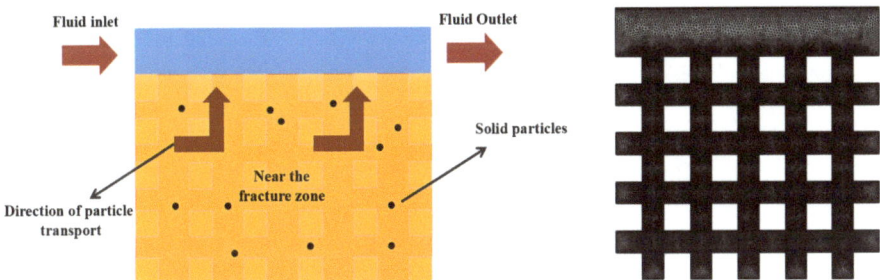

Figure 2. Simulation using geometric models and mesh division results.

2.3.2. Theoretical Model

Considering the actual movement of fluids and clay particles within the fractures and flushing zones, the model accounts for the flow of fluids within the fractures, the carrying force of fluids on particles, and factors such as the gravity and buoyancy of clay particles. The specific model is as follows:

(1) Flow of fluid within the fracture

During the sand flushing process, the flow of fluid within the fracture is generally laminar. The flow of fluid can be represented by Equations (1) and (2):

$$\rho \frac{\partial u}{\partial t} + \rho(u \cdot \nabla)u = \nabla \cdot [-p_l + \mu(\nabla \mu + (\nabla \mu)^T] + F + \rho g \quad (1)$$

where P—the movable cemented clay microparticle diameter, m; F—the volume force exerted on the fluid, N/m^3; g—gravity acceleration, N/kg; p_l—pressure of fluid inside the fracture, Pa; T—temperature, K; t—non-dimensional time; u—fluid flow velocity within the fracture, m/s; ρ—fluid density, kg/m^3; and μ—fluid viscosity, Pa·s.

$$\rho \nabla \cdot (u) = 0 \quad (2)$$

(2) The carrying force of fluid on particles

$$F_D = \frac{1}{\tau_P} m_P (u' - v) \quad (3)$$

$$\tau_P = \frac{\rho_p d_p^2}{18\mu} \quad (4)$$

where F_D—the carrying force of fluid on movable cemented clay microparticles, N; m_P—Mass of movable cemented clay microparticles, kg; τ_P—time corresponding to the velocity of movable cemented clay microparticles, m/s; and ρ_p—density of movable cemented clay microparticles, kg/m^3.

(3) Inter-particle collision force.

$$m_{p1}v_1 + m_{p2}v_2 = m_{p1}v_1' + m_{p2}v_2' \qquad (5)$$

where m_{p1}, m_{p2}—Mass of movable cemented clay microparticles, kg; v_1, v_2—velocity of movable cemented clay microparticles before mutual collision, m/s; and v_1', v_2'—velocity of movable cemented clay microparticles after mutual collision, m/s.

Force of gravity and buoyancy acting on the particles:

$$F_g = \frac{\pi d_p^3}{6} g \frac{\rho_p - \rho}{\rho_p} \qquad (6)$$

$$FG = mpg \qquad (7)$$

where F_g—buoyancy acting on movable cemented clay microparticles, N; F_G—gravity acting on movable cemented clay microparticles, N.

(4) Initial conditions

Initial conditions for the fluid:

$$p_i = p + p_h \qquad (8)$$

$$p_i = pgh \qquad (9)$$

where p_i—pressure caused by the gravitational effect on the fluid, Pa.

Initial conditions for particles:

$$q = q_0 \qquad (10)$$

$$v = v_0 \qquad (11)$$

(5) Boundary conditions

Fluid at the crack surface:

$$u_1 = 0 \qquad (12)$$

where u_1—fluid flow velocity at the wall position, m/s.

Rebound of movable cemented clay microparticles at the crack surface:

$$v = v_c - 2(n \cdot v_c)n \qquad (13)$$

where v—velocity of movable cemented clay microparticles, m/s; v_c—velocity of movable cemented clay microparticles before colliding with the crack surface, m/s; and n—number of collisions of movable cemented clay microparticles.

2.3.3. Boundary Condition

The flow velocity of the fluid under strong flushing conditions was set at 0.05 m/s, with pore throat size of 4 μm, rock particle size of 5 μm × 5 μm, model size of 50 μm × 50 μm, 500 clay particles, release time of 0 s, clay particle density of 2200 kg/m^3, diameter of 2 μm, and the injected fluid is water with a density of 1000 kg/m^3. The simulation time was set to 5000 μs with a time step of 1 μs.

3. Results and Discussion

3.1. Analysis of Microscopic Structural Characteristics of Near-Fracture Pore Channels

Six core samples from the strong flushing zone, low-saturation oil zone, and deep matrix zone were selected for microscopic structure observation. For rocks at different locations, the internal pore structure was observed through scanning electron microscopy (SEM). The observed pore types in the core samples were recorded, as shown in Table 1. Samples 1 and 2 belong to the core of strong erosion zone, samples 3 and 4 belong to the core of low saturation oil, and samples 5 and 6 belong to the core of matrix penetration zone.

Table 1. Data table of pore parameters and pore types in core samples.

Core ID	Pore Parameters		Pore Type
	Pore Size (μm)	Maximum Pore Size (μm)	
1#	10~70	200	Predominantly dissolution pores, with secondary intergranular pores and intercrystalline pores.
2#	5~50	70	
3#	10~80	200	Predominantly intergranular pores, with secondary dissolution pores.
4#	20~80	100	
5#	20~60	100	Mainly composed of intergranular pores, with secondary dissolution pores.
6#	10~70	120	

The changes in pore structure size and types of the core under the scanning electron microscope (SEM) are shown in Figure 3. By observing with SEM, the pore size of the core can be roughly inferred. The micro-pore structure inside the rock will undergo significant changes under different flushing conditions. A comparison of each core sample allows observation of different positions within the core.

Figure 3. Microscopic pore structure of rock at different positions. (a) 1#: 0.5 cm from the fracture; (b) 2#: 1.5 cm from the fracture; (c) 3#: 2.5 cm from the fracture; (d) 4#: 3.5 cm from the fracture; (e) 5#: Matrix rock; (f) 6#: Matrix rock.

Comparing sample 1# with samples 5# and 6#, it is evident that the cement content of the rock in the near-fracture area significantly decreases due to the influence of fluid flushing. This indicates that although there is a fracture structure in the formation, the cementation of the rock near the fracture is washed away and carried by the high-intensity water drive flushing. The gaps between rock particles increase, causing clay particles to move with the injected fluid, forming a high-permeability zone centered around the fracture. Comparing rock samples 1# to 4#, it is observed that the closer to the fracture, the stronger the flushing effect. This is because the fluid in the near-fracture area is less affected by the flow resistance, and as the fluid passes through the fracture, it also extends to the surrounding area. This leads to a larger displacement of the fluid near the fracture end, resulting in a stronger flushing effect on the matrix.

In summary, during the water flooding process, clay minerals undergo physical and chemical reactions due to differences in acidity and alkalinity between the injected water and formation water. Eventually, they decompose, and the partially decomposed clay mineral fragments are transported to the surface with the injected water, leading to a reduction in the total amount of clay minerals. To verify this hypothesis, rocks from different locations were selected, and their mineral composition changes were analyzed using X-ray diffraction. The X-ray diffraction patterns for rocks from different locations are shown in Figure 4. Additionally, based on the X-ray diffraction patterns, the mineral compositions of rocks at different locations were analyzed, and the results are presented in Table 2.

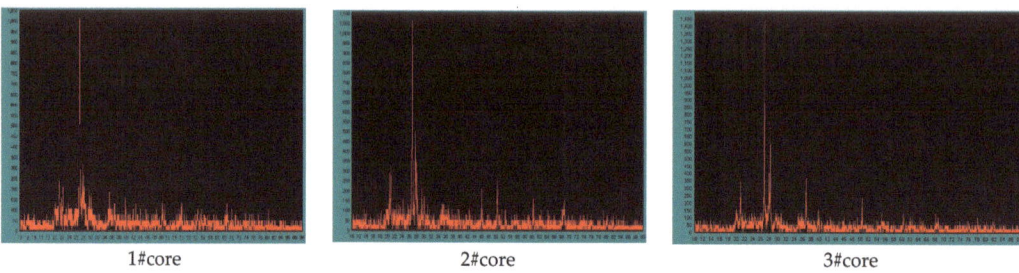

Figure 4. The X-ray diffraction patterns of cores from different locations.

Table 2. Mineral composition of cores from different locations.

Core ID	The Mineral Composition (%)					
	Clay Minerals	Quartz	Calcite	Dolomite	Potassium Feldspar	Hematite
1#	19.4	34.1	38	1.5	2.8	4.2
2#	23	31.4	29.4	6	7.8	2.4
3#	30.5	28	23.4	10	7.2	0.9

According to Table 2, it can be observed that the clay mineral content in sample 1# is relatively low, at 19.4%; and the content of quartz and calcite is higher, both above 30%. In sample 2#, the clay mineral content is 23%, significantly higher than in 1#, while the content of quartz and calcite is notably lower. This indicates that the clay minerals, mainly composed of cementitious materials, have been severely flushed in the near-fracture area. Comparing with sample 3#, the clay mineral content is even higher, indicating that the closer to the fracture, the more severe the flushing effect. As the clay on the pore walls is peeled off and transported with the injected fluid, some throats are transformed into pores, further increasing the throat diameter of the pores in the water flow channel, shortening the flow path, and leading to a decrease in tortuosity and an increase in permeability.

3.2. Analysis of Microscopic Pore Size Distribution Characteristics of Fractured Pore Channels

The loss of clay minerals can lead to significant changes in the pore structure and pore throat size of rocks within the formation. Therefore, the use of constant-rate mercury injection experiments was employed to analyze the variation patterns in pore structure at different locations in the rock cores. Based on the constant-rate mercury injection results, curves depicting pore throat radius against the frequency of pore throat distribution were plotted, as shown in Figure 5.

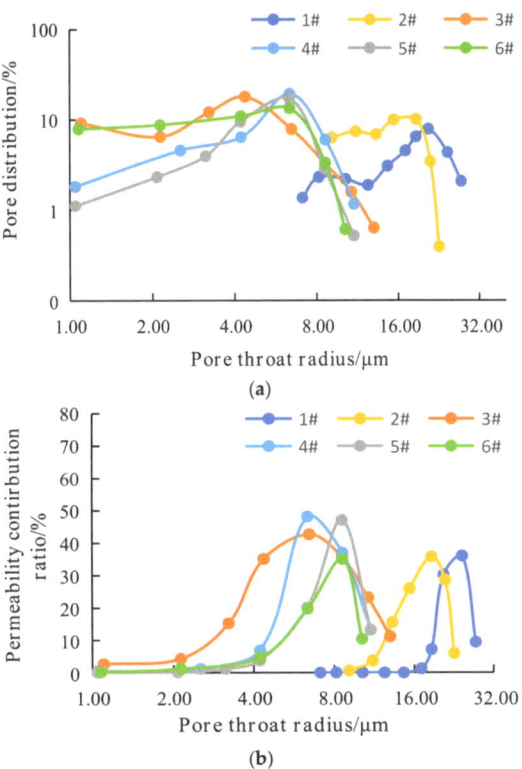

Figure 5. Mercury injection results of the rock core. (**a**) Curve of pore throat radius distribution frequency; (**b**) Curve of pore throat radius contribution to permeability.

The results from Figure 5 reveal that the permeability of rock cores 1# to 6# varies, and there are significant differences in pore radius distribution. For rock cores with lower permeability, the pore radius distribution tends to concentrate in the low-value zone, with a narrow spread range and a higher peak in the curve. Conversely, for higher permeability rock cores, the pore radius distribution extends towards the high-value zone, with a wider spread range and a lower peak in the curve.

It is evident that the pore radius constrains the permeability of the rock core. Rock cores with smaller permeability exhibit symmetric shapes in the contribution rate curve. Different pore radii significantly impact the permeability, with a higher permeability resulting in a wider span in the contribution rate curve, and the peak shifts towards the high-value zone.

Further analysis of the pore structure variations in different locations of the rock cores is presented in Table 3. From Table 3, it can be observed that the permeability, porosity, and average pore radius of the rock cores at different distances from the fracture edge have undergone significant changes. As the distance from the fracture edge decreases, parameters such as permeability, porosity, and average pore radius of the rock increase significantly. The reason for this phenomenon is that the rock framework of the reservoir is damaged under the prolonged scouring effect of injected water. Previously, the weak points with relatively fragile particle support were opened up, leading to an increase in the number of large channels and improved connectivity. At a distance of 0.5 cm from the fracture, the permeability of the rock core can reach 2533.24×10^{-3} μm^2, porosity can reach 33.77%, and the average pore radius is 20.32 μm. This indicates that after a long period of water flooding development, the properties of the matrix rock near the fracture in the formation will also undergo changes, significantly increasing the permeability, forming a high-permeability zone with a fractured structure. However, when the distance from the fracture reaches 2.5 cm, the permeability, porosity, and pore radius of the rock show no significant changes compared to the matrix rock. This suggests that, influenced by factors such as injection intensity, the area of the high-permeability zone is limited. After the formation of large channels, the supporting part of the rock framework particles is mostly destroyed, resulting in a huge difference in pore sizes.

Table 3. Constant-speed mercury injection experiment data.

ID	Distance from the Crack (cm)	Core Position	Permeability ($10^{-3} \mu m^2$)	Porosity (%)	Relative Permeability Coefficient	Peak State	Skewness	Mean Pore Throat Radius (μm)
1#	0.5	High-impact Scour Zone	2533.24	33.77	0.23	2.39	0.71	20.32
2#	1.5		1315.66	32.04	0.45	2.26	0.71	18.25
3#	2.5	Low Saturation Oil Zone	673.02	31.52	0.63	1.45	0.81	9.48
4#	3.5		537.99	31.19	0.39	2.28	0.69	8.35
5#	—	Matrix Flow Zone	506.34	29.36	0.41	2.16	0.72	8.11
6#	—		521.55	28.91	0.26	1.81	0.68	8.44

3.3. Residual Oil Characteristics Analysis in Rock under Strong Jetting State

According to the study in Section 3.1, under strong jetting conditions, the clay minerals cemented in the rock within the formation are severely flushed, and fine particles are easily carried away by the strong entraining fluid. This leads to an increase in the pores of the rocks near the fracture. To further investigate the characteristics of the fracture-jetting zone and the low-saturation oil area in high-water-cut oil reservoirs, an analysis of the microscopic residual oil changes in the rock under strong jetting conditions was conducted. The appearance of the rock cores at different distances from the fracture was observed, as shown in Figure 6.

Figure 6. The surface structures of the rock cores at different distances from the fracture. (a) 1#; (b) 2#; (c) 3#; (d) 4#; (e) 5#; (f) 6#.

Through the observation of the cross-sectional appearance of the cores in Figure 6, it can be seen that the residual oil content in cores 1# and 2# is relatively less compared to cores 3# to 6#. However, a small amount of residual oil can still be observed. Cores 1# and 2# are located in the strong flushing zone, while the other cores are located in the matrix permeation zone and the low-saturation oil zone. It is inferred that cores 1# and 2# may have experienced a reduction in residual oil content due to the high-intensity injection, making it difficult to observe residual oil on the rock surface.

To quantitatively study the characteristics of residual oil in different regions, microscopic observations of residual oil are needed for analysis. The comparative microscopic observations of cores from different regions are shown in Figure 7.

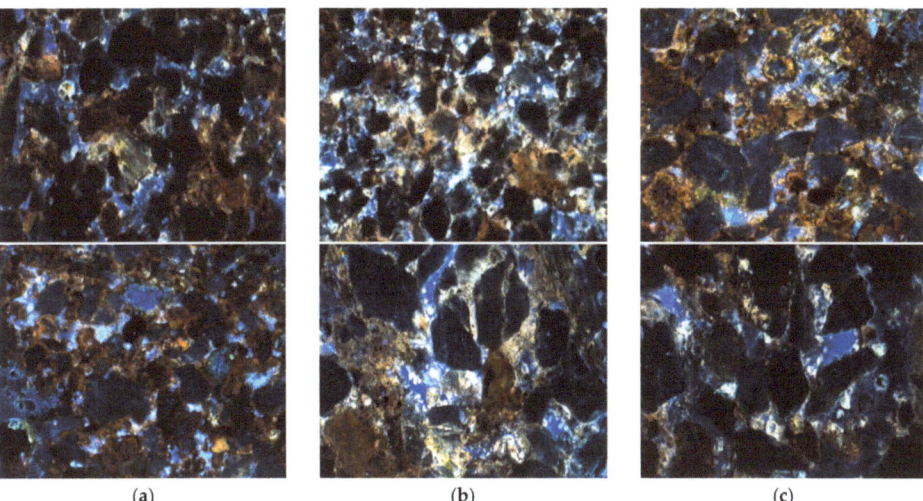

Figure 7. Distribution of remaining crude oil in different locations of the rock core. (**a**) Strong flushing zone; (**b**) Low saturation oil zone; (**c**) Matrix permeation zone.

The microscopic observations of different regions of the rock cores reveal that the matrix permeability zone has a higher crude oil content with a significant amount of blue water phase, indicating a higher residual oil content. In the low-saturation oil zone, there is a considerable variation in oil phase content, with an increase in the blue area. The water phase is observed in the form of droplets on the rock surface, resulting in a lower residual oil content compared to the matrix permeability zone. Observations of the intense flushing zone show that due to the flushing of the internal cementitious materials, the blue water phase areas have formed continuous sheets, creating high-permeability flow channels. The injected fluid flows through these channels, avoiding areas with lower permeability, further exacerbating the generation of low-efficiency or ineffective cycling bands.

In order to further quantitatively study the distribution of residual oil inside the rock cores at different locations, the results were summarized and analyzed, resulting in the distribution of residual oil content at various locations within the rock cores, as shown in Figure 8.

From Figure 8, it can be observed that, for the matrix core, it is mainly dominated by throat and cluster oil. This is because the oil inside the throat remains stable without displacement, resulting in a higher oil content. Clustered oil is formed during the saturation process where oil and water shear each other in the porous medium. Film oil occurs after displacement, where some oil is adsorbed on the rock surface due to adsorption, contributing to the lower content of film oil in the matrix core.

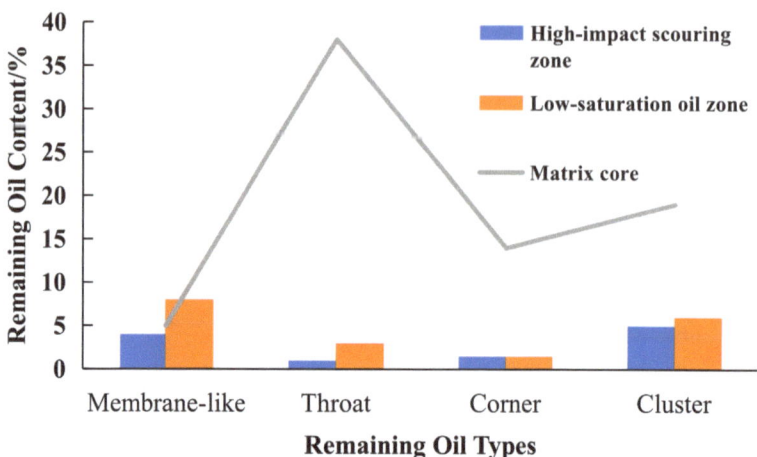

Figure 8. Distribution of different types of remaining crude oil in cores at different locations.

In the strong flushing zone and low saturation oil area, the distribution of oil in cores is primarily composed of film and clustered residual oil, with less content of throat-type and corner-type residual oil. For the low saturation area, the content of throat, corner, and clustered oil inside the rock significantly decreases due to the impact of water flooding. The adsorption effect between the rock and crude oil during water flooding results in a substantial increase in the content of film oil. In the strong flushing zone, the content of various types of oil is very low. This is attributed to the intense flushing effect of water flow, causing different types of oil inside the rock to be carried away under the strong drag force. Compared to other regions, there is almost no oil present in the strongly flushed area.

3.4. The Formation Mechanism and Influencing Factors of Fracture-Flush Zone-Low Saturation Area

Based on the results of the previous section, the formation mechanism of the fracture-flush zone-low saturation area in high-water content oil reservoirs was preliminarily determined. However, due to limited experimental conditions, it was challenging to accurately describe the influencing factors in the formation of the fracture-flush zone-low saturation area in high-water content oil reservoirs. Therefore, in this section, finite element simulation methods are employed to simulate the fluid flow and clay particle migration characteristics within the fracture-flush zone and the flush zone-low saturation area. The established porous media simulation and grid partitioning are illustrated in Figure 9.

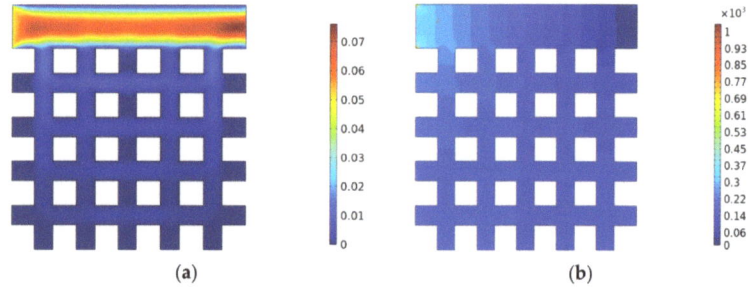

Figure 9. Variations in fluid flow velocity and pressure during the flushing process. (**a**) Variations in flow velocity during the flushing process; (**b**) Changes in pressure during the flushing process.

Through simulation, it can be assumed that the fluid velocity and pressure remain stable during the flushing process, with relatively small variations over time. Additionally, from Figure 9, it can be observed that the fluid velocity at the crack end is relatively high, reaching up to 0.075 m/s, while the fluid velocity at the far end of the crack is relatively small, with local fluid velocity in surrounding dead pores being 0 m/s. The significant difference in fluid velocity between different regions indicates that the flow of fluid at the crack end affects the flow state of the fluid near the crack end.

The simulation has confirmed that the flow of fluid in the porous medium affects the flow state of the fluid near the crack end. It is necessary to further analyze the movement of clay particles near the crack end. The movement patterns of clay particles near the crack end at different times are illustrated in Figure 10.

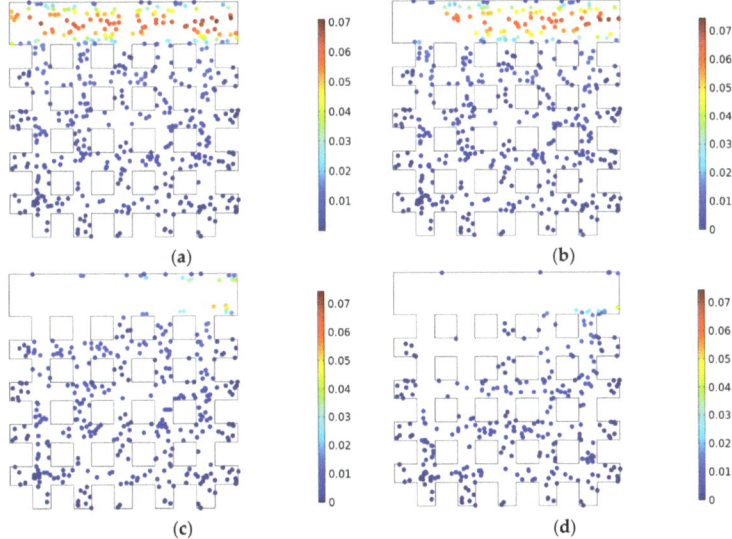

Figure 10. The distribution patterns of clay particles near the crack end at different times. (**a**) 0 μs; (**b**) 208 μs; (**c**) 500 μs; (**d**) 5000 μs.

From Figure 10, it can be observed that with the increase in simulation time, the distribution state of clay particles near the crack end undergoes significant changes. The clay particles near the crack end are carried away by the incoming fluid, flowing out through the exit crack. Moreover, the closer to the entrance, the stronger the fluid's carrying capacity. In contrast, the clay particles in the pores located at the far end of the crack show no significant changes. This indicates that the flow of fluid within the crack does have an impact on poorly cemented clay particles, causing them to detach from the originally stable cemented surface and enter the crack. This ultimately results in an increase in pore size near the crack end, forming a high-permeability zone.

To further quantitatively study the movement characteristics of clay particles near the crack end, the impact of different parameters such as flow velocity, fluid viscosity, and clay particle density on the carrying capacity was analyzed, as shown in Figure 11.

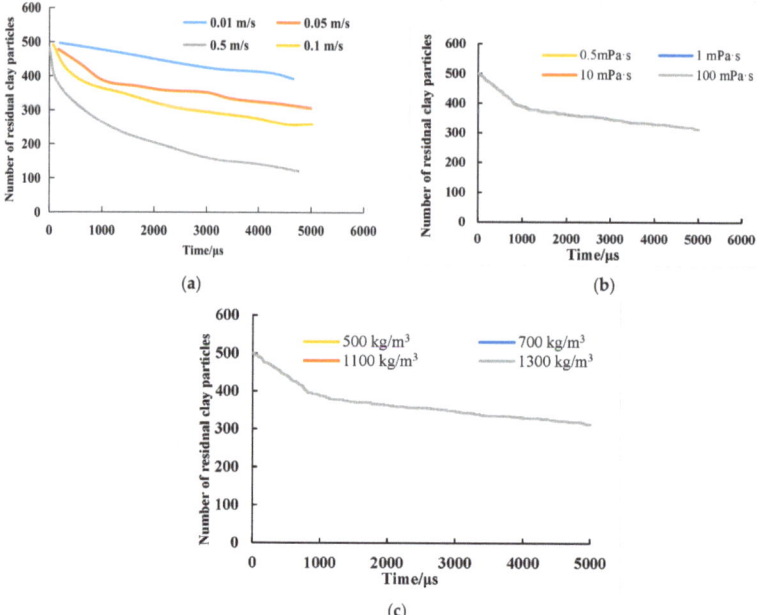

Figure 11. Impact of different parameters on the number of residual clay particles. (**a**) Variation in the number of residual clay particles under different flow velocities; (**b**) change in the number of residual clay particles under different fluid viscosities; (**c**) change in the number of residual clay particles under different particle densities.

From Figure 11, it can be observed that under strong flushing conditions, the number of residual clay particles is mainly influenced by flow velocity and flushing time. In other words, the greater the flushing intensity, the faster the water flow velocity, and the longer the flushing time, the fewer residual clay particles near the fracture. However, the viscosity of the flushing fluid and the density of clay particles have no effect on the flushing efficiency, and the number of residual clay particles remains unchanged.

In summary, the formation mechanism of high-water content oil reservoirs, including fractures, flushing zones, and low-saturation zones, can be outlined as follows (Figure 12): Under the influence of strong water flow, the fluid velocity within the fractures is high, subsequently driving the flow of fluid in the surrounding pores near the fractures. On one hand, this reduces the saturation of oil near the fractures. Simultaneously, under the strong drag force, the cementing materials in the rock and fine clay minerals are also carried away, leading to the formation of high-permeability bands near the fractures. The pore structure and throat size significantly increase. Additionally, under the action of the flushing zone's liquid flow, the oil within the rocks at the far end of the fractures is also carried away by the drag force. This results in a decrease in oil saturation, reduced flow resistance, and ultimately the formation of the fracture-flushing zone-low saturation oil region.

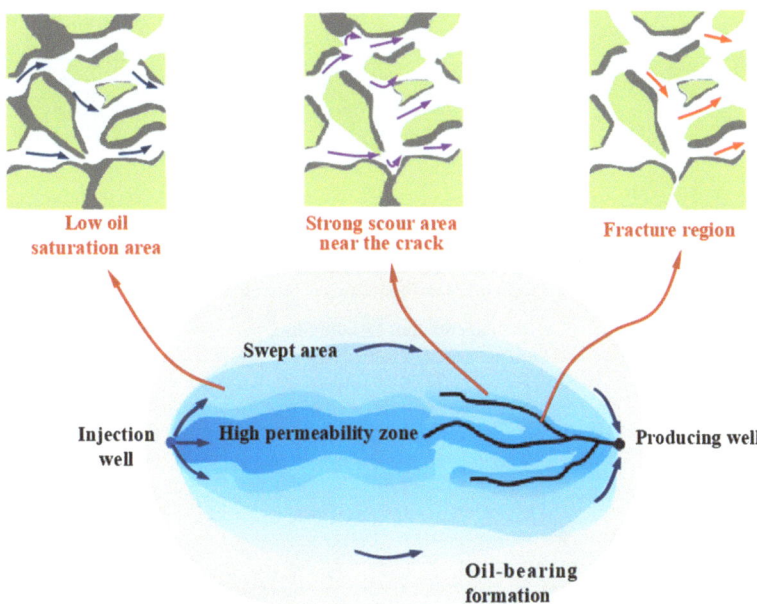

Figure 12. Characteristics of high-water content fracture-type pore zones.

4. Conclusions

(1) It was found that the pore structure and mineral content in different regions are different by simulating the formation of the dominant porous channel of fracture type in the laboratory. After long-term water flooding development, the intense water flushing near the fracture washes away the cementing materials of the rocks, causing changes in the properties of the matrix rocks near the fracture. The gaps between rock particles increase, and clay particles are transported to the surface with the injected fluid. The permeability capacity significantly increases, forming a high-permeability band with a fractured structure. This is also one of the reasons for the formation of fractured pores.

(2) The remaining oil content is different in different areas, and the remaining oil types are mainly throat and cluster oil. In the low-saturation region, the content of pores, corners, and clustered oil within the rocks significantly decreases, while the content of lamellar oil increases substantially. Clay minerals are not severely flushed, and there is almost no original oil present in the intensely flushed area.

(3) The lower the core permeability, the more concentrated the distribution of throat radius in the low-value zone, with a narrower distribution range and higher peak values. Comparing core parameters at different distances from the fracture, it is observed that after a certain period of water flooding, the properties of the matrix rocks near the fracture change, and the permeability significantly increases, forming a high-permeability band with a fractured structure. However, when the distance from the fracture reaches 2.5 cm, the parameters such as permeability, porosity, and throat radius of the rock do not change significantly compared to the matrix rocks, indicating that the area of the high-permeability band is limited, influenced by factors such as injection intensity.

(4) Mathematical models of fluid flow and clay particle migration in fissure-scour zone and scour zone-low saturation zone were established. Under intense flushing conditions, the number of residual clay particles is primarily influenced by flow velocity and flushing time. In other words, the greater the flushing intensity, the faster the water flow, and the longer the flushing time, the fewer clay particles are left near the

fracture end. The viscosity of the flushing liquid and the density of clay particles have no significant impact on the flushing effect.

Author Contributions: Conceptualization, N.Z.; formal analysis, T.L.; supervision, D.Y.; validation, G.C.; writing—original draft, N.Z. All authors have read and agreed to the published version of the manuscript.

Funding: This research was funded by the National Natural Science Foundation of China (No. 51574089).

Data Availability Statement: The data that support the findings of this study are available from the corresponding author upon reasonable request.

Conflicts of Interest: The authors declare no conflict of interest.

References

1. Chen, C.; Han, X.; Yang, M.; Zhang, W.; Wang, X.; Dong, P. A New Artificial Intelligence Recognition Method of Dominant Channel Based on Principal Component Analysis. In Proceedings of the SPE/IATMI Asia Pacific Oil & Gas Conference and Exhibition of the Society of the Petroleum Engineers, Bali, Indonesia, 29–31 October 2019; Paper SPE-196295-MS.
2. Skjaerstein, A.; Tronvoll, J.; Santarelli, F.J.; Jøranson, H. Effect of Water Breakthrough on Sand Production: Experimental and Field Evidence. In Proceedings of the SPE Annual Technical Conference and Exhibition of the Society of the Petroleum Engineers, San Antonio, TX, USA, 5–8 October 1997; Paper SPE-38806-MS.
3. Zhang, M.; Fan, J.; Zhang, Y.; Liu, Y. Study on the relationship between the water cutting rate and the remaining oil saturation of the reservoir by using the index percolating saturation formula with variable coefficients. *J. Pet. Explor. Prod. Technol.* **2020**, *10*, 3649–3661. [CrossRef]
4. Bridge, J.S. The interaction between channel geometry, water flow, sediment transport and deposition in braided rivers. *Geol. Soc. Lond. Spec. Publ.* **1993**, *75*, 13–71. [CrossRef]
5. Sugiura, S.; Nakajima, M.; Iwamoto, S.; Seki, M. Interfacial tension driven monodispersed droplet formation from microfabricated channel array. *Langmuir* **2001**, *17*, 5562–5566. [CrossRef]
6. Monty, J.P.; Hutchins, N.; Ng, H.; Marusic, I.; Chong, M. A comparison of turbulent pipe, channel and boundary layer flows. *J. Fluid Mech.* **2009**, *632*, 431–442. [CrossRef]
7. Trallero, J.L.; Sarica, C.; Brill, J.P. A Study of Oil/Water Flow Patterns in Horizontal Pipes. *SPE Prod. Facil.* **1997**, *12*, 165–172. [CrossRef]
8. Hoefner, M.L.; Fogler, H.S. Pore evolution and channel formation during flow and reaction in porous media. *AIChE J.* **1988**, *34*, 45–54. [CrossRef]
9. Fulcher, R.A., Jr.; Ertekin, T.; Stahl, C.D. Effect of Capillary Number and Its Constituents on Two-Phase Relative Permeability Curves. *J. Pet. Technol.* **1985**, *37*, 249–260. [CrossRef]
10. Abbaszadeh-Dehghani, M.; Brigham, W.E. Analysis of well-to-well tracer flow to determine reservoir layering. *J. Pet. Technol.* **1984**, *36*, 1753–1762. [CrossRef]
11. Alabert, F.G.; Modot, V. Stochastic models of reservoir heterogeneity impact on connectivity and average permeabilities. In Proceedings of the SPE Annual Technical Conference and Exhibition, Washington, DC, USA, 4–7 October 1992; SPE-24893-MS.
12. Draper, N.; Smith, H. *Applied Regression Analysis*; Wiley: New York, NY, USA, 1998.
13. Gentil, P. The Use of Multilinear Regression Models in Patterned Waterfloods. Master's Thesis, University of Texas, Austin, TX, USA, 2005.
14. Hird, K.B.; Dubrule, O. Quantification of Reservoir Connectivity for Reservoir Description Applications. *SPE Res. Eval. Eng.* **1995**, *1*, SPE-30571-MS. [CrossRef]
15. Hocking, R. *Technometrics*; Wiley: New York, NY, USA, 1983.
16. Abbaszadeh-Dehghani, M. Analysis of Unit Mobility Ratio Well-to-Well Tracer Flow to Determine Reservoir Heterogeneity. Ph.D. Thesis, Stanford University, Stanford, CA, USA, 1983.
17. Coutinho, A.; Dias, C.M.; Alves, J.L.D.; Landau, L.; Loula, A.F.D.; Malta, S.M.C.; Castro, R.G.S.; Garcia, E.L.M. Stabilized methods and post-processing techniques for miscible displacements. *Comput. Methods Appl. Mech. Eng.* **2004**, *193*, 1421–1436. [CrossRef]
18. Brigham, W.E.; Maghsood, A.D. Tracer Testing for Reservoir Description. *J. Pet. Technol.* **1987**, *39*, 519–527. [CrossRef]
19. Garland, T.M. Selective Plugging of Water Injection Wells. *Soc. Pet. Eng.* **1966**, *18*, 1550–1560. [CrossRef]
20. Vatkins, Y.P.R. How to Diagnose a Thief Zone. *Soc. Pet. Eng.* **1973**, *25*, 839–840.
21. Chan, K.S.; Yortsos, Y.C. Transport of Particulate Suspensions in Porousmedia Model Formulation. *AIChE J.* **1987**, *33*, 1636–1643.
22. Naides, C.; Paris, J.M. Identification of Water-Bearing Layers with Resistivity Image Logs. In Proceedings of the SPE Latin American Caribbean Petroleum Engineering Conference, Rio de Janeiro, Brazil, 20–23 June 2005.
23. Yousef, A.A.; Gentil, F. A Capacitance Model to Interwell Connectivity Form Production and Injection Rate Fluctuations. In Proceedings of the SPE Annual Technical Conference and Exhibition, Dallas, TX, USA, 9–12 October 2005.
24. Felsenthal, M. A Statistical Study of Some Vaterflood Parameters. *Soc. Pet. Eng.* **1979**, *31*, 1303–1304.

25. Izgec, B.; Kabir, S. Identification arid Characterization of High-Conductive Layers in VPaterfloods. *Soc. Pet. Eng.* **2009**, *14*, 113–119.
26. Shawket, G.; Younes, B.; Moutaz, S. Thief Zones and Effectiveness of Water Shut-Off Treatments under Variable Levels of Gravity and Reservoir Heterogeneity in Carbonate Reservoirs. In Proceedings of the SPE EUROPEC/EAGE Annual Technical Conference and Exhibition, Barcelona, Spain, 14–17 June 2010. SPE-131055-MS.
27. Sadeghnejad, S.; Masihi, M.; Shojaei, A.; Pishvaie, M.; King, P.R. Field Scale Characterization of Geological Formations Using Percolation Theory. *Transp. Porous Media* **2012**, *92*, 357–372. [CrossRef]
28. Ding, S.; Jiang, H.; Liu, G.; Sun, L.; Zhao, L. Determining the Levels and Parameters of Thief Zone Based on Automatic History Matching and Fuzzy Method. *J. Pet. Sci. Eng.* **2016**, *138*, 138–152. [CrossRef]
29. Xue, L.; Liu, P.; Zhang, Y. Status and Prospect of Improved Oil Recovery Technology of High Water Cut Reservoirs. *Water* **2023**, *15*, 1342. [CrossRef]
30. Feng, G.; Zhou, Y.; Yao, W.; Liu, L.; Feng, Z.; Yi, Y.; Feng, Y. Countermeasures to decrease water cut and increase oil recovery from high water cut, narrow-channel reservoirs in Bohai Sea. *Geofluids* **2021**, *2021*, 6671098. [CrossRef]
31. Wang, Y.; Jiang, H.; Wang, Z.; Diwu, P.; Li, J. Study on the Countermeasures and Mechanism of Balanced Utilization in Multilayer Reservoirs at Ultra-High Water Cut Period. *Processes* **2023**, *11*, 3111. [CrossRef]
32. Tan, J.; Liu, Y.-X.; Li, Y.-L.; Liu, C.-Y.; Mou, S.-R. Study on oil displacement efficiency of offshore sandstone reservoir with big bottom water. *J. Pet. Explor. Prod. Technol.* **2021**, *11*, 3289–3299. [CrossRef]

Disclaimer/Publisher's Note: The statements, opinions and data contained in all publications are solely those of the individual author(s) and contributor(s) and not of MDPI and/or the editor(s). MDPI and/or the editor(s) disclaim responsibility for any injury to people or property resulting from any ideas, methods, instructions or products referred to in the content.

Article

Wellbore Integrity Analysis of a Deviated Well Section of a CO_2 Sequestration Well under Anisotropic Geostress

Xiabin Wang [1], Shanpo Jia [2,*], Shaobo Gao [3,*], Long Zhao [4], Xianyin Qi [2] and Haijun He [5]

1. State Key Laboratory of Offshore Oil Exploitation, CNOOC, Beijing 100028, China; wangxb35@cnooc.com.cn
2. Bohai Rim Energy Research Institute, Northeast Petroleum University, Qinhuangdao 066004, China; yixianyin001@163.com
3. College of Civil Engineering & Architecture, Northeast Petroleum University, Daqing 163318, China
4. School of Environmental & Safety Engineering, Liaoning Petrochemical University, Fushun 113001, China; long.zhao78@foxmail.com
5. PetroChina Daqing Oilfield Production Engineering & Research Institute, Daqing 163453, China; hehj@petrochina.com.cn
* Correspondence: jiashanporsm@163.com (S.J.); shaobo.gao@foxmail.com (S.G.)

Abstract: On the basis of "Carbon Peak and Carbon Neutral" goals, carbon sequestration projects are increasing in China. The integrity of cement sheaths, as an important factor affecting carbon sequestration projects, has also received more attention and research. When CO_2 is injected into the subsurface from sequestration wells, the cement sheath may mechanically fail due to the pressure accumulated inside the casing, which leads to the sealing of the cement sheath failing. The elasticity and strength parameters of the cement sheath are considered in this paper. The critical bottom-hole injection pressures of inclined well sections under anisotropic formation stresses at different depths were calculated for actual carbon-sealing wells in the X block—the CO_2 sequestration target block. The sensitivity factors of the critical bottom-hole injection pressure were also analyzed. It was found that the cement sheath damage criterion was tensile damage. The Young's modulus and tensile strength of the cement sheath are the main factors affecting the mechanical failure of the cement sheath, with Poisson's ratio having the second highest influence. An increase in the Young's modulus, Poisson's ratio, and tensile strength of the cement sheath can help to improve the mechanical stability of cement sheaths in CO_2 sequestration wells. This model can be used for the design and evaluation of cement in carbon sequestration wells.

Keywords: carbon sequestration; deviated well; cement sheath; the critical bottom-hole injection pressure; mechanical stability

Citation: Wang, X.; Jia, S.; Gao, S.; Zhao, L.; Qi, X.; He, H. Wellbore Integrity Analysis of a Deviated Well Section of a CO_2 Sequestration Well under Anisotropic Geostress. *Energies* 2024, 17, 3290. https://doi.org/10.3390/en17133290

Academic Editor: Hossein Hamidi

Received: 29 May 2024
Revised: 28 June 2024
Accepted: 30 June 2024
Published: 4 July 2024

Copyright: © 2024 by the authors. Licensee MDPI, Basel, Switzerland. This article is an open access article distributed under the terms and conditions of the Creative Commons Attribution (CC BY) license (https://creativecommons.org/licenses/by/4.0/).

1. Introduction

With the rapid development in the construction of global carbon sequestration facilities, the carbon sequestration capacity is increasing [1]. Carbon dioxide capture and storage (CCS) in deep geological formations is an effective strategy to mitigate severe climate change. CCS is the process of capturing CO_2 from the atmosphere or other carbon sources and injecting it into deep geological formations suitable for storage [2]. Examples include depleted oil, gas reservoirs, and deep saline aquifers. The temperature and pressure of deep saline aquifers generally reach supercritical conditions for CO_2, which is sequestered in a supercritical state during its formation and thus permanently isolated from the atmosphere [3].

The integrity of the wellbore is critical in the CO_2 storage process. If the integrity of the wellbore is jeopardized, CO_2 leakage can be easily caused. The CO_2 injected into the saline formation under the action of saline multiphase fluids and the cement sheath undergoes chemical dissolution and leaching reactions, resulting in the corrosion of the cement sheath [4]. This increases the permeability of the cement sheath and reduces the compressive strength of the cement sheath [5]. In addition, the internal pressure of the casing also has an important effect on the failure of the cement sheath. When the casing pressure is high, the casing–cement sheath assembly may become damaged. The cement sheath body may become tensile- or shear-damaged due to reaching the yield limit, which will lead to the failure of the wellbore's integrity [6].

During long-term CO_2 sequestration processes, maintaining the wellbore integrity (including casing, cement sheath, and near-wellbore formation) is important to maintain the normal operation of the sequestered wells [7]. In terms of experimental studies on cement sheath failure, L. Connell et al. [8] investigated cement sheath sealing at the cement–formation interface and derived a relationship between the cement erosion rate and the water flow rate. Agata Lorek et al. [9] experimentally illustrated that cement corrosion is more severe at the cement–formation interface. Wu Zhiqiang et al. [10] concluded that the improvement of cement quality and the increase in the effective sealing length of the cement sheath can effectively enhance the sealing integrity of the cement sheath interface through indoor experiments on hydraulic sealing integrity. In a theoretical study and numerical simulation of cement sheath sealing, Ai et al. [11] evaluated cement sheath stress integrity by using a dynamic stress integrity model of cement sheath while taking into account changes in temperature and formation pore pressure. Yan Tie et al. [12] analyzed the effects of formation parameters and cement sheath parameters on the sealing capacity of the cement sheath and determined the mechanical parameters of the cement sheath under effective CO_2 sealing conditions. Song Li et al. [13] used the computational software FLAC to simulate the temperature change, fluid pressure, and solid deformation at the casing–cement sheath–formation interface and determined that no leakage would occur in the CO_2 sealing when the internal pressure of the wellbore was between 8 MPa and 750 MPa.

Zheng et al. considered the effects of cement composition [14], casing eccentricity [15], and elliptical geometry [16] on zonal isolation and provided a risk assessment workflow for abandoned wells in any region [17]. In addition, Zheng et al. also studied the influence of the formation creep effect on the integrity of cement sheaths [18]. Artificial neural networks (ANNs) are used to predict the fatigue failure of cement sheaths. A more convenient and accurate model for predicting the fatigue failure of cement sheaths under cyclic pressure and temperature changes was proposed [19,20]. Chen et al. used statistical methods to analyze the accuracy of the proposed concrete strength criteria, providing a reference for the selection of concrete strength criteria under complex stress states [21]. Zhou et al. [22] used a dynamic multiphase flow simulator to evaluate the effectiveness and suitability of using a subsea capping stack to respond to a CO_2 well blowout. Gao Deli et al. [23] established a calculation method for the fracture parameters of radial and interfacial cracks in the cement sheath of the wellbore based on the continuous dislocation distribution method and the virtual crack closure technique. It was established that the thermal expansion coefficient of cement has an important influence on the fracture parameters of cracks. The increased temperature difference between the wellbore fluid and the formation will lead to an increased risk of cement sheath interface failure. Li Q et al. [24] established a thermo-hole-elastic coupling model for the rock damage process and investigated the stress state and damage around the wellbore after CO_2 injection. Wang Dian et al. [25] established a cement sheath–formation numerical model based on the cohesive unit method and evaluated the influence of the cement slurry system and cement quality on the leakage risk. Li et al. [26] considered an analytical model with anisotropic formation stress and isotropic inner casing pressure.

In most of the above models, isotropic formation stresses are applied on the outer boundary of the cement sheath, and anisotropic formation stresses are not considered. Although Li et al. considered the anisotropic formation stress case, they calculated the cement sheath failure in the straight section. In this paper, the critical bottom-hole injection pressure in the case of cement sheath failure in inclined well sections under anisotropic formation stress is investigated. This can provide a reference for actual production.

2. Analytical Modeling

The computational model in this paper was used to evaluate the integrity of the cement sheath in an inclined section of the well by taking into account the anisotropic formation stresses and the interaction between the cement sheath and the formation. This model can be used to analyze critical bottom-hole injection pressures of cement sheaths before shear or tensile damage occurs. This section describes the derivation of the computational model.

2.1. Calculation of Cement Sheath Stress

After the borehole is cased and cemented, the forces and displacements are continuous at the cement sheath/formation interface, the second interface, assuming that the cement sheath is in close contact with the surrounding formation rock [26]. The far-field principal stress acts in the x-direction as σ_1^∞ and in the y-direction as σ_2^∞. Under the influence of the cement sheath, the circumferential, radial, and tangential stresses at the cement sheath–formation interface can be calculated from Equations (1)–(3).

$$\sigma_{\theta fc} = \frac{1}{2}(\sigma_1^\infty + \sigma_2^\infty)\left[1 + B\left(\frac{r_{CO}}{r}\right)^2\right] - \frac{1}{2}(\sigma_1^\infty - \sigma_2^\infty) \times \left[1 - 3C\left(\frac{r_{CO}}{r}\right)^4\right]\cos 2\theta \quad (1)$$

$$\sigma_{rfc} = \frac{1}{2}(\sigma_1^\infty + \sigma_2^\infty)\left[1 - B\left(\frac{r_{CO}}{r}\right)^2\right] + \frac{1}{2}(\sigma_1^\infty - \sigma_2^\infty) \times \left[1 - 2A\left(\frac{r_{CO}}{r}\right)^2 - 3C\left(\frac{r_{CO}}{r}\right)^4\right]\cos 2\theta \quad (2)$$

$$\tau_{r\theta fc} = -\frac{1}{2}(\sigma_1^\infty - \sigma_2^\infty)\left[1 + A\left(\frac{r_{CO}}{r}\right)^2 + 3C\left(\frac{r_{CO}}{r}\right)^4\right]\sin 2\theta \quad (3)$$

where $A = \frac{2(1-\beta_{cf})}{\beta_{cf}k_f + 1}$, $B = \frac{(k_c - 1) - \beta_{cf}(k_f - 1)}{2\beta + (k_c - 1)}$, $C = \frac{(\beta_{cf} - 1)}{\beta_{cf}k_f + 1}$, $\beta_{cf} = \frac{E_c}{E_f}$, $k_c = 3 - 4v_c$, and $k_f = 3 - 4v_f$.

s_{qfc} is the circumferential stress at the formation/cement sheath interface, MPa; s_{rfc} is the radial stress between the formation and the cement sheath, MPa; t_{rqfc} is the shear stress between the formation and the cement sheath, MPa; σ_1^∞ is the far-field principal stress in the x-direction, MPa; σ_2^∞ is the far-field principal stress in the y-direction, MPa; r_{co} is the outer radius of the cement sheath, m; r is the radius at any point, m; q is the angle between any point and the x-positive direction; Ec is Young's modulus of the cement sheath, GPa; E_f is the Young's modulus of the formation, GPa; v_c is the Poisson's ratio of the cement sheath; v_f is the Poisson's ratio of the formation; and b_{cf} is the stiffness ratio, defined as the ratio of the Young's modulus of the cement sheath to that of the formation.

The far-field principal stresses, σ_1^∞ and σ_2^∞, in the x- and y-directions in the above equation are the principal stresses in a straight well section. When the well section is an inclined well, it is necessary to transform the formation stress by coordinate transformation. The transformation equations are shown in Equations (4) and (5).

$$\begin{bmatrix} \sigma_{xx} & \sigma_{xy} & \sigma_{xz} \\ \sigma_{yx} & \sigma_{yy} & \sigma_{yz} \\ \sigma_{zx} & \sigma_{zy} & \sigma_{zz} \end{bmatrix} = [L]\begin{bmatrix} \sigma_H & & \\ & \sigma_h & \\ & & \sigma_v \end{bmatrix}[L]^T \quad (4)$$

$$L = \begin{bmatrix} \cos\varphi\cos\Omega & \cos\varphi\sin\Omega & -\sin\varphi \\ -\sin\varphi & \cos\Omega & 0 \\ \sin\varphi\cos\Omega & \sin\varphi\sin\Omega & \cos\varphi \end{bmatrix} \quad (5)$$

where j is the well inclination angle, °; W is the azimuth, °; s_{ij} (i, j = x, y, z) is the principal stress (i = j) and shear stress (i ≠ j) in different directions after the coordinate transformation, MPa; s_H is the horizontal maximum principal stress, MPa; s_h is the horizontal minimum principal stress, MPa; and s_v is the vertical principal stress, MPa.

Both the casing and the cement sheath alone can be considered a hollow cylinder system. The composite wellbore system consisting of the casing and cement sheath can be regarded as a composite cylinder composed of two different materials. On the inside of the cement sheath; the inner radius of the cement sheath, r_{ci}, is equal to the outer radius of the casing, r_{so}. On the outside of the cement sheath, the radial stress, σ_{rco}, is equal to the formation's radial stress, σ_{rfi}. The modified lame solution in Equations (6) and (7) can be solved to find the stress state in the composite wellbore system.

$$\sigma_{rw} = \frac{(r_{co}^2 \sigma_{rco} - r_{si}^2 P_i)}{r_{co}^2 - r_{si}^2} + \frac{r_{co}^2 r_{si}^2 (P_i - \sigma_{rco})}{(r_{co}^2 - r_{si}^2) r^2} \quad r_{si} \leq r \leq r_{co} \tag{6}$$

$$\sigma_{\theta w} = \frac{(r_{co}^2 \sigma_{rco} - r_{si}^2 P_i)}{r_{co}^2 - r_{si}^2} - \frac{r_{co}^2 r_{si}^2 (P_i - \sigma_{rco})}{(r_{co}^2 - r_{si}^2) r^2} \quad r_{si} \leq r \leq r_{co} \tag{7}$$

where s_{rw} is the radial stress in a composite wellbore system, MPa; r_{co} is the outer radius of the cement sheath, m; s_{rco} is the radial stress at the outer edge of the cement sheath, MPa; r_{si} is the inner radius of the casing, m; P_c is the casing pressure, MPa; and s_{qw} is the circumferential stress in the composite wellbore system, MPa.

As mentioned earlier, it is assumed that the casing and the cement sheath together form a composite cylinder. The cement sheath is firmly bonded to the outside of the casing. The first and second cemented surfaces have an outer casing–inner cement sheath interface bonding stress, σ_{sc}, and an outer cement sheath–inner formation interface bonding stress, σ_{cf}, respectively (Figure 1). Considering the influence of these two kinds of bonding stresses, the radial and circumferential stresses of the cement sheath can be calculated using Equations (8) and (9).

$$\sigma_{rcs} = \frac{\left(r_{co}^2 \sigma_{cf} - r_{ci}^2 \sigma_{sc}\right)}{r_{co}^2 - r_{ci}^2} + \frac{r_{co}^2 r_{ci}^2 (\sigma_{sc} - \sigma_{cf})}{(r_{co}^2 - r_{ci}^2) r^2} \tag{8}$$

$$\sigma_{\theta cs} = \frac{\left(r_{co}^2 \sigma_{cf} - r_{ci}^2 \sigma_{sc}\right)}{r_{co}^2 - r_{ci}^2} - \frac{r_{co}^2 r_{ci}^2 (\sigma_{sc} - \sigma_{cf})}{(r_{co}^2 - r_{ci}^2) r^2} \tag{9}$$

where s_{rcs} is the radial stress between the cement sheath and the casing affected by the bonding stresses, MPa; s_{cf} is the bonding stress between the cement sheath and the formation, MPa; s_{sc} is the bonding stress between the cement sheath and the casing, MPa; r_{ci} is the distance of the inside of the cement sheath from the center of the casing, m; and s_{qcs} is the circumferential stress between the cement sheath and the casing affected by the bonding stresses, MPa.

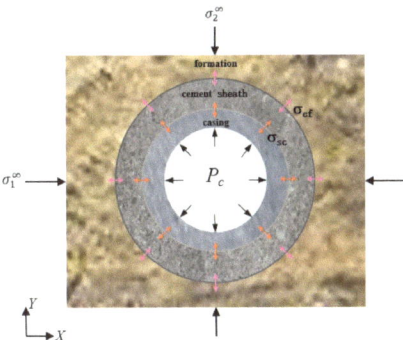

Figure 1. Bonding stress state of a composite wellbore system.

The final stress state of the cement sheath is mainly composed of the following three parts: (1) the radial external stress at the interface of the outer cement sheath and the inner formation produced by the influence of the far-field stress in two directions; (2) the radial external stress at the interface of the outer casing and the inner cement sheath produced by the influence of the pressure of the inner casing; and (3) the initial state of the stress induced by the cement sheath itself under the influence of the bonding stress on the two cemented surfaces. According to the principle of superposition, the sum of the stresses generated by $\sigma_1^\infty, \sigma_2^\infty, P_i, \sigma_{sc}$, and σ_{cf} was calculated as in Equations (6)–(9), which led to the total radial stresses, σ_r, and circumferential stresses, σ_θ, of the cement sheath. The final stress state of the cement sheath can be expressed by Equations (10) and (11).

$$\sigma_r = \sigma_{rw} + \sigma_{rcs} = \frac{(r_{co}^2 \sigma_{rco} - r_{si}^2 P_i)}{r_{co}^2 - r_{si}^2} + \frac{r_{co}^2 r_{si}^2 (P_i - \sigma_{rco})}{(r_{co}^2 - r_{si}^2) r^2} + \frac{\left(r_{co}^2 \sigma_{cf} - r_{ci}^2 \sigma_{sc}\right)}{r_{co}^2 - r_{ci}^2} + \frac{r_{co}^2 r_{ci}^2 \left(\sigma_{sc} - \sigma_{cf}\right)}{(r_{co}^2 - r_{ci}^2) r^2} \quad (10)$$

$$\sigma_\theta = \sigma_{\theta w} + \sigma_{\theta cs} = \frac{(r_{co}^2 \sigma_{rco} - r_{si}^2 P_i)}{r_{co}^2 - r_{si}^2} - \frac{r_{co}^2 r_{si}^2 (P_i - \sigma_{rco})}{(r_{co}^2 - r_{si}^2) r^2} + \frac{\left(r_{co}^2 \sigma_{cf} - r_{ci}^2 \sigma_{sc}\right)}{r_{co}^2 - r_{ci}^2} - \frac{r_{co}^2 r_{ci}^2 \left(\sigma_{sc} - \sigma_{cf}\right)}{(r_{co}^2 - r_{ci}^2) r^2} \quad (11)$$

where $r_{ci} \leq r \leq r_{co}$.

2.2. Establishment of Failure Criteria

The critical bottom-hole injection pressure, P_{imax}, was evaluated using tensile and shear damage criteria. As shown in Equation (12), the circumferential stress of the cement sheath must be lower than the tensile strength of the cement sheath when evaluated using the tensile damage criterion.

$$-\sigma_\theta < \sigma_{tensile} \quad (12)$$

where s_q is the circumferential stress of the cement sheath, MPa, and $s_{tensile}$ is the tensile strength of the cement sheath, MPa.

To evaluate the performance of the cement sheath, the failure index FI was defined. The failure index FI for the tensile damage criterion is defined as shown in Equation (13). The judgment criterion is that if $FI_{tensile}$ is greater than 1, it means that tensile damage occurs at that position in the cement sheath. Otherwise, the cement sheath will not be considered to have experienced tensile failure.

$$FI_{tensile} = \frac{-\sigma_\theta}{\sigma_{tensile}} \quad (13)$$

where $FI_{tensile}$ is the failure index for the tensile damage criterion.

The shear damage criterion used in this paper was the Mohr–Coulomb theory, and the shear damage criterion was evaluated as shown in Equation (14).

$$|\sigma_r| \leq \frac{2\sigma_{cohesion} \cos \varphi}{1 - \sin \varphi} + |\sigma_\theta| \frac{1 + \sin \varphi}{1 - \sin \varphi} \quad (14)$$

where $s_{cohesion}$ is the cohesion of the cement sheath, MPa, and j is the angle of the internal friction of the cement sheath, °.

The FI of the shear damage is defined as shown in Equation (15). The judgment criterion is that if FI_{shear} is greater than 1, shear damage will occur at that location in the cement sheath. Conversely, the cement sheath will not undergo shear damage if it is less than 1.

$$FI_{shear} = \frac{|\sigma_r|}{\left(\frac{2\sigma_{cohesion} \cos \varphi}{1 - \sin \varphi} + |\sigma_\theta| \frac{1 + \sin \varphi}{1 - \sin \varphi}\right)} \quad (15)$$

where FI_{shear} is the failure index for the shear damage criterion.

3. Instance Validation

The X gas field is located in the southwestern part of the Bozhong Depression in the Bohai Bay Basin. It is a dorsal tectonic zone characterized by a ridge in a depression sandwiched between the southwestern sub-depression and the southern sub-depression of the Bohai Depression [27]. The main part of the block has several long-axis backslopes, which are basically divided by near-north–east- and north–east-oriented faults [28]. Block X is influenced by the right-wing strike–slip rupture. The clamping area formed a north–south-oriented tensile stress, which in turn formed an east–west- and north–east-oriented normal fault system [29].

The Guantao Formation in the X gas field is a braided river deposit [30], with a burial depth of 2000 m~2900 m and a thickness of 510 m~650 m in the sand body and a stable distribution and spread throughout the whole area. The CO_2 injection wellbore frame is shown in Figure 2. The cement sheath between the production casing and the intermediate casing returns to the top of the reservoir. The upper part of the Guantao Formation is the Minghuazhen Formation, and the longitudinal drilling rate of sandstone in the Minghuazhen Formation is 10%~20%. The spread of mudstone is relatively stable in the whole area, with a thickness of 30 m~40 m. This study mainly focuses on the reservoirs of the Neoproterozoic Guantao reservoir.

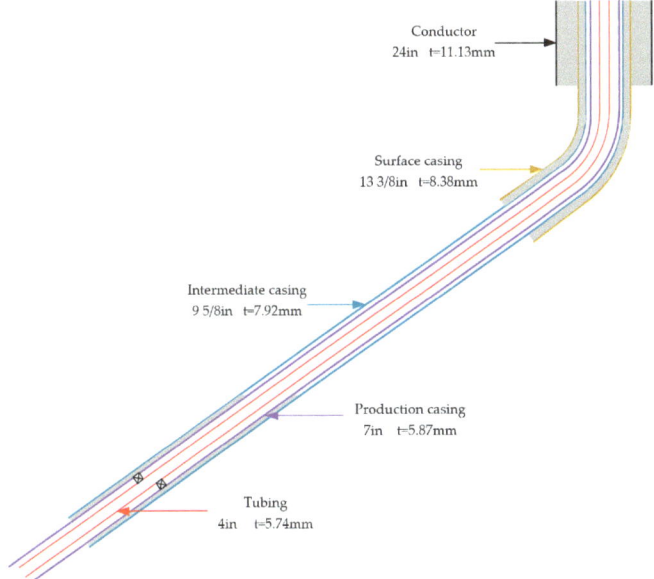

Figure 2. CO_2 injection wellbore frame.

The maximum bottom displacement of the CO_2 injection well in the X gas field is 2704.37 m. The CO_2 storage is planned for 100 years. The planar wave range is 1700 m, which is about 800 m away from the water source well. This CO_2 injection well has a large displacement and slope. The well's parameters are shown in Table 1. The CO_2 injection well consists of a straight section and an inclined section. The measured depths of the straight section and inclined section are 775.88 m and 3262.06 m, respectively. The inclined section has an inclination angle of 56 degrees. The true vertical depth of the CO_2 injection zone ranges from 2400 m to 2600 m.

Table 1. CO_2 injection well's wellbore parameters.

Type	Measured Depth/m	Inclination Angle/°	Vertical Depth/m	Bottom Hole Displacement/m	Target Horizon
CO_2 injection well	4037.94	56	2600	2704.37	Guantao formation

Table 2 lists the depth and dimensional information for the tubing and each casing. Table 3 lists the mechanical property information for the cement sheath and the formation. There are no tensile strength, internal friction angle, or cohesion data for the cap shale and reservoir sandstone.

Table 2. CO_2 injection well casing's outer diameter and measured depth.

Type	Outer Diameter /mm	Wall Thickness /mm
Conductor	609.60	11.13
Surface casing	339.73	8.38
Intermediate casing	244.48	7.92
Production casing	177.80	5.87
Tubing	101.60	5.74

Table 3. Mechanical properties of cement and formation rock.

Type	Young's Modulus/GPa	Poisson Ratio	Tensile Strength /MPa	Internal Frictional Angle/°	Cohesion /MPa
Cement	11	0.12	2.49	31.67	9.16
Cap rock shale	40	0.15	—	—	—
Reservoir sandstone	23	0.29	—	—	—

During CO_2 injection, when the casing pressure below the packer reaches a critical value, the cement sheath will be damaged and its wellbore integrity will be lost. The critical pressure at this time is called the critical bottom-hole injection pressure. Due to the lack of ground stress data in the CO_2 injection well, the vertical principal stress and the horizontal maximum and minimum principal stresses adopted in this paper are the average values of the measured ground stress in the three adjacent wells in the target block, respectively. At 2400 m, 2500 m, and 2600 m, the vertical principal stress values were 49.20 MPa, 51.49 MPa, and 53.79 MPa, respectively. The maximum horizontal principal stress values were 48.42 MPa, 50.32 MPa, and 52.69 MPa, respectively. The minimum horizontal principal stress values were 39.71 MPa, 40.36 MPa, and 42.70 MPa, respectively.

It should be noted that after the conversion of the maximum and minimum principal stresses in the horizontal direction to the maximum and minimum principal stresses in the inclined section, shear stresses are actually generated. When calculating the critical bottom-hole injection pressure in the inclined section, the effect of shear stress is not considered in this paper. The bonding stress between the cement sheath and the formation, σ_{cf}, and the bonding stress between the cement sheath and the casing, σ_{sc}, were selected according to relevant test data and are, respectively, 0.3 MPa and 0.2 MPa.

In judging the damage of the cement sheath, it is assumed that the shear damage and tensile damage of the cement sheath are independent. That is to say, the tensile damage of the cement sheath is not considered when shear damage occurs, and the shear damage of the cement sheath is not considered when tensile damage occurs. The damage to the cement sheath was determined by which critical value of the shear failure index and tensile failure index was reached first.

Generally speaking, the damage risk of the cement sheath in the reservoir section is higher than that in the cap section, so the sandstone in the reservoir section is chosen for the calculation. As shown in Figure 3, the radial and circumferential stress distributions of the cement sheath at true vertical depths of 2400 m, 2500 m, and 2600 m in the Guantao reservoir show that the radial stress in the y-direction of the cement sheath is obviously larger than that in the x-direction, which is due to the fact that the far-field principal stress in the y-direction obtained by the transformation of the diagonal cross-section, σ_2^∞, is larger than that in the x-direction by about 5.5 MPa. Relatively, the circumferential stress in the x-direction is larger than that in the y-direction. The maximum values of the radial stress at the three depths are 23.75 MPa, 23.85 MPa, and 25.45 MPa, respectively, and the maximum values of the circumferential stress are 12.05 MPa, 11.50 MPa, and 12.60 MPa, respectively.

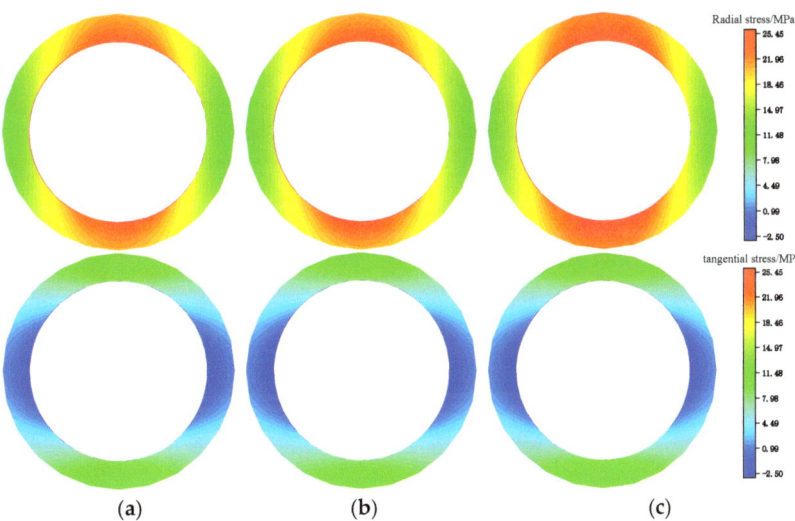

Figure 3. Radial and tangential stress distributions of cement sheath at different vertical depths: (**a**) 2400 m; (**b**) 2500 m; and (**c**) 2600 m.

From Figure 4, it can be seen that the shear failure indices of the cement sheath at the three depths range from 0.288 to 0.540, which is not enough for shear damage to occur. The tensile failure indices of the inner side of the cement sheath in the x-direction are all very close to 1, which reaches the critical failure state. The critical bottom-hole injection pressures at 2400 m, 2500 m, and 2600 m are 80.37 MPa, 83.57 MPa, and 87.39 MPa, respectively.

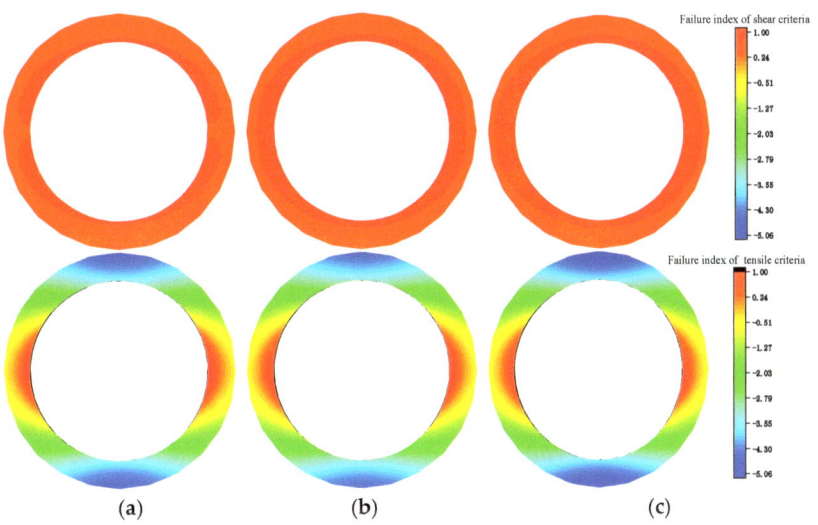

Figure 4. Distribution of failure index for shear and tensile criteria for cement sheath at different vertical depths: (**a**) 2400 m; (**b**) 2500 m; and (**c**) 2600 m.

For the inclined section of the CO_2 injection well in the X gas field, the tensile damage of the cement sheath at the sandstone in the reservoir section occurs before the shear damage under the consideration of the anisotropic formation stress. Therefore, tensile damage can be used as the main damage criterion for this CO_2 injection well.

If the inclined well is replaced by a straight well, the critical injection pressures at 2400 m, 2500 m, and 2600 m could be 67.66 MPa, 68.80 MPa, and 72.62 MPa, respectively. Tensile failure also occurs. It can be seen that the critical bottom-hole injection pressure of the vertical well is about 14 MPa lower on average than that of the inclined well. This may be because when calculating the inclined well, the far-field principal stress, σ_2^∞, in the y-direction is larger than that of σ_1^∞ in the x-direction; therefore, the calculated circumferential stress is smaller, and the critical bottom-hole injection pressure of the inclined well is larger.

4. Results and Discussion

4.1. Effect of Young's Modulus

Sensitivity analyses were conducted for sandstone at depths of 2400 m, 2500 m, and 2600 m, respectively, so as to evaluate the effect of the Young's modulus of the cement sheath on the critical bottom-hole injection pressure. Figure 5 shows that the Young's modulus has a positive effect on the critical bottom-hole injection pressure. For sandstone at a depth of 2600 m, the critical bottom-hole injection pressure increased from 87.39 MPa to 91.26 MPa when the Young's modulus of the cement sheath increased from 11 GPa to 35 GPa, which is a 4.43% increase in the critical bottom-hole injection pressure. For other depths, the pattern of change between the critical bottom-hole injection pressure and the Young's modulus of the cement sheath is basically the same as that of sandstone at a depth of 2600 m.

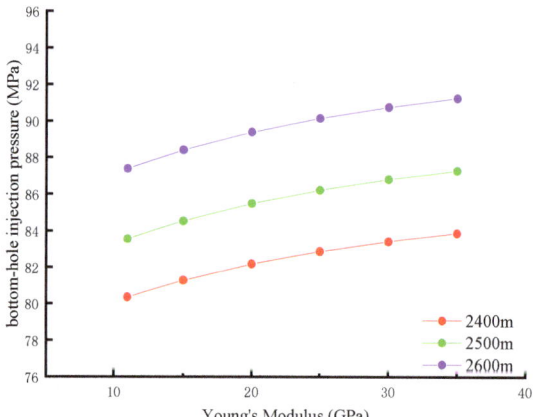

Figure 5. Effect of Young's modulus on critical bottom-hole injection pressure.

In general, the higher the Young's modulus of the cement sheath, the more reliable the well. Borehole cement with a high Young's modulus is preferred for new well construction and design. The Young's modulus is recommended to be between 20 and 40 GPa.

4.2. Effect of Poisson's Ratio

Figure 6 shows the effect of Poisson's ratio on the critical bottom-hole injection pressure. With an increase in Poisson's ratio, there is a slight increase in the critical bottom-hole injection pressure.

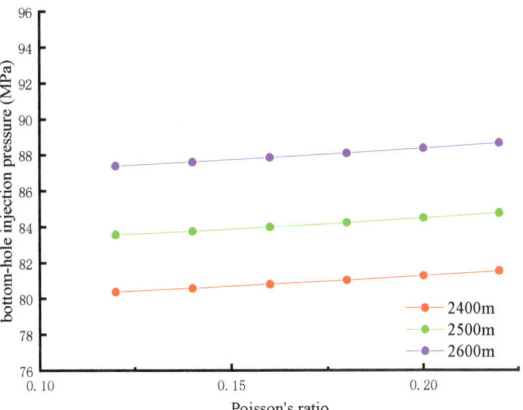

Figure 6. Influence of Poisson's ratio on critical bottom-hole injection pressure.

Similarly, for the sandstone at 2600 m, the critical bottom-hole injection pressure increased from 87.39 MPa to 88.66 MPa when Poisson's ratio for the cement sheath increased from 0.12 to 0.22, and the critical bottom-hole injection pressure increased by 1.45%. For other depths, the pattern of change between the critical bottom-hole injection pressure and Poisson's ratio for the cement sheath is basically the same as that of sandstone at a depth of 2600 m.

Overall, Poisson's ratio has a small positive effect on the critical bottom-hole injection pressure. When cementing CO_2 sequestration wells, cement with a larger Poisson's ratio can be prioritized. Poisson's ratio is recommended to be between 0.15 and 0.25.

4.3. Effect of Tensile Strength

The tensile strength of the cement sheath is an important factor affecting the tensile damage of the cement sheath. The effect of the cement tensile strength on the critical bottom-hole injection pressure is given in Figure 7. The tensile strength of the cement sheath has a significant positive effect on the critical bottom-hole injection pressure.

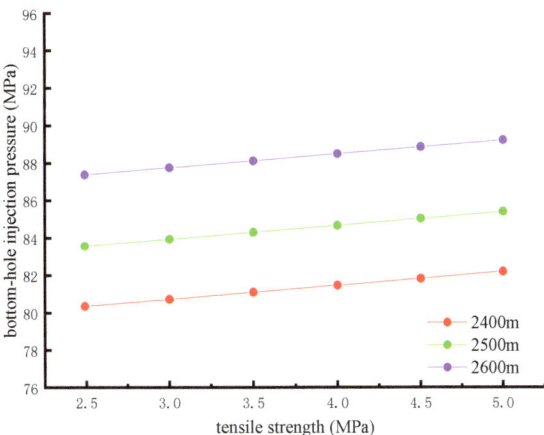

Figure 7. Influence of the tensile strength of the cement sheath on the critical bottom-hole injection pressure.

The magnitude of the shear failure index is mainly controlled by the internal friction angle, φ, and the cohesion, $\sigma_{cohesion}$, of the cement sheath. The magnitude of the tensile failure index is mainly controlled by the tensile strength of the cement. In this model, the shear failure index is much smaller than the tensile failure index under the same conditions; the damage to the cement sheath is controlled by the tensile failure criterion. Therefore, in this paper, the influence of the friction angle, φ, and cohesion, $\sigma_{cohesion}$, of the cement sheath on the critical bottom-hole injection pressure was not analyzed. Only the tensile strength of the cement sheath was considered.

For sandstone at a depth of 2600 m, the critical bottom-hole injection pressure increases from 87.39 MPa to 89.23 MPa when the cement sheath tensile strength increases from 2.49 MPa to 5 MPa, which is a 2.11% increase in the critical bottom-hole injection pressure. For other depths, the pattern of change between the critical bottom-hole injection pressure and cement sheath tensile strength is basically the same as that of sandstone at a depth of 2600 m.

5. Conclusions

This paper provides a calculation scheme for the critical bottom-hole injection pressure in the reservoir of an inclined section of a CO_2 sequestration well under an anisotropic stratigraphic stress state. The following conclusions were obtained:

1. The cement sheath in the reservoir section of this work area underwent tensile damage first. The cement sheath may have undergone shear damage during high casing pressure levels. The cement sheath damage criterion of such injection wells is controlled by tensile damage. It is worth noting that the tensile failure of the cement sheath does not occur first in all working conditions. The order of failure will change with different working conditions, such as different formation and cement mechanical parameters and wellbore frames.
2. The Young's modulus and tensile strength of the cement sheath are the main factors affecting the critical bottom-hole injection pressure. Poisson's ratio has a smaller effect. An increase in the Young's modulus, Poisson's ratio, and tensile strength of the cement sheath will increase the critical bottom-hole injection pressure. The Young's modulus for cement is recommended to be between 20 and 40 GPa. Poisson's ratio is recommended to be between 0.15 and 0.25.
3. The deeper the reservoir section, the greater the critical bottom-hole injection pressure.

Author Contributions: Investigation, methodology, validation, and writing—original draft, X.W.; formal analysis, S.J.; data curation and writing—review and editing, S.G.; data curation, L.Z.; supervision, X.Q.; visualization, H.H. All authors have read and agreed to the published version of the manuscript.

Funding: This research received no external funding.

Data Availability Statement: Data are not available as the relevant dataset for oil and gas field companies was used.

Conflicts of Interest: Author Xiabin Wang was employed by the company CNOOC. The remaining authors declare that the research was conducted in the absence of any commercial or financial relationships that could be construed as a potential conflict of interest.

References

1. Zhang, Z.; Tang, Y.; Wu, Z. Exergy analysis based on CO_2 capture and selective exhaust gas recirculation in a natural gas combined cycle power plant. *J. South China Norm. Univ. Nat. Sci. Ed.* **2023**, *55*, 1–8. [CrossRef]
2. Vafaie, A.; Cama, J.; Soler, J.M.; Kivi, I.R.; Vilarrasa, V. Chemo-hydro-mechanical effects of CO_2 injection on reservoir and seal rocks: A review on laboratory experiments. *Renew. Sustain. Energy Rev.* **2023**, *178*, 113270. [CrossRef]
3. Chaparro, M.C.; Klose, T.; Hirsch, A.; Schilling, F.; Butscher, C.; Blum, P. Modelling of wellbore cement alteration due to CO_2-rich brine interaction in a large-scale autoclave experiment. *Int. J. Greenh. Gas Control* **2021**, *110*, 103428. [CrossRef]
4. Zhang, Z.; Bai, M.; Chen, Q. Influencing factors of corrosion behavior of carbon dioxide storage wellbore. *Corros. Prot.* **2021**, *42*, 54–57+61. [CrossRef]
5. Zhang, J.; Xu, M.; Zhu, J.; Wang, G.; Ma, S. Corrosion of oilwell cement by carbon dioxide. *J. Chin. Ceram. Soc.* **2007**, *12*, 1651–1656. [CrossRef]

6. Guan, J.; Liao, H.; Lin, Z.; Xie, S.; Zhao, X. Test device for evaluating cement sheath seal integrity under cyclic loading. *China Pet. Mach.* **2021**, *45*, 48–53. [CrossRef]
7. Shi, Y.; Guan, Z.; Xi, C.; Miao, Z.; Fu, C. An analytical method for the calculation of allowable internal casing pressure based on the cement sheath integrity analysis. *Nat. Gas Ind.* **2017**, *37*, 89–93. [CrossRef]
8. Connell, L.; Down, D.; Lu, M.; Hay, D.; Heryanto, D. An investigation into the integrity of wellbore cement in CO_2 storage wells: Core flooding experiments and simulations. *Int. J. Greenh. Gas Control* **2015**, *37*, 424–440. [CrossRef]
9. Lorek, A.; Labus, M.; Bujok, P. Wellbore cement degradation in contact zone with formation rock. *Environ. Earth Sci.* **2016**, *75*, 499. [CrossRef]
10. Wu, Z.; Yue, J.; Li, Q.; Cao, Y.; Geng, Y.; Liu, S. Experimental study on the hydraulic seal integrity evaluation of casing-cement sheath bonding interface. *China Offshore Oil Gas* **2018**, *30*, 129–134.
11. Ai, C.; Li, J.; Li, Z.; Zhang, Z.; Chen, D. Research on cement sheath stress integrity of CO_2 buried well in the process of injection. *Sci. Technol. Eng.* **2013**, *13*, 2057–2061. [CrossRef]
12. Yan, T.; Li, Y.; Li, J.; Feng, F.; Bai, M. The study on cement sheath sealing effect and its performance parameters of CO_2 buried well. *China Sci. Technol. Eng.* **2014**, *14*, 25–29. [CrossRef]
13. Song, L.; Tan, Y. A two-dimensional model analysis of casing-cement-strata in geological storage of CO_2. *Chin. J. Appl. Mech.* **2015**, *32*, 288–293+356.
14. Zheng, D. Geological Analysis and Cement Composition for Improving Zonal Isolation. Ph.D. Thesis, The University of Tulsa, Tulsa, OK, USA, 2023.
15. Zheng, D.; Ozbayoglu, E.; Baldino, S.; Miska, S.; Liu, Y. The influence of casing eccentricity on zonal isolation. In Proceedings of the ARMA 23241 at the 57th US Rock Mechanics/Geomechanics Symposium, Atlanta, GA, USA, 25–28 June 2023.
16. Zheng, D.; Miska, Z.; Ozbayoglu, E.; Zhang, Y. The influence of elliptical-geometry wellbore on zonal isolation. In Proceedings of the Paper Presented at the 56th U.S. Rock Mechanics/Geomechanics Symposium, Santa Fe, NM, USA, 26 June 2022. [CrossRef]
17. Zheng, D.; Turhan, C.; Wang, N.; Ashok, P.; van Oort, E. Prioritizing wells for repurposing or permanent abandonment based on generalized well integrity risk analysis. In Proceedings of the SPE/IADC Drilling Conference and Exhibition, Galveston, TX, USA, 27 February 2024; p. D021S018R001.
18. Zheng, D.; Miska, S.; Ozbayoglu, E. The influence of formation creeping on wellbore integrity. In Proceedings of the Paper presented at the SPE 2021 Symposium Compilation, Virtual, 26 November 2021. [CrossRef]
19. Zheng, D.; Ozbayoglu, E.; Miska, S.; Liu, Y. Cement sheath fatigue failure prediction by ANN-based model. In Proceedings of the Paper Presented at the Offshore Technology Conference, Houston, TX, USA, 2–5 May 2022. [CrossRef]
20. Zheng, D.; Ozbayoglu, E.; Miska, S.; Liu, Y. Cement sheath fatigue failure prediction by support vector machine based model. In Proceedings of the Paper Presented at the SPE Eastern Regional Meeting, Wheeling, WV, USA, 18–20 October 2022. [CrossRef]
21. Chen, X.; Zheng, D.; Wu, X.; Li, C. A review of three common concrete multiaxial strength criteria from 2010 to 2020. *Arch. Computat. Methods Eng.* **2023**, *30*, 811–829. [CrossRef]
22. Zhou, L.; Upchurch, E.R.; Liu, Y.; Anfinsen, B.; Hashemian, Y.; Zhao, G. Evaluating subsea capping stack usage for CO_2 blowouts. In Proceedings of the Paper Presented at the Offshore Technology Conference, Houston, TX, USA, 6–9 May 2024. [CrossRef]
23. Gao, D.; Dou, H.; Dong, X. Research progress in wellbore cement sheath integrity under conditions of CO_2 injection and storage. *J. Yan'an Univ. Nat. Sci. Ed.* **2022**, *41*, 1–9+17.
24. Li, Q.; Jing, M. Thermo-poroelastic coupling analysis of rock damage around wellbore due to CO_2 injection. *Chin. J. Rock Mech. Eng.* **2013**, *32*, 2205–2213.
25. Wang, D.; Li, J.; Liu, P.; Liu, X.; Lian, W.; Lu, Z. Simulation study of sealing integrity in abandoned wells within CO_2 sequestration block. *Drill. Fluid Complet. Fluid* **2023**, *40*, 384–390. [CrossRef]
26. Li, B.; Guo, B.; Li, H.; Shi, Y. An analytical solution to simulate the effect of cement/formation stiffness on well integrity evaluation in carbon sequestration projects. *J. Nat. Gas Sci. Eng.* **2015**, *27*, 1092–1099. [CrossRef]
27. Zhang, G.; Cheng, Q.; Zhang, L.; Liu, W.; Zhao, Y.; Wu, H.; Shen, C. Calculation of 3D reservoir rock mechanical parameters of metamorphic rock reservoirs in the Bozhong 19-6 gas field of the Bohai Bay Basin and their significance. *Earth Sci.* **2018**.
28. Zhao, T.; Wang, F.; Peng, J.; Wang, B. Influence of CO_2 injection on phase behavior of oil and gas reservoirs and analysis of gas condensate reservoir genesis: A case study of BZ19-6 gas condensate reservoir. *China Offshore Oil Gas* **2023**, *35*, 24–34. [CrossRef]
29. Yang, X.; Liu, C.; Wang, F.; Li, G.; Feng, D.; Yang, T.; He, Z.; Su, J. Distribution and origin of overpressure in the Paleogene Dongying Formation in the southwestern sub-sag, Bozhong Sag, Bohai Bay Basin. *Oil Gas Geol.* **2024**, *45*, 96–112.
30. Li, X.; Qin, R. Method of fracture characterization and productivity prediction of 19-6 buried-hill fractured reservoirs, Bohai Bay Basin. *Earth Sci.* **2023**, *48*, 475–487.

Disclaimer/Publisher's Note: The statements, opinions and data contained in all publications are solely those of the individual author(s) and contributor(s) and not of MDPI and/or the editor(s). MDPI and/or the editor(s) disclaim responsibility for any injury to people or property resulting from any ideas, methods, instructions or products referred to in the content.

Article

Recognition of Artificial Gases Formed during Drill-Bit Metamorphism Using Advanced Mud Gas

Janaina Andrade de Lima Leon [1,*], Henrique Luiz de Barros Penteado [1], Geoffrey S. Ellis [2], Alexei Milkov [3,†] and João Graciano Mendonça Filho [4]

1. Brazilian Petrol S/A—Petrobras, Rio de Janeiro 20031-912, Brazil; hpenteado@petrobras.com.br
2. U.S. Geological Survey (USGS), Denver, CO 80225, USA; gsellis@usgs.gov
3. Department of Geology and Geological Engineering, Colorado School of Mines, Golden, CO 80401, USA
4. Department of Geological Sciences, Federal University of Rio de Janeiro, Rio de Janeiro 21941-916, Brazil; graciano@geologia.ufrj.br
* Correspondence: janainaleon@petrobras.com.br
† Deceased author.

Abstract: Drill-bit metamorphism (DBM) is the process of thermal degradation of drilling fluid at the interface of the bit and rock due to the overheating of the bit. The heat generated by the drill when drilling into a rock formation promotes the generation of artificial hydrocarbon and non-hydrocarbon gas, changing the composition of the gas. The objective of this work is to recognize and evaluate artificial gases originating from DBM in wells targeting oil accumulations in pre-salt carbonates in the Santos Basin, Brazil. For the evaluation, chromatographic data from advanced mud gas equipment, drilling parameters, drill type, and lithology were used. The molar concentrations of gases and gas ratios (especially ethene/ethene+ethane and dryness) were analyzed, which identified the occurrence of DBM. DBM is most severe when wells penetrate igneous and carbonate rocks with diamond-impregnated drill bits. The rate of penetration, weight on bit, and rotation per minute were evaluated together with gas data but did not present good correlations to assist in identifying DBM. The depth intervals over which artificial gases formed during DBM are recognized should not be used to infer pay zones or predict the composition and properties of reservoir fluids because the gas composition is completely changed.

Keywords: drill-bit metamorphism; advanced mud gas logging; drilling parameters; Santos Basin; pre-salt

Citation: Leon, J.A.d.L.; Penteado, H.L.d.B.; Ellis, G.S.; Milkov, A.; Filho, J.G.M. Recognition of Artificial Gases Formed during Drill-Bit Metamorphism Using Advanced Mud Gas. *Energies* **2024**, *17*, 4383. https://doi.org/10.3390/en17174383

Academic Editors: Mofazzal Hossain, Yongjun Deng, Lin Chen, Kai Wang and Jie Wu

Received: 14 March 2024
Revised: 6 July 2024
Accepted: 19 July 2024
Published: 2 September 2024

Copyright: © 2024 by the authors. Licensee MDPI, Basel, Switzerland. This article is an open access article distributed under the terms and conditions of the Creative Commons Attribution (CC BY) license (https:// creativecommons.org/licenses/by/ 4.0/).

1. Introduction

Mud gas logging is the process of measuring the chemical composition, recognition of artificial gases, and concentrations of gases released from the returned drilling mud [1–3]. Mud gas logging helps recognize potential well safety issues (e.g., gas kicks), detect possible pay zones and the depths of petroleum–water contacts, predict the type and composition of reservoir fluid, and identify potential reservoir compartmentalization [4–7]. In general, mudlogging is a valuable evaluation technique that has evolved since the advent of rotary drilling in 1939 [8]. It provides continuous real-time information that can be used for drilling process optimization and detecting downhole drilling problems [9–11].

This technology, mud gas logging, has been used in the petroleum industry for decades [12–14] and has continuously and significantly improved over time [15,16]. Petroleum companies use advanced mud gas logging (AMG) to collect data during drilling, including measuring molecular compositions of gases and making real-time corrections to eliminate the effects of gas recycling and various drilling artifacts [3,7,17–21].

Drill-bit metamorphism (DBM) is the process of thermal degradation of drilling fluid at the drill bit and rock interface due to drill-bit overheating. The heat generated by the bit when drilling into rock formations promotes the generation of artificial gas, changing the gas composition and the drill cuttings that return to the surface [22–26]. Bowen and Aurousseau [27] first documented that the heat produced by the friction of drilling can significantly affect (fuse) cored rocks. Taylor [28] introduced the term "bit metamorphism" into the literature after describing severely altered cuttings commonly accompanied by the increase in gas volumes in wells from the German North Sea. DBM commonly occurs when drillers use impregnated drill bits or polycrystalline diamond compact (PDC) bits [29], together with drilling fluid containing oil-based mud. Because of these two factors, high temperature (from ~350 °C to 1200 °C) results from both in situ formation temperature and the friction of the bit with the rocks during rotary drilling [22,25,30,31]. Mud temperature is an important factor to be evaluated as it impacts both the generation of artificial gas and the stability of the well [32,33].

DBM generates carbon monoxide (CO) and alkenes (C_nH_{2n}, including ethene (C_2H_4), propene (C_3H_6), and butene (C_4H_8)) that are not present in the geological formations and natural petroleum fluids [24,25]. DBM also generates alkanes (methane (CH_4 or C_1) to pentanes (C_5H_{12} or C_5)), aromatics (benzene), hydrogen (H_2), carbon dioxide (CO_2), hydrogen sulfide (H_2S), and carbonyl sulfide (COS), and these artificial compounds contaminate the indigenous compounds present in geological formations and reservoirs before drilling [24,26,34–36]. Artificial gases associated with DBM may be generated by Fischer–Tropsch reactions [37], pyrolysis of the oil-based mud during drilling, corrosion of the drill pipe, alteration of the gas already present in the drilled rock, and by a combination of these processes [23,36]. DBM increases the total gas amount and the concentrations of specific compounds changing gas ratios and affecting the isotopic composition of gases [23,24], all of which may lead to false identification of pay zones and incorrect interpretation of reservoir fluid types and properties [26].

According to Wenger [24] DBM can be observed in several ways including in cutting samples, in the composition and isotopic analysis of mud gases, and in the analysis of fluid-inclusion volatiles. In wells where DBM occurs, mud gas analysis shows that gases typically not found in natural gases such as alkenes, CO_2, CO, and COS are generated during DBM.

When drilling a well, conventional mud gas or advanced mud gas analysis can be used to identify gas during drilling. Conventional mud gas uses a chromatograph and is unable to distinguish alkanes from alkenes as the retention times for alkanes and alkenes are very similar. Because advanced mud gas uses a mass spectrometer to identify drilling gas and the spectrometer differentiates gas based on molecular weight, alkanes can be easily distinguished from alkenes with this technique that is currently being used [38]. This differentiation is important for identifying DBM and therefore advanced mud gas will be used as the basis of this work.

Ref. [39] developed specific DBM research work by carrying out controlled experiments in the laboratory, using rock samples 0.5 m long. This experiment was able to simulate drilling conditions up to 5 km. In this experiment, the generation of alkanes, alkenes, CO, and H_2 was observed through the extraction, analysis, and sampling of mud gas during drilling. The alkanes, alkenes, CO, and H_2 that were generated during the experiment responded to a gradual increase in WOB or RPM generated through the cracking of the drilling fluid due to the increase in temperature at the cutter–rock interface.

As companies drill more wells in deep and hot reservoirs [40–42], which often have low permeability [43,44] and are characterized by low rates of penetration (ROP), the occurrences of DBM have become more common. Still, there are few published case studies that describe how to recognize artificial gases generated during DBM using advanced mudlogging data. Here, we discuss occurrences of DBM in wells that targeted oil accumulations in pre-salt carbonate reservoirs in the Santos Basin offshore Brazil [45]. Fifty wells were evaluated in the Santos Basin, targeting pre-salt carbonate reservoirs, and this is an academic innovation, as such a broad study had never been carried out with so many wells, totaling more than 66 km of well drilled using AMG for a study focused on identifying DBM. It was possible to carry out a robust analysis of the presence, absence, and intensity of DBM.

We present a workflow to identify artificial gases formed during the DBM using advanced mud gas logging data and estimate the intensity of the DBM as a function of drill-bit types, penetrated lithologies, and drilling parameters. During the drilling of these wells, we did not have real-time isotopic data, nor a methodology to assist in the decontamination of artificial gases.

2. Methodology
2.1. Overview

The dataset studied is available at the Brazilian regulatory agency "Brazilian National Agency for Petroleum, Natural Gas and Biofuels—ANP". These data include the molecular composition of the mud gas and drilling parameters from 50 wells drilled between 2009 and 2020 by Petrobras in the Santos Basin (Figure 1). Mud gas analyses were performed during drilling by service providers using equipment for advanced mud gas analysis (Figure 2). Three service providers for advanced gas analysis were employed, but the names of wells (Table 1), the service providers, and the names of the equipment were coded for confidentiality reasons: MUDX with the GASX equipment (37 wells), MUDY with the GASY equipment (11 wells), and MUDZ with the GASZ equipment (two wells).

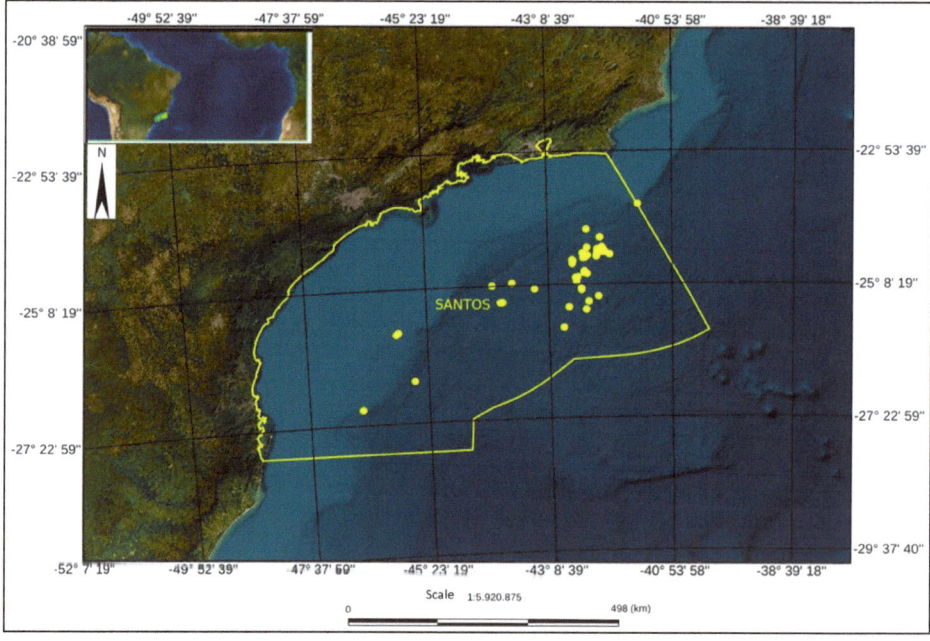

Figure 1. Regional map of the Santos Basin, showing the location of wells that were drilled using advanced mud gas analysis.

The evaluation to recognize the DBM in the wells started with the analysis of the gas ratio plots. Initially, several ratios were calculated along each well: $C_1/(C_1 + C_2 + C_3 + i\text{-}C_4 + n\text{-}C_4 + i\text{-}C_5 + n\text{-}C_5)$, called dryness [46], C_2/C_1, C_2/C_3, C_2C/C_3, $C_3/n\text{-}C_4$, $i\text{-}C_4/i\text{-}C_5$, $n\text{-}C_4/n\text{-}C_5$, $i\text{-}C_4/n\text{-}C_4$, $i\text{-}C_5/n\text{-}C_5$, and ethene/(ethene+ethane) [24]. Then, these ratios were plotted in scatter plots as a function of depth so that it was possible to identify the patterns of variation of the gaseous compounds. We identified that C_2/C_3, C_2/C_1, ethene/(ethene+ethane), and dryness ratios were most relevant for the evaluation of DBM in the wells. For each well, evaluations were performed in the same manner using mud gas data, lithologies, and drill-bit types as displayed in three panels (Figures 3–5). For Figures 3–5, panel "a" represents the general graph of the well with the indication of lithologies, bit changes, formation tops, and gas ratios. The panel "b" shows the ratios C_2/C_1, dryness, and ethene/(ethene+ethane). Panel "c" shows concentrations (chromatography) of C_1–C_5 gases and benzene (DELTA data), and panel "d" shows the normalized alkanes.

Table 1. Survey of wells and lithostratigraphic units for which advanced gas analysis was performed. Lithostratigraphic units color coding: post-salt in green, evaporites in purple, pre-salt in blue and orange colors, and the economic basement of the basin in red. Blank cells indicate the absence of formation in the well or interval in which advanced mud gas analysis was not acquired. The letters IGN in the cells indicate the presence of igneous rock.

Wells	Marambaia	Juréia	Itajaí-Açu	Itanhaém	Guarujá	Ariri	Barra Velha	Itapema	Piçarras	Camboriú
D3			x			x	x IGN	x IGN		
E3			x	x	x	x	x IGN			
F4		x	x	x	x	x				x
H1	x		x	x		x				
H4				x	x	x	x	x	x	
I3			x	x		x	x			
I4			x	x	x IGN	x IGN	x			
J1	x		x	x	x	x	x IGN			
J2		x	x							
J3			x							
A3						x IGN	x			
A4						x	x	x	x IGN	
B1						x	x	x	x	
B2						x	x IGN			
B5						x	x			
D4						x	x	x	x	x
D5						x IGN	x IGN	x		
E1						x	x IGN			
E2						x IGN	x IGN			
E4						x	x	x	x	
F1						x	x	x		
F2						x	x	x		
F3						x	x IGN	x		
F5						x	x IGN			
G1						x	x IGN			
G2						x	x			
G4						x	x			
G5						x	x	x	x IGN	
H2						x	x			
H3						x	x	x		
H5						x	x	x	x	x
I1						x	x	x	x	
I2						x	x			
I5						x				
J5						x	x			
A1							x IGN			
A2							x IGN	x		x
A5							x	x		x
B4							x	x		
C1							x	x	x	
C2							x	x		
C3							x	x	x	
C4							x	x	x	
C5							x	x		
D1							x	x		x
D2							x			
G3							x IGN	x		
J4							x IGN			
E5								x		
B3										x

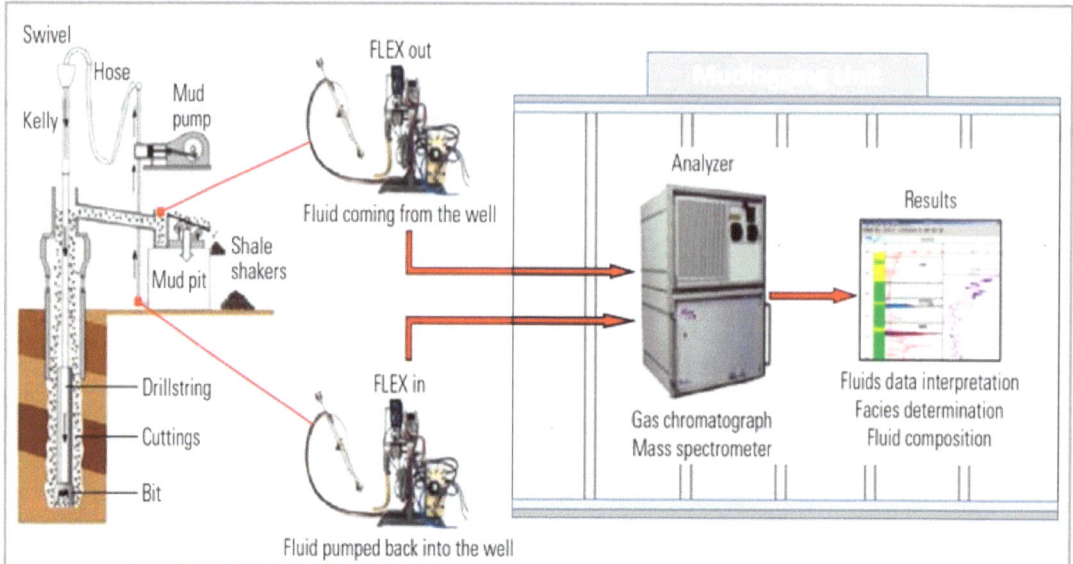

Figure 2. Generalized process of advanced mud gas extraction and analysis (modified from Ablard et al., 2012 [3]). The schematic illustrates well drilling and mud circulation, positioning of mud extraction probes at the IN and OUT along the mudflow line, and subsequent analysis of the gas inside the mudlogging unit by gas chromatograph and mass spectrometer.

We recognized artificial mud gases generated by DBM based on the following criteria:

- ✓ A gas chromatography response from a reservoir without DBM shows a normal distribution of the gaseous fractions ($C_1 > C_2 > C_3 > C_4 > C_5$; [12,16]). However, in the presence of DBM, the chromatography order is reversed ($C_2 > C_1$ in the case of extreme metamorphism), or the values tend to be very close ($C_2 \approx C_1$ in the case of moderate metamorphism).
- ✓ The concentrations of benzene tend to increase in the presence of DBM and may be higher than C_5 depending on the intensity of the metamorphism.
- ✓ During the occurrence of DBM, the average values for C_2 (ethane + ethene) normalized to the C_1 to C_5 range normally are higher than 20%, reaching very high values (up to 70%) in extreme cases.
- ✓ The ratio C_2/C_3 tends to present high dispersion in the presence of DBM and low dispersion when there is no DBM.
- ✓ The C_2/C_1 ratio tends to be greater than 0.2, with high dispersion and greater than dryness and ethene/(ethene+ethane) ratio values depending on the intensity of DBM.
- ✓ The ethene/(ethene+ethane) ratio is close to 1 in the presence of DBM and has lower values or is close to zero in the absence of DBM [24].
- ✓ Dryness tends to be close to 1 in intervals without DBM and close to 0 in intervals with DBM because DBM generates many wet (C_2+) gases.

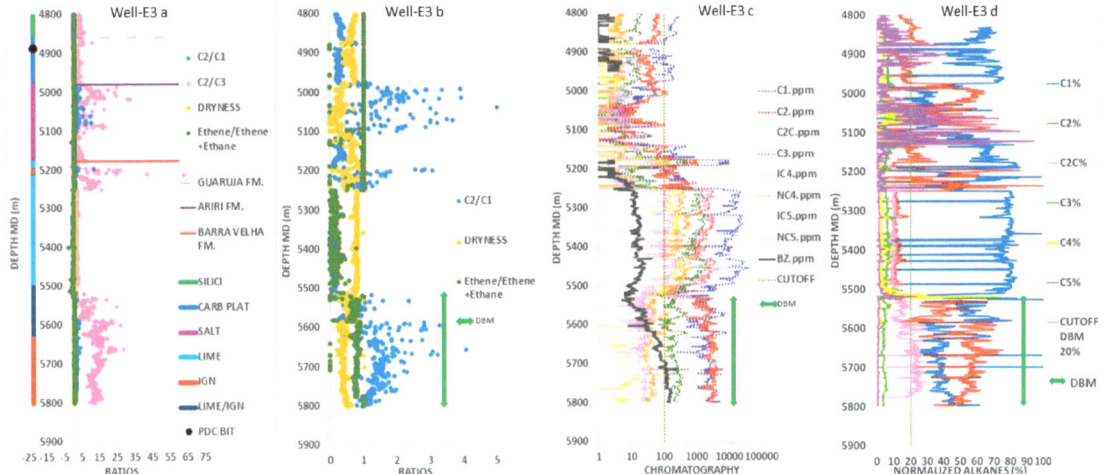

Figure 3. Mud gas logs for well E3 are divided into four parts including formation tops, lithology, and gas ratios (Well-E3 a panel), ratios C_2/C_1, dryness, and ethene/(ethane+ethene) (Well-E3 b panel), gas chromatography (Well-E3 c panel), and normalized alkanes (Well-E3 d panel). See Table 2 for mnemonics of lithological types. Interval with drill-bit metamorphism marked with green arrow. From 5500 until the end of the well, changes were observed in the gas curves, mainly in the igneous rock interval caused by drill-bit metamorphism. In Well-E3 b, we observed an increase in the C_2/C_1 curve and a decrease in dryness causing the inversion of these two curves. In Well-E3 c, an increase in C_2 is also observed, overlapping C_1 from 5500 m to the end of the well, and in Well-E3 d, the relative percentage of ethane is greater than that of methane depending on the increase in ethylene.

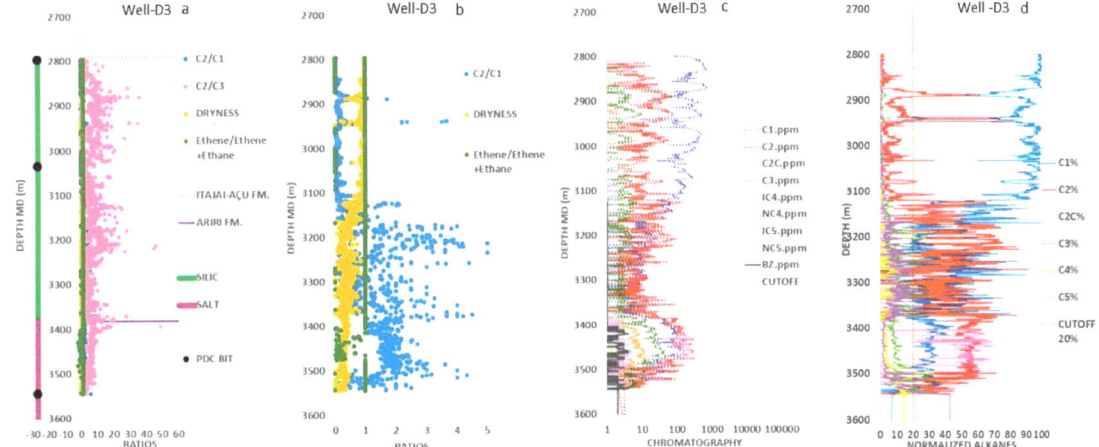

Figure 4. Mud gas logs for well D3 are divided into four parts including formation tops, lithology, and gas ratios (Well-D3 a panel), ratios C_2/C_1, dryness, and ethene/(ethane+ethene) (Well-D3 b panel), and gas chromatography (Well-D3 c panel) and normalized alkanes (Well-D3 d panel). See Table 2 for mnemonics of lithological types.

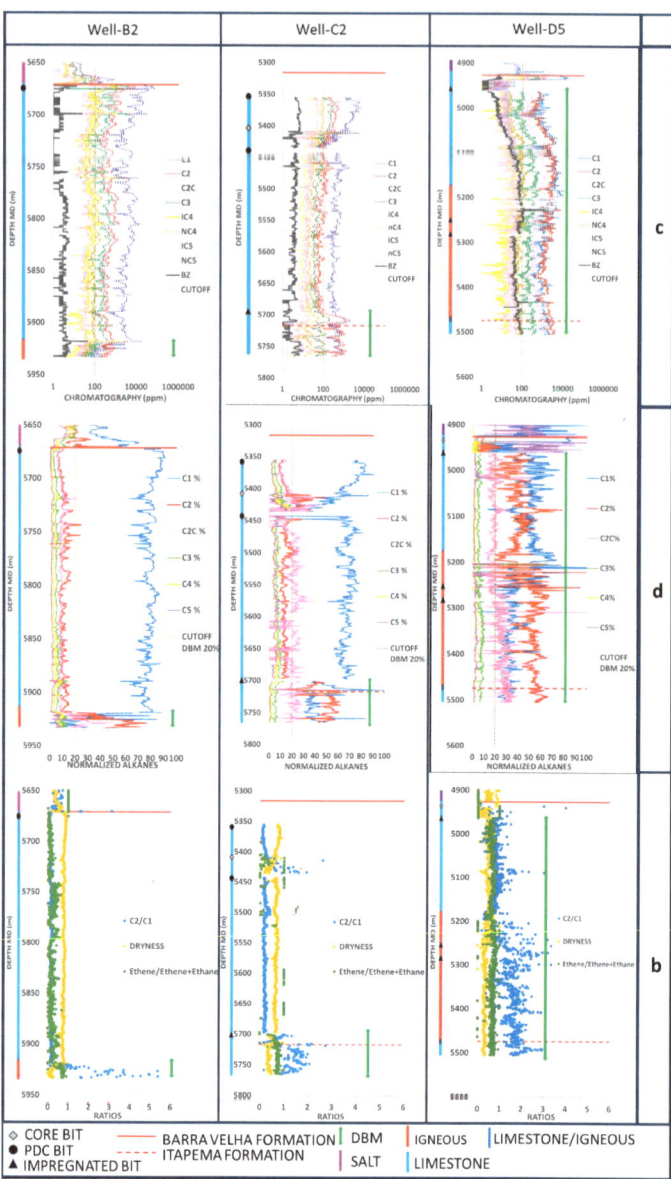

Figure 5. The panel is separated into three different wells. Well-B2 (c, d, and b), Well-C2 (c, d, and b), and Well-D5 (c, d, and b) are represented on all the graphs that identify the drill-bit metamorphism in wells B2, C2, and D5. For the three wells, the chromatographic distribution graphs of alkanes (Well-B2 c, Well-C2 c, and Well-D5 c), the concentration of normalized alkanes from C_1 to C_5 (Well-B2 d, Well-C2 d, and Well-D5 d), and ratios (Well-B2 b, Well-C2 b, and Well-D5 b) were evaluated. Comparison between the gas chromatography of wells B2 (Well-B2 c—without drill-bit metamorphism until 5918 m and with drill-bit metamorphism when started the igneous rock), well C2 (Well-C2 c—with drill-bit metamorphism in the interval below 5700 m after changing from PDC to impregnated drill), and well D5 (Well-D5 c—with drill-bit metamorphism throughout the well drilled with the impregnated drill). Interval with drill-bit metamorphism marked with green arrow.

Table 2. Lithology groups were used in the evaluation charts. The table includes the lithologies, groups, mnemonics, and the color used in each lithology group.

Type of Lithology		Mnemonic	Color
Lithology	Group Name		
Sedimentary after salt	Siliciclastic	SILICI	
Post-salt carbonate platform	Carbonate platform	CARB PLAT	
Salt – Halite and soluble salt	Salt	SALT	
Pre-salt carbonate	Limestone	LIME	
Shale, siltite, mudstone, laminite	Calcilutite	CALCI	
Unidentified igneous	Unidentified igneous	IGN	
Diabase/Basalt	Diabase	DIABA OR BASA	
Carbonate pre-salt intercalated with igneous	Limestone/Igneous	LIME/IGN	
Pre-salt carbonate intercalated with calcilutite	Limestone/Calcilutite	LIME/CALCI	

2.2. Tools and Data Source

The equipment for advanced mud gas analysis can analyze gases during well drilling and identify several compounds such as alkanes (methane (CH_4 or C_1), to octane (C_8H_{18} or C_8)), aromatics (benzene and toluene), and cycloalkanes (cyclohexane and methylcyclohexane). The analyses provide quantitative data for the lighter hydrocarbons (from C_1 to pentanes, C_5) and qualitative data for the heavier (C_6+) hydrocarbons [3].

The wells that used GASX and GASY were analyzed with a mass spectrometer. With this equipment, it is possible to detect single-bonded ethane (C_2H_6) and double-bonded ethene (C_2H_4) through a channel with the presence of ethane and ethene (C_2) together and another channel called "corrected C_2—(C_2C)" where only the presence of pure ethane is detected. Ethene concentrations were estimated indirectly by subtracting these channels (ethene = $C_2 - C_2C$) because the advanced gas analysis equipment did not have an ethylene analyzer available at the time the wells were drilled. The C_2C channel is only presented when there are relatively high concentrations of C_2C values (>200 ppm). All evaluation and interpretation of this work were performed considering C_2C above 200 ppm.

In the advanced mud gas analysis, there are two gas extractors in the mudflow line [3]. One is positioned to measure mud gas in the drilling fluid injected into the well (IN) and the other one measures mud gas after complete circulation of the drilling fluid in the well (OUT; Figure 2). This facilitates extraction and quantification of the gas that is incorporated into the drilling fluid at the entrance of the well (IN) and the gas that is returning from the well (OUT). The calculated difference ("DELTA" data) between the measurements in the OUT and IN extractors (DELTA = OUT − IN) removes the effect of the recycled gas in the mud from the concentration measurements and only accounts for the gas that was released during drilling. The gas reading at the IN and OUT is a great advantage of the advanced gas analysis when compared to the conventional gas analysis.

The drilling fluid is heated (80 °C or 90 °C depending on the service provider) both at the inlet and at the outlet before being extracted. As a result, the gas dissolved in the fluid is released more easily and with greater efficiency because the increase in temperature leads to an increase in the solubility [43] (especially in the heavy fractions), which is another great advantage of advanced mud gas analysis [3].

During gas extraction, hydrocarbons tend to stick to the drilling fluid and are therefore difficult to extract, causing an extraction deficit in gas analysis. Advanced gas analysis equipment also has the limitation of keeping some parameters stable. This difficulty in stabilization also causes an extraction deficit in gas analysis. To compensate for these extraction difficulties at the end of well drilling, mudlogging companies normally calculate the Extraction Efficiency Correction (EEC) and apply that correction to the DELTA data from C_1 to C_5. This correction occurs by multiplying a different coefficient for each of the alkanes (C_1, C_2, C_3, C_4, and C_5). When this coefficient is identified for each alkane, it will be applied throughout all depths of the well. This coefficient varies for each of the alkanes, being progressively larger from C_1 to C_5, according to the difficulty of extraction. This coefficient is used to compensate for both the extraction deficit, which is inherent to the equipment's capacity, and for the characteristics of the drilling fluid. Each mudlogging company has its criteria for calculating the correction factor and they apply a unique value for each of the alkanes. The EEC factor can be recalculated for each compound at the end of each drilling phase or when the drilling fluid changes configuration. Extraction coefficients are not calculated for C_6+ heavy hydrocarbons.

2.3. Data Preparation and Processing

Initially, compositional data from advanced mud gas analysis were collected for each of the 50 wells. All data were cataloged, organized, and evaluated for quality. Gas data with zero values, values greater than one million, values consecutively repeated, fractionated values, and the data acquired before the start of the mud return were excluded. Data in which the unit of measurement was not parts per million (ppm), such as a unit of total gas, a unit of heavy gas, and percentage data, were also disregarded. At the end of this process, all invalid and spurious data were removed. For each of the 50 wells, the drilling data (type of bit used and bit change depths), type of drilling fluid, the tops of lithostratigraphic formations, lithologies, the final and initial depths of advanced mud-gas analysis, the identification of the company providing the advanced gas analysis service, and drilling parameters (rate of penetration—ROP, weight on bit—WOB, and rotations per minute—RPM) were recorded.

After organizing these data, several gas ratios were calculated. Then, these ratios were plotted versus measured depth MD (m) for each well to identify the variation patterns of the gaseous compounds. The top of each lithostratigraphic unit and the depth of each bit change were marked on these plots. In addition, plots of gas concentrations (C_1, C_2, C_2c, C_3, $i-C_4$, $n-C_4$, $i-C_5$, $n-C_5$, and benzene, using DELTA data versus depth MD (m)) and plots of drilling parameters (ROP, WOB, and RPM versus depth MD (m)), were prepared.

During the drilling of the wells, geological information such as rock samples, gas samples, well logs, and drilling parameters were used to identify the lithostratigraphic formations of the wells. This information was recorded in a document called a geological monitoring profile, from which the top and bottom information of the lithostratigraphic formation used in this work were taken. According to [47], the Santos Basin comprises the following lithostratigraphic formations: Marambaia (Paleocene sandstones, siltstones, and shales), Juréia (Late Cretaceous siltstones, shales, and diamictites), Itajaí-Açu (Late Cretaceous siltstones, shale, and diamictites), Itanhaém (Albian/Cenomanian dark laminated shales and calcilutites), Guarujá (Albian shales, calcilutites, calcirrudites, and calcarenites), Ariri (Neoaptian halite and anhydrite), Barra Velha (Aptian stromatolitic limestones and microbial laminites), Itapema (Neobarremian and Eoaptian calcirrudites and shales), Piçarras (Barremian conglomerates and polymictic sandstones), and Camboriú (Early Cretaceous basalts and diabase) Formations.

In all wells, the top and bottom of each lithostratigraphic formation were identified, and the results of this survey are compiled in Table 1. Some wells penetrated igneous rocks with different thicknesses interspersed with siliciclastic and carbonate rocks. Seven wells penetrated the Camboriú Formation, which is the economic basement of the Santos Basin.

The predominant lithologies were identified for each lithostratigraphic unit (Table 2). The "siliciclastic" lithology group includes all sedimentary lithologies from the Marambaia, Juraí-Açu, and Itanhaém formations. The "carbonate platform" group includes all lithologies of the Guarujá Formation, whereas the "salt" group includes all lithologies of the Ariri Formation. The "limestone" group includes the Barra Velha, Itapema, and Piçarras formations. All other lithologies with low porosity were included in the "calcilutite" group.

For some wells that crossed igneous rocks, it was not possible to identify the rock type in detail, and therefore the rocks were classified only as "igneous". In the "diabase" group, igneous rocks that have been identified as diabase or basalt are included. Finally, two groups were created in which the carbonate lithologies are interspersed with igneous or fine low-porosity lithologies. These intercalations of carbonates with igneous or fine rocks are layers with thicknesses <10 m. Thus, the groups "limestone/igneous" and "limestone/calcilutite" were created. These two last groups were created because these thin lithologies are difficult to visually recognize on the lithology profiles of the wells, but they can affect the gas response.

3. Results and Discussion

Among 50 wells that penetrated the pre-salt section in the Santos Basin, 33 wells had clear evidence of DBM in some parts of the wells. Figure 3 represents an evaluation model of the well E3, showing the presence of DBM (based on all the criteria) in the section below 5500 m up to 5800 m (e.g., $C_2 > C_1$ in ppm, with C_2 normalized > 20%). In the Well-E3 b panel, we observe between 5500 and 5800 m the inversion of the curves and dispersion of the ratios within the limestone/igneous lithological group (pre-salt carbonate rocks with thin (centimeters to <10 m) intercalations of igneous rocks). The limestone and igneous intercalations make it difficult to advance the drilling, promoting more friction and thus generating artificial gas, even before drilling the thick package of igneous rock at the end of the well. The normalized alkane chart values were calculated by normalizing the advanced gas analysis ranges of each well from C_1 to C_5 in percentage terms.

Figures 3 and 4 show a typical DBM signature in the Ariri Formation (salt) section. This formation has low porosity and permeability, causing the gas background to be normally below 100 ppm. Sometimes, methane and ethane are present in higher concentrations during salt drilling, while the other hydrocarbons have low concentrations. This characteristic of the formation (all alkanes are below 100 ppm) can generate a false DBM signature that is possibly an artifact of the low gas concentration (<100 ppm).

Similar low gas anomalies also sometimes occur in siliciclastic rocks drilled in the post-salt interval. In Figure 4, these low concentration values (<100 ppm) associated with the inversion of the dryness curve and dispersion of the gas ratios occur both in the post-salt interval (Itajaí-Açu Formation) and at the top of the Ariri Formation. The inversion of the C_2/C_1 and dryness curves and a larger dispersion of the C_2/C_3 and C_2/C_1 ratios are observed below 3100 m just when the gaseous anomaly decreases.

3.1. Influence of Drill-Bit Type and Lithologies on DBM

At Petrobras, well drilling is carried out with different bits depending on the objective and drilled interval. Until the present work, a broad study had not been carried out correlating the intensity of the DBM with the drills used to drill wells in the pre-salt section.

We found that different types of drill bits and lithology influence the presence and severity of DBM. Therefore, wells were separated according to the type of drill bit that was more common (PDC bits, diamond-impregnated bits, and tricone bits) and according to the main lithologies (igneous, pre-salt carbonate, salt, and post-salt).

Different intensities of DBM were observed during the drilling of pre-salt carbonate rocks with different bit types. Figure 5 shows a comparison between the chromatography of three wells with different amounts of artificial DBM-generated gases in the pre-salt interval.

The well B2 (Figure 5, Well-B2 c) was drilled with a PDC bit in the Barra Velha Formation (limestones). Based on the normal chromatographic distribution of $C_1 > C_2 > C_3 > C_4 > C_5$, C_2 normalized = 10%, low dispersion (C_2/C_1 and C_2/C_3), dryness close to 1, and low values of benzene, in this case, the PDC bit did not generate artificial gases until the depth of 5918 m. The drill bit started to drill the igneous rock (5918 m until 5933 m), and all the criteria for identifying DBM were observed (e.g., $C_2 > C_1$ in ppm with C_2 normalized = 46.8% range average 5918/5933).

In well C2 (Figure 5, Well-C2 c), the mud gas data were acquired starting a few meters below the top of the Barra Velha Formation and drilled with a PDC drill bit to a depth of ~5700 m without DBM. Then, the PDC bit was changed to a diamond-impregnated bit. After changing the bit (5700 m until 5765 m), all the criteria were observed, suggesting the occurrence of DBM (e.g., $C_2 > C_1$ in ppm with C_2 normalized = 47.9% range average 5700/5765).

The well D_5 (Figure 5, Well-D5 c) was drilled with an impregnated drill bit from the top of the Barra Velha Formation to the end of the well. This drill bit generated moderate DBM in the section below the depth of 4965 m until 5180 m (based on all the criteria, e.g., $C_2 = C_1$ in ppm with C_2 normalized = 39.6% range average 4965/5180) in the pre-salt carbonate and generated extreme metamorphism in the igneous rock interval 5180 m until 5505 m (e.g., C_2 normalized = 51.3% range average 5180/5505). The extreme DBM response was recognized by observing the inversion of the chromatographic curves ($C_2 > C_1$) and significant concentrations of benzene (benzene $> C_5$).

As a result of the integrated evaluation of the drill bits, lithologies, and the responses of gas ratios, we found that only the PDC and impregnated drill bits resulted in DBM. Artificial gases generated by DBM were observed in 92% of the cases in which an impregnated bit was used to drill igneous rocks. DBM also often occurs when the PDC drill bit cuts through an igneous rock (73%). As for the pre-salt carbonates, the impregnated drill bit caused DBM in 71% of the cases and the PDC drill bit caused DBM much less often (24% of cases).

The occurrences of DBM and associated artificial gases in the salt section are rare. The gas background is normally low (<100 ppm) when the salt section is drilled because of its relative ease for drilling, given its characteristic behavior as a seal rock, and a lack of indigenous gases. For the 50 wells that were evaluated, only PDC or tricone bits were used to drill the Ariri Formation. DBM was observed in only 8% of cases when the PDC was used to drill the salt, normally associated with anhydrite drilling.

Igneous rocks were penetrated in 23 wells with impregnated or PDC bits. We observed that the intensity of DBM was much higher in the igneous rocks than in sedimentary rocks. DBM in the igneous rocks generated abundant artificial gases (both alkanes and alkenes), with strong inversions of the dryness and ethene/(ethene+ethane) curves, high values, and high dispersion of C_2/C_1, inversion of normal chromatography $C_2 > C_1$, C_2/C_3 dispersion, and the concentrations of benzene exceeding the concentrations of C_4 and C_5 (Figure 3). Rocks that have a high degree of cohesion and low permeability, such as igneous rocks and silicified carbonate rocks, for example, may have a high generation of artificial gas since the drill will have greater difficulty drilling these rocks. Rocks with a high degree of resistance associated with small grain sizes can also influence drilling. The smaller the grain, the more cohesive the rock and the more difficult it is to drill.

We have not observed DBM when the tricone drilling bits were used for drilling. In all cases observed, the presence of DBM impacts the gas data, compromising the interpretation. Knowledge about the types of drill bits associated with DBM is quite applicable during the drilling operation so that well projects can plan and avoid using drill bits that produce more DBM in the intervals of interest or the objective of the well. Therefore with this knowledge, the evaluation of advanced gas analysis during drilling can be conducted with the best possible reliability.

3.2. Influence of Drilling Parameters on DBM

Drilling parameters, such as ROP, were evaluated to understand how or if they influence the DBM. Of the 50 wells evaluated, 13 showed DBM in the entire section drilled in the pre-salt section and 15 wells did not show DBM in any drilled interval. These 28 wells were separated into 2 groups of "wells with DBM and wells without DBM", and the correlations between the drilling parameters ROP, WOB, RPM, and the DBM proxies C_2/C_1, dryness, and ethene/ethene+ethane were evaluated.

According to information on the influence of lithology on the DBM, for each group of wells, the lithologies were grouped as follows: igneous rock, limestone, calcilutite, post-salt sedimentary rock, and salt.

For the two groups of wells, the following plots were made: ROP versus C_2/C_1, ROP versus ethene/ethene+ethane, ROP versus dryness, WOB versus C_2/C_1, WOB versus ethene/ethene+ethane, WOB versus dryness, RPM versus C_2/C_1, RPM versus ethene/ethene+ethane, and RPM versus dryness. Figure 6 shows some of the correlations that were evaluated for the two groups of wells (ROP versus C_2/C_1, WOB versus ethene/ethene+ethane, and RPM versus dryness) and the respective R^2 values separated by lithology. All correlations in both groups had $R^2 < 0.4$, indicating little to no influence of drilling parameters on DBM generation or low linear correlation between drilling parameters (ROP, WOB, and RPM) and the DBM (C_2/C_1, dryness, and ethene/ethene+ethane).

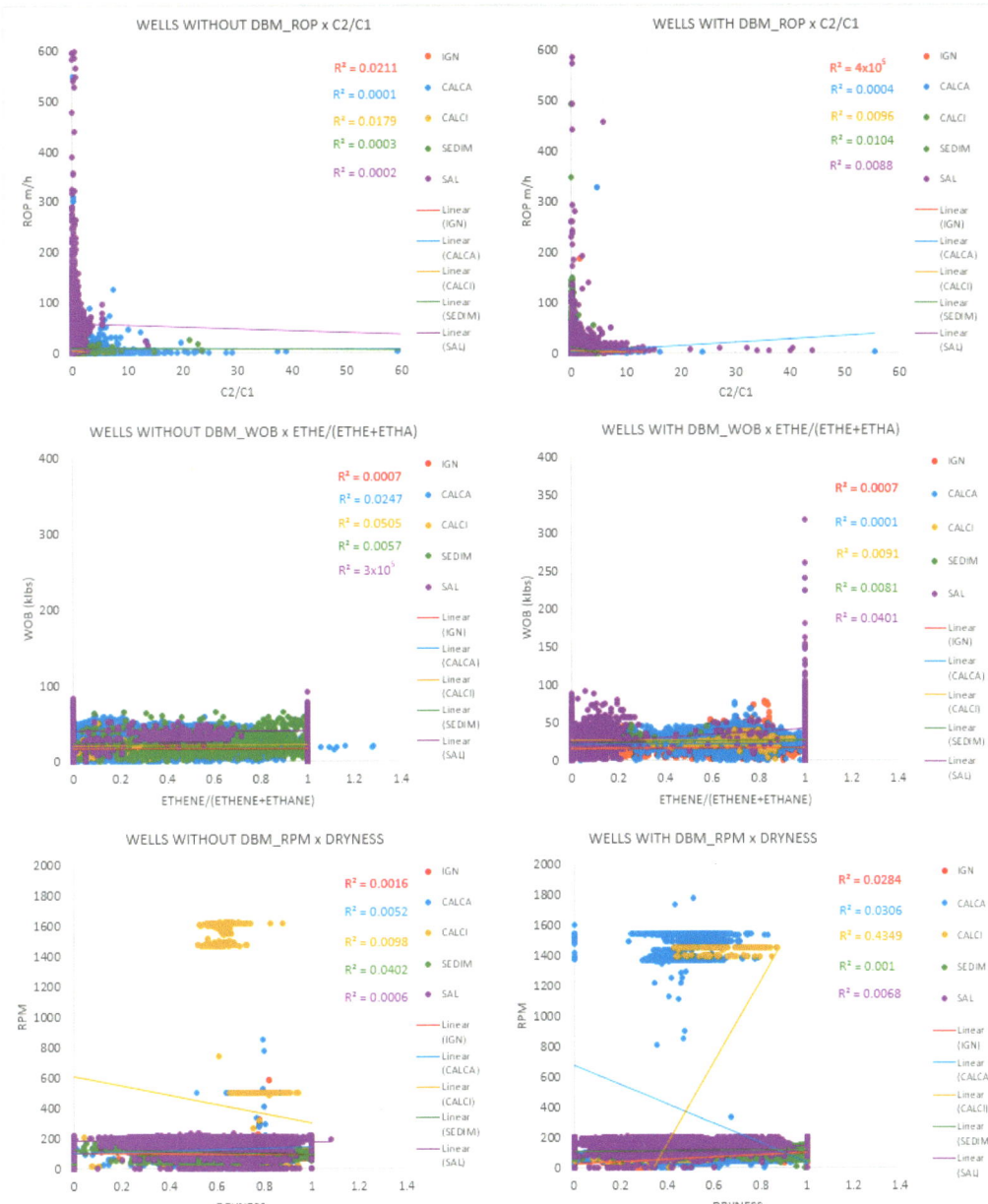

Figure 6. Correlations between the drilling parameters and the gas ratios that were used for the identification of DBM considering the groups of wells with and without DBM, separated by lithology (ROP x C_2/C_1, WOB x ethene/ethene+ethane, and RPM x dryness).

4. Conclusions

DBM was recognized using advanced mud gas data in numerous wells drilled in the Santos Basin. DBM was most common and intense when igneous rocks and pre-salt carbonate rocks were drilled with impregnated bits. PDC bits also often result in DBM, especially when used to drill through igneous rocks.

As a result of the integrated evaluation of the drill bits, lithologies, and the responses of gas ratios, we found that only the PDC and impregnated drill bits resulted in DBM. Artificial gases generated by DBM were observed in 92% of the cases in which an impregnated bit was used to drill igneous rocks. DBM also often occurs when the PDC drill bit cuts through an igneous rock (73%). As for the pre-salt carbonates, the impregnated drill bit caused DBM in 71% of the cases and the PDC drill bit caused DBM much less often (24% of cases).

When drilling with impregnated or PDC bit, it is necessary to carefully evaluate for the presence of DBM. The identification of DBM during the drilling of exploration wells is essential. DBM generates artificial gases and changes the chromatography, gas dryness, and other gas ratios, which can potentially lead to incorrect interpretations of the origin and maturity of hydrocarbon gases.

We did not find discernable relationships between drilling parameters and the geomechanical properties that indicate DBM. The presence or absence of DBM is more correlated with the type of lithology and drill-bit properties.

For the 50 wells in this work (wells drilled between 2009 and 2020), the intervals where artificial gas was generated by DBM should not be used to interpret the composition of gases in the reservoir. As of yet, no methodology has been developed to decontaminate or separate gas components generated by DBM from indigenous gases coming from reservoir rocks. The application of analysis tools like machine learning or artificial intelligence may aid in this effort. New technologies are advancing, and it may be possible for companies providing AMG services to apply a methodology based on algorithms to decontaminate DBM.

Author Contributions: Conceptualization, J.A.d.L.L. and H.L.d.B.P.; Methodology, H.L.d.B.P., G.S.E. and J.G.M.F.; Formal analysis, J.A.d.L.L.; Investigation, J.A.d.L.L.; Data curation, J.A.d.L.L.; Writing—original draft preparation, J.A.d.L.L.; Writing—review and editing, H.L.d.B.P., G.S.E., A.M. and J.G.M.F.; Visualization, A.M.; Supervision, H.L.d.B.P., G.S.E. and J.G.M.F. All authors have read and agreed to the published version of the manuscript.

Funding: The authors J.A.d.L.L. and H.L.d.B.P. are employees of the company Brazilian Petrol S/A–Petrobras, which acquired all the data that were analyzed in this manuscript. This paper was part of J.A.d.L.L.'s Ph. and was funded by Brazilian Petrol S/A-Petrobras.

Data Availability Statement: The dataset studied is public and can be requested. It is currently available at the Brazilian regulatory agency "Brazilian National Agency for Petroleum, Natural Gas and Biofuels—ANP". Interested parties should contact the ANP "Technical Data Superintendence" (SDT) via email at helpdesk@anp.gov.br.

Acknowledgments: We are grateful to Petrobras for supporting and providing fundamental data and materials for the development and execution of this work. We thank the U.S. Geological Survey Program and the Department of Geology and Geological Engineering, Colorado School of Mines, for supporting and stimulating this type of research. We express our gratitude to Alexei Milkov (in memoriam) for accepting to be part of this research, for his guidance, and for his great commitment to helping us. Any use of trade, firm, or product names is for descriptive purposes only and does not imply endorsement by the U.S. Government.

Conflicts of Interest: The authors J.A.d.L.L. and H.L.d.B.P. are employees of the company Brazilian Petrol S/A–Petrobras. The remaining authors declare that the research was conducted in the absence of any commercial or financial relationships that could be construed as a potential conflict of interest.

References

1. Kennedy, K.F. *Hydrocarbon Well Logging Recommended Practice: Society of Professional Well Log Analysts*; The Society: Lancaster, MA, USA, 1983.
2. Whiticar, M.J. Correlation of natural gases with their sources. In *The Petroleum System—From Source to Trap*; Magoon, L., Dow, W., Eds.; AAPG: Tulsa, OK, USA, 1994; Memoir 60; pp. 261–284.
3. Ablard, P.; Bell, C.; Cook, D.; Fornasier, I.; Poyet, J.P.; Sharma, S.; Fielding, K.; Lawton, L.; Haines, G.; Herkommer, M.A.; et al. The expanding role of mud logging. *Oilfield Rev.* **2012**, *24*, 24–41.
4. Elshahawi, H.; Venkataramanan, L.; McKinney, D.; Flannery, M.; Mullins, O.C.; Hashem, M. Combining continuous fluid typing, wireline formation tester, and geochemical measurements for an improved understanding of reservoir architecture. In Proceedings of the SPE 2006, Annual Technical Conference and Exhibition, San Antonio, TX, USA, 24–27 September 2006. [CrossRef]
5. Liew, Y.Y.; Fornasier, I.; Hartman, A. Integration of mud gas analysis with conventional logs to aid fluid typing in thinly bedded, argillaceous clastic reservoirs. In Proceedings of the SPWLA 51st Annual Logging Symposium, Perth, Australia, 19–23 June 2010. Paper Number: SPW-LA-2010-80310.
6. Yang, T.; Arief, I.H.; Niemann, M.; Houbiers, M.; Meisingset, K.K.; Martins, A.; Froelich, L.A. machine learning approach to predict Gas Oil Ratio based on advanced mud gas data. In Proceedings of the SPE 2019, Europec featured at 81st EAGE Conference and Exhibition, London, UK, 3–6 June 2019. SPE-195459-MS. [CrossRef]
7. Arief, I.H.; Yang, T. Real time reservoir fluid log from advanced mud gas data. In Proceedings of the SPE Annual Technical Conference and Exhibition, Virtual, 27–29 October 2020. [CrossRef]
8. Varhaug, M. The Defining Series, Mud logging. Available online: https://www.slb.com/resource-library/oilfield-review/defining-series/defining-mud-logging (accessed on 14 April 2024).
9. Elmgerbi, A.M.; Ettinger, C.P.; Tekum, P.M.; Thonhauser, G.; Nascimento, A. Application of Machine Learning Techniques for Real Time Rate of Penetration Optimization. In Proceedings of the SPE/IADC Middle East Drilling Technology Conference and Exhibition, Abu Dhabi, United Arab Emirates, 25–27 May 2021. [CrossRef]
10. Elmgerbi, A.; Chuykov, E.; Thonhauser, G.; Nascimento, A. Machine Learning Techniques Application for Real-Time Drilling Hydraulic Optimization. In Proceedings of the International Petroleum Technology Conference, Dhahran, Saudi Arabia, 21–23 February 2022. [CrossRef]
11. Elmgerbi, A.; Thonhauser, G. Holistic autonomous model for early detection of downhole drilling problems in real-time. *Process Saf. Environ. Prot.* **2022**, *164*, 418–434. [CrossRef]
12. Pixler, B.O. Formation evaluation by analysis of hydrocarbon ratios. *J. Pet. Technol.* **1969**, *21*, 665–670. [CrossRef]
13. Haworth, J.H.; Sellens, M.; Whittaker, A. Interpretation of hydrocarbon shows using light (C_1–C_5) hydrocarbon gases from mud-log data. *AAPG Bull.* **1985**, *69*, 1305–1310.
14. Wright, A.C. Estimation of gas/oil ratios and detection of unusual formation fluids from mud logging gas data. In Proceedings of the SPWLA 1996, 37th Annual Logging Symposium, New Orleans, LA, USA, 16–19 June 1966.
15. De Pazzis, L.L.; Delahaye, T.R.; Besson, L.J.; Lombez, J.P. New gas logging system improves gas shows analysis and interpretation. In Proceedings of the SPE 1989, Annual Technical Conference and Exhibition, San Antonio, TX, USA, 8–11 October 1989. [CrossRef]
16. Kandel, D.; Quagliaroli, R.; Segalini, G.; Barraud, B. Improved integrated reservoir interpretation using the Gas While Drilling (GWD) Data. In Proceedings of the SPE 2000, European Petroleum Conference, Paris, France, 24–25 October 2000. [CrossRef]
17. McKinney, D.; Flannery, M.; Elshahawi, H.; Stankiewicz, A.; Clarke, E.; Breviere, J.; Sharma, S. Advanced mud gas logging in combination with wireline formation testing and geochemical fingerprinting for an improved understanding of reservoir architecture. In Proceedings of the SPE 2007, Annual Technical Conference and Exhibition, Anaheim, CA, USA, 11–14 November 2007. [CrossRef]
18. Loermans, T.; Bradford, C.; Kimour, F.; Karoum, R.; Meridji, Y.; Kasprzykowski, P.; Bondabou, K.; Marsala, A. Advanced mud logging aids (AML) formation evaluation and drilling, yields precise hydrocarbon fluid composition. In Proceedings of the SPE 2011, Middle East Oil and Gas Show and Conference, Manama, Bahrain, 25–28 September 2011. [CrossRef]
19. Kyi, K.K.; Lynn, C.S.; Haddad, S.; Chouya, S.; Wa, W.W.; Gligorijevic, A. Integration of downhole fluid analysis and advanced mud gas logging reduces uncertainty in reservoir evaluation. In Proceedings of the International Petroleum Technology Conference, Doha, Qatar, 19–22 January 2014. IPTC-17485-MS. [CrossRef]
20. Dashti, J.; Al-Awadi, M.; Al-Ajmi, B.; Rao, S.; Shoeibi, A.; Estarabadi, J.; Nadiminti, A. Use of advanced mud gas chromatography for reservoir quality prediction while drilling. In Proceedings of the International Petroleum Technology Conference, Bangkok, Thailand, 14–16 November 2016. IPTC-18624-MS. [CrossRef]
21. Blue, D.; Blakey, T.; Rowe, M. Advanced mud logging: Key to safe and efficient well delivery. In Proceedings of the Offshore Technology Conference 2019, Houston, TX, USA, 6–9 May 2019. [CrossRef]
22. Graves, W. Bit-generated rock textures and their effect on evaluation of lithology, porosity, and shows in drill-cutting samples. *AAPG Bull.* **1986**, *70*, 1129–1135.
23. Faber, E.; Gerling, P.; Dunke, I. Gaseous hydrocarbons of unknown origin found while drilling. *Org. Geochem.* **1987**, *13*, 875–879. [CrossRef]

24. Wenger, L.M.; Pottorf, R.J.; Macleod, G.; Otten, G.; Dreyfus, S.; Justwan, H.; Wood, E.S. Drill-bit metamorphism: Recognition and impact on show evaluation. In Proceedings of the SPE 2009, Annual Technical Conference and Exhibition, New Orleans, LA, USA, 4–7 October 2009. [CrossRef]
25. Qubaisi, K.; Torlov, V.; Fernandes, A.; Bajsair, A. Using drill bit metamorphism to aid in formation evaluation of tight gas reservoirs. In Proceedings of the International Petroleum Technology Conference 2020, Dhahran, Saudi Arabia, 13–15 January 2020. [CrossRef]
26. Strapoć, D.; Fornasier, I.; Di Santo, I. Drill-bit metamorphism: Impact awareness, monitoring technology, mitigation attempts. In Proceedings of the OMC Med Energy Conference and Exhibition, Ravenna, Italy, 28–30 September 2021. Paper Number: OMC-2021-208.
27. Bowen, N.L.; Aurousseau, M. Fusion of sedimentary rocks in drill-holes. *Bull. Geol. Soc. Am.* **1923**, *34*, 431–448. [CrossRef]
28. Taylor, J.C.M. Bit metamorphism, illustrated by lithological data from German North Sea wells. In *Petroleum Geology of the Southeastern North Sea and the Adjacent Onshore Areas: (The Hague, 1982)*; Kaasschieter, J.P.H., Reijers, T.J.A., Eds.; Geologie en Mijnbouw; Springer: Berlin/Heidelberg, Germany, 1983; Volume 62, pp. 211–219.
29. Kerr, C.J. PDC drill bit design and field application evolution. *J. Pet. Technol.* **1988**, *40*, 327–332. [CrossRef]
30. Ortega, A.; Glowka, D.A. Frictional heating and convective cooling of polycrystalline diamond drag tools during rock cutting. *SPE J.* **1984**, *24*, 121–128. [CrossRef]
31. Kennedy, L.A.; Spray, J.G. Frictional melting of sedimentary rock during high-speed diamond drilling: An analytical SEM and TEM investigation. *Tectonophysics* **1992**, *204*, 323–337. [CrossRef]
32. Qingchao, L.; Jin, L.; Shiming, W.; Ying, G.; Xiaoying, H.; Qiang, L.; Yuanfang, C.; Zhuo, D.; Xianzhong, L.; Xiaodong, Z. Numerical insights into factors affecting collapse behavior of horizontal wellbore in clayey silt hydrate-bearing sediments and the accompanying control strategy. *Ocean. Eng.* **2024**, *294*, 11702. [CrossRef]
33. Li, Q.; Wang, F.; Wang, Y.; Bai, B.; Zhang, J.; Lili, C.; Sun, Q.; Wang, Y.; Forson, K. Adsorption behavior and mechanism analysis of siloxane thickener for CO_2 fracturing fluid on shallow shale soil. *J. Mol. Liq.* **2023**, *376*, 121394. [CrossRef]
34. Jeffrey, A.W.A.; Kaplan, I.R. Hydrocarbons and inorganic gases in the Gravberg-1 well, Siljan Ring, Sweden. *Chem. Geol.* **1988**, *71*, 237–255. [CrossRef]
35. Zimmer, M.; Erzingeer, J. On the geochemistry of gases in formation and drilling fluids—Results from KTB. *Sci. Drill.* **1995**, *5*, 101–109.
36. Strapoć, D.; Ammar, M.; Abolins, N.; Gligorijevic, A. Key role of regearing mud gas logging for natural H_2 exploration. In Proceedings of the SPWLA 2022, 63rd Annual Logging Symposium, Stavanger, Norway, 10–15 June 2022. [CrossRef]
37. Fischer, F.; Tropsch, H. The synthesis of petroleum at atmospheric pressures from gasification products of coal. *Brennstoff-Chemie* **1926**, *7*, 97–104.
38. Reitsma, M.; Harlow, C. Real-time detection of drill bit metamorphism for accurate interpretation of hydrocarbon shows. 2016. In Proceedings of the International Conference and Exhibition, Barcelona, Spain, 3–6 April 2016. [CrossRef]
39. Strapoć, D.; Gonzalez, D.; Gerbaud, L. Controlled drill bit metamorphism (DBM) using indoor rig floor experiments. In Proceedings of the 30th International Meeting on Organic Geochemistry (IMOG 2021), Montpellier, France, 12–17 September 2021; Volume 2021, p. 1. [CrossRef]
40. DeBruijn, G.; Skeates, C.; Greenaway, R.; Harrison, D.; Parris, M.; James, S.; Mueller, F.; Ray, S.; Riding, M.; Temple, L.; et al. High-pressure, high-temperature technologies. *Oilfield Rev.* **2008**, *8*, 46–60.
41. Goodwin, N.R.J.; Abdullayev, N.; Javadova, A.; Volk, H.; Riley, G. Diamondoids and basin modelling reveal one of the world's deepest petroleum systems, South Caspian basin, Azerbaijan. *J. Pet. Geol.* **2020**, *43*, 133–149. [CrossRef]
42. Zhu, G.; Milkov, A.V.; Li, J.; Xue, N.; Chen, Y.; Hu, J.; Li, T.; Zhang, Z.; Chen, Z. Deepest oil in Asia: Characteristics of petroleum system in the Tarim basin, China. *J. Pet. Sci. Eng.* **2021**, *199*, 108246. [CrossRef]
43. Al-Mahrooqi, S.; Mookerjee, A.; Walton, W.; Scholten, S.; Archer, R.; Al-Busaidi, J.; Al-Busaidi, H. Well logging and formation evaluation challenges in the deepest well in the Sultanate of Oman (HPHT tight sand reservoirs). In Proceedings of the SPE 2011, Middle East Unconventional Gas Conference and Exhibition, Muscat, Oman, 31 January–2 February 2011. [CrossRef]
44. Carcione, E.; Easow, I.; Chiniwala, B. Alkenes detection from drill bit metamorphism and real-time geochemical elemental analysis on drill cuttings aids drilling optimization and geo-steering in tight unconventional laterals. In Proceedings of the SPE/AAPG/SEG Unconventional Resources Technology Conference 2017, Austin, TX, USA, 24–26 July 2017. [CrossRef]
45. Souza, I.V.A.F.; Ellis, G.S.; Ferreira, A.A.; Guzzo, J.V.P.; Díaz, R.A.; Albuquerque, A.L.S.; Amrani, A. Geochemical characterization of natural gases in the pre-salt section of the Santos Basin (Brazil) focused on hydrocarbons and volatile organic sulfur compounds. *Mar. Pet. Geol.* **2022**, *144*, 105763. [CrossRef]
46. Stahl, W.J. Carbon and nitrogen isotopes in hydrocarbon research and exploration. *Chem. Geol.* **1977**, *20*, 121–149. [CrossRef]
47. Milani, E.J.; Rangel, H.D.; Bueno, G.V.; Stica, J.M.; Winter, W.R.; Caixeta, J.M.; Neto, O.C.P. Bacias sedimentares Brasileiras—*Cartas estratigráficas*. *Bol. Geociências Petrobras* **2007**, *15*, 183–205.

Disclaimer/Publisher's Note: The statements, opinions and data contained in all publications are solely those of the individual author(s) and contributor(s) and not of MDPI and/or the editor(s). MDPI and/or the editor(s) disclaim responsibility for any injury to people or property resulting from any ideas, methods, instructions or products referred to in the content.

Article

A New Prediction Model of Annular Pressure Buildup for Offshore Wells

Renjun Xie [1,2,*] and Laibin Zhang [1]

[1] Department of Safety Engineering, College of Safety and Ocean Engineering, China University of Petroleum (Beijing), Beijing 102249, China
[2] CNOOC Research Institute Ltd., Beijing 100028, China
* Correspondence: xierj@cnooc.com.cn

Abstract: Subsea wellheads and Christmas trees are commonly utilized in deepwater oil and gas development. However, the special structure of subsea wellheads makes it difficult to monitor casing–casing annular pressure buildup, which in turn poses a greater risk to the integrity of the wellbore. In order to analyze the effect of changes in the casing-free section and the sealed section on the variation in annulus volume, a new annular pressure buildup model of casing-cement sheath-formation deformation was established and verified according to the elastic deformation theory. Furthermore, the influence of casing deformation on annulus pressure buildup was analyzed. Results indicate that the error of annulus pressure buildup predicted by the multi-string mechanical model proposed in this paper that considers the deformation of the casing sealing section is approximately 13% lower than the one that does not consider this factor. This paper provides guidance for the design of casing strings in deepwater oil and gas wells, ensuring safe production.

Keywords: deepwater oil and gas well; annular pressure buildup; casing sealing section deformation; annulus volume change

Citation: Xie, R.; Zhang, L. A New Prediction Model of Annular Pressure Buildup for Offshore Wells. *Appl. Sci.* **2024**, *14*, 9768. https://doi.org/10.3390/app14219768

Academic Editor: Tiago Miranda

Received: 13 August 2024
Revised: 18 October 2024
Accepted: 24 October 2024
Published: 25 October 2024

Copyright: © 2024 by the authors. Licensee MDPI, Basel, Switzerland. This article is an open access article distributed under the terms and conditions of the Creative Commons Attribution (CC BY) license (https://creativecommons.org/licenses/by/4.0/).

1. Introduction

Annular pressure buildup is caused by fluid expansion due to temperature rise, which in turn leads to an increase in pressure in the annular space between the layers of the casing [1–3]. For onshore and shallow water wells, annular pressure buildup can be reduced by wellhead pressure relief. However, for deepwater wells where the wellhead is located underwater, the annular pressure cannot be released or reduced [4,5]. Therefore, the accurate prediction of annular pressure buildup is crucial for the casing design and safe production of deepwater oil and gas wells [6–8].

After several hours of production in Well A-2 of the Marlin Oilfield in the Gulf of Mexico, the high-temperature fluid caused the annular pressure to rise, resulting in the collapse of the production casing [9–12]. During the drilling of Well Pompano A-31 in the Gulf of Mexico, the high circulating temperature of the drilling fluid caused the high annular pressure, resulting in the collapse of the 16″ casing [13,14].

A. Adams [15] studied the design and strength analysis of multi-layer tubing-casing and proposed the use of SLA (Service Life Analysis) based on the wellbore string system to check the overall loading of wellbore strings. The method takes into account the additional loads generated by the thermal expansion of annular fluid for the first time. M.A. Goodman [16] applied the SLA casing design method to study and analyze the casing strength of HPHT wells in the field. A.S. Halal and R.F. Mitchell [17] took into account the fluid nonlinear PTV relationship and elastic deformation of the casing and performed a rigorous mathematical derivation. Lubinski [18] proposed that the annular pressure buildup is important to the wellbore. The thermal expansion of confined annulus fluid can cause packer leakage, and a high wellbore profile temperature can cause tubing bending. Klementich [19] and Adams [15], respectively, proposed that the annular fluids generate annular pressure buildup through being heated. Oudeman [20] proposed that the annular pressure buildup is mainly caused by the thermal expansion/compression characteristics of the fluid trapped in the annulus, the deformation of the casing, and the migration of the fluid into the annulus, and established a calculation model for annular pressure buildup. In previous studies, only the impact of casing-free section deformation on annular pressure bulldup was investigated. However, the impact of casing sealing section deformation on annular pressure buildup has not been studied [21–25]. B. Moe [26] simplified the calculation of casing deformation in the PVT equation and gave suggestions on how to prevent annular pressure. In 2006, Deng Yuanzhou [27] proposed an iterative model to predict annular pressure. Considering the coupling effect of annular pressure and volume, the calculation result of the iterative method is 10% smaller than that of the traditional method, mainly due to the influence of annular volume change. Huang Xiaolong et al. [28] established a calculation model of annulus pressure by using the plane strain equation and analyzed the influences of reservoir temperature, bottom-hole temperature, fluid properties, production flow rate, and well mechanism on annulus pressure buildup model for the wellbore–reservoir system. The annular temperature calculated by this model was more accurate than that calculated by the traditional semistable method. In 2009 and 2010, A.R. Hasan [23] proposed two models, semi-steady state and transient, to calculate the annular pressure in the production process, in which the thermal expansion of the fluid and the leakage effect of the fluid were strictly considered.

In order to illustrate the influence of casing sealing section deformation on annular pressure buildup, in this paper, a coordination deformation model of casing-cement sheath-formation was established, and then a new multi-string annular pressure buildup model was proposed. The influence of casing sealing section deformation on annular pressure was analyzed. The results show that the error of the annular pressure buildup model established in this paper, which considers the deformation of the sealed casing, is about 13% lower than that predicted without considering this factor and is closer to the measured value in the field. This study has important practical significance for guiding the casing string design and safety production of deepwater oil and gas wells.

2. Source and Formation Mechanism of Annular Pressure

2.1. Thermally Induced Annular Pressure

At present, the sources that cause the increase in annular pressure between casing mainly include the following three categories:

(1) Sustained Casing Pressure

Sustained casing pressure refers to the pressure phenomenon that persists in the casing annulus after the wellhead is relieved through the blowout valve. According to the statistics of the U.S. Department of Mines, sustained casing pressure mainly exists in the jacket annulus, which is mainly caused by casing leakage, packer failure, and poor cementing quality. Once sustained casing pressure is created, it is usually necessary to replace the oil casing to eliminate that part of the annular pressure;

(2) Operator-Imposed Pressure

The annulus pressure rise caused by operation and construction mainly comes from special production processes, such as gas lifting, heavy oil thermal recovery, and annulus pressure monitoring and pressure test measures. Therefore, the management of this part of annulus pressure needs to select appropriate process technology and manage it in accordance with relevant company standards;

(3) Thermally Induced Pressure

Due to the relatively high temperature of the oil and gas extracted from the formation, the high-temperature laminar flow will continue to lose heat to the low-temperature medium around the wellbore during the upward migration along the wellbore, which causes the thermal expansion of the closed fluid medium in the casing and forms the trap pressure. This part of pressure is common in the development of deep-water, high-temperature oil and gas fields. Because pressure relief cannot be achieved through the wellhead, casing collapse accidents caused by annular thermal effect are prone to occur in the initial stage of production, threatening wellbore safety and integrity. In this paper, the annular pressure caused by the thermal effect is studied and analyzed.

2.2. Formation Mechanism of Annular Pressure

For the study of the formation mechanism of annular pressure buildup caused by the thermal effect, it can first be reduced to a simple physical model; in a closed structural space filled with fluid, how much pressure will be generated in the container after the annulus fluid is heated? As can be seen from Figure 1, due to the limitation of the rigid structure volume, the annulus fluid cannot expand freely after heating, so a certain force will be generated on the inner wall of the container, which the annulus pressure buildup required to be solved. According to Newton's third law, the inner wall of the container will also exert the same amount of reaction force on the annular fluid; that is, when the volume of the annular fluid expands, it is also affected by the volume compression effect, and the two are coupled. Therefore, on the premise that the integrity of the container structure is not damaged, the annular pressure rise problem in the confined space is mainly related to the thermal expansion performance of the fluid itself and the change of the annulus volume.

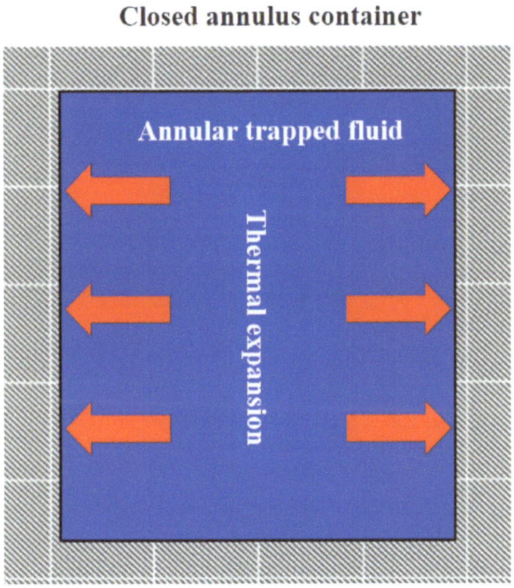

Figure 1. Schematic of single sealed annular pressure buildup.

3. Wellbore Annulus Volume Model

As shown in Figure 2, the cementing operation is carried out after the casing is run in the wellbore, but the cement does not return to the wellhead. In the process of displacement change, all parts of the medium at the first and second cementation planes are always tightly cemented. According to the cement return height outside the casing, the casing can be divided into a free section and a sealed section. The free section of the casing is subjected to internal and external liquid pressure. The sealing section of the casing is subjected to internal and external hydraulic pressure as well as cement ring pressure. The change in wellbore annulus volume takes into account the deformation of the free section of the casing and the deformation of the sealing section [29–32]. After the cementing operation, due to the cementing effect of cement slurry, the casing, cement ring, and formation in the sealing section become a tightly connected elastic body, which forms the outer surface of the wellbore formation system and plays the role of sealing formation fluid and preventing gas channeling. Under the coupling of temperature and pressure, the medium in each part of the wellbore formation system will change. The elastomer composed of casing, cement ring, and formation in the sealing section will produce elastic deformation under the effect of thermal stress and inner annular pressure and expand outward in the radial direction of the wellbore.

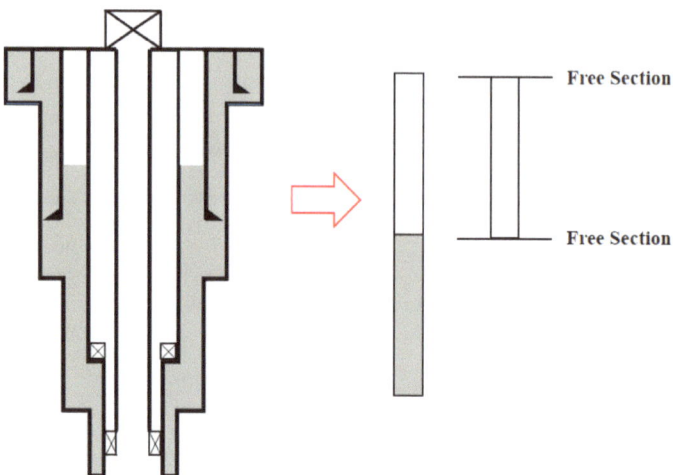

Figure 2. Schematic diagram of wellbore structure.

3.1. Deformation of Casing-Free Section

The deformation of the casing-free section consists of two parts: one is caused by thermal expansion, and the other is caused by stress. Assuming the total radial strain of the free section of casing is ε_r, then the strain due to the stress part is $\varepsilon_r - \alpha T$. Therefore, the following occurs:

$$\begin{cases} \varepsilon_r - \alpha T = \frac{1}{E}[\sigma_r - \nu(\sigma_\theta + \sigma_z)] \\ \varepsilon_\theta - \alpha T = \frac{1}{E}[\sigma_\theta - \nu(\sigma_z + \sigma_r)] \\ \varepsilon_z - \alpha T = \frac{1}{E}[\sigma_z - \nu(\sigma_r + \sigma_\theta)] \end{cases} \quad (1)$$

under planar conditions, $\varepsilon_z = 0$;

Where σ_r is radial stress, MPa; σ_θ is circumferential stress, MPa; σ_z is axial stress, MPa; ε_r is the radial strain, m; ε_θ is circumferential stress, m; ε_z is the axial stress, m; E is the elastic modulus of elastomer, MPa; ν is Poisson's ratio of an elastic body, dimensionless; α is the expansion coefficient of the elastomer, $°C^{-1}$; T is the temperature change, $°C$.

Due to the difference between the subsea wellhead and the land wellhead, the deformation ability of the deepwater casing in the longitudinal direction is limited after cementing. According to the theory of thermoelasticity, the wellbore formation system can be regarded as the two-dimensional plane strain problem of a multi-layer composite thick-walled barrel for displacement and stress analysis and solution. The geometric equation of the plane strain problem in polar coordinates is as follows:

$$\begin{cases} \varepsilon_r = \frac{\partial v}{\partial r} \\ \varepsilon_\theta = \frac{v}{r} + \frac{1}{r}\frac{\partial v}{\partial \theta} \\ \varepsilon_z = \frac{1}{r}\frac{\partial v}{\partial \theta} + \frac{\partial v}{\partial r} - \frac{v}{r} \end{cases} \quad (2)$$

The wellbore formation system belongs to the axisymmetric plane strain problem, which only considers the deformation along the radial direction, so the geometric equation is simplified as follows:

$$\begin{cases} \varepsilon_r = \dfrac{d\partial_v}{d_r} \\ \varepsilon_\theta = \dfrac{v}{r} \\ \varepsilon_z = 0 \end{cases} \tag{3}$$

In polar coordinates, the physical equation of plane strain considering temperature change is as follows:

$$\begin{cases} \varepsilon_r = \dfrac{1+v}{E}[(1-v)\sigma_r - v\sigma_\theta] + (1+v)\alpha T \\ \varepsilon_\theta = \dfrac{1+v}{E}[(1-v)\sigma_{r\theta} - v\sigma_r] + (1+v)\alpha T \end{cases} \tag{4}$$

Therefore, the state equation of stress and strain derived from the formula is as follows:

$$\begin{cases} \sigma_r = \dfrac{E}{(1+v)(1-2v)}[(1-v)\varepsilon_r + v\varepsilon_\theta - (1+v)\alpha T] \\ \sigma_\theta = \dfrac{E}{(1+v)(1-2v)}[(1-v)\varepsilon_r + v\varepsilon_r - (1+v)\alpha T] \end{cases} \tag{5}$$

Because the wellbore formation system belongs to axisymmetric plane strain problem, the influence of physical force can be ignored. At the same time, the tangential stress is zero, and the magnitude of the circumferential stress is independent of the rotation Angle, so the formula can be simplified as follows:

The plane stress satisfies the equilibrium equation:

$$\frac{d\sigma_r}{dr} + \frac{\sigma_r - \sigma_\theta}{r} = 0 \tag{6}$$

The boundary conditions are as follows:

$$\begin{cases} \sigma_r(a) = -q_1 \\ \sigma_r(b) = q_2 \end{cases} \tag{7}$$

Substituting the constitutive equation expressed by stress into the equilibrium Equation (6):

$$\frac{d^2u}{dr^2} + \frac{1}{r}\frac{du}{dr} - \frac{u}{r^2} = \frac{1+v}{1-v}\alpha\frac{dT}{dr} \tag{8}$$

According to the boundary Equation (7), the radial displacement formula of the casing-free section can be obtained:

$$u(r) = u_q(r) + u_T(r) \tag{9}$$

In the formula, $u_q(r)$ is the change in displacement due to stress, m and $u_T(r)$ is the change in displacement of the elastomer due to temperature, m.

Substituting $r = a$ and $r = b$ into Equation (9), respectively, the displacements of the inner and outer boundaries can be obtained:

$$\begin{cases} u_1 u(a) = c_{11}q_1 + c_{12}q_2 + u_T(a) \\ u(b) = c_{21}q_1 + c_{22}q_2 + u_T(b) \end{cases} \tag{10}$$

The expression of each coefficient is as follows:

$$\begin{cases} c_{11} = \frac{b^2+(1-2v)a^2}{b^2-a^2}\frac{(1+v)a}{E} \\ c_{12} = \frac{2(1-v)b^2}{b^2-a^2}\frac{(1+v)a}{E} \\ c_{21} = \frac{2(1-v)ab}{b^2-a^2}\frac{(1+v)a}{E} \\ c_{22} = \frac{(1-2v)b^2+a^2}{b^2-a^2}\frac{b}{a}\frac{(1+v)a}{E} \end{cases} \quad (11)$$

3.2. Deformation of Casing Sealing Section

The deformation of the casing sealing section must consider the deformation of the external cement sheath and the formation under the influence of thermal expansion and stress. As shown in Figure 3, assuming that the cement sheath and the formation are regarded as a stressed whole, the cement sheath formation can be divided into a hot area affected by temperature and a cold area not affected by temperature according to the temperature influence range. The inner radius of the casing is r_0, with an outer radius of r_1. The outer radius of the cement ring is r_2. The area where the formation is affected by the temperature is the hot area with a radius of b, and the region where the temperature cannot be swept is the cold area with a radius of c.

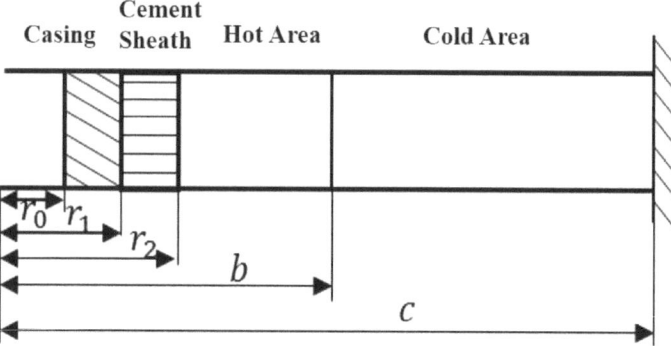

Figure 3. Schematic diagram of casing-cement sheath formation system.

For casing deformation, Equation (10) can also be used for its calculation. The pressure in the casing is q_0, the pressure at the cementation surface of the casing and cement sheath is q_1, the pressure at the cementation surface of the cement sheath and the formation hot zone is q_2, and the pressure at the interface of the hot and cold areas is q_3.

When $r = r_0$, the $q = q_0$, $r = r_1$, $q = -q_1$, the radial deformation of the casing is as follows:

$$\begin{cases} u_{1i} = c^1_{1i}q_0 - c^2_{1i}q_1 + u_{T1i} \\ u_{1o} = c^1_{1o}q_0 - c^2_{1o}q_1 + u_{T1o} \end{cases} \quad (12)$$

When $r = r_1$, the $q = q_1$, $r = r_2$, $q = -q_2$, the radial deformation of the cement sheath is as follows:

$$\begin{cases} u_{2i} = c^1_{2i}q_1 - c^2_{2i}q_2 + u_{T2i} \\ u_{2o} = c^1_{2o}q_1 - c^2_{2o}q_2 + u_{T2o} \end{cases} \quad (13)$$

When $r = r_2$, the $q = q_2$, $r = b$, $q = -q_3$, the radial deformation of the formation hot area is as follows:

$$\begin{cases} u_{3i} = c_{3i}^1 q_2 - c_{3i}^2 q_3 + u_{T3i} \\ u_{3o} = c_{3o}^1 q_2 - c_{3o}^2 q_3 + u_{T3o} \end{cases} \quad (14)$$

When $r = b$, the $\sigma_1 = q_3$, considering the assumption that the formation is infinite, the radial deformation of the formation's cold area is as follows:

$$u_{4i} = \frac{q_3}{k_b} \quad (15)$$

where

$$k_b = \frac{(1-2v)\left(\frac{c^2}{b^2}\right)+1}{(1-2v)\left[\left(\frac{c^2}{b^2}\right)-1\right]} \frac{E}{(1+v)b} \quad (16)$$

According to the principle of coordinated deformation, when $r = a$ or $r = b$, the deformation at the junction of the hot and cold zone meets Equations (14) and (15) at the same time, and the deformation of the casing sealing section can be obtained by combining Equation (12) with Equation (16).

4. Multi-String Annular Pressure Buildup Model

The annulus pressure buildup is the increase in fluid expansion pressure in the sealed annulus between each layer of casing due to temperature rise. The typical calculation model is as follows:

$$\Delta p = \left(\frac{\partial p}{\partial T}\right)_{m,V} \cdot \Delta T + \left(\frac{\partial p}{\partial V_{ann}}\right)_{m,T} \cdot \Delta V_{ann} + \left(\frac{\partial p}{\partial m}\right)_{V,T} \cdot \Delta m \quad (17)$$

The first term on the right side of the equation represents the pressure due to the thermal expansion effect, and the second term represents the pressure due to the annular deformation caused by the thermal effect or load of the casing. The third term is the pressure due to annular fluid leakage or reservoir fluid overflow into the annulus.

In the typical three-annulus well structure, the tubing and production casing form the A annulus, the production casing and intermediate casing form the B annulus, and the intermediate casing and surface casing form the C annulus. When the production parameters of oil and gas wells and the annular drilling fluid system are determined, according to the calculation formula of annular pressure buildup, the main factor affecting annular pressure buildup is the change of annular volume. The annular pressure buildup is affected by the pressure of the adjacent annulus; therefore, the casing annulus pressure buildup values are solved as a system.

Taking the A annulus as an example, the changes of annulus volume mainly include the following: ① radial deformation of tubing; ② the radial deformation of the casing-free section (related to the B annulus pressure); ③ the radial deformation of the casing sealing section.

The volume change caused by the radial deformation of the tubing is as follows:

$$\Delta V_1 = \int_0^{L_t} \pi \left((r_{To} + u_c(r_{To}))^2 - r_{To}^2 \right) dx \quad (18)$$

The volume change caused by the radial deformation of the casing-free section (related to the B annulus pressure) is as follows:

$$\Delta V_2 = \int_0^{L_{ps}} \pi \left((r_{p1i} + u_s(r_{p1i}))^2 - r_{p1i}^2 \right) dx \quad (19)$$

The volume change caused by the radial deformation of the casing sealing section is as follows:

$$\Delta V_3 = \int_0^{L_{pc}} \pi\left(\left(r_{p1i} + u_c(r_{p1i})\right)^2 - r_{p1i}^2\right) dx \tag{20}$$

In the process of calculating the change of the A annulus volume, the influence of the B annular pressure on the A annulus volume needs to be considered.

The deformation of annulus B and C can be obtained using the same method. The change of the B annulus volume is affected by the pressure of annulus A and C, and the change of the C annulus volume is affected by the B annulus pressure.

5. Solution

A, B, and C annular pressure buildup is affected by each other and needs to be solved iteratively by using multiple strings. The three annular pressure buildup calculation formulas are transformed to form a set of ternary quadratic equations:

$$\begin{cases} f_1 = f(P_1^2, P_2^2) \\ f_2 = f(P_1^2, P_2^2, P_3^2) \\ f_3 = f(P_2^2, P_3^2) \end{cases} \tag{21}$$

where P_1 is the A annular pressure, MPa, P_2 is the B annular pressure, MPa, and P_3 is the C annular pressure, MPa.

The ternary quadratic equations are solved by Newton's iterative method. The solution of $f(P_n)$ is obtained by establishing the equation $f(P_n)$, taking the derivative $f'(P_n)$ of the equation, and then, according to Newton's iteration principle.

$$P_{n+1} = P_n - \frac{f(P_n)}{f'(P_n)} \tag{22}$$

$$Df = \begin{bmatrix} \frac{df_1}{dP_1} & \frac{df_1}{dP_2} & \frac{df_1}{dP_3} \\ \frac{df_2}{dP_1} & \frac{df_2}{dP_2} & \frac{df_2}{dP_3} \\ \frac{df_3}{dP_1} & \frac{df_3}{dP_2} & \frac{df_3}{dP_3} \end{bmatrix} \tag{23}$$

Therefore, the calculation steps of annular pressure buildup of multi-string are as follows:

(1) Based on the wellbore structure and the production parameters, the temperature profile of the well is solved, and then the temperature change of the fluid in the sealed section, the free section, and the annulus of each layer of the case is obtained;
(2) Obtain ΔV of the annular volume change according to the wellbore structure;
(3) Establish a set of equations for calculating the annular pressure buildup;
(4) Setting an initial pressure matrix;
(5) Calculate the derivative matrix DF of the pressure equation;
(6) Obtain the annular pressure buildup matrix according to Newton's iteration principle;
(7) Comparing the pressure obtained in step (6) with the initial value and repeating steps (4) to (6) if the error is not within the range;
(8) Obtain the annular pressure buildup value of each annular space.

6. Model Validation

To verify the model established in the paper, an example well was selected. The wellbore structure of the example well is shown in Table 1 and Figure 4 The high-precision temperature and pressure sensors were placed on the outer wall of the 9 5/8″ casing at a depth of 190 m. The casing was properly cemented, and the cement was returned to 2438 m. This well was further drilled to a depth of 3627 m, and a 7″ tailpipe was installed.

Table 1. Well structure.

Name	OD (in)	Line Weight (lbm/ft)	Depth (ft)	TOC (ft)
Conductor	30	200	377	
Surface casing	20	133	3025	374
Intermediate casing	13 3/8	72	8258	3525
Producer casing	9 5/8	53.5	10,456	8000
Liner	7	38	11,900	
Tubing	3 1/2	9.2	11,000	

Figure 4. Wellbore structure diagram.

6.1. Well Structure

After the test was completed, the well was abandoned, and the 9 5/8″ casing with sensors was retrieved. The three testing processes clearly demonstrated the changes in temperature and pressure data during well flushing, flow testing, and pressure rise.

As the temperature rose from 17 °C to 65 °C, the pressure increased from 1.2 MPa to 25 MPa. The internal pressure resistance strength of the 13 3/8″ P110 72 lbs/ft casing was 51 MPa, and the external compression resistance strength of the 9 5/8″ P110 72 lbs/ft casing was 54.7 MPa. The density of the drilling fluid was 1.3 g/cm^3. Due to the pressure of the liquid column, the pressure increased by 26 MPa from the sensor position to the mud return position at 2328 m.

6.2. Annular Pressure Buildup Calculation

According to casing and annulus fluid temperature changes, annular pressure buildup was iteratively solved for the well, and annular pressure buildup simulation analysis was carried out in three cases:

(1) Only the expansion and compression effect of the liquid is considered;

(2) Considering the expansion and compression effect of liquid and the deformation of the casing-free section;

(3) Consider the expansion and compression effect of the liquid, the deformation of the casing-free section, and the sealing section.

The annular pressure buildup values are shown in Figure 5.

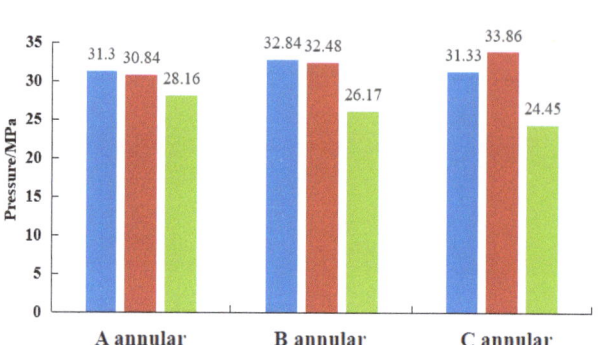

C1: Only the expansion and compression effect of the liquid is considered
C2: Considering the expansion and compression effect of liquid and the deformation of the casing free section
C3: Consider the expansion and compression effect of the liquid, the deformation of the casing free section and sealing section

Figure 5. Annular pressure buildup comparison diagram.

As can be seen from Figure 5, for annulus A and B, C1 and C2 were basically equal because there was little difference in deformation of the free section of the casing in the inner and outer layers and little change in annulus volume.

For the C annulus, only the inner casing had a free section of pressure, and the annulus volume was reduced; therefore, C2 was greater than C1.

If the deformation of the sealed section is considered, it can be seen from the figure that the C3 values were all lower than the C1 and C2 values, and the predicted value of the trap pressure was smaller than that without considering the deformation of the sealing section.

Due to the different annular cement return heights, the sealing section had different influences on the trap pressure of A, B, and C and had the greatest influence on the C annular pressure buildup and the least influence on the A annular pressure buildup.

Specifically, the annular trap pressure predicted by the case of B annulus C3 was better than that predicted by the case of C1 and C2, which decreased by 14.9% and 13.6%, respectively. The predicted annular trap pressure of the C annulus C3 was 13.2% and 14.8%, which was lower than that of C1 and C2, respectively.

6.3. Wellbore Temperature Profile

The average temperature and average temperature changes in the annulus when the temperature sensor reached 30 °C are shown in Figures 6 and 7.

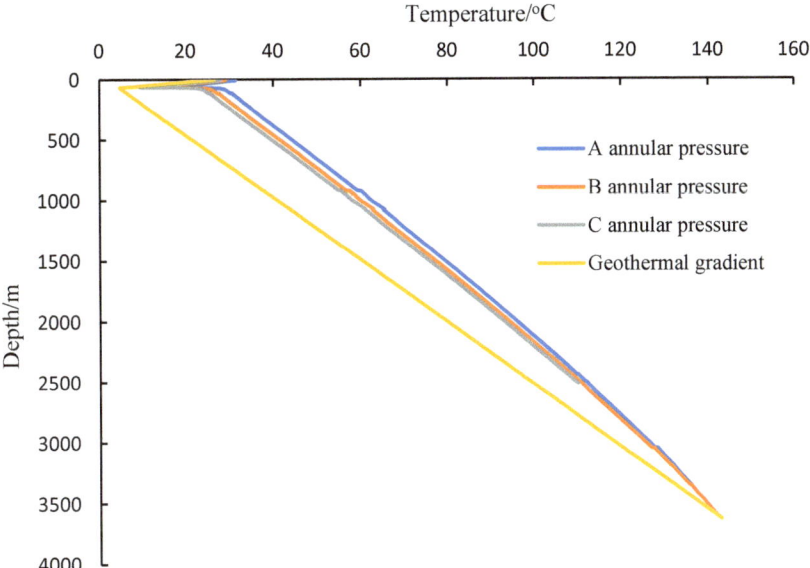

Figure 6. Temperature profile.

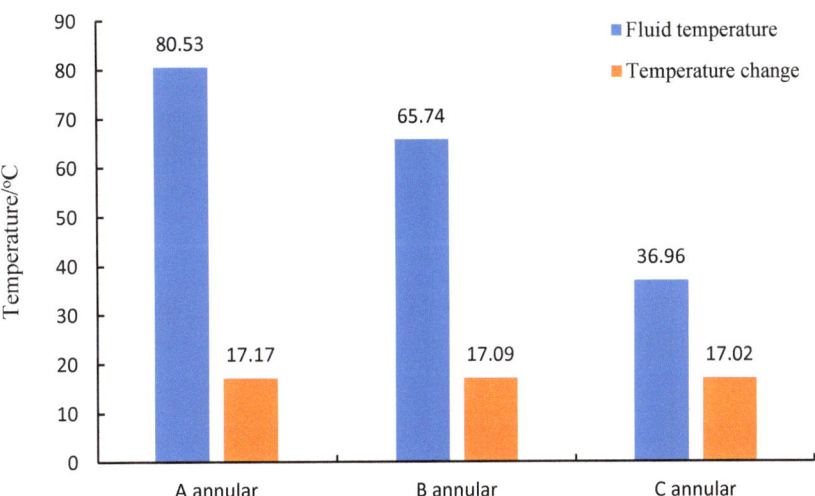

Figure 7. Average temperature and average temperature variation in the annulus.

The wellbore temperature profile and the average temperature and average temperature changes in the annulus when the temperature sensor reached 40 °C are shown in Figures 8 and 9.

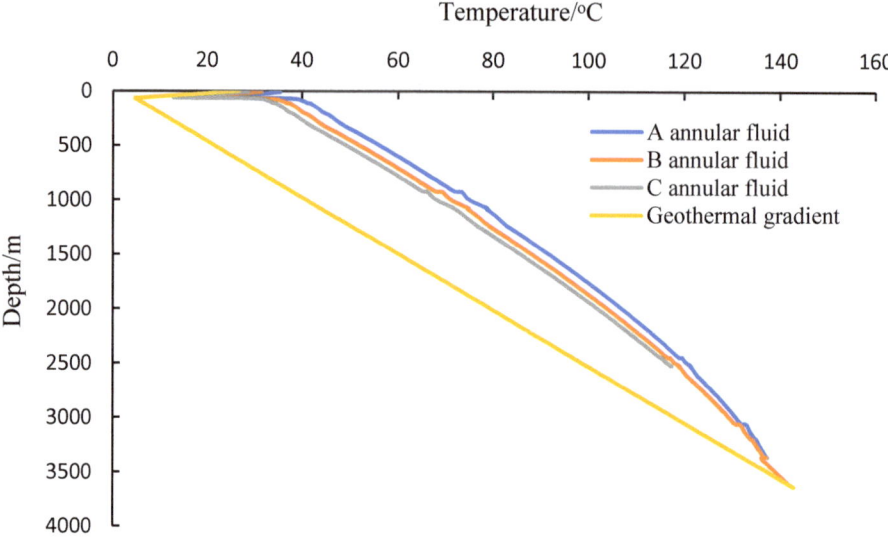

Figure 8. Temperature profile.

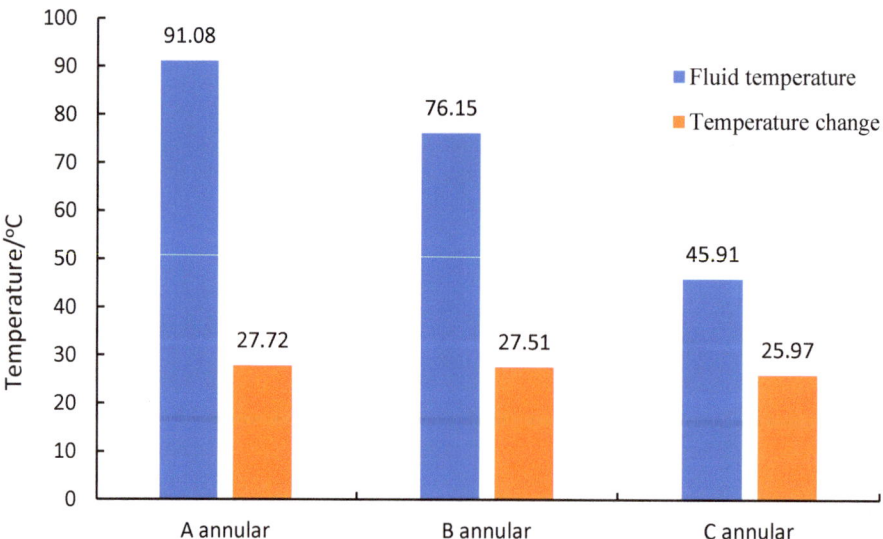

Figure 9. Average temperature and average temperature variation in the annulus.

The wellbore temperature profile and the average temperature and average temperature changes in the annulus when the temperature sensor reached 50 °C are shown in Figures 10 and 11.

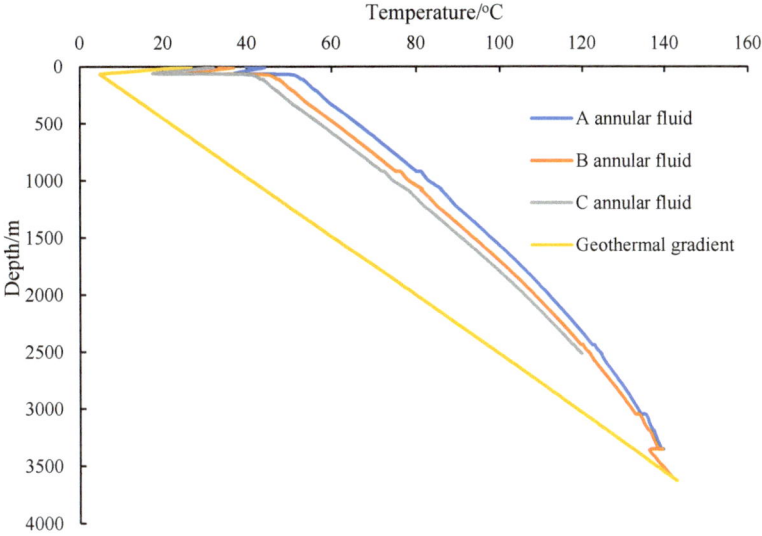

Figure 10. Temperature profile.

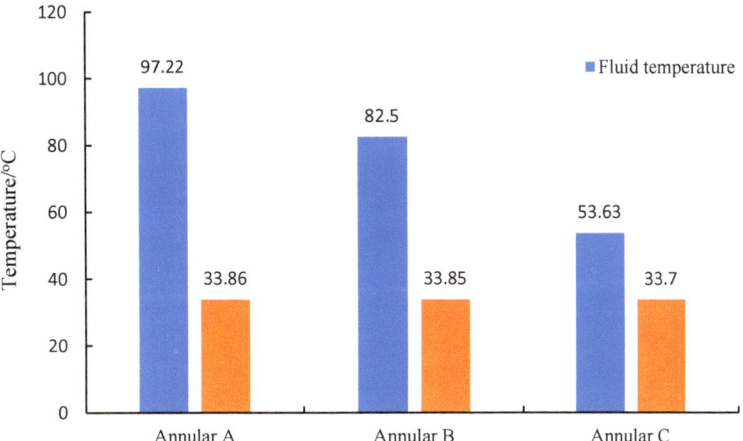

Figure 11. Average temperature and average temperature variation in the annulus.

6.4. Comparison Between Calculated and Measured Values

Based on the average temperature and temperature changes of the annulus at temperature points of 30 °C, 40 °C, and 50 °C, combined with the experimentally measured thermal and physical parameters of the annulus fluid, the annulus pressure buildup at each temperature point can be determined. As shown in Figure 12, Wellcat software (The version is EDM5000.17.2 belonged to Schlumberger, USA) was also used to compare it with the measured data and data calculated by the model in this paper.

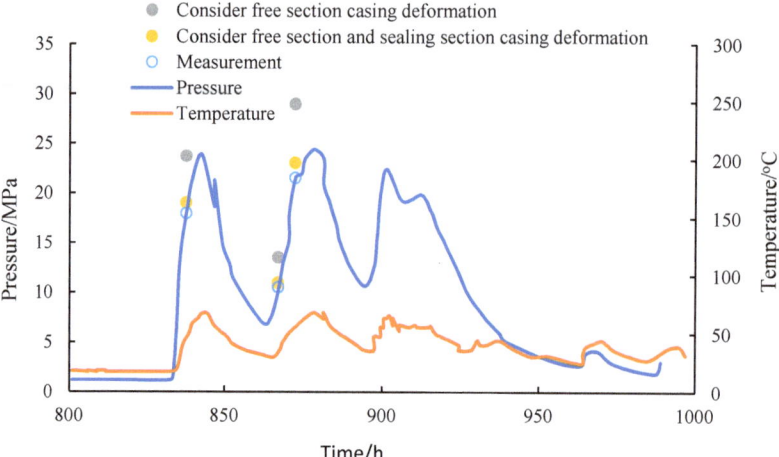

Figure 12. Comparison of measured pressure and temperature with those calculated by model in this paper.

According to Table 2, the calculated values of Wellcat were larger than the measured values, with an error of 10–30%. By comparison, the error calculated using the model established in this article was within 10%, which was smaller than that calculated by Wellcat, thus verifying the accuracy of the model.

Table 2. Comparison between calculated and measured values.

Temperature (°C)	Measured Pressure (MPa)	Wellcat Calculated Value (MPa)	Error (%)	The Calculated Values of This Model (MPa)	Error (%)
30	10.52	13.51	28.42	11.00	4.56
40	17.98	23.7	31.81	19.07	6.06
50	21.54	28.94	34.35	23.06	7.05

7. Results and Discussion

7.1. Wellbore Temperature

In this section, the example well in Section 6.1 was used for analysis. The well parameters are described in Section 6.1. The average temperature of the annulus at different production times was calculated and is shown in Figure 13. The increase in temperature is the main factor causing annulus pressure buildup, and the annulus pressure buildup increases as the temperature rises. The temperature profile of the annular fluid is related to the production rate and production time of oil and gas wells. As shown in Figure 11, the temperature of the annular fluid increased rapidly in the early stage of production, then slowly increased and finally stabilized. The temperature of the annular fluid changed significantly in the early stage of production. The temperature profile of the annular fluid slowly increased with the production of oil and gas wells, as shown in Figure 14.

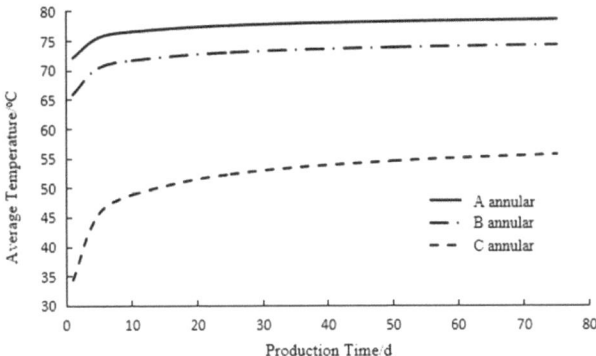

Figure 13. Average temperature of the annulus at different production times.

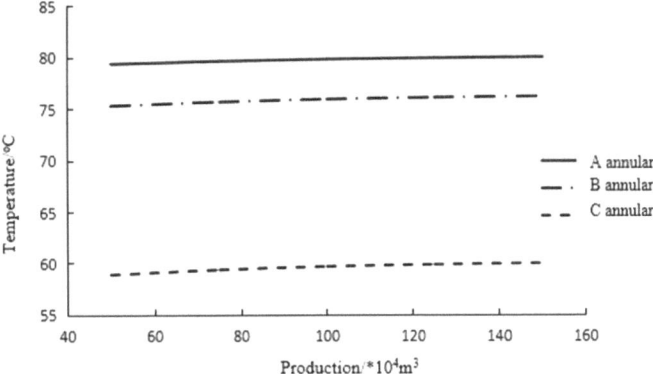

Figure 14. Temperature profile of annulus at different production rates.

In the early stage of production, due to the unstable temperature distribution in the wellbore, there were significant changes in annulus pressure buildup, as shown in Figure 15. Within the first 20 days of production, the pressure in the B annulus was the highest. As the temperature gradually stabilized due to the location of the C annulus fluid at the wellhead, the average temperature of the annulus fluid changed greatly, and the C annulus pressure was the highest. Although the fluid temperature in the A annulus was the highest, the temperature change in the A annulus was minimal, resulting in the lowest pressure formed in the A annulus.

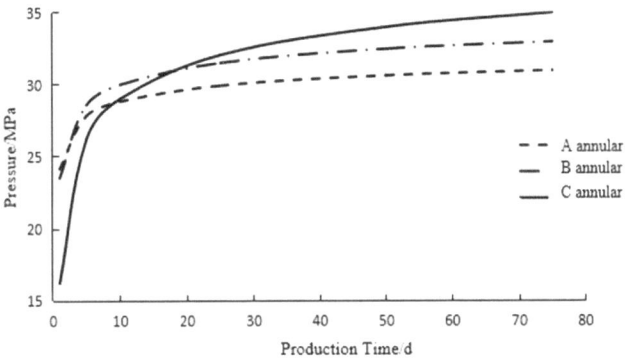

Figure 15. Annular pressure at different production times.

With the increase in wellbore production, the annular pressure gradually increased. Due to the maximum temperature change in the C annulus, the annular pressure was the highest, while the temperature change in the A annulus was the smallest, and the annular pressure was the smallest, as shown in Figure 16.

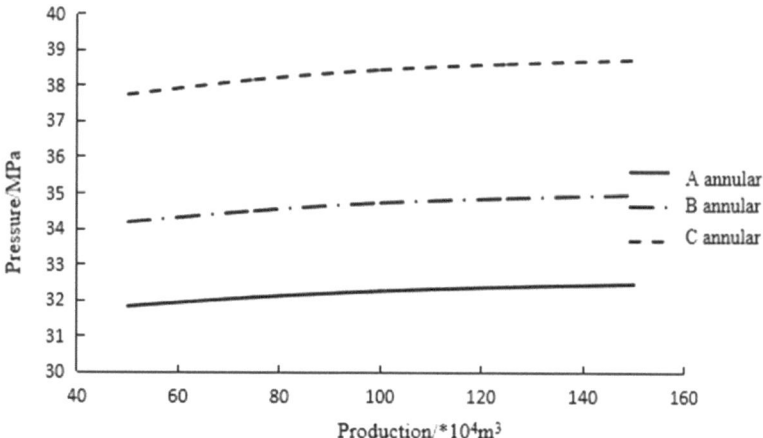

Figure 16. Annular pressure at different production rates.

7.2. Thermal Properties Parameters of Drilling Fluid

The expansion coefficient and compression coefficient of the annular fluid vary with different temperatures. Table 3 shows the expansion coefficient and compression coefficient of the annular fluid at different temperatures. As the temperature of the annular fluid increased from 34.42 °C to 55.71 °C, both the expansion coefficient and compression coefficient of the fluid increased with temperature, and the increase in the expansion coefficient was greater than that of the compression coefficient. In the process of calculating the pressure, the expansion coefficient and compression coefficient should be selected according to different temperatures. Using a constant coefficient may lead to significant errors. Figure 17 shows the annular pressure calculated using the expansion coefficient and compression coefficient, which vary with temperature, as well as a constant coefficient. The constant coefficient was used for the expansion coefficient and the compression coefficient at 55.71 °C, and the maximum error of the annular pressure calculated using the constant coefficient was 11.58%.

Table 3. Fluid expansion coefficient and compression coefficient at different temperatures.

Average Temperature of the Annulus (°C)	Annular Temperature Difference (°C)	Expansion Coefficient ($\times 10^{-6}$ °C^{-1})	Compression Coefficient ($\times 10^{-6}$ MPa^{-1})
34.42	16.08	355.15	451.03
45.72	27.38	397.12	460.75
48.94	30.6	409.09	463.52
52.39	34.05	421.90	466.49
54.57	36.23	430.00	468.36
55.71	37.37	434.24	469.34

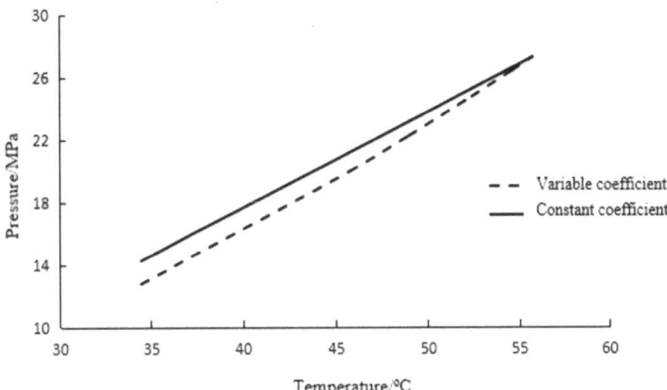

Figure 17. The influence of fluid expansion coefficient and compression coefficient on annular pressure.

7.3. Casing Parameters

The elastic modulus of the casing varies with temperature, and different types of pipes experience varying changes in elastic modulus due to temperature fluctuations. These changes have an impact on the radial deformation of the casing and the volume change of the annulus, thereby affecting the annulus pressure. Table 4 shows the elastic modulus of three common types of pipes at different temperatures. Similarly, Poisson's ratio and elastic modulus can also affect casing deformation and annulus pressure.

Table 4. The elastic modulus of the casing varies with temperature.

Type of Casing	20 °C	200 °C	250 °C	300 °C	350 °C
N80 (MPa)	205.94	165.73	149.06	136.31	117.68
P110 (MPa)	205.94	184.37	177.50	172.60	166.71
TP90H (MPa)	—	170.75	163.25	155.75	148.25

As shown in Figure 18, annulus pressure increased linearly and rapidly with the increase in casing elastic modulus. On the contrary, the higher the Poisson's ratio, the lower the annular pressure, but the smaller the variation of annular pressure.

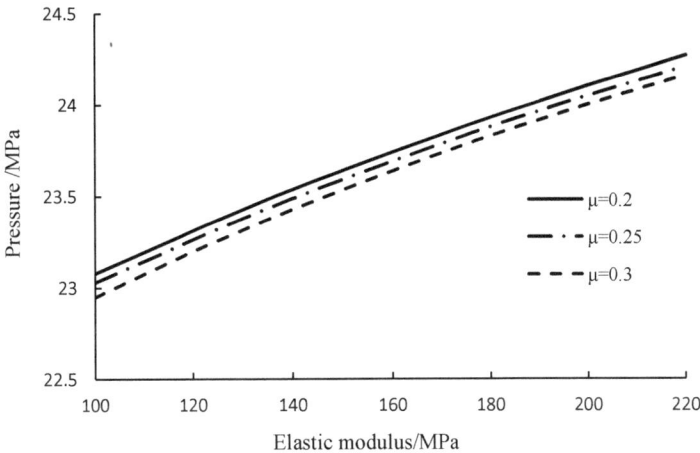

Figure 18. Relationship between elastic modulus of casing and pressure.

7.4. Cement Sheath

As shown in Figure 19, the larger the elastic modulus of the cement sheath, the smaller the radial displacement of the sealing section casing under compressive and thermal loads. This results in a smaller change in annular volume, leading to a higher pressure. The Poisson's ratio of the cement sheath had a smaller impact on the annular pressure.

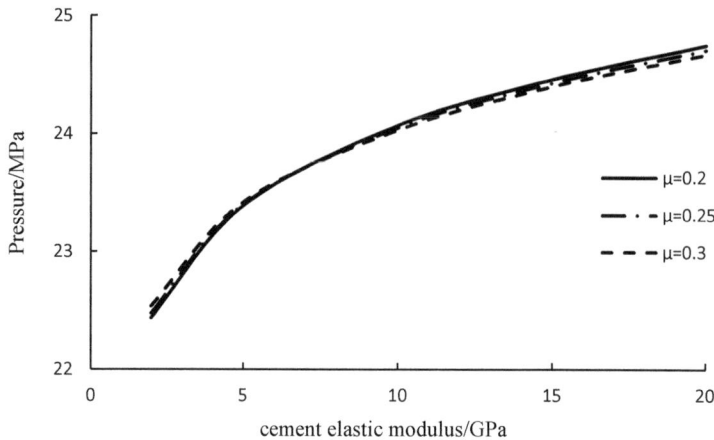

Figure 19. Relationship between elastic modulus of cement sheath and pressure.

7.5. Release of Annular Pressure

In deepwater oil and gas wells, the special structure of the wellhead device only allows the A annulus pressure to be released through the wellhead pressure relief valve, while the B and C annulus pressure buildup cannot be released through the wellhead. Due to the mutual influence of A, B, and C annulus pressure buildup, when A annulus is depressurized, the B and C annulus pressure buildup would be affected correspondingly. Figure 20 shows the effect of the A annulus on the annulus pressure buildup of the B and C annulus during pressure release. When the A annulus was depressurized, the annulus pressure buildup in the B and C annulus decreased, and the annular pressure buildup in the B annulus decreased faster.

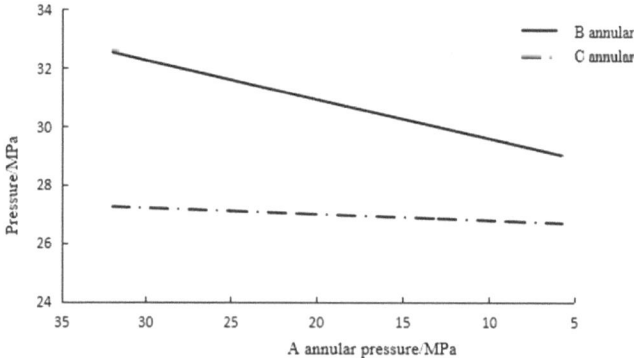

Figure 20. The impact of A annulus pressure release on B and C annulus pressures.

8. Conclusions

Based on elastic deformation theory, a new deformation model of the casing sealing section is established, and a new prediction method for multi-string annular pressure buildup is developed. The annular pressure buildup predicted by considering the deformation of the casing sealing section is about 13% lower than that predicted by not considering this factor, which is closer to the measured value. Based on the obtained results, the following conclusions can be drawn.

(1) The annular pressure increases with the increase in the elastic modulus of the casing while decreasing with the increase in Poisson's ratio.

(2) Based on elastic deformation theory, the deformation model of the casing sealing section is established, and the prediction method of multi-string annular pressure buildup is improved. The predicted annular pressure buildup value considering casing sealing section deformation is about 13% lower than that without considering this factor. This also shows that the predicted annular pressure buildup value considering casing sealing section deformation is lower than that without considering casing sealing section deformation and is more consistent with reality.

(3) Temperature rise is the most important factor causing the annular pressure buildup, and the annulus pressure buildup increases with the increase in temperature. The annular fluid temperature profile is related to oil and gas well production and production time. The annulus fluid temperature increases rapidly at the initial stage of production, then slowly increases and finally becomes stable. The annular fluid temperature profile increases slowly with the increase in oil and gas production.

(4) Reducing the pressure in the A annulus also releases the annulus pressure buildup in the B and C annulus, while the annulus pressure buildup in the B annulus decreases faster.

Author Contributions: Conceptualization, R.X. and L.Z.; methodology, R.X.; software, R.X.; validation, R.X. and L.Z.; formal analysis, R.X.; investigation, R.X.; resources, L.Z.; data curation, R.X.; writing—original draft preparation, R.X.; writing—review and editing, L.Z.; visualization, R.X.; supervision, L.Z.; project administration, R.X.; funding acquisition, R.X. All authors have read and agreed to the published version of the manuscript.

Funding: This work is supported by the Major technology projects of China National Offshore Oil Corporation Limited (KJGG-2022-17-05).

Institutional Review Board Statement: Not applicable.

Informed Consent Statement: Not applicable.

Data Availability Statement: The original contributions presented in the study are included in the article, further inquiries can be directed to the corresponding author.

Conflicts of Interest: Author Renjun Xie was employed by the company CNOOC Research Institute Ltd. The remaining authors declare that the research was conducted in the absence of any commercial or financial relationships that could be construed as a potential conflict of interest. The authors declare that this study received funding from China National Off-shore Oil Corporation Limited. The funder was not involved in the study design, collection, analysis, interpretation of data, the writing of this article or the decision to submit it for publication.

References

1. Mainguy, M.; Innes, R. Explaining sustained "A"-annulus pressure in major north sea high-pressure/high-temperature fields. *SPE Drill. Complet.* **2019**, *34*, 71–80. [CrossRef]
2. Sathuvalli, U.B.; Pilko, R.M.; Gonzalez, R.A.; Pai, R.M.; Sachdeva, P.; Suryanarayana, P.V. Design and performance of annular-pressure-buildup mitigation techniques. *SPE Drill. Complet.* **2017**, *32*, 168–183. [CrossRef]
3. Ma, T.; Tang, Y.; Chen, P.; He, Y. Mitigation of annular pressure buildup for deepwater wells using a recovery relief method. *Energy Sci. Eng.* **2019**, *7*, 1727–1747. [CrossRef]
4. da Veiga, A.P.; Martins, I.O.; Barcelos, J.G.; Ferreira, M.V.D.; Alves, E.B.; da Silva, A.K.; Hasan, A.R.; Barbosa, J.R., Jr. Predicting thermal expansion pressure buildup in a deepwater oil well with an annulus partially filled with nitrogen. *J. Pet. Sci. Eng.* **2022**, *208*, 109275. [CrossRef]
5. Liu, Z.C.; Samuel, R.; Gonzales, A.; Kang, Y.F. Modeling and simulation of annular pressure buildup APB management using syntactic foam in HP/HT deepwater wells. In Proceedings of the SPE Deepwater Drilling and Completions Conference, Galveston, TX, USA, 14–15 September 2016; SPE: Richardson, TX, USA, 2016.

6. Azzola, J.H.; Tselepidakis, D.; Pattillo, P.D.; Richey, J.F.; Tinker, S.J.; Miller, R.; Segreto, S.J. Application of Vacuum–Insulated Tubing to Mitigate Annular Pressure Buildup. *SPE Drill. Complet.* **2007**, *22*, 46–51. [CrossRef]
7. Mwang'Ande, A.W.; Liao, H.; Zeng, L. Mitigation of annulus pressure buildup in offshore gas wells by determination of top of cement. *J. Energy Resour. Technol.* **2019**, *141*, 102901. [CrossRef]
8. Guan, Z.-C.; Zhang, B.; Wang, Q.; Liu, Y.-W.; Xu, Y.-Q.; Zhang, Q. Design of thermal-insulated pipes applied in deepwater well to mitigate annular pressure build-up. *Appl. Therm. Eng.* **2016**, *98*, 129–136. [CrossRef]
9. Bradford, D.W.; Fritchie, D.G.; Gibson, D.H.; Gosch, S.W.; Pattillo, P.D.; Sharp, J.W.; Taylor, C.E. Marlin Failure Analysis and Redesign; Part 1—Description of Failure. In Proceedings of the IADC/SPE Drilling Conference, Dallas, TX, USA, 26–28 February 2002. SPE 74528.
10. Ellis, R.C.; Fritchie, D.G.; Gibson, D.H.; Gosch, S.W.; Pattillo, P.D. Marlin Failure Analysis and Redesign; Part 2—Redesign. In Proceedings of the IADC/SPE Drilling Conference, Dallas, TX, USA, 26–28 February 2002. SPE 74529.
11. Gosch, S.W.; Horne, D.J.; Pattillo, P.D.; Sharp, J.W.; Shah, P.C. Marlin Failure Analysis and Redesign; Part 3—VIT Completion with Real-Time Monitoring. In Proceedings of the IADC/SPE Drilling Conference, Dallas, TX, USA, 26–28 February 2002. SPE 74530.
12. Vargo, R.F., Jr.; Payne, M.; Faul, R.; LeBlanc, J.; Griffith, J.E. Practical and Successful Prevention of Annular Pressure Buildup on the Marlin Project. In Proceedings of the SPE Annual Technical Conference and Exhibition, San Antonio, TX, USA, 29 September–2 October 2002. SPE 77473.
13. Pattillo, P.D.; Cocales, B.W.; Morey, S.C. Analysis of an annular pressure buildup failure during drill ahead. *SPE Drill. Complet.* **2006**, *21*, 242–247. [CrossRef]
14. Ansari, M.T. Evaluation of Annular Pressure Buildup (APB) During WCD Blowout of Deepwater Hp/Ht Wells. Master's Thesis, Texas A&M University, College Station, TX, USA, 2014.
15. Adams, A. How to Design for Annulus Fluid Heat-Up. In Proceedings of the SPE Annual Technical Conference and Exhibition, Dallas, TX, USA, 6–9 October 1991. SPE 22871.
16. Goodman, M.A.; Halal, A.S. Case study: HPHT casing design achieved with multistring analysis. In Proceedings of the SPE Annual Technical Conference and Exhibition, Houston, TX, USA, 3–6 October 1993. SPE 26322.
17. Halal, A.S.; Mitchell, R.F. Casing design for trapped annular pressure buildup. *SPE Drill. Complet.* **1994**, *9*, 107–114. [CrossRef]
18. Lubinski, A.; Althouse, W.S.; Logan, J.L. Helical buckling of tubing sealed in packers. *J. Pet. Technol.* **1962**, *14*, 655–670. [CrossRef]
19. Klementich, E.F.; Jellison, M.J. A service life model for casing strings. *SPE Drill. Eng.* **1986**, *1*, 141–152. [CrossRef]
20. Oudeman, P.; Bacarreza, L.J. Field Trial Results of Annular Pressure Behavior in a High-Pressure/High-Temperature Well. *SPE Drill. Complet.* **1995**, *10*, 1084–1088. [CrossRef]
21. Maiti, S.; Gupta, H.; Vyas, A.; Kulkarni, S.D. Evaluating precision of annular pressure buildup (APB) estimation using machine-learning tools. *SPE Drill. Complet.* **2022**, *37*, 93–103. [CrossRef]
22. Oudeman, P.; Kerem, M. Transient behavior of annular pressure build-up in HP/HT wells. *SPE Drill. Complet.* **2006**, *21*, 234–241. [CrossRef]
23. Hasan, A.R.; Izgec, B.; Kabir, C.S. Sustaining production by managing annular-pressure buildup. *SPE Prod. Oper.* **2010**, *25*, 195–203. [CrossRef]
24. Yin, F.; Gao, D.L. Improved calculation of multiple annuli pressure buildup in subsea HPHT wells. In Proceedings of the IADC/SPE Asia Pacific Drilling Technology Conference, Bangkok, Thailand, 25–27 August 2014; SPE: Richardson, TX, USA, 2014.
25. Kang, Y.; Gonzales, A.; Liu, Z.; Samuel, R. Modeling and Simulation of Annular Pressure Buildup APB in a Deepwater Wellbore with Vacuum-Insulated Tubing. In Proceedings of the SPE/IADC Drilling Conference and Exhibition, The Hague, The Netherlands, 14–16 March 2017; SPE: Richardson, TX, USA, 2017.
26. Moe, B.; Erpelding, P. Annular Pressure Buildup. What Is It and What to do About It. *Deep. Technol.* **2000**, *21*, 2–4.
27. Deng, Y.; Chen, P.; Zhang, H. Iterative method for calculating confined annular pressure in oil and gas wells. *Offshore Oil* **2006**, *26*, 93–96.
28. Huang, X.; Yan, D.; Tian, R. Analysis of deepwater casing annular-void ring closure pressure calculation and control technology. *China Offshore Oil Gas* **2014**, *26*, 65–69.
29. Liu, J.; Fan, H.; Peng, Q.; Deng, S.; Kang, B.; Ren, W. Research on the prediction model of annular pressure buildup in subsea wells. *J. Nat. Gas Sci. Eng.* **2015**, *27*, 1677–1683. [CrossRef]
30. Wang, H.; Zhang, H.; Li, J.; Sun, T. Study on annular pressure buildup phenomenon in subsea wells considering the effect of cement. *Energy Sci. Eng.* **2022**, *10*, 81–95. [CrossRef]
31. Gao, D.; Feng, Q.; Zheng, H. On a method of prediction of the annular pressure buildup in deepwater wells for oil & gas. *Comput. Model. Eng. Sci. (CMES)* **2012**, *89*, 1–15.
32. Dong, G.; Chen, P. A review of the evaluation methods and control technologies for trapped annular pressure in deepwater oil and gas wells. *J. Nat. Gas Sci. Eng.* **2017**, *37*, 85–105. [CrossRef]

Disclaimer/Publisher's Note: The statements, opinions and data contained in all publications are solely those of the individual author(s) and contributor(s) and not of MDPI and/or the editor(s). MDPI and/or the editor(s) disclaim responsibility for any injury to people or property resulting from any ideas, methods, instructions or products referred to in the content.

Article

Numerical Simulation on the Influence of the Distribution Characteristics of Cracks and Solution Cavities on the Wellbore Stability in Carbonate Formation

Jingzhe Zhang [1], Rongrong Zhao [1,*], Hongyi An [1], Wenhao Li [1], Yuxin Geng [2], Xiangyu Fan [3] and Qiangui Zhang [4,*]

1. Exploration Utility Department, CNPC Southwest Oil and Gas Field Company, Chengdu 610095, China; zhangjingzhe@petrochina.com.cn (J.Z.); anhongyi@petrochina.com.cn (H.A.); liwhao@petrochina.com.cn (W.L.)
2. Shanxi CBM Exploration and Development, CNPC Huabei Oil Field Company, Changzhi 046000, China; ggynl1999@163.com
3. School of Geoscience and Technology, Southwest Petroleum University, Chengdu 610500, China; 199931010004@swpu.edu.cn
4. Petroleum Engineering School, Southwest Petroleum University, Chengdu 610500, China
* Correspondence: zhao_rr@petrochina.com.cn (R.Z.); qgzhang@swpu.edu.cn (Q.Z.); Tel.: +86-028-8303-7050 (Q.Z.)

Abstract: The development of cracks and solution cavities in carbonate reservoirs can notably reduce the rock's mechanical properties, leading to a severe wellbore collapse problem during drilling operations. To clarify the influence of the characteristics of cracks and solution cavities on the wellbore stability in the Dengying Formation carbonate reservoir in the Gaoshiti–Moxi area of Sichuan, the mechanical properties of carbonate rock were analyzed. Then, the influences of the attitude and width of cracks, the size and quantity of solution cavities, and their connectivity on wellbore stability were studied using FLAC3D 6.00 numerical simulation software. Our results show the following: (1) The cracks and solution cavities in the Dengying Formation carbonate rock cause significant differences in the rock's mechanical properties. (2) The equivalent drilling fluid density of collapse pressure (ρ_c) considering the effects of cracks and solution cavities is 6.4% higher than without these effects, which is in good accordance with engineering practice. Additionally, cracks play a more significant role than solution cavities in affecting the wellbore stability. (3) When the orientation of a crack is closer to the direction of maximum horizontal stress, and the dip angle and width of the crack increase, the stress and deformation at the intersection of the crack and wellbore gradually increase, and correspondingly, ρ_c also increases. (4) The stress and displacement of various points around the solution cavities gradually increase with the increases in diameter and quantity of solution cavities, and ρ_c also increases. (5) Compared with the situation where cracks and solution cavities are not interconnected, the stress disturbance area around the wellbore is larger, and ρ_c is greater when cracks and solution cavities are interconnected.

Keywords: wellbore stability; dengying formation; carbonate rocks; cracks and solution cavities characteristics; numerical simulation

1. Introduction

Wellbore instability, which is the key to impeding quality and efficient drilling, is the focus of attention in the petroleum engineering field [1–6]. In particular, carbonate rocks are tricky for oil and gas drilling because they are full of unpredictable cracks and solution cavities [7–10]. These features can disrupt the pressure distribution and weaken the rock, leading to instability issues of the wellbore in the drilling process [11,12]. The instability of the wellbore in carbonate formation has perplexed engineers and researchers, even though this problem has been on the research radar for a while now [13–15], especially since there is still a lot we do not know about how these cracks and solution cavities affect the wellbore stability.

It is a common understanding that cracks and solution cavities play a key role in affecting the wellbore stability in carbonate formation. Undoubtedly, the cracks can affect the rock's mechanical properties and fluid flow in the formation, resulting in instability of the wellbore. Yan et al. [16], Zhang et al. [17], Liu et al. [18], and Karatela et al. [19] pointed out that the crack's properties, such as strength and friction, are crucial for the wellbore stability. Cui et al. [20] discovered that the weak plane effect is significant, and the borehole wall is prone to collapse along the crack. Liu et al. [21] believe that the higher risk of wellbore instability may correspond to the formation with relatively high permeability, porosity, and gas saturation. Zhang et al. [22] found that cracks would promote rock hydration and then weaken the stability of the wellbore. On the other hand, the size and distribution of formation pores have a significant impact on wellbore stability. The size and distribution of pores can greatly affect the effectiveness of mud in supporting the wellbore [23]. Pore throats can affect the flow of drilling fluid in the formation and change the wellbore stability condition [24,25]. Moreover, pores can also impact the pore pressure distribution and ultimately result in a different wellbore collapse condition [26]. Researchers have tried various ways to evaluate the stability of wellbores in the formation of cracks and solution cavities. For example, Gentzis et al. [27] utilized lithological descriptions to predict rock strength parameters and subsequently introduced a wellbore stability evaluation method based on finite element theory. Based on a Monte Carlo probability analysis and the Mogi-Coulomb failure criterion, Al-Ajmi et al. [28,29] proposed a method for evaluating the wellbore collapse pressure of a carbonate reservoir. Karatela et al. [19] and Karatela and Taheri [30] established a discrete element numerical model to study the wellbore stability of fractured shale formation. Guo et al. [31] proposed a generalized H-B criterion model based on acoustic data for evaluating the wellbore stability of multi-set fractured limestone formation. Liu et al. [32] employed the fractal method to fit the functional relationship between the cohesion, internal friction angle, and fractal dimension of micro-fractured rocks and then developed a wellbore stability evaluation method for fractured carbonate formation. Based on the thermo-poroelastic medium theory, Gomar et al. [33] proposed a wellbore stability evaluation method for porous formation. These studies attempted to obtain a method to evaluate the wellbore stability of carbonate formation, considering the effect of cracks and solution cavities. Still, many of these studies did not fully address how unevenly distributed these cracks and solution cavities can be.

As mentioned above, although scholars have recognized that the development of cracks and solution cavities is a key factor influencing the wellbore stability of carbonate formation, and some methods for evaluating its wellbore stability were obtained, there is currently a lack of systematic research on the influence of the distribution characteristics of cracks and solution cavities on the wellbore stability in carbonate formation. In the present work, the characteristics of cracks and solution cavities, the mechanics properties of the carbonate rock in the Ordovician Dengying carbonate reservoir in the Gaoshiti–Moxi area of the Sichuan Basin are analyzed using log interpretation, electron microscope scanning experiments, and rock triaxial experiments. Then, employing the FLAC3D numerical simulation software, the influence of the distribution characteristics of cracks and solution cavities on the wellbore stability in carbonate formation is studied considering the orientation and dip angle of the crack, size, and quantity of solution cavities, connectivity of crack and solution cavity. These results can provide guidance for the drilling design in the carbonate formation with developed cracks and solution cavities.

2. Reservoir Rock Characteristics and Engineering Problems

2.1. Geological Features of the Dengying Formation Carbonate Reservoir

The Dengying Formation carbonate reservoir of the Gaoshiti–Moxi area in the Sichuan Basin sits within the Gaoshiti-Anpingdian-Moxi latent tectonic zone, depicted in Figure 1. This formation is located in the eastern part of the ancient Anyue–Suining–Tongnan uplift [34–36] and is characterized by its low and gentle structure. Shaped by intersecting structural lines running north to south and east to west, the area is marked by moderate elevations and a scarcity of faults. Notably, the seismic top structure here is more pronounced and enclosed than the cold top structure. The Dengying reservoir, known for its antiquity, exhibits notable plasticity and density, reinforced by strong compaction strength. Dominated by fractured-vuggy reservoirs, it is defined by its fragmented nature and limited capacity to withstand formation pressure. These geological features often lead to issues such as formation collapse, blockages, and stuck drill strings during drilling operations, presenting considerable challenges to the natural gas exploration and development efforts in the area.

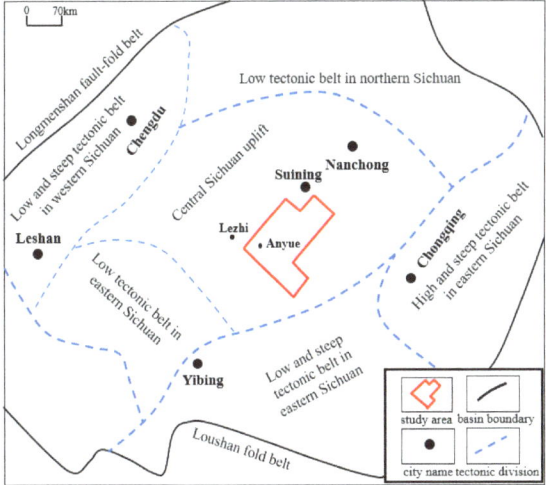

Figure 1. Reservoir geographical location and geological feature map.

2.2. Rock Cracks and Solution Cavities Characteristics

The rock samples were collected from the Dengying Formation at depths ranging from 4959 m to 5481 m of Well GS10 in the Gaoshiti–Moxi area of the Sichuan Basin using the core drill bit, as shown in Figure 2. These rock samples were obtained when the drilling operation reached the gas-producing reservoir. The dissolution pores in the rock of the Dengying Formation exhibit a variety of shapes, predominantly irregular or nearly circular. The distribution of these pores is notably uneven, often displaying band-like, honeycomb, patchy, and scattered arrangements. The formation is rich in solution cavities, with an average density of 26.45 per meter. Horizontal cracks are more common, followed by inclined ones. A large number of small cracks are also presented, with quantity densities ranging from 0.86 to 8.26 per meter. Notably, in the solution cavities and cracks, we observe fillings of quartz, dolomite, and asphalt. Additionally, some cracks are observed to be associated with solution cavities, indicating a complex interplay of geological processes.

2.3. Rock Mechanical Parameters

The triaxial compression tests on rock samples were conducted using a specialized apparatus (GCTS Ltd., RTR–1500, Tempe, AZ, USA), adhering to the method specified in the "Regulation for testing the physical and mechanical properties of rock—Part 20: Test for determining the strength of rock in triaxial compression" (Chinese Standard, Standard ID: DZ/T 0276.20–2015) [37]. The tests were carried out in accordance with the procedures detailed by Zhang et al. [38], using the samples extracted from the 5182 m to 5187 m depth interval of Well GS10, as shown in Figure 2. The results of these tests are listed in Table 1. A notable observation is the increase in rock compressive strength and elastic modulus with elevated confining pressure. However, the variation in the Poisson's ratio with confining pressure is not evident. A significant disparity in compressive strength is observed across different rock samples tested under identical confining pressure conditions. For instance, at a confining pressure of 50 MPa, the highest compressive strength recorded is 53.63% greater than the lowest. This substantial variation emphasizes the profound impact of cracks and solution cavities on the mechanical characteristics of carbonate rock, thus highlighting the distinct variability of the rock's mechanical parameters.

Figure 2. Rock samples collected from the Dengying Formation: (**a**) Brownish-gray solution cavities in dolomite at a depth from 5480.27 m to 5480.41 m; (**b**) Dark gray dolomite with solution cavities at a depth from 5460.12 m to 5460.22 m; (**c**) Light gray dolomite, at a depth from 4970.32 m to 4970.41 m; (**d**) Blackish-gray muddy dolomite, at a depth from 4959.81 m to 4959.89 m.

Table 1. Rock parameters of the carbonate rock tested by the triaxial compression tests.

No.	Compressive Strength/MPa	Elastic Modulus/GPa	Poisson's Ratio	Confining Pressure/MPa
1	143.61	33.45	0.23	0
2	264.20	46.32	0.32	50
3	415.38	53.55	0.24	60
4	174.99	41.10	0.22	0
5	402.31	48.84	0.29	50
6	390.08	50.06	0.26	60
7	197.03	36.20	0.23	0
8	358.30	49.30	0.27	50
9	383.27	50.01	0.23	60
10	253.11	38.41	0.24	0
11	261.86	49.34	0.21	50
12	391.34	49.19	0.28	60

2.4. Wellbore Instability Problem

A critical wellbore collapse incident occurred at a depth ranging from 5182 m to 5187 m in Well GS10, which is located in the Gaoshiti–Moxi area of the Sichuan Basin. The target reservoir in this area is the Dengying Formation. The Dengying Formation was designed as the fifth opening section of Well GS10 with a drill bit diameter of 127 mm. This issue arose while using an oil-based drilling fluid with a density of 1.14 g/cm^3, which was initially selected to prevent clay mineral hydration. However, this density decision did not factor in the presence of cracks and solution cavities, leading to mechanical instability. As a remedial measure, the density of the drilling fluid was increased to 1.24 g/cm^3, which effectively resolved the issue, allowing for the successful completion of the drilling operation in this section. Subsequent imaging logging analysis highlighted the extensive development of cracks and solution cavities in the formation. As depicted in Figure 3, notable features included two large cracks with a dip angle of 19.81° and a maximum width of 10 mm and three solution cavities, each with an average diameter of 10 mm. According to the cutting logging result based on the lithologic difference analysis, the maximum width of the crack is bigger than 30 mm, and the largest diameter of the solution cavity is approximately 900 mm. Therefore, the analysis concluded that these developed cracks and solution cavities were the key contributors to the wellbore instability due to the mechanical properties of the carbonate rocks decreasing significantly when the cracks and solution cavities appear in the rock. This indicates that the influence of the distribution characteristics of cracks and solution cavities on the wellbore stability in carbonate formation is worth further studying to optimize the drilling design on the new wells in this area.

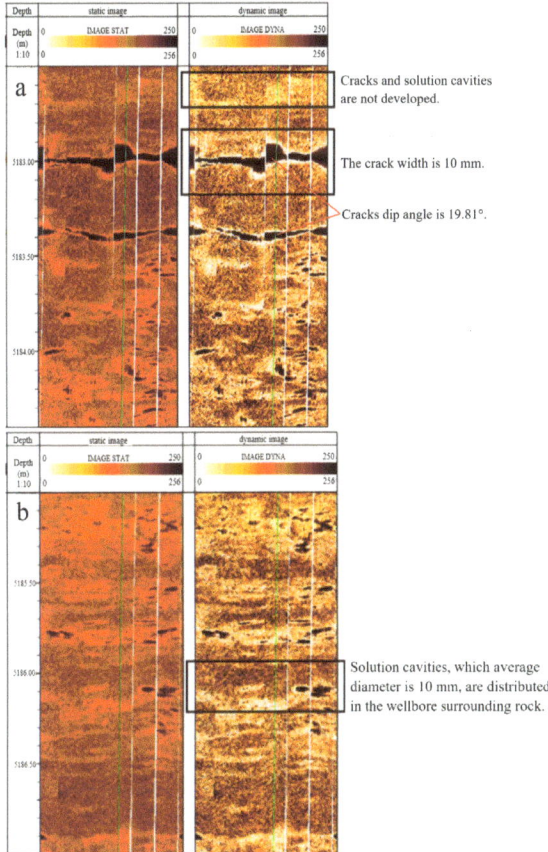

Figure 3. Distribution of cracks and solution cavities in the 5182–5187 m section of Well GS10 in the Gaoshiti–Moxi area of the Sichuan Basin: (**a**) 5182–5185 m section; (**b**) 5184.5–5187.5 m section.

3. Methodology and Model Parameters
3.1. Numerical Simulation Scheme

According to the real-life scenario of cracks and solution cavities characteristics in the carbonate formation of the Gaoshiti–Moxi area of the Sichuan Basin (Figure 3), three factors are considered for evaluating the influence of cracks and solution cavities on the wellbore stability, including cracks characteristics, solution cavities characteristics and connectivity between cracks and solution cavities. Based on the formation reality stated in Section 3.1, the parameters setting of these factors in the numerical simulation is simplified through an arithmetic progression to obtain the favorable rules regarding the influence of cracks and solution cavities characteristics on the wellbore stability, as shown in Table 2. The total stresses are employed to determine the stability of the wellbore. This is because the rock strength parameters, which are used for the numerical simulation, are calculated by using logging data. Additionally, there is another reason for using total stress. Considering the actual situation where a properly formed mud cake exists on the borehole wall, It is assumed that the gas in the formation does not flow. The equivalent drilling fluid density of collapse pressure (ρ_c) is calculated for these different conditions to analyze these factors affecting the wellbore stability. The model designs of the angle of the crack strike and the maximum horizontal principal stress direction (β), the dip angle of the crack (θ), the solution cavity diameter (D), and the solution cavities quantity (N) are shown in Figures 4–7, respectively.

Table 2. Parameters setting of the analytical factors for the numerical simulation.

Factor	Parameter	Parameters Setting	Other Parameters
Cracks characteristics	β	0°, 15°, 30°, 45°, 60°, 75°, 90°	L_f = 10 mm, θ = 90°
	θ	0°, 15°, 30°, 45°, 60°, 75°, 90°	L_f = 10 mm, β = 90°
	L_f	8 mm, 12 mm, 16 mm, 25 mm, 30 mm	β = 90°, β = 90°
Solution cavities characteristics	D	152 mm, 304 mm, 457 mm, 609 mm, 762 mm, 914 mm	L = 300 mm
	N	1 (Unilateral distribution), 2 (Symmetrical distribution), 4 (Symmetrical distribution around the circumference), 8 (Symmetrical distribution around the circumference with two groups), 16 (Symmetrical distribution around the circumference with four groups)	L = 300 mm, D = 152 mm
Connectivity between cracks and solution cavities	Connectivity	Situation I: The cracks are not connected to the four solution cavities. Situation II: the cracks are connected to the four solution cavities.	L_f = 10 mm, D =1 52 mm

β is the angle of the crack strike and the maximum horizontal principal stress direction, °; θ is the dip angle of the crack, °; L_f is the crack width, mm; D is the solution cavities diameter, mm; N is the solution cavities quantity, Integer; L is the distance from the solution cavities to the wellbore wall, mm.

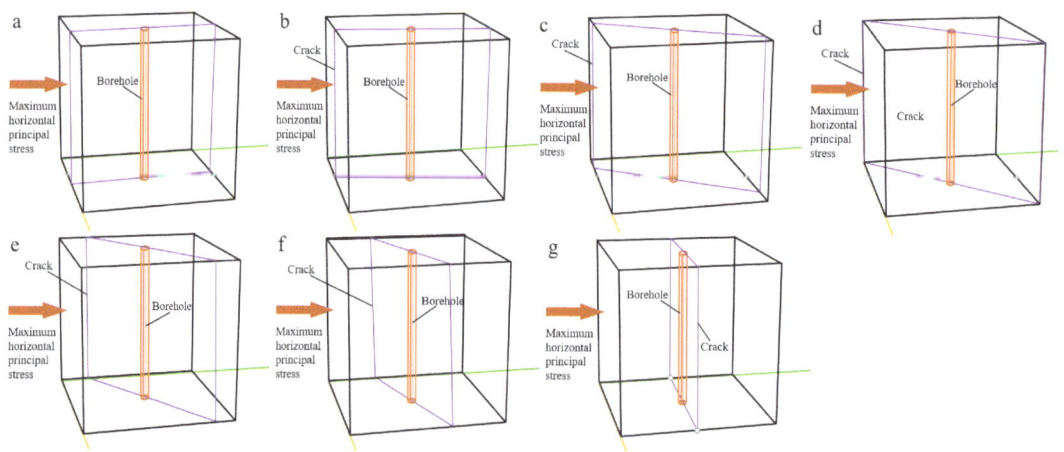

Figure 4. Model design of the angle of the crack strike and the maximum horizontal principal stress direction (β): (a) β = 0°; (b) β = 15°; (c) β = 30°; (d) β = 45°; (e) β = 60°; (f) β = 75°; (g) β = 90°.

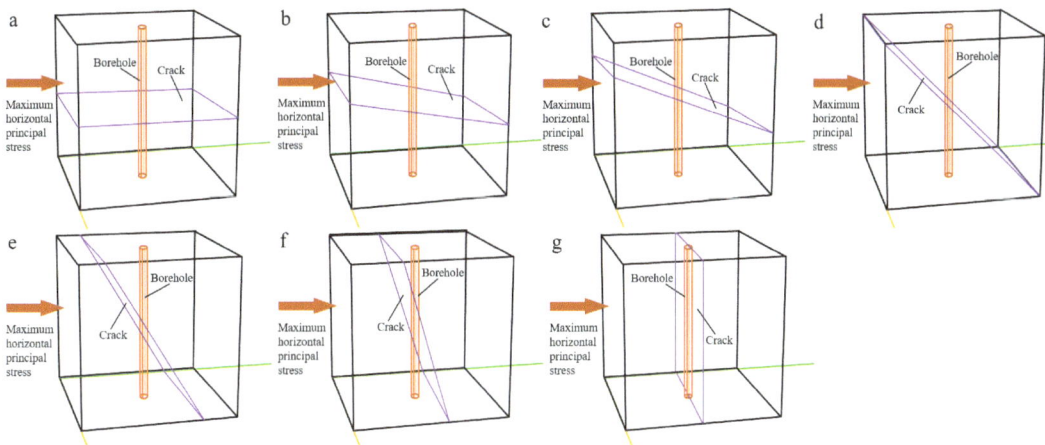

Figure 5. Model design of the dip angle of the crack (θ): (**a**) $\theta = 0°$; (**b**) $\theta = 15°$; (**c**) $\theta = 30°$; (**d**) $\theta = 45°$; (**e**) $\theta = 60°$; (**f**) $\theta = 75°$; (**g**) $\theta = 90°$.

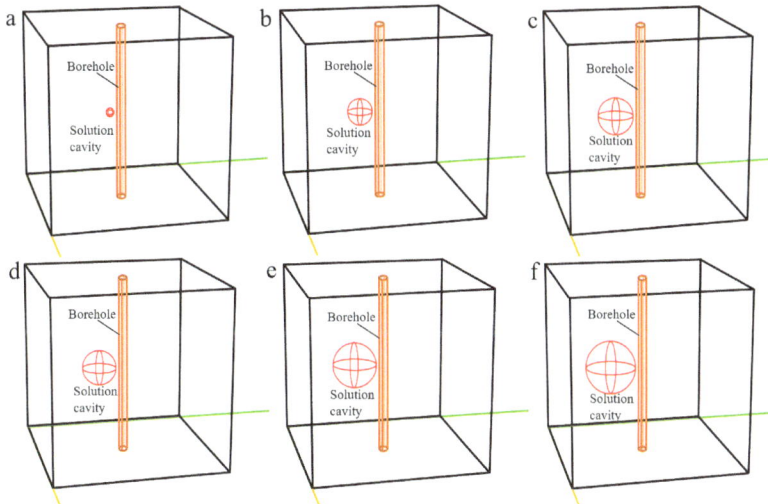

Figure 6. Model design of the solution cavity diameter (D): (**a**) $D = 152$ mm; (**b**) $D = 304$ mm; (**c**) $D = 457$ mm; (**d**) $D = 609$ mm; (**e**) $D = 762$ mm; (**f**) $D = 914$ mm.

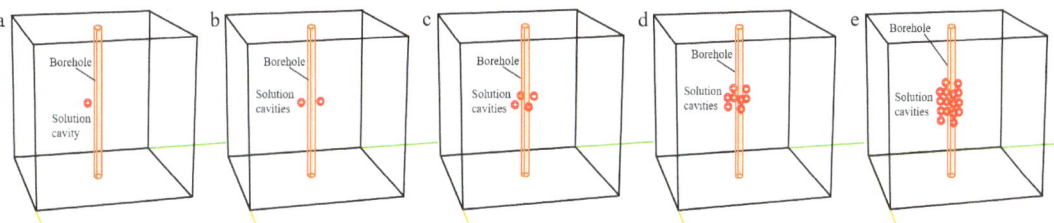

Figure 7. Model design of the solution cavities quantity (N): (**a**) $N = 1$; (**b**) $N = 2$; (**c**) $N = 4$; (**d**) $N = 8$; (**e**) $N = 16$.

3.2. Numerical Simulation Method

The software FLAC3D, which is adept at calculating equivalent stress by assessing the equivalent plastic strain, was employed for the numerical simulations. It is vital to determine the stress state after the plastic yield of the surrounding rock. When the wellbore rock's plastic yield value surpasses its critical threshold, the convergence of the calculation should be judged. If the calculation cannot converge, it signals a state of instability and the potential failure of the surrounding rock.

Following the factors and parameters detailed in Table 2, the numerical simulation to assess the influence of cracks and solution cavities on wellbore stability comprises several stages:

(1) The stratum model containing the cracks and solution cavities according to the parameters setting as shown in Table 2 is created, the model parameters are set, and initial stress and pore pressure are applied to the model. The model then undergoes a calculation phase to achieve stress balance.

(2) Within this stratum model, a borehole is simulated. Since the main concern is the deformation of the strata around the wellbore, the inside of the wellbore is set as a certain pressure boundary condition considering the supporting effect of drilling fluid on the wellbore wall. To ensure the stability of the rock surrounding the wellbore, a high fluid column pressure is applied in the borehole at first.

(3) The fluid column pressure is gradually decreased. The point at which the wellbore collapses with a fluid column pressure (P_c) is determined by the onset of plastic deformation without a convergence result in the surrounding rock, and the Mohr–Coulomb criterion is used to justify this plastic deformation. From this, the equivalent drilling fluid density of collapse pressure (ρ_c) is calculated by dividing P_c by the formation depth (H).

The numerical simulation flow chart is illustrated in Figure 8.

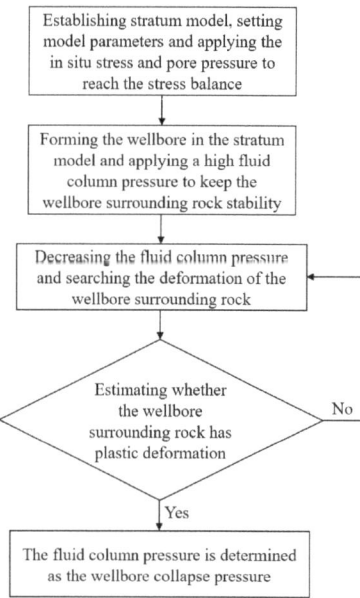

Figure 8. Numerical simulation flow chart.

3.3. Numerical Simulation Model

The numerical simulation model is designed as a 3-m cube featuring a vertical wellbore at its center to replicate an actual wellbore, as illustrated in Figure 9a. In this setup, the four lateral sides of the model are fixed horizontally, while vertically, they are subject to free boundary conditions. Normal stress is applied to the model's outer boundaries to maintain a balance between the internal and external forces. Given that the stresses in all three directions exceed 120 MPa, the gravitational effect on the cube is considered negligible. The mesh layout of the model is displayed in Figure 9b. A more accurate simulation result can be obtained when a denser mesh is set compared to when a sparse mesh is set for numerical simulation; thus, a denser mesh is generated around the wellbore area. After establishing the numerical simulation model, as shown in Figure 9, the rock, cracks, and solution cavities are generated by changing the material parameters of the designated area in the numerical simulation model according to the numerical simulation scheme, as introduced in Section 3.1.

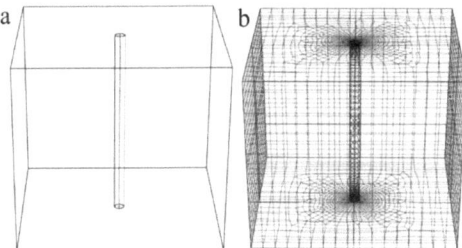

Figure 9. Numerical simulation model (**a**) and mesh generation (**b**).

3.4. Model Parameters

By utilizing the well logging data for the 5182 m–5187 m section of Well GS10 (as shown in Figure 10), the pore pressure and three in situ stresses (overburden pressure, minimum principal stress, and maximum principal stress) were determined. The pore pressure was calculated using the Eation method [39], and the three in situ stresses were calculated using Huang's model [40]. The parameters of the numerical simulation are presented in Table 3. Due to the discrepancy in stress conditions between rock core tests and the actual formation, rock parameters were first calculated using logging data and then corrected with test results. Therefore, the rock strength parameters in Table 3 are determined by total stresses since effective stress cannot be determined through well logging interpretation.

Table 3. Parameters for numerical simulation.

Parameter	Value
Maximum horizontal stress/MPa	154.08
Minimum horizontal stress/MPa	122.49
Vertical stress/MPa	142.58
Formation pressure/MPa	65.00
Elastic modulus of rock/GPa	49.52
Poisson's ratio of rock	0.28
Cohesion of rock/MPa	20.52
Internal friction angle of rock/°	39.60
Elastic modulus of the fillings in cracks/GPa	25.68
Poisson's ratio of the fillings in cracks	0.23
Cohesion of the fillings in cracks/MPa	7
Internal friction angle of the fillings in cracks/°	25.90
Elastic modulus of the fillings in solution cavities/GPa	3.85
Poisson's ratio of the fillings in solution cavities	0.13
Cohesion of the fillings in solution cavities/MPa	7
Internal friction angle of the fillings in solution cavities/°	3.20

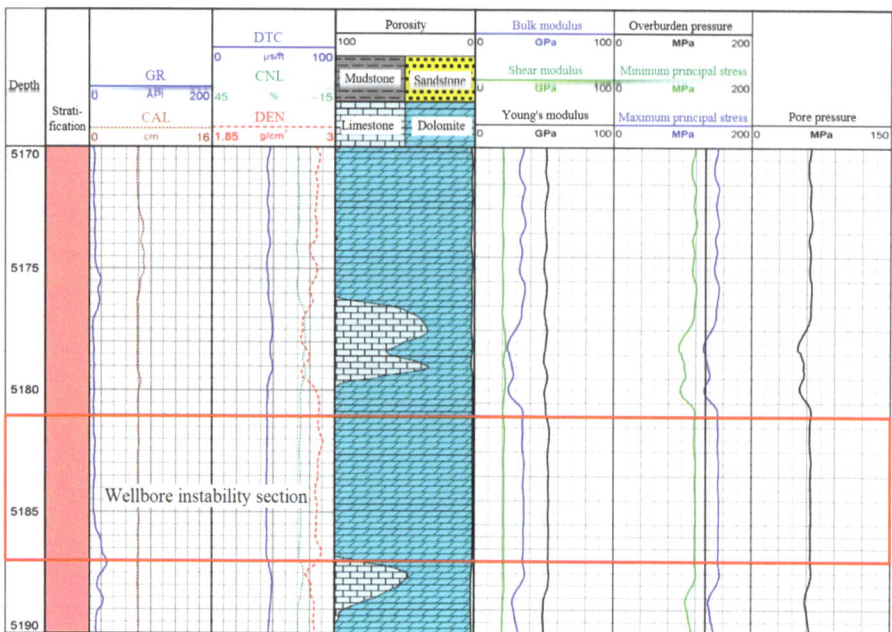

Figure 10. Geomechanical parameters of the 5170–5190 m section of Well GS10 obtained from well logging interpretation.

4. Results and Analysis

4.1. Verification of the Numerical Simulation Results

Reflecting the imaging log data for the 5182–5187 m section of Well GS10, the numerical model was designed to explore three situations: one without the influence of cracks and solution cavities, another considering only the presence of cracks, and the third one encompassing both cracks and solution cavities. In the latter two situations, the cracks and solution cavities were determined based on the imaging logging chart, as depicted in Figure 3. Two cracks with a dip angle of 19.81° were designed to cross the well in the model. The width of one crack was 10 mm, while the width of the other was 3 mm. Ten solution cavities were designed around the well, with diameters ranging from 3 mm to 20 mm and an average value of 10 mm. The third case was consistent with the actual situation on the site. Following the methodology mentioned in Section 3, the equivalent drilling fluid densities of collapse pressure (ρ_c) were calculated for these varied conditions, as shown in Table 4. Notably, ρ_c considering the impacts of both cracks and solution cavities increases by 6.42% compared to ignoring cracks and solution cavities. Furthermore, ρ_c considering both cracks and solution cavities increases by 1.75% compared with the value for only considering cracks, and ρ_c considering only cracks increases by 4.59% compared with the value for ignoring cracks and solution cavities. This indicates that cracks play a more significant role than solution cavities in affecting the wellbore stability.

From a practical standpoint, as discussed in Section 3.1, the initial employment of an oil-based drilling fluid with a density of 1.14 g/cm^3, which overlooked the effects of cracks and solution cavities, resulted in significant wellbore collapse. Adjusting the drilling fluid density to 1.24 g/cm^3 subsequently alleviated this problem. This practical situation corresponds well with our simulation findings in Table 4, reinforcing the validity of our numerical approach. By accounting for both cracks and solution cavities, this method offers a more accurate and dependable means for assessing wellbore stability in the Dengying Formation carbonate reservoir in the Gaoshiti–Moxi area of the Sichuan Basin.

Table 4. Equivalent drilling fluid densities of collapse pressure for the three situations.

Situation	Description	Result/g·cm^{-3}
Without considering cracks and solution cavities	The formation is considered as the intact rock	1.09
Only considering cracks	Two cracks with a dip angle of 19.81° are designed in the formation model. The width of one crack is 10 mm, while the width of the other is 3 mm	1.14
Considering cracks and solution cavities	Two cracks with a dip angle of 19.81° and ten solution cavities are designed in the formation model. The width of one crack is 10 mm, while the width of the other is 3 mm. The diameters of the solution cavities range from 3 mm to 20 mm, with an average value of 10 mm. This situation is consistent with the actual situation on the site	1.16

4.2. Influence of Cracks Characteristics on the Wellbore Stability

4.2.1. Influence of Crack Orientation on the Wellbore Stability

Figures 11 and 12 present the stress and displacement patterns around the wellbore, varying with the angle (β) between the crack strike and the maximum horizontal principal stress direction, as shown in Table 2. The observed trend is the following: with a decrease in β, both the stress and displacement at all points around the wellbore gradually increase. The lowest stress and displacement values are recorded when β equals 90°. For instance, at $\beta = 0°$, the stress and displacement at point A are measured at 125.21 MPa and 1.53 mm, respectively. In contrast, at $\beta = 90°$, these values drop to 105.61 MPa and 1.25 mm, representing decreases of 15.65% and 18.30%, respectively. This pattern suggests that when cracks align parallel to the maximum horizontal principal stress, the stress and deformation at the crack-wellbore intersection are at their peak. Conversely, these values progressively reduce as β increases. This can be attributed to the following reasons: At $\beta = 0°$, the maximum horizontal stress causes the cracks near the wellbore to undergo tensile expansion, leading to concentrated stress and increased tensile displacement in the area. However, at $\beta = 90°$, the maximum horizontal stress tends to close the cracks, resulting in lesser stress concentration and tensile displacement compared to the former situation.

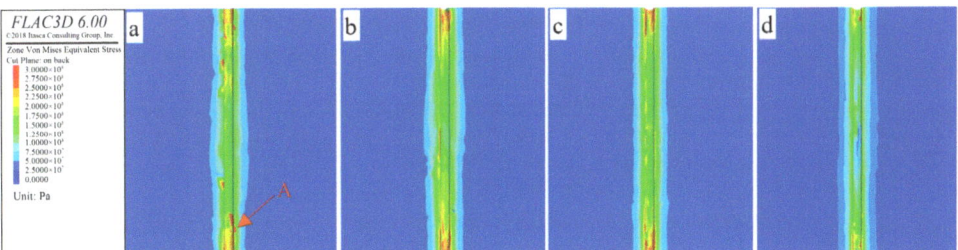

Figure 11. Stress nephogram around the wellbore at different β: (**a**) $\beta = 0°$; (**b**) $\beta = 30°$; (**c**) $\beta = 60°$; (**d**) $\beta = 90°$.

Figure 13 showcases the correlation between the equivalent drilling fluid densities of collapse pressure (ρ_c) and the angle (β) between the crack strike and the maximum horizontal principal stress direction. A notable trend emerges from the graph: ρ_c linearly diminishes with an increase in β. Specifically, ρ_c is 1.26 g/cm^3 at $\beta = 0°$ but drops to 1.06 g/cm^3 at $\beta = 90°$, marking a substantial reduction of 15.87%. When considered alongside the previously discussed stress and displacement data, this trend suggests a clear rule: in situations where cracks run parallel to the maximum horizontal stress, vertical cracks intersecting the wellbore are prone to tensile expansion. This expansion significantly contributes to the weakening and potential failure of the rock formation surrounding the wellbore, thus resulting in the highest collapse pressure under these conditions.

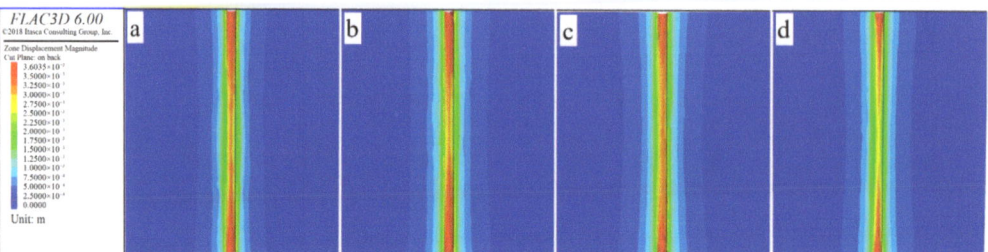

Figure 12. Displacement nephogram around the wellbore at different β: (**a**) $\beta = 0°$; (**b**) $\beta = 30°$; (**c**) $\beta = 60°$; (**d**) $\beta = 90°$.

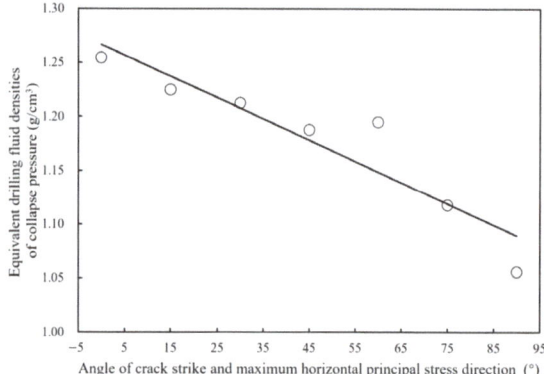

Figure 13. Curve between the equivalent drilling fluid density of collapse pressure and the angle of the crack strike and maximum horizontal principal stress direction.

4.2.2. Influence of Crack Dip Angle on the Wellbore Stability

Figures 14 and 15 illustrate the variations in stress and displacement around the wellbore in response to different crack dip angles (θ). Analyzing these figures leads to two significant observations: First, an increase in θ results in both stress and displacement at points around the wellbore. For example, at $\theta = 0°$, the stress and displacement at point A are recorded at 82.58 MPa and 0.85 mm, respectively. At a higher angle of $\theta = 60°$, these values increase to 102.56 MPa and 1.08 mm, indicating increases of 24.08% and 27.05%, respectively. The rationale behind this trend is that as θ increases, more points along the crack surface come closer to the wellbore. This proximity weakens the integrity of the surrounding rock layers, preventing the formation of a uniformly stressed, arch-like structure around the wellbore. Consequently, areas where cracks intersect the wellbore experience heightened stress concentration and larger displacement, as shown in Figures 14 and 15.

Figure 16 presents the relationship between the equivalent drilling fluid density of collapse pressure (ρ_c) and the crack dip angle (θ). The graph in Figure 16 illustrates that ρ_c exhibits a linear increase as θ rises. For instance, ρ_c is calculated at 1.08 g/cm³ at $\theta = 0°$, and it escalates to 1.41 g/cm³ at $\theta = 90°$, representing a substantial increase of 30.56% from the initial value. Correlating this with the stress and displacement changes depicted in Figures 14 and 15, it becomes evident that an increase in θ adversely affects the integrity of the rock surrounding the wellbore. This weakening of the rock structure makes the wellbore more prone to damage, thus leading to an increase in ρ_c with the rising θ.

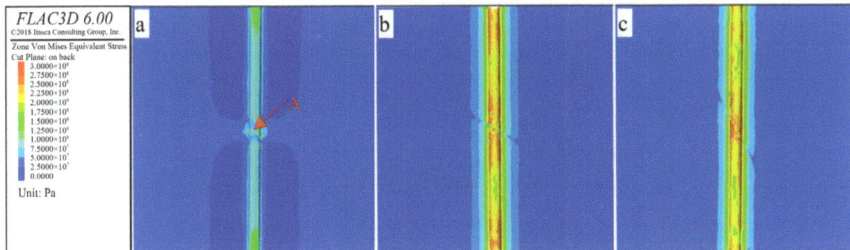

Figure 14. Stress nephogram around the wellbore for different crack dip angles: (**a**) $\theta = 0°$; (**b**) $\theta = 30°$; (**c**) $\theta = 60°$.

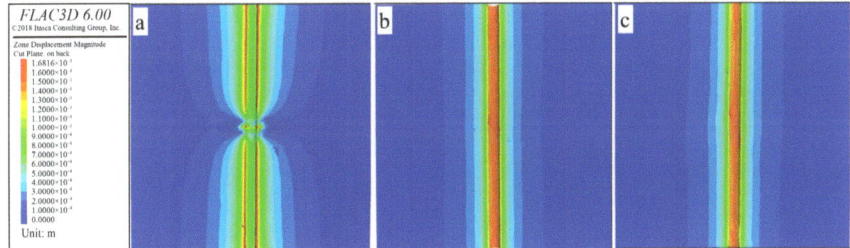

Figure 15. Displacement nephogram around the wellbore for different crack dip angles: (**a**) $\theta = 0°$; (**b**) $\theta = 30°$; (**c**) $\theta = 60°$.

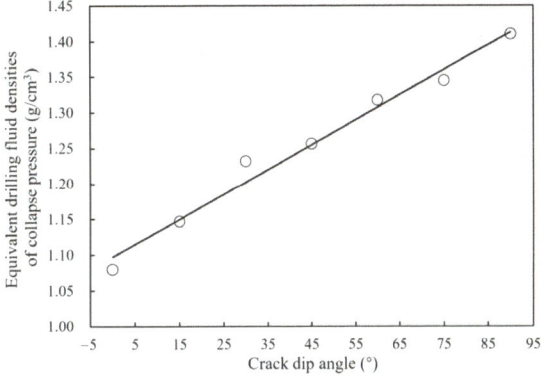

Figure 16. Curve between the equivalent drilling fluid densities of collapse pressure and crack dip angles.

4.2.3. Influence of Crack Width on the Wellbore Stability

Figures 17 and 18 illustrate the variations in stress and displacement around the wellbore concerning various crack widths (L_f). Notably, both the stress and the displacement at different locations around the wellbore increase gradually as L_f increases. For example, when L_f is 8 mm, the stress and displacement values at point A are 108.23 MPa and 1.27 mm, respectively. However, when L_f is 30 mm, the stress and displacement values at point A rise to 146.32 MPa and 1.52 mm, respectively. This represents a substantial increase of 35.13% in stress and 19.68% in displacement compared to the previous values. This trend underscores the fact that an expansion in the crack width leads to a reduction in the stability of the surrounding rock, consequently resulting in elevated stress and displacement levels at various points around the wellbore.

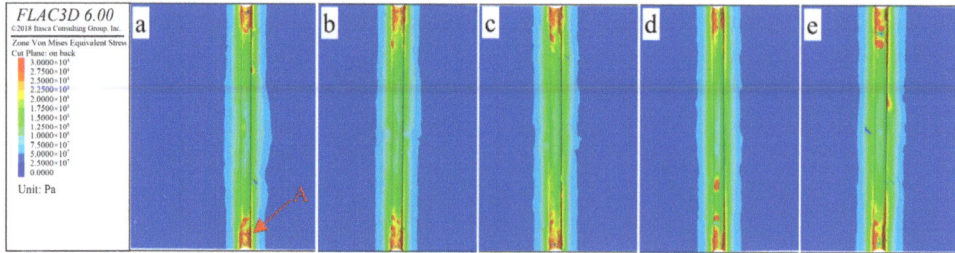

Figure 17. Stress nephogram around the wellbore for different crack widths: (**a**) L_f = 8 mm; (**b**) L_f = 12 mm; (**c**) L_f = 16 mm; (**d**) L_f = 25 mm; (**e**) L_f = 30 mm.

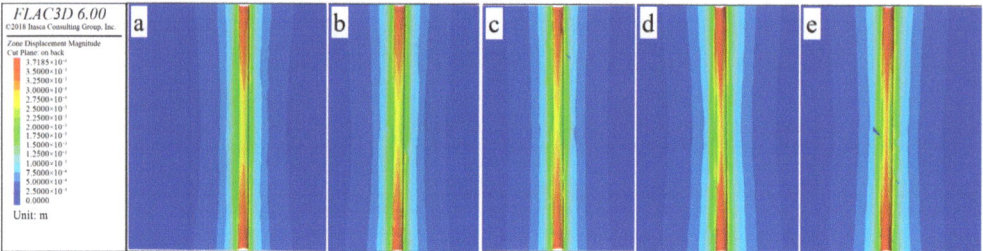

Figure 18. Displacement nephogram around the wellbore for different crack widths: (**a**) L_f = 8 mm; (**b**) L_f = 12 mm; (**c**) L_f = 16 mm; (**d**) L_f = 25 mm; (**e**) L_f = 30 mm.

Figure 19 presents the relationship between the equivalent drilling fluid densities of collapse pressure (ρ_c) and crack widths (L_f). It is evident from the figure that ρ_c exhibits a linear increase with the increase in L_f. For example, when L_f is 8 mm, ρ_c is measured as 1.06 g/cm^3, while for L_f = 30 mm, ρ_c increases to 1.33 g/cm^3, representing a 25.47% increment compared to the previous value. This observed increase can be attributed to the reduced stability of the rock surrounding the wellbore as L_f increases, resulting from stress concentration near the cracks, as demonstrated in Figure 17. Consequently, ρ_c exhibits an upward trend with the growth of L_f.

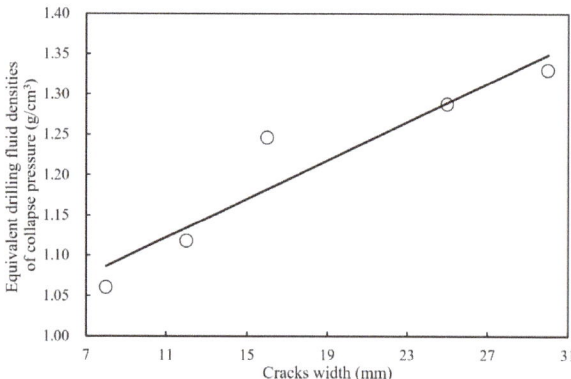

Figure 19. Curve between the equivalent drilling fluid densities of collapse pressure and crack widths.

4.3. Influence of Solution Cavities Characteristics on the Wellbore Stability

4.3.1. Influence of Solution Cavities Size on the Wellbore Stability

Figures 20 and 21 present the variations in stress and displacement around the wellbore for various solution cavity diameters (D). These figures reveal a consistent trend: stress and displacement at different locations around the wellbore gradually increase with an increase in D. For instance, when D is 152 mm, the stress and displacement values at point A are measured as 201.03 MPa and 3.52 mm, respectively. However, with D increased to 914 mm, the stress and displacement at point A rise to 256.32 MPa and 4.03 mm, respectively. This represents a significant increase of 28.01% in stress and 14.48% in displacement compared to the previous values.

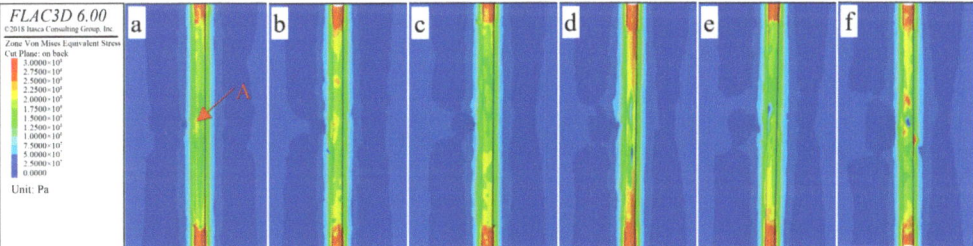

Figure 20. Stress nephogram around the wellbore for different solution cavities diameter: (**a**) D = 152 mm; (**b**) D = 304 mm; (**c**) D = 457 mm; (**d**) D = 609 mm; (**e**) D = 762 mm; (**f**) D = 914 mm.

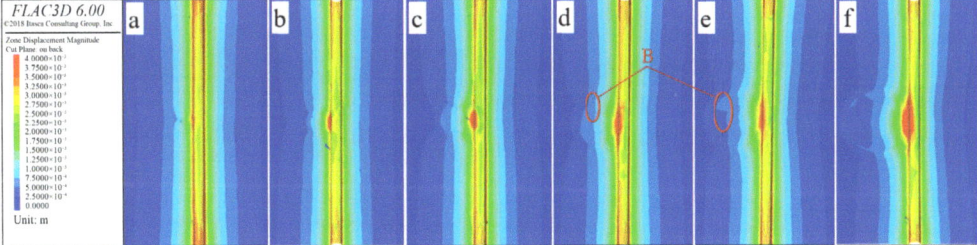

Figure 21. Displacement nephogram around the wellbore for different solution cavities diameter: (**a**) D = 152 mm; (**b**) D = 304 mm; (**c**) D = 457 mm; (**d**) D = 609 mm; (**e**) D = 762 mm; (**f**) D = 914 mm.

This phenomenon can be attributed to the lower mechanical strength of the fillings within the solution cavities compared to the surrounding rock. In such cases, stress tends to concentrate on the solution cavities wall due to the limited supporting capacity of the fillings, as exemplified by the conditions observed at point A in Figure 20. Furthermore, this stress concentration amplifies with the enlargement of D, leading to increased deformation in the solution cavities area, as shown in Figure 21.

Figure 22 presents the relationship between the equivalent drilling fluid densities of collapse pressure (ρ_c) and solution cavities diameter (D). The figure reveals an initial linear increase in ρ_c with the growth of D, and ρ_c for the situation of D = 609 mm is increased by 14.12% compared with the situation of D = 152 mm. However, when D exceeds 609 mm, there is no significant change in ρ_c. This phenomenon can be ascribed to the fact that when the size of the solution cavities reaches a certain value, for example, D = 609 mm, it may cause wellbore collapse and damage to the surrounding rock mass, thereby forming a substantial cavity. As shown in Figure 21, the area (area B) with a displacement larger than 3.5 mm for D = 609 mm increases sharply compared to that for D = 457 mm. In this case, the main factor affecting wellbore stability would be the damage to the rock between the solution cavity and the borehole, and this cavity may become large enough to minimize the impact of further increases in solution cavity size on ρ_c. Therefore, the linear relationship between the diameter of solution cavities and ρ_c rises to 609 mm and then becomes flat.

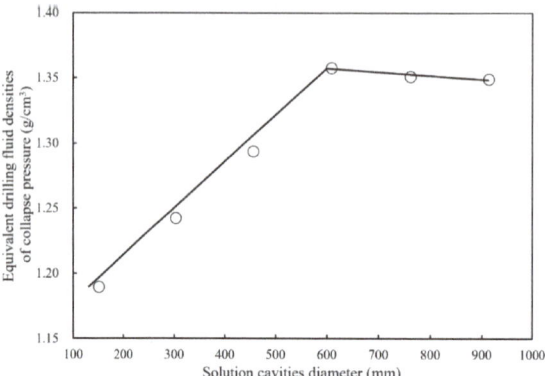

Figure 22. Curve between the equivalent drilling fluid densities of collapse pressure and solution cavities diameter.

4.3.2. Influence of Solution Cavities Quantity on Wellbore Stability

Figures 23 and 24 illustrate the variations in stress and displacement distribution around the wellbore for varying solution cavity quantities (N). Notably, these figures illustrate a consistent trend: as solution cavities quantity (N) increases, both stress and displacement around the wellbore gradually rise. For example, when $N = 1$, the stress and displacement values at point A are measured as 165.42 MPa and 2.78 mm, respectively. However, with an increase in N to 16, the stress and displacement at point A increase to 178.62 MPa and 2.92 mm, respectively. This marks a notable increase of 7.98% in stress and 5.03% in displacement compared to the previous values.

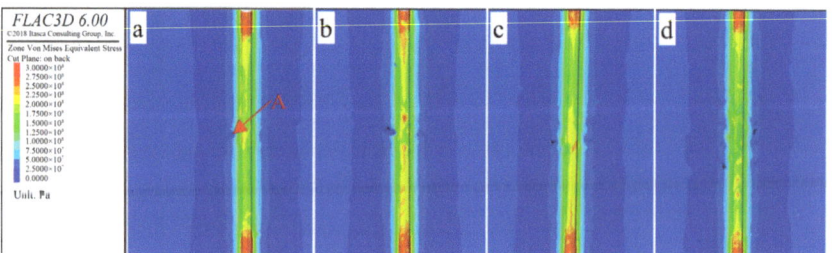

Figure 23. Stress nephogram around the wellbore for different solution cavities quantities: (**a**) $N = 1$; (**b**) $N = 4$; (**c**) $N = 8$; (**d**) $N = 16$.

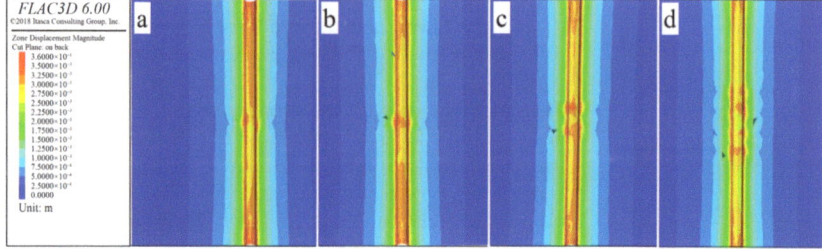

Figure 24. Displacement nephogram around the wellbore for different solution cavities quantities: (**a**) $N = 1$; (**b**) $N = 4$; (**c**) $N = 8$; (**d**) $N = 16$.

This trend can be attributed to the fact that an increase in the quantity of solution cavities near the wellbore can compromise the integrity of the rock, significantly reducing the stability of the surrounding rock mass. Consequently, stress and displacement around the wellbore, particularly in proximity to the solution cavities, increase as N grows.

Figure 25 presents the relationship between the equivalent drilling fluid densities of collapse pressure (ρ_c) and the number of solution cavities (N). Figure 25 reveals an approximately linear increase in ρ_c with the growth of N. For instance, when $N = 1$, ρ_c is measured at 1.18 g/cm^3, whereas for $N = 16$, ρ_c increases to 1.31 g/cm^3. This represents a notable increase of 11.02% compared to the previous value. This trend indicates that wellbore stability would significantly decrease with an increase in the number of solution cavities. This decrease in stability is attributed to the compromised integrity of the rock surrounding the wellbore, as mentioned above.

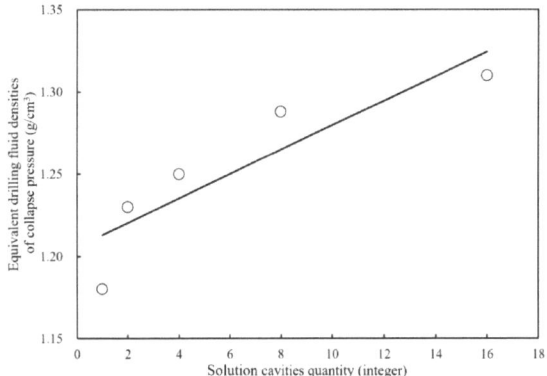

Figure 25. Curve between the equivalent drilling fluid densities of collapse pressure and solution cavities quantities.

4.4. Influence of Connectivity Between Cracks and Solution Cavities on the Wellbore Stability

Figures 26 and 27 illustrate the variations in stress and displacement distribution around the wellbore under different conditions of cracks and solution cavities connectivity. Figure 22 shows the following: (1) The disturbed stress area around the wellbore is 45.23% larger for the condition of connected cracks and solution cavities compared to the condition of disconnected cracks and solution cavities. (2) The stress around the wellbore is higher when the cracks and solution cavities are connected. For instance, when the cracks and solution cavities are disconnected, the stress at point A is measured as 222.25 MPa. However, when the cracks and solution cavities are connected, the stress at point A increases to 278.62 MPa, representing a 25.36% increase compared to the previous condition. As for Figure 27, it is observed that the connectivity of cracks and solution cavities has an insignificant impact on the displacement of various points around the wellbore. This may be because the connectivity of cracks and solution cavities does not significantly affect the integrity of the surrounding rock, resulting in minimal differences in rock deformation around the wellbore. However, when cracks extend through the solution cavities into the formation, deformation of the solution cavities might lead to stress release, which would increase the influence of cracks on the stress distribution around the wellbore. Consequently, stress disturbance for the condition of connected cracks and solution cavities would expand compared to the condition of disconnected cracks and solution cavities.

The equivalent drilling fluid densities of collapse pressure (ρ_c) under the conditions of disconnected and connected cracks and solution cavities are 1.202 g/cm^3 and 1.362 g/cm^3, respectively. It reveals that ρ_c for the condition of connected cracks and solution cavities increases by 13.33% compared to that for the condition of disconnected cracks and solution cavities. This emphasizes the importance of carefully surveying the connectivity of cracks and solution cavities during drilling design. It also highlights the need to pay greater attention to carbonate formation with connected cracks and solution cavities during the drilling process in such formation.

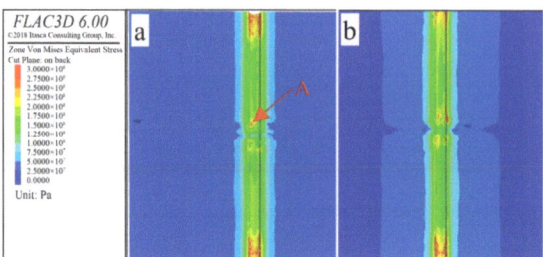

Figure 26. Stress nephogram around the wellbore for (**a**) disconnected cracks and solution cavities; (**b**) connected cracks and solution cavities.

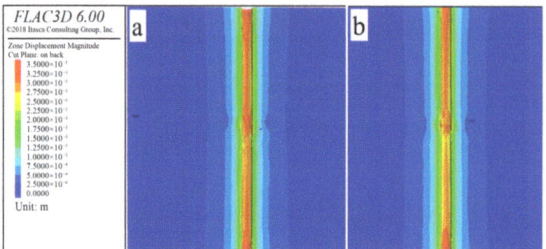

Figure 27. Displacement nephogram around the wellbore for (**a**) disconnected cracks and solution cavities; (**b**) connected cracks and solution cavities.

4.5. Sensitivity Analysis of Cracks and Solution Cavities Characteristics on the Wellbore Stability

Figure 28 shows the percentage of the maximum change in the equivalent densities of collapse pressure under the circumstances of variations in crack orientation, crack dip angle, crack width, solution cavities size, solution cavities quantity, and the connectivity between cracks and solution cavities. As shown in Figure 28, cracks have a much more significant impact on wellbore stability than solution cavities. When comparing the influencing factors such as the orientation, dip angle, and width of cracks, the wellbore stability is most sensitive to the crack dip angle, followed by the crack width and crack orientation. For example, the equivalent drilling fluid densities of collapse pressure (ρ_c) for a crack dip angle (θ) of 90° is 30.55% higher than that for $\theta = 0°$; ρ_c for a crack width (L_f) of 8 mm is 25.47% higher than that for $L_f = 30$ mm; and ρ_c for a crack orientation (β) of 0° is 15.87% higher than that for $\beta = 90°$. Regarding the influencing factor of the solution cavity characteristics, the solution cavity size has the greatest impact on the wellbore stability, followed by the connectivity between cracks and solution cavities and the solution cavities quantity. However, the difference in the influence among these solution cavities characteristics factors is smaller than that of the cracks characteristics factors. It should be noted that the sensitivity analysis of cracks and solution cavities characteristics on wellbore stability here is based on the numerical simulation conditions and parameter settings, as shown in Section 3.2, which is in line with the real-life scenario of cracks and solution cavities characteristics in the carbonate formation in the Gaoshiti–Moxi area of the Sichuan Basin. The results of the sensitivity analysis may be different if other formations or geological conditions are studied.

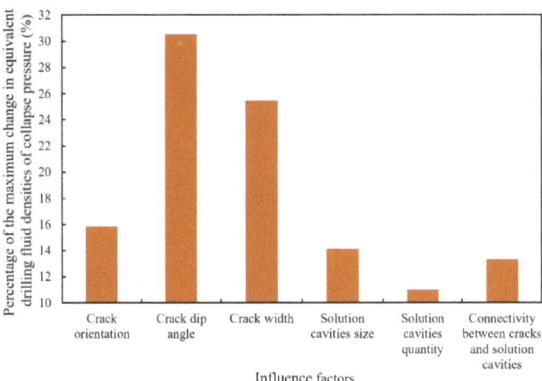

Figure 28. Percentage of the maximum change in the equivalent drilling fluid densities of collapse pressure for different influence factors.

As previously discussed, the presence of cracks and solution cavities in carbonate formation significantly diminishes the wellbore stability. In the case of vertical wells, cracks oriented closer to the direction of maximum horizontal stress can lead to reduced wellbore stability. Furthermore, the increased dip angle and width of cracks, along with the greater quantity and size of solution cavities, including connected cracks and solution cavities, all contribute to wellbore instability in carbonate formation. Consequently, a meticulous survey of cracks and solution cavities is essential during the drilling design phase for carbonate formation. This survey should encompass the relationship between cracks and in situ stress, crack dip angle and width, solution cavity diameters and quantity, as well as the connectivity of cracks and solution cavities. A comprehensive analysis of the characteristics of cracks and solution cavities should then be integrated to evaluate the wellbore stability. Ultimately, this approach enables the development of a reliable drilling design that incorporates appropriate drilling fluid densities and excellent drilling fluid properties. Such a design is crucial for ensuring the safety of drilling operations in carbonate formation.

5. Conclusions

(1) Cracks and solution cavities are prevalent in the Dengying Formation carbonate reservoir in the Gaoshiti–Moxi area of the Sichuan Basin. Horizontal cracks are the most predominant, followed by inclined cracks, with an average crack density ranging from 0.86 to 8.26 per meter. Additionally, there is a multitude of solution cavities, with an average development density of 26.45 per meter. The extensive development of cracks and solution cavities contributes to the discrete nature of the rock's mechanical parameters.

(2) Taking into account the influence of cracks and solution cavities, the equivalent drilling fluid density for collapse pressure (ρ_c) in the 5182–5187 m section of Well GS10 is determined to be 1.16 g/cm^3. This represents a 6.42% increase compared to the value obtained without considering the effect of cracks and solution cavities, which aligns well with engineering practice. In addition, cracks play a more significant role than solution cavities in affecting the wellbore stability.

(3) As the angle of the crack strike deviates further from the direction of maximum horizontal principal stress, stress and displacement at all points around the wellbore progressively increase, resulting in a linear increase in ρ_c. Similarly, as the dip angle and width of cracks increase, both stress and displacement at the intersection of cracks and the wellbore gradually intensify, leading to a linear increase in ρ_c.

(4) As the solution cavities diameter (D) increases, the stress and displacement at various points around the wellbore gradually intensify, and initially, ρ_c increases linearly.

However, when D exceeds 609 mm, there is no significant change in ρ_c. Additionally, as the number of solution cavities increases, both stress and displacement around the wellbore progressively rise, and ρ_c increases approximately in a linear trend.

(5) The disturbed stress area around the wellbore increases by 15.23% when cracks and solution cavities are connected compared to when disconnected. Furthermore, the stress around the wellbore is higher when cracks and solution cavities are connected than when disconnected; ρ_c for the situation when cracks and solution cavities are connected increases by 13.33% compared to when disconnected.

Author Contributions: Conceptualization, Q.Z.; Methodology, J.Z. and R.Z.; Project administration, R.Z.; Validation, H.A.; Formal analysis, W.L. and Q.Z.; Investigation, J.Z. and Y.G.; Resources, R.Z. and X.F.; Data curation, R.Z. and W.L.; Writing-original draft, J.Z. and Y.G.; Writing—review and editing, H.A. and Q.Z.; Funding acquisition, X.F. and Q.Z.; Supervision, X.F. All authors have read and agreed to the published version of the manuscript.

Funding: This research was funded by the National Natural Science Foundation of China (Grant Nos. 42172313 and 51774246) and the Natural Science Foundation of Sichuan Province (Grant Nos. 2024NSFSC0203, 2022NSFSC0185 and 2023NSFSC0938).

Institutional Review Board Statement: Not applicable.

Informed Consent Statement: Not applicable.

Data Availability Statement: The data presented in this study are openly available in [Mendeley Data] at [https://doi.org/10.17632/t6bmr8thbc.1].

Acknowledgments: The authors would like to express their gratitude to the editors and anonymous reviewers for their constructive comments on the draft paper. The authors are also grateful to Xingyu Mou, Qianwen Mo, Kang Ding, Xiwei Wang, and Yang Feng for their contributions to data curation, investigation, and reviewing.

Conflicts of Interest: Jingzhe Zhang, Rongrong Zhao, Hongyi An and Wenhao Li are currently employed by CNPC Southwest Oil and Gas Field Company. Yuxin Geng is currently employed by CNPC Huabei Oil Field Company. The remaining authors declare that the research was conducted in the absence of any commercial or financial relationships that could be construed as a potential conflict of interest.

References

1. Zhang, Q.; Yao, B.; Fan, X.; Li, Y.; Fantuzzi, N.; Ma, T.; Chen, Y.; Zeng, F.; Li, X.; Wang, L. A failure criterion for shale considering the anisotropy and hydration based on the shear slide failure model. *Int. J. Min. Sci. Technol.* **2023**, *33*, 447–462. [CrossRef]
2. Zhang, M.; Fan, X.; Zhang, Q.; Yang, B.; Zhao, P.; Yao, B.; He, L. Influence of multi-planes of weakness on unstable zones near wellbore wall in a fractured formation. *J. Nat. Gas Sci. Eng.* **2021**, *93*, 104026. [CrossRef]
3. Hao, Y. Effect of stratum properties on wellbore stability. *Chem. Technol. Fuels Oil* **2015**, *51*, 556–563. [CrossRef]
4. Allawi, R.H. Chemical and mechanical model to analysis wellbore stability. *Pet. Sci. Technol.* **2023**, *42*, 3062–3084. [CrossRef]
5. Younessi, A.; Rasouli, V. A fracture sliding potential index for wellbore stability analysis. *Int. J. Rock Mech. Min.* **2010**, *47*, 927–939. [CrossRef]
6. Qiu, Y.; Ma, T.; Peng, N.; Liu, Y.; Liu, J.; Ranjith, P.G. Wellbore stability analysis of inclined wells in transversely isotropic formations accounting for hydraulic-mechanical coupling. *Geoenergy Sci. Eng.* **2023**, *224*, 211615. [CrossRef]
7. Zhang, Q.; Yan, Z.; Fan, X.; Li, Z.; Zhao, P.; Shuai, J.; Jia, L.; Liu, L. Field monitoring and identification method for overflow of fractured-vuggy carbonate reservoir. *Energies* **2023**, *16*, 2399. [CrossRef]
8. Yang, X.; Liu, Z.; Zhang, H.; Xiong, Q.; Liu, H. An analysis on wellbore collapse of open hole completion in carbonate formation. *Appl. Mech. Mater.* **2012**, *220*, 220–223. [CrossRef]
9. Martyushev, D.A.; Ponomareva, I.N.; Chukhlov, A.S.; Davoodi, S.; Osovetsky, B.M.; Kazymov, K.P.; Yang, Y.F. Study of void space structure and its influence on carbonate reservoir properties: X-ray microtomography, electron microscopy, and well testing. *Mar. Pet. Geol.* **2023**, *151*, 106192. [CrossRef]
10. Li, W.Q.; Mu, L.; Zhao, L.; Li, J.; Wang, S.; Fan, Z.; Shao, D.; Li, C.; Shan, F.; Zhao, W.; et al. Pore-throat structure characteristics and its impact on the porosity and permeability relationship of Carboniferous carbonate reservoirs in eastern edge of Pre-Caspian Basin. *Pet. Explor. Dev.* **2020**, *47*, 1027–1041. [CrossRef]
11. Liu, H.; Cui, S.; Men, Y.; Han, X. Dynamic analysis of wellbore stress field and wellbore stability in carbonate reservoir production process. *Arab. J. Geosci.* **2019**, *12*, 585. [CrossRef]

12. Liu, P.; Jiang, L.; Tang, B.; Ren, K.; Huang, M.; Geng, C. Residual oil distribution pattern in a fault-solution carbonate reservoir and countermeasures to improve oil development effectiveness. *Geofluids* **2022**, *2022*, 2147200. [CrossRef]
13. Ren, Q.; Jin, Q.; Feng, J.; Li, M. Simulation of stress fields and quantitative prediction of fractures distribution in upper Ordovician biological limestone formation within Hetianhe field, Tarim Basin, NW China. *J. Pet. Sci. Eng.* **2019**, *173*, 1236–1253. [CrossRef]
14. Sun, Y.; Cheng, Y.; Hao, Y.; Meng, M.; Dai, X.; Wang, Y.; Dong, S.; Peng, G. Analysis of borehole stability for inclined wells in fractured carbonate formation. *Fresen Environ. Bull.* **2019**, *28*, 3893–3899.
15. Liu, H.; Cui, S.; Meng, Y.; Han, X.; Liu, T.; Fan, Y.; Tao, Y.; Yu, A. A new method for wellbore stability evaluation based on fractured carbonate reservoir rock breaking degree. *Arab. J. Geosci.* **2021**, *14*, 647. [CrossRef]
16. Yan, X.; You, L.; Kang, Y.; Li, X.; Xu, C.; She, J. Impact of drilling fluids on friction coefficient of brittle gas shale. *Int. J. Rock Mech. Min.* **2018**, *106*, 144–152. [CrossRef]
17. Zhang, L.; Yan, X.; Yang, X.; Zhao, X. An analytical model of coal wellbore stability based on block limit equilibrium considering irregular distribution of cleats. *Int. J. Coal Geol.* **2015**, *152*, 147–158. [CrossRef]
18. Liu, H.; Liu, T.; Meng, Y.; Han, X.; Cui, S.; Yu, A. Experimental study and evaluation for borehole stability of fractured limestone formation. *J. Pet. Sci. Eng.* **2019**, *180*, 130–137. [CrossRef]
19. Karatela, E.; Taheri, A.; Xu, C.; Stevenson, G. Study on effect of in-situ stress ratio and discontinuities orientation on bore-hole stability in heavily fractured rocks using discrete element method. *J. Pet. Sci. Eng.* **2016**, *139*, 94–103. [CrossRef]
20. Cui, S.; Liu, H.; Meng, Y.; Zhang, Y.; Tao, Y.; Zhang, X. Study on fracture occurrence characteristics and wellbore stability of limestone formation. *J. Pet. Sci. Eng.* **2021**, *204*, 108783. [CrossRef]
21. Liu, J.; Ma, T.; Fu, J.; Peng, N.; Qiu, Y.; Liu, Y.; Gao, J. Fully coupled two-phase hydro-mechanical model for wellbore stability analysis in tight gas formations considering the variation of rock mechanical parameters. *Gas Sci. Eng.* **2023**, *115*, 205023. [CrossRef]
22. Zhang, Q.; Ran, J.; Fan, X.; Yang, B.; Zhao, P.; Chen, Y.; Huang, P.; Zhang, M.; He, L. Mechanical properties of basalt, tuff and breccia in the Permian system of Sichuan Basin after water absorption–implications for wellbore stability analysis. *Acta Geotech.* **2023**, *18*, 2059–2080. [CrossRef]
23. Andabily, Z.; Ebrahim, M. Management of Wellbore Stability by Controlling Physical and Chemical Properties of Drilling Muds. Ph.D. Dissertation, University of New South Wales, Kensington, Australia, 1997.
24. Akhtarmanesh, S.; Shahrabi, M.J.A.; Atashnezhad, A. Improvement of wellbore stability in shale using nanoparticles. *J. Pet. Sci. Eng.* **2013**, *112*, 290–295. [CrossRef]
25. Li, J.; Ma, Y.; Liu, Y.; Liu, Q. Numerical simulation for the effect of mud cake time-dependent properties on wellbore stability. *Rock Mech. Rock Eng.* **2024**, *57*, 581–596. [CrossRef]
26. Gholilou, A.; Far, P.B.; Vialle, S.; Madadi, M. Determination of safe mud window considering time-dependent variations of temperature and pore pressure: Analytical and numerical approaches. *J. Rock Mech. Geotech. Eng.* **2017**, *9*, 900–911. [CrossRef]
27. Gentzis, T.; Deisman, N.; Chalaturnyk, R.J. A method to predict geomechanical properties and model well stability in horizontal boreholes. *Int. J. Coal Geol.* **2009**, *78*, 149–160. [CrossRef]
28. Al-Ajmi, A.M.; Al-Harthy, M.H. Probabilistic wellbore collapse analysis. *J. Pet. Sci. Eng.* **2010**, *74*, 171–177. [CrossRef]
29. Al-Ajmi, A. Wellbore Stability Analysis Based on a New True-Triaxial Failure Criterion. Ph.D. Dissertation, KTH Royal Institute of Technology, Stockholm, Sweden, 2006.
30. Karatela, E.; Taheri, A. Three-dimensional hydro-mechanical model of borehole in fractured rock mass using discrete element method. *J. Nat. Gas Sci. Eng.* **2018**, *53*, 263–275. [CrossRef]
31. Guo, J.; She, C.; Liu, H.; Cui, S. Rock mechanical properties and wellbore stability of fractured dolomite formations. *ACS Omega* **2023**, *8*, 35152–35166. [CrossRef]
32. Liu, H.; Cui, S.; Meng, Y.; Han, Z.; Yang, M. Study on rock mechanical properties and wellbore stability of fractured carbonate formation based on fractal geometry. *ACS Omega* **2022**, *7*, 43022–43035. [CrossRef]
33. Gomar, M.; Goodarznia, I.; Shadizadeh, S.R. Transient thermo-poroelastic finite element analysis of borehole breakouts. *Int. J. Rock Mech. Min.* **2014**, *71*, 418–428. [CrossRef]
34. Zhao, W.; Wang, Z.; Jiang, H.; Fu, X.; Xie, W.; Xu, A.; Shen, A.; Shi, S.; Huang, S.; Jiang, Q. Exploration status of the deep Sinian strata in the Sichuan Basin: Formation conditions of old giant carbonate oil/gas fields. *Nat. Gas Ind. B* **2020**, *7*, 462–472. [CrossRef]
35. Zhou, L. Optimization and field application of reservoir stimulation of the Sinian Deng 4 member in the Sichuan Basin. *Nat. Gas Ind. B* **2022**, *9*, 277–288. [CrossRef]
36. Wang, L.; He, Y.; Peng, X.; Deng, H.; Liu, Y.; Xu, W. Pore structure characteristics of an ultradeep carbonate gas reservoir and their effects on gas storage and percolation capacities in the Deng IV member, Gaoshiti-Moxi Area, Sichuan Basin, SW China. *Mar. Pet. Geol.* **2020**, *111*, 44–65. [CrossRef]
37. DZ/T 0276.20–2015; Ministry of Land and Resources of the People's Republic of China Regulation for the Physical and Mechanical Propertier of Rock–Part 20: Test for Determining the Strength of Rock in Triaxial Compression. Industry Standard of Geology and Mineral Resources of the People's Republic of China: Beijing, China, 2015.
38. Zhang, Q.; Fan, X.; Chen, P.; Ma, T.; Zeng, F. Geomechanical behaviors of shale after water absorption considering the combined effect of anisotropy and hydration. *Eng. Geol.* **2020**, *269*, 105547. [CrossRef]
39. Fan, X.; Gong, M.; Zhang, Q.; Wang, J.; Bai, L.; Chen, Y. Prediction of the horizontal stress of the tight sandstone formation in eastern Sulige of China. *J. Pet. Sci. Eng.* **2014**, *113*, 72–80. [CrossRef]

40. Fan, X.; Zhang, Q.; Duan, M.; Yang, Y.; Duan, Q.; Chen, X.; Lv, D. Mathematical model for calculating the horizontal principal stress of a faulted monoclinal structure in southwest Qaidam Basin, China. *AAPG Bull.* **2019**, *103*, 2697–2729. [CrossRef]

Disclaimer/Publisher's Note: The statements, opinions and data contained in all publications are solely those of the individual author(s) and contributor(s) and not of MDPI and/or the editor(s). MDPI and/or the editor(s) disclaim responsibility for any injury to people or property resulting from any ideas, methods, instructions or products referred to in the content.

Article

Research on Key Parameters of Wellbore Stability for Horizontal Drilling in Offshore Hydrate Reservoirs

Zhengfeng Shan [1,2], Xiansi Wang [1,2], Zeqin Li [1,3], Zhenggang Gong [1,3], Nan Ma [1,3], Jianbo Zhang [1,3], Zhiyuan Wang [1,3,*] and Baojiang Sun [1,3]

1. School of Petroleum Engineering, China University of Petroleum (East China), Qingdao 266580, China; zhangjianbo2@163.com (J.Z.)
2. Drilling Division, CNPC Offshore Engineering Co., Ltd., Tianjin 300280, China
3. National Engineering Research Center of Oil & Gas Drilling and Completion Technology, Qingdao 266580, China
* Correspondence: wangzy1209@126.com

Abstract: The South China Sea has abundant reserves of natural gas hydrates, and if developed effectively, it can greatly alleviate the pressure on the energy supply in China. But the hydrate reservoirs in the sea area are loose, shallow, porous, and have poor mechanical properties. During the drilling process, the invasion of drilling fluid into this kind of reservoir is likely to induce mass decomposition of gas hydrate and, in turn, a significant reduction in mechanical strength around the wellbore as well as instability of the wellbore. In this study, in light of the engineering background of exploratory wells at the South China Sea, a temperature and pressure field model in a gas hydrate reservoir at sea during open circuit drilling was established, and then, based on this model, a comprehensive model for the stability analysis of the well drilled in the hydrate reservoir at sea was constructed, both of them with errors of less than 10%. With these two models, the effects of different drilling parameters on wellbore stability were investigated. The gas and liquid produced by the decomposition of hydrates in the formation will increase the pore pressure in the formation, thereby reducing the effective stress in the formation. The closer the formation is to the wellbore, the more thorough the decomposition of hydrates in the formation and the greater the effective plastic strain. Keeping all other conditions constant, the increase in drilling fluid invasion pressure and temperature, as well as reservoir permeability, will lead to a decrease in the mechanical strength of the formation around the wellbore and an expansion of the wellbore yield zone. The results can provide a theoretical reference for the stability analysis at sea.

Citation: Shan, Z.; Wang, X.; Li, Z.; Gong, Z.; Ma, N.; Zhang, J.; Wang, Z.; Sun, B. Research on Key Parameters of Wellbore Stability for Horizontal Drilling in Offshore Hydrate Reservoirs. *Appl. Sci.* **2024**, *14*, 10922. https://doi.org/10.3390/app142310922

Keywords: hydrate; horizontal well drilling; formation mechanical strength; wellbore stability; prediction model

Academic Editors: Dino Musmarra and Andrea L. Rizzo

Received: 5 September 2024
Revised: 20 October 2024
Accepted: 14 November 2024
Published: 25 November 2024

Copyright: © 2024 by the authors. Licensee MDPI, Basel, Switzerland. This article is an open access article distributed under the terms and conditions of the Creative Commons Attribution (CC BY) license (https://creativecommons.org/licenses/by/4.0/).

1. Introduction

Natural gas hydrate discovered in nature is usually white and decomposes quickly when heated [1–5]. During the drilling process, the drilling fluid circulating in the wellbore would exchange heat with the interface of the reservoir near the well and filter through the well wall into the formation [6–8]. Meanwhile, the phase transformation of hydrate would greatly weaken the original mechanical strength of the formation [9–11]. Moreover, hydrate reservoirs are mostly argillaceous siltstone with weak mechanical properties. These two factors coming together will make the risk of wellbore instability rise sharply in the process of hydrate reservoir drilling [12–16]. During the drilling process of horizontal wells in marine hydrate reservoirs, there is dynamic heat and mass transfer between the wellbore and the reservoir, as well as multiphase flow in the reservoir under fluid thermal solid coupling, which is a complex process of multi-physical field coupling. Therefore, revealing the heat and mass transfer mechanisms between the wellbore and reservoir, the degradation characteristics of formation mechanics, the coupling laws of multiple physical fields in the near wellbore zone, and conducting stability analysis of horizontal well drilling in hydrate layers are of great significance for the safe and efficient exploitation of hydrates.

During the drilling process of the horizontal well in the hydrate reservoir at sea, the drilling fluid invading the reservoir could make the solid hydrate particles decompose into gas and liquid in two phases. Consequently, the mechanical properties of the hydrate reservoir may change significantly, and the stress distribution state of the reservoir may change accordingly. So, an accurate assessment of the variations of hydrate reservoir mechanical properties during the drilling process is an important prerequisite for analyzing the wellbore stability. Yun et al. [10] (2007) investigated the effects of reservoir rock type, hydrate saturation, and effective confining pressure on the mechanical strength of hydrate reservoir by rock triaxial tests. Hyodo et al. [11] (2013) carried out stiffness experiments on samples of water and gas-saturated hydrate deposits and found that the stiffness and strength of hydrate deposits saturated with gas were significantly higher than those saturated with water. Through experiments, Lei et al. [17] reached the findings that in the formation of argillaceous siltstone, hydrate particles, when formed, would occupy the original space of the rock particles and be distributed in uneven micro-fractures in a belt or lenticular zone. This finding is very similar to the results of laboratory X-CT scanning and field logging. But in the sandy deposits with hydrate, things are completely different, either in the form of hydrate filling the pores or in the way of rock cementation. Therefore, argillaceous siltstone and sandy sediments containing gas hydrate differ widely in mechanical properties, porosity, and permeability. Hyodo et al. [18] (2014) investigated the shear strength and deformation characteristics of hydrate deposits during hydrate decomposition by depressurization and heating. Researchers at home and abroad have carried out a lot of studies on the mechanical properties of rocks containing hydrate, but few studies have covered the variation patterns of mechanical characteristics of the argillaceous siltstone reservoirs with weak cementation in the South China Sea under the disturbance by wellbore temperature and pressure in the drilling process.

In the process of drilling and mining of hydrate reservoirs, once the wellbore loses stability, many difficult problems could arise. As a result, researchers have conducted numerous studies on the stability of wellbores in hydrate formation. After simplifying the multi-phase flow caused by the hydrate, Freij-Ayoub et al. [19] established a wellbore stability evaluation model by coupling temperature, pressure, and mechanical fields. By extending the Biot consolidation model, Jin et al. [20] simulated the mechanical response characteristics of hydrate reservoirs in the process of horizontal well drilling and production, and the results provided a theoretical reference for the study of the stability of horizontal wellbores in hydrate reservoirs during drilling and production. Based on the theory of overloading and secondary loading surface, Sun et al. [21] built a mechanical constitutive model of a hydrate reservoir and simulated the formation settlement characteristics during the depressurization extraction of hydrate; moreover, they simulated the mechanical behavior of an argillaceous siltstone hydrate reservoir in the first year of depressurization extraction. Li et al. [22] considered the evolution of reservoir mechanical strength, wellbore temperature, and pressure parameters in depth; established a wellbore stability analysis model; and analyzed the stability evolution law of the wellbore area during shallow seabed drilling. Wang et al. [23] analyzed the influence of different factors on the stability of a horizontal wellbore by establishing a multi-physics coupling model between a drilling wellbore and marine natural gas hydrate reservoir. Wang et al. [24] derived analytical solutions of temperature, seepage, and mechanical parameters during hydrate reservoir drilling by coupling multiple fields. Most existent studies took vertical wells as research objects, but few studies covered the stability of horizontal wells during drilling, failing to reveal the strong coupling effect between the wellbore and hydrate reservoir in depth.

In summary, current research on the stability of hydrate wellbores is mostly focused on vertical wells, while there is relatively little research on the stability of wellbores during the drilling process of hydrate horizontal wells. Moreover, research on hydrate reservoir drilling still regards the wellbore and reservoir as two relatively independent physical models, and has not deeply revealed the strong coupling characteristics between the wellbore and hydrate reservoir. Therefore, it is necessary to establish a model coupling heat, fluid, solid, and chemical fields of the near-well zone, which takes the impact of drilling fluid loss on the reservoir into consideration in light of the drilling conditions, and to realize the analysis of wellbore stability in the drilling process by numerical simulation software-COMSOL. In this study, a model of horizontal well in the hydrate reservoir at sea during open-circuit drilling was established, and, on this basis, a comprehensive model for stability analysis was constructed, both with errors of less than 10%. This study can provide theoretical references for stability analysis of a horizontal wellbore and control of safety pressure in the wellbore during drilling at sea areas.

2. The Model of the Hydrate Reservoir at Sea Around the Horizontal Well During Drilling

Generally, shallow and hydrate reservoirs are mostly argillaceous siltstone, loose and porous, small in thickness, and average in mechanical properties. The whole horizontal well drilling process is complex, and needs to consider the influence of multi-physical fields to establish a coupled model around the wellbore.

2.1. Fluid–Solid Coupling Seepage Equation

During the drilling of a hydrate horizontal well, the drilling fluid may invade into the hydrate stratum and cause decomposition of hydrate. This is a complex coupling process of fluid and solid affected by multiple factors. The underlying assumptions are as follows:

(1) The gas in the pores and pores in the formation were compressible, and the formation water was slightly compressible. The deformations of rock skeleton were all small in displacement.
(2) The heat exchange in the formation only affected the decomposition of hydrate and had no effect on the properties of the rock skeleton and formation fluid.
(3) The rock skeleton and reservoir hydrate did not change in density in the whole process.
(4) The generalized Darcy's law was applicable to the flow of all fluids in the formation.

2.1.1. Continuity Equation

Rock skeleton:

$$\nabla \cdot (1-\phi)v_s + \frac{\partial(1-\phi)}{\partial t} = 0 \tag{1}$$

Hydrate:

$$\frac{\partial(\phi S_h \rho_h)}{\partial t} + \nabla \cdot (\phi S_h \rho_h v_s) = -m_h \tag{2}$$

Water phase:

$$\frac{\partial(\phi S_w \rho_w)}{\partial t} + \nabla \cdot (\phi S_w \rho_w v_w) = m_w \tag{3}$$

Gas phase:

$$\frac{\partial(\phi S_g \rho_g)}{\partial t} + \nabla \cdot (\phi S_g \rho_g v_g) = m_g \tag{4}$$

where φ denotes the formation porosity, dimensionless; v_g, v_w, and v_s denote the actual velocities of gas, water, and rock skeleton, respectively, m/s; S_g, S_w, and S_h denote the pore saturations of gas, water, and hydrate, respectively; ρ_g, ρ_w, and ρ_h denote the densities of gas, water, and hydrate, respectively, kg/m^3; m_h denotes the decomposition rate of hydrate, kg/s; m_g denotes the formation rate of natural gas per unit volume of stratum, kg/s; and m_w denotes the formation rate of water per unit volume of stratum, kg·s^{-1}.

2.1.2. Darcy's Law of Two-Phase Seepage

Water phase:
$$\phi S_w v_{rw} = -\frac{kk_{rw}}{\mu_w}(\nabla P_w - \rho_w g) \quad (5)$$

Gas phase [25]:
$$\phi S_g v_{rg} = -\frac{kk_{rg}}{\mu_g}(\nabla P_g - \rho_g g) \quad (6)$$

v_{rw} and v_{rg} denote the relative seepage velocity of water and gas to the rock skeleton, m·s^{-1}; μ_g and μ_w represent the gas phase viscosity and dynamic viscosity of the water phase, mPa·s; P_g and P_w denote the seepage pressure of the gas phase and water phase, respectively, MPa; k_{rg} and k_{rw} denote the relative permeability of the gas phase and water phase, respectively, m^2; and k denotes the absolute permeability of the hydrate reservoir, m^2.

From the continuity equation and two-phase seepage equation, the fluid–solid coupling control equation of the hydrate reservoir was obtained:

$$\frac{\partial(\phi \rho_g S_g)}{\partial t} - \nabla \cdot \left(-\frac{k_{rg} k \rho_g}{\mu_g}(\nabla P_g - \rho_g g)\right) + \nabla \cdot (\phi \rho_g S_g v_s) = m_g \quad (7)$$

$$\frac{\partial(\phi \rho_w S_w)}{\partial t} - \nabla \cdot \left(-\frac{k_{rw} k \rho_w}{\mu_w}(\nabla P_w - \rho_w g)\right) + \nabla \cdot (\phi \rho_w S_w v_s) = m_w \quad (8)$$

2.2. Energy Conservation Equation

Without considering the Joule Thomson effect, the energy conservation equation is

$$((1-\phi)\rho_s C_s + \phi S_h \rho_h C_h + \phi S_w \rho_w C_w + \phi S_g \rho_g C_g)\frac{\partial T}{\partial t} = \nabla \cdot (K_e \nabla T) - \nabla \cdot [(\phi S_g C_g \rho_g v_g + \phi S_w C_w \rho_w v_w)T] - m_h \Delta H_D + Q_{in} \quad (9)$$

where ρ_s denotes the density of the rock skeleton, kg·m^{-3}; H_s denotes the enthalpy of the rock skeleton, J; H_h denotes the enthalpy of the hydrate, J; H_w denotes the enthalpy of the water, J; H_g denotes the enthalpy of the methane gas, J; ρ_h denotes the hydrate density, kg·m^{-3}; K_e denotes the effective thermal conductivity of the hydrate reservoir, W·m^{-1}·K^{-1}; C_w denotes the specific heat of the water, J·kg^{-1}·K^{-1}; C_h denotes the specific heat of the hydrate, J·kg^{-1}·K^{-1}; C_s denotes the specific heat of the rock skeleton, J·kg^{-1}·K^{-1}; and ΔH_D denotes the reaction heat of the hydrate decomposition, J·mol^{-1}, calculated by the model proposed by Selim (1989) [26] in this work:

$$\Delta H_D = \begin{cases} 215.59 \times 10^3 - 394.945, & 248K < T < 273K \\ 446.12 \times 10^3 - 132.638, & 273K \leq T < 298K \end{cases} \quad (10)$$

2.3. Equation of Solid Field of the Reservoir

2.3.1. Balance Equation of the Rock Skeleton

The balance equation of the rock skeleton solid field based on the effective stress is [27]

$$\frac{\partial \sigma_{xx}}{\partial x} + \frac{\partial \tau_{xy}}{\partial y} + \frac{\partial \tau_{zx}}{\partial z} - \alpha \frac{\partial P}{\partial x} + f_x = 0 \quad (11)$$

$$\frac{\partial \sigma_{yy}}{\partial y} + \frac{\partial \tau_{xy}}{\partial x} + \frac{\partial \tau_{yz}}{\partial z} - \alpha \frac{\partial P}{\partial y} + f_y = 0 \quad (12)$$

$$\frac{\partial \sigma_{zz}}{\partial z} + \frac{\partial \tau_{xz}}{\partial x} + \frac{\partial \tau_{yz}}{\partial y} - \alpha \frac{\partial P}{\partial z} + f_z = 0 \quad (13)$$

$$P = P_g S_g + P_w S_w \quad (14)$$

where σ_{ij} denotes effective stress component of the rock, MPa; α is the Biot coefficient, 1 in this study; P denotes the equivalent pore pressure, MPa; and f_i denotes the component of volume force, MPa.

The equation in tensor form is as follows:

$$\sigma_{ij,j} + f_i - (\alpha P \delta_{ij})_{,j} = 0 \tag{15}$$

δ_{ij}, Kronecker symbol, $\delta_{ij} = [1\ 1\ 1\ 0\ 0\ 0]^T$.

2.3.2. Geometric Equation of Reservoir Skeleton

For the hydrate reservoir, since the deformations of the rock skeleton are all small, it is expressed by the following equation [3,4]:

$$\begin{Bmatrix} \varepsilon_x \\ \varepsilon_y \\ \varepsilon_z \\ \gamma_{xy} \\ \gamma_{yz} \\ \gamma_{zx} \end{Bmatrix} = \begin{bmatrix} \frac{\partial u}{\partial x} \\ \frac{\partial v}{\partial y} \\ \frac{\partial w}{\partial z} \\ \frac{\partial u}{\partial y} + \frac{\partial v}{\partial x} \\ \frac{\partial v}{\partial z} + \frac{\partial w}{\partial y} \\ \frac{\partial w}{\partial x} + \frac{\partial u}{\partial z} \end{bmatrix} = \begin{bmatrix} \frac{\partial}{\partial x} & 0 & 0 \\ 0 & \frac{\partial}{\partial y} & 0 \\ 0 & 0 & \frac{\partial}{\partial z} \\ \frac{\partial}{\partial y} & \frac{\partial}{\partial x} & 0 \\ 0 & \frac{\partial}{\partial z} & \frac{\partial}{\partial y} \\ \frac{\partial}{\partial z} & 0 & \frac{\partial}{\partial x} \end{bmatrix} \begin{Bmatrix} u \\ v \\ w \end{Bmatrix} \tag{16}$$

It is expressed in tensor form as follows:

$$\varepsilon_{ij} = \frac{1}{2}(u_{i,j} + u_{j,i}) \tag{17}$$

where ε_{ij} represents the strain tensor and u represents the displacement component, m.

2.3.3. Constitutive Equation of Reservoir Skeleton

In this study, the constitutive relation of the rock skeleton is expressed by the elastic–plastic constitutive equation [28]:

$$d\sigma_{ij} = D_{ijkl} d\varepsilon_{kl} \tag{18}$$

where σ_{ij} denotes the stress increment; D_{ijkl} denotes the elastic–plastic matrix tensor; and ε_{ij} denotes the strain tensor.

The Mohr Coulomb criterion is commonly used in engineering to determine the yield of rocks. The Mohr Coulomb criterion holds that the shear failure of rocks on a surface is the result of the combined action of shear stress and normal stress on that surface, rather than solely the influence of shear stress. When a rock experiences the most unfavorable combination of shear stress and normal stress, it fails. The schematic diagram of rock failure is shown in Figure 1.

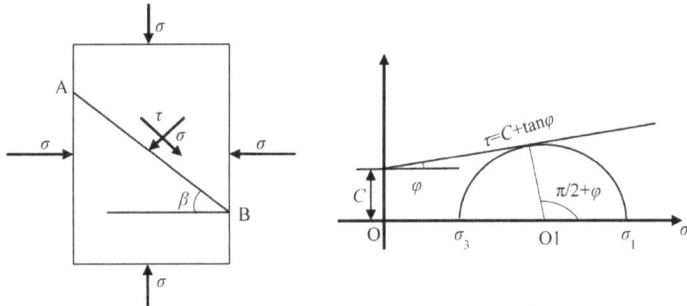

Figure 1. Rock failure diagram of Mohr Coulomb criterion.

2.4. Kinetic Equations for Hydrate Decomposition

The chemical reaction equation of gas hydrate decomposition is [29]

$$CH_4 nH_2O \rightarrow CH_4 + nH_2O \tag{19}$$

In which n denotes hydration number, dimensionless, and is taken as 5.75.

The gas production rate of the hydrate decomposition is calculated by the equations [30]

$$m_g = k_d M_g A_s (P_e - P) \tag{20}$$

$$k_d = k_0 e^{-\frac{\Delta E_a}{RT}} \tag{21}$$

$$A_s = \phi S_h \left(\frac{\phi_{wg}^3}{2K}\right)^{1/2} \tag{22}$$

From the rate of the gas production and chemical decomposition equation of the hydrate, the rate of hydrate decomposition and the rate of water production were derived:

$$m_h = m_g \frac{nM_w + M_g}{M_g} \tag{23}$$

$$m_w = m_g \frac{nM_w}{M_g} \tag{24}$$

where m_h denotes the rate of the hydrate decomposition, kg·s^{-1}; k_d denotes the kinetic reaction rate of the hydrate, mol·m^{-2}·Pa^{-1}·s^{-1}; A_s denotes the specific surface area of the hydrate particles, m^{-1}; ϕ denotes the absolute porosity of the hydrate reservoir, dimensionless; ϕ_{wg} denotes the effective porosity of the hydrate reservoir, dimensionless; P_e denotes the phase equilibrium pressure of the hydrate, MPa.

2.5. Model Definite Solution Conditions

The boundary conditions for this model are as follows.
Boundary flow continuity [31]:

$$-\vec{n} \frac{\rho_g k k_{rg}}{\mu_g} (\nabla P_g + \rho_g g) - q_g = 0 \tag{25}$$

$$-\vec{n} \frac{\rho_w k_{rw} k}{\mu_w} (\nabla P_w + \rho_w g) - q_w = 0 \tag{26}$$

where n denotes the unit normal vector at the flow boundary and q_g and q_w denote the flow rate of gas and liquid per unit area at the flow boundary, respectively, kg/m^3/s.

Constant pressure boundary conditions:

$$P_B = f_P(x, y, z, t) \tag{27}$$

where $f_p(x, y, z, t)$ is the pressure function at time t at a point (x, y, z) on boundary B, MPa.

The initial conditions of the seepage field are as follows:

$$P(x, y, z)|_{t=0} = P_1(x, y, z) \tag{28}$$

$$S|_{t=0} = S_1(x, y, z) \tag{29}$$

$P_1(x, y, z)$ is the pressure function, MPa and $S_1(x, y, z)$ is the saturation function, dimensionless.

The initial conditions of the temperature field are as follows:

$$T(x, y, z)|_{t=0} = T_1(x, y, z) \tag{30}$$

$T_1(x, y, z)$ is the temperature field function, K.

3. Numerical Solution Method and Model Validation

3.1. Mesh Division

In the process of solving the model, it is necessary to calculate the changes in fluid state at various spatial nodes inside the wellbore, so the model needs to be meshed. Divide the grid into quadrilateral grids. In order to improve the convergence, computational accuracy, and efficiency of the model, a mapping mesh partitioning model is adopted, which includes 6 vertex elements, 160 boundary elements, 1050 domain elements, and a minimum element mass of 0.9833. In addition, the grid was appropriately densified using a distribution function in the near wellbore area, as shown in Figure 2.

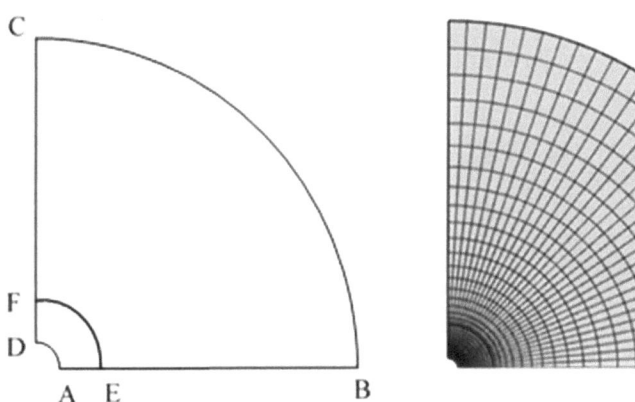

Figure 2. Grid diagram of horizontal well model.

Length of any grid segment:

$$\Delta z_j = z_j - z_{j+1} = \Delta z \quad (31)$$

Total number of grids:

$$N = \text{int}\left(\sum_j z/\Delta z\right) \quad (32)$$

Any node location:

$$z_{j+1} = z_j + \Delta z \quad (33)$$

Among them, j = 1, 2, 3, 4...N.

3.2. The Numerical Solution Method

Horizontal well drilling in a hydrate reservoir at the sea is a complex coupling process. Moreover, considering the coupling between the wellbore and reservoir at the same time makes the problem more complicated. Based on the established coupling model around the horizontal well at the sea area, COMSOL software (the software version is 5.5) was used in this work to realize the simulation of stability of the horizontal wellbore in a hydrate reservoir by coupling the wellbore and reservoir conditions.

In order to better study the influence of fluid–solid coupling on wellbore stability, the following assumptions are made in this study:

(1) The pores and gases in the formation are compressible, and the formation water is considered as a slightly compressible fluid. The deformation generated by the rock skeleton is small displacement deformation;

(2) The heat exchange occurring in the formation only affects the decomposition of hydrates and has no effect on the rock skeleton and formation fluid properties;

(3) The density of rock skeleton and reservoir hydrates remains unchanged throughout the entire process;
(4) Hydrate reservoirs are uniformly isotropic;
(5) The generalized Darcy's law is applicable to the flow of all fluids in geological formations;
(6) The relative velocity of hydrates to the rock skeleton is zero, and the relative velocity between the gas–liquid phases is zero.

3.3. Validation of the Model

In light of the well drilling process in the shallow gas hydrate reservoir in deep water, the stability of the horizontal well was simulated, with the horizontal section of the well assumed to be 400 m long and all the physical parameters of the simulated reservoir assumed to be isotropic. The basic parameters used in the simulations are shown in Table 1. The two-dimensional model of the hydrate reservoir was established as a quarter circle with a radius of 4 m, where the horizontal well diameter is 0.1143 m.

Table 1. Basic drilling parameters.

Parameter	Unit	Value	Parameter	Unit	Value
Depth	m	1420–1464	Horizontal section length	m	400
Absolute porosity	-	0.4	Poisson's ratio	-	0.35
Initial hydrate saturation	-	0.33	Absolute permeability	m^2	4.935×10^{-14}
Sea level temperature	°C	25	Pore pressure	MPa	14.508
Internal friction angle	°	12	Initial temperature of reservoir	°C	13.373
Vertical depth	m	1435.1	Minimum horizontal principal stress	MPa	15.47
Sea water depth	m	1235	Ultimate horizontal principal stress	MPa	15.47
Elastic modulus	MPa	$S_h = 0:32$	Vertical stress	MPa	16.36
Cohesion	MPa	$S_h = 0:0.1$	Temperature of injected drilling fluid	°C	25

The model was divided by mapping grids, which included 6 vertex units, 160 boundary units, and 1050 domain units and had a minimum unit mass of 0.9833. In addition, the grids near the wellbore were properly refined by the distribution function.

In the simulations by COMSOL Multiphysics, the boundary conditions were as follows:

(1) AD was the boundary of the drilling fluid invading the hydrate reservoir, the temperature and pressure of the drilling fluid invading the reservoir were set, the volume fraction of the liquid phase was set at 1, and the boundary load condition of the solid field was set as the drilling fluid invading pressure.
(2) AB and CD were both set as symmetrical boundaries, the solid field was set as roller supporting, with the normal displacement of 0.
(3) BC was the outer boundary, which was set as the outflow boundary condition in the porous medium phase transfer and the porous medium heat transfer modules, as the boundary with the constant pressure of the initial pressure of the hydrate reservoir in the Darcy law module, and as the boundary with the load condition of the initial pressure in the solid field.

The proposed model was verified by using the drilling data of the SH2 vertical well in the hydrate reservoir of the Shenhu Sea area. In the layers at 200.1 m and 217.5 m below the mud line, the hydrate saturations calculated by the model are in good agreement with those from the field logging data, so these two reservoir layers were taken as the main research objects. The basic data are shown in Table 2.

Table 2. Basic data of drilling fluid invading the hydrate reservoir.

Depth of Seabed/m	Formation Pressure/MPa	Formation Temperature/°C	Hydrate Saturation
200.1	14.508	13.357	0.33
217.5	14.689	14.178	0.27
Drilling fluid pressure/MPa	Drilling fluid temperature/°C	Horizontal stress/MPa	Vertical stress/MPa
14.771	14.35	15.47	16.36
14.975	15.2	15.74	16.70

The effects of drilling fluid invasion in these two layers were simulated by COMSOL software. The basic data are shown in Table 3. The simulation results showed that the wellbore at these two layers expanded in some degrees, and the yield radii of the wellbore at these two layers after drilling fluid invasion were 0.00998 m and 0.01191 m, respectively, as shown in Figure 3. By comparing with the measured caliper data of the wellbore from logging, the errors at these two points were 9.78% and 5.99%, respectively, which verified the accuracy in simulating the horizontal well drilled in the hydrate reservoir at sea. Figure 4 shows the details.

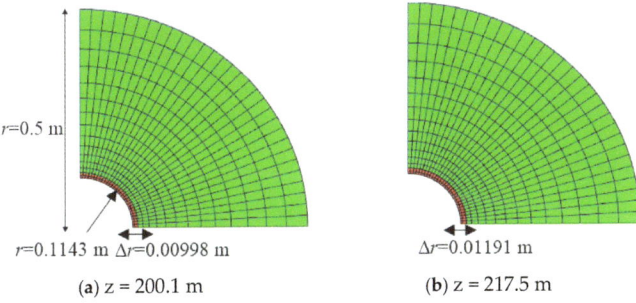

Figure 3. Distributions of wellbore yield area.

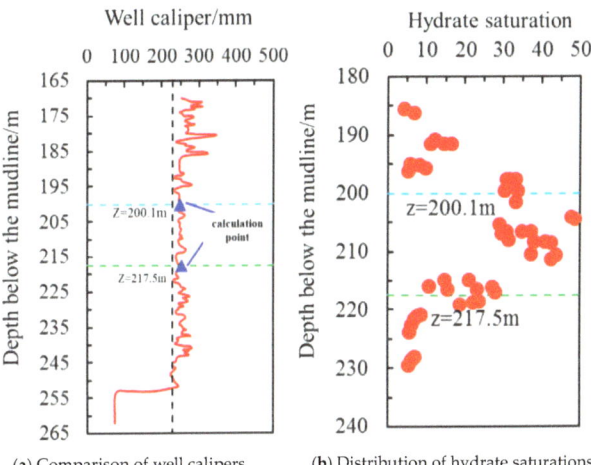

Figure 4. Comparison of well caliper data from logging and simulation.

Table 3. Comparison of wellbore enlargement values from logging and from simulation.

Serial No.	Depth of Seabed/m	Measured Well Caliper Enlargement/mm	Simulated Well Caliper Enlargement/mm	Error/%
1	200.1	18.181	19.96	9.78
2	217.5	22.473	23.82	5.99

4. Results and Discussion

This section presents the variation patterns of pressure and temperature, hydrate saturation, porosity, permeability, volume strain, and effective plastic strain of the formation at the cross-section of 200 m of the horizontal section from simulation in the drilling process.

4.1. Sensitivity Analysis of Basic Parameters

4.1.1. Distribution Pattern of Formation Pressure

The drilling fluid infiltrated into the formation constantly and the pore pressure in the formation increased gradually. When the two reached a balance, the drilling fluid stopped infiltrating. Figure 5 shows the distribution nephograms of the reservoir pressure and near-wellbore zone pressure at different moments; Figure 6 shows the distribution of the formation isobaric lines and the distribution of formation fluid streamlines 20 h after the drilling invasion; and Figure 7 shows the distribution patterns of the pressure around the well at different moments. As can be seen from the graph, with the increase in time, the wellbore pressure continues to propagate further into the formation, the pressure front continues to advance, and the pore pressure of the formation continues to increase, and in the early stage of drilling fluid invasion, due to the large pressure difference, the speed of pressure wave propagation is fast. After entering the middle and late stages of invasion, the pressure difference gradually decreases, the seepage effect weakens, and the speed of pressure propagation slows down.

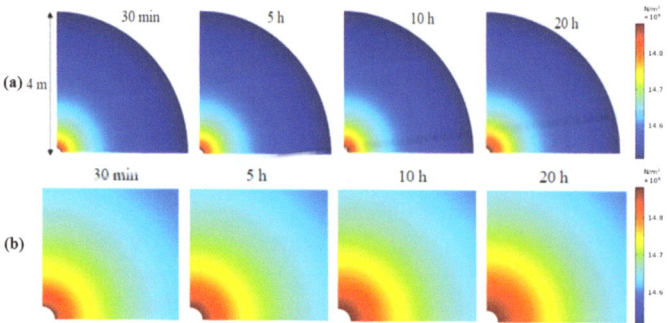

Figure 5. Variations of formation pressure and pressure in near wellbore zone over time. (**a**) reservoir pressure (**b**) near-wellbore zone pressure.

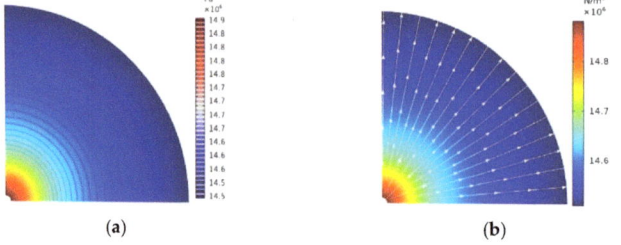

Figure 6. Formation isobaric lines and fluid streamlines 20 h into the drilling. (**a**) Distribution of formation isobaric lines, (**b**) Distribution of formation fluid streamlines.

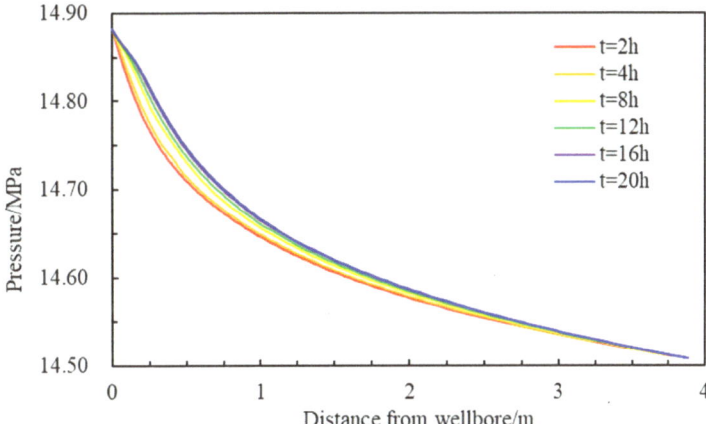

Figure 7. Curves of pressure around the wellbore.

4.1.2. Distribution Pattern of Formation Temperatures

High-temperature drilling fluids intrude into the formation under differential pressure, and the in situ temperature gradually increases after heat exchange with the drilling fluids. Once the temperature and pressure conditions of the hydrate phase equilibrium state are broken, the solid hydrate begins to decompose. Since heat transfer occurs throughout the drilling fluid intrusion into the hydrate reservoir, it is crucial to find out the pattern of reservoir temperature change after drilling fluid intrusion. Figure 8 shows the distribution cloud of the reservoir temperature and near-well zone temperature at different times. Figure 9 shows the distribution pattern of the temperature in the near-well zone at different times. From the graph, it can be seen that, with the continuous invasion of drilling fluid, the temperature inside the formation gradually increases. However, the propagation of temperature has a certain lag compared to pressure. The rate of temperature increase in the near wellbore zone is relatively slow, and the range of temperature changes in the formation is small. After 30 min, the pressure has already propagated to a range of 4 m around the wellbore. The temperature disturbance caused by the invasion of drilling fluid within 20 h only has a significant impact on the range of about 0.5 m around the wellbore. Therefore, the risk of wellbore instability is highest within a range of 0.5 m around the wellbore.

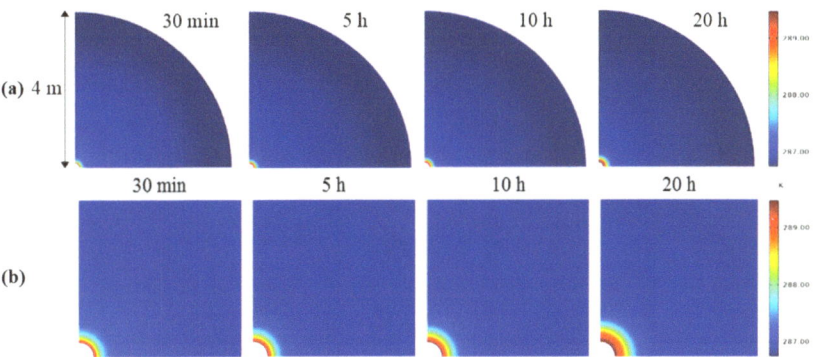

Figure 8. Variations of formation temperature and near-wellbore zone temperature over time. (**a**) reservoir temperature (**b**) near-well zone temperature.

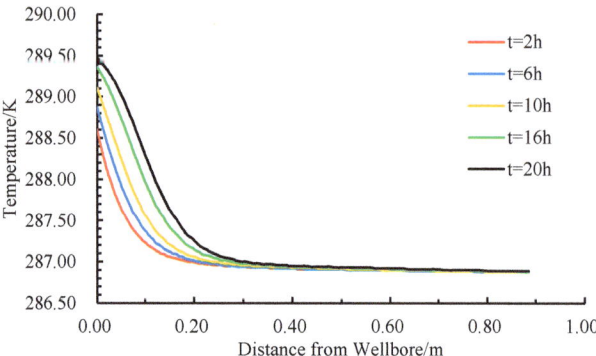

Figure 9. Curves of temperatures of the near-wellbore zone.

4.1.3. Distribution Pattern of Hydrate Saturation

As can be seen in Figures 10–12, with the extension of the invasion time of the drilling fluid, the decomposition of the hydrate in the formation constantly increased in degree and range, and the hydrate decomposition front constantly advanced. The decomposition front divided the distribution of the hydrate saturation into two regions. From the decomposition front to the far end of the formation, the hydrate had not decomposed. From the wellbore to the decomposition front, the decomposition rate was faster, and the closer to the wellbore, the more sufficient the heat exchange, and the higher the degree of hydrate decomposition was. Due to the hysteresis of heat transfer of drilling fluid, the pressure front is always ahead of the hydrate decomposition front, that is, the pressure wave propagation range is always greater than the hydrate decomposition range.

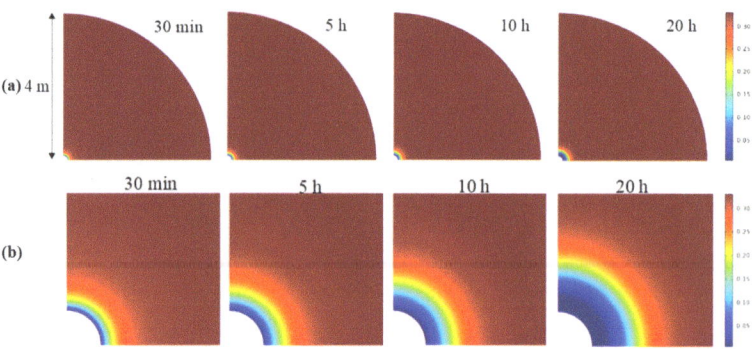

Figure 10. Variations of hydrate saturations in the formation and near-wellbore zone with time. (a) formation (b) near-wellbore zone.

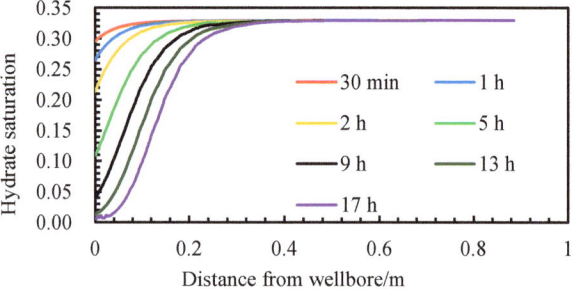

Figure 11. Distributions of hydrate saturations in the near-wellbore zone.

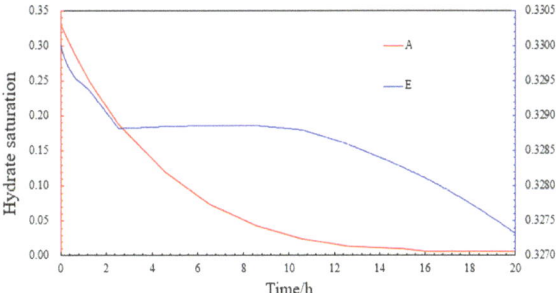

Figure 12. Variations of hydrate saturation at two points around the well with time.

4.1.4. Distribution Patterns of Porosity and Permeability

Figure 13 shows the distribution nephograms of formation-effective porosity and formation-effective permeability at different times after the drilling fluid invaded the formation. Figure 14 shows the distribution patterns of effective porosity and effective permeability in the near-wellbore zone at different times. The distribution characteristics of effective porosity and effective permeability are very similar to those of hydrate saturation of the formation.

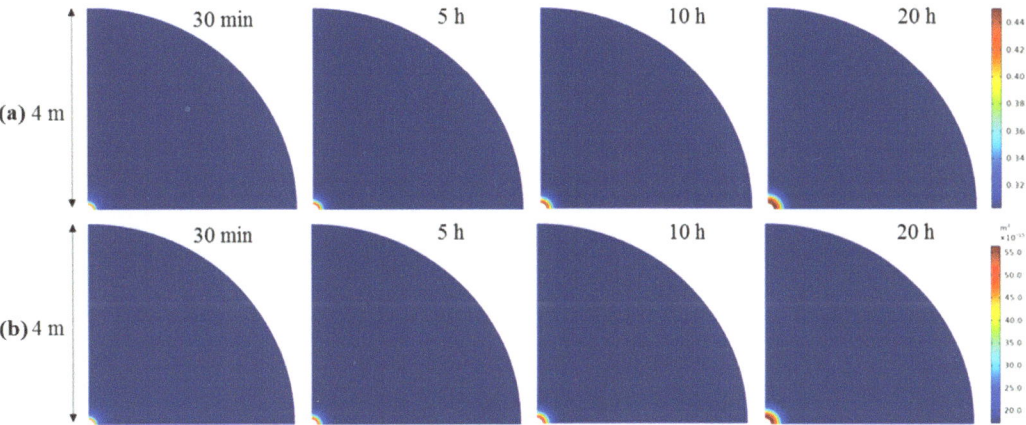

Figure 13. The distribution nephograms of formation-effective porosity and formation-effective permeability at different times. (**a**) formation-effective porosity (**b**) formation-effective permeability.

Similarly, the distribution of effective porosity and permeability can be divided into two regions, from the wellbore to the decomposition front, as a lot of hydrate decomposed, the effective porosity and permeability rapidly increased, and from the decomposition front to the far end of the formation, they basically remained almost undisturbed. Therefore, decomposition is the main factor that affects the porosity and permeability. In the near-wellbore zone, the combination of the fluid–solid coupling effect with the hydrate decomposition effect led to the development of fractures in the reservoir massively, making the effective porosity and permeability there significantly higher than the undisturbed formation. A total of 20 h after the drilling fluid invaded the formation, the ultimate effective porosity of the formation was 0.45, about 1.13 times the absolute porosity of the formation, and the ultimate effective permeability of the formation was 5.64×10^{-14} m^2, about 1.14 times of the absolute permeability of the formation.

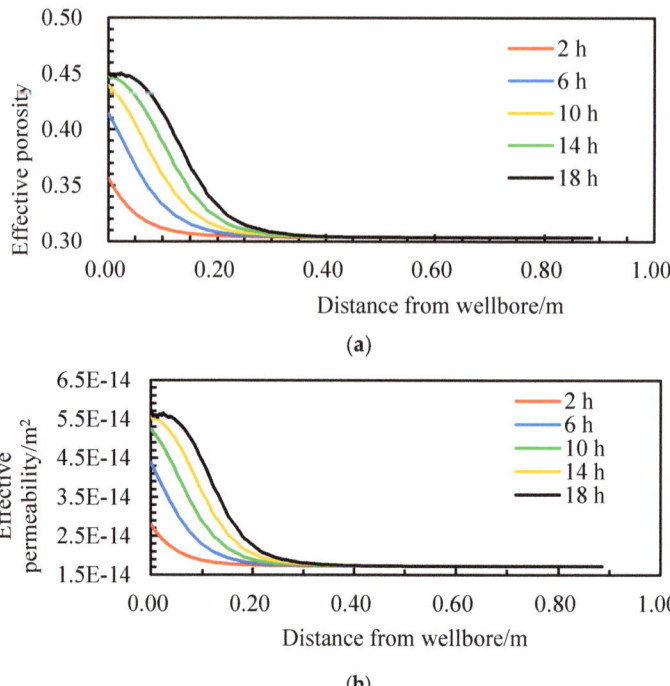

Figure 14. The distribution patterns of effective porosity and effective permeability. (**a**) Distributions of effective porosity in the near-wellbore zone, (**b**) Distributions of effective permeability in the near-wellbore zone.

4.2. Sensitivity Analysis

This section presents our study of the key parameters affecting the stability of horizontal wellbores in hydrate reservoirs, and the simulation parameters are shown in Table 4.

Table 4. Values of key parameters for horizontal wellbore stability analysis.

Pressure of Injected Drilling Fluid/MPa	Temperature of Injected Drilling Fluid/°C	Absolute Porosity	Absolute Permeability/m^2	Initial Hydrate Saturation
14.6	15	0.3	2.961×10^{-14}	0.33
14.882	16.32	0.4	4.935×10^{-14}	0.4
15	17	0.5	6.909×10^{-14}	0.5
15.2	18	0.6	8.883×10^{-14}	0.6

4.2.1. Pressure

From Figures 15 and 16, with the increase in the pressure of the injected drilling fluid, the reservoir was more and more sufficient and had larger and larger ranges impacted. Therefore, the hydrate decomposition became more thorough, and the front of the hydrate decomposition enlarged. Meanwhile, the accelerated invasion of drilling fluid made the elastic modulus and cohesion of the stratum drop sharply, and the range of the elastic modulus and cohesion drop in the reservoir expanded, which together made the mechanical properties of the hydrate reservoir with weak cementation intrinsically decline further.

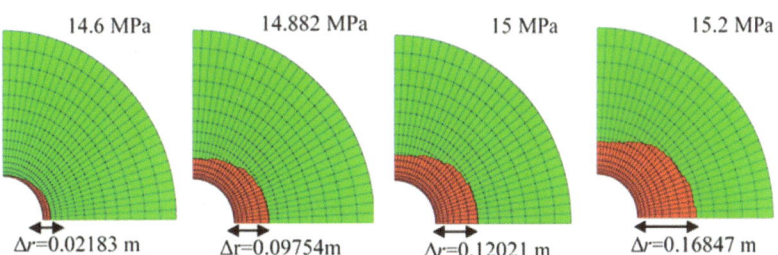

Figure 15. Variations of wellbore yield area under different pressures of injected drilling fluid.

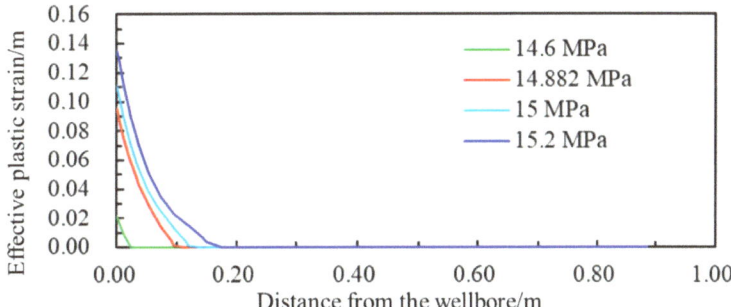

Figure 16. Distributions of effective plastic strain.

Therefore, for the formation of a hydrate reservoir with poor mud cake, the density of drilling fluid should be lowered as much as possible.

4.2.2. Temperature

Figures 17 and 18 show that the decomposition region of hydrate gradually expands as the temperature of the injected drilling fluid increases, and the modulus of elasticity and cohesion of the formation also gradually decreases as a result of hydrate decomposition, and the range of the effective plastic strain around the well expands, with a corresponding increase in the plastic yield region of the wellbore. In addition, the expansion of the hydrate decomposition region also weakened the mechanical properties of the formation over a larger area around the wellbore and increased the extent of wellbore collapse. The final yield radius increased from 0.0527m to 0.14245 m (almost tripled) after 20 h of drilling fluid intrusion when the temperature of the injected drilling fluid was increased to 18 °C. The final diameter expansion reached 124.6%.

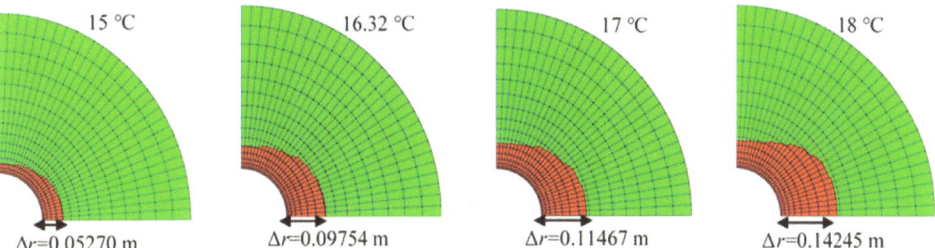

Figure 17. Variations of wellbore yield zone at different temperatures of injected drilling fluid.

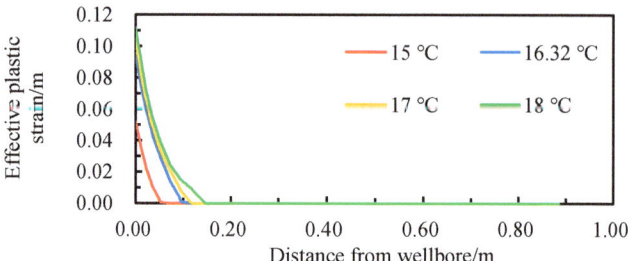

Figure 18. Distributions of effective plastic strain around wellbore at different times.

4.2.3. Reservoir Porosity

Figure 19 shows the variations of hydrate saturation, elastic modulus, and cohesion of the reservoir and wellbore yield area within 0.5 m of the wellbore 20 h after the drilling fluid invasion with other parameters remaining constant (assuming the reservoir porosity changed, but the formation mechanical parameters did not). Figure 20 shows the distributions of hydrate saturation and effective plastic strain around the wellbore 20 h after the drilling fluid invasion, respectively. We can see from the simulation results that the hydrate decomposition ranges at different reservoir porosities differed to some extent; with the increase in porosity of the hydrate reservoir, the hydrate decomposition range near the wellbore gradually reduced, the range of the formation with elastic modulus and cohesion drop also decreased, the range of effective plastic strain around the well also became smaller, and the region of the wellbore with plastic yield also shrunk accordingly. The increase in reservoir porosity makes the plastic yield area decrease. The risk of wellbore instability in the hydrate reservoir with higher porosity will be much greater than that in the reservoir with lower porosity after the complete decomposition of the hydrate.

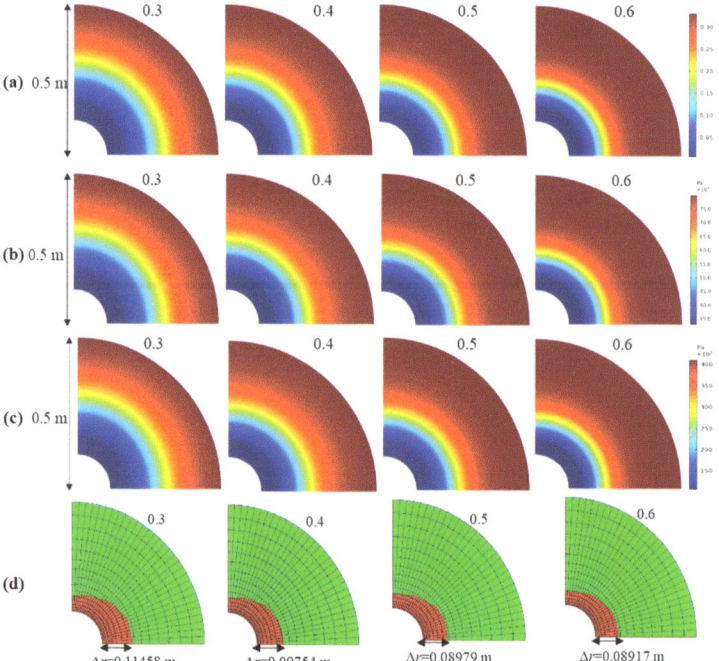

Figure 19. The variations of hydrate saturation and elastic modulus, and cohesion of the reservoir and wellbore yield area after the drilling fluid invasion. (**a**) hydrate saturation, (**b**) elastic modulus, (**c**) cohesion of the reservoir, (**d**) cohesion of the wellbore yield area.

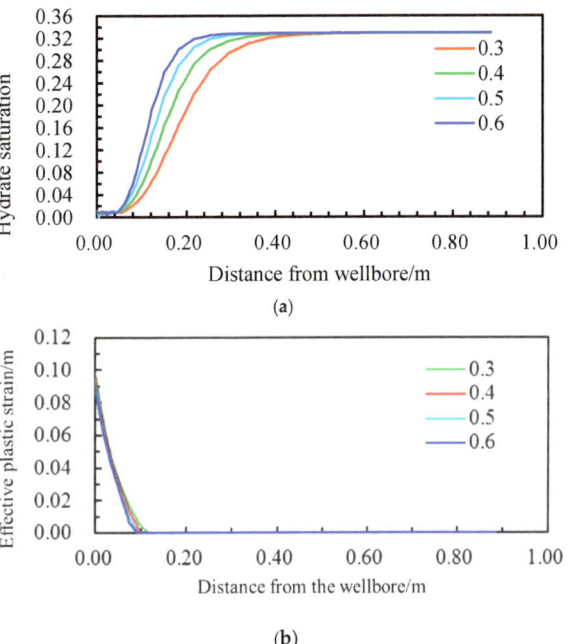

Figure 20. The distributions of hydrate saturation and effective plastic strain. (**a**) Distributions of hydrate saturation around the wellbore, (**b**) Distributions of effective plastic strain around wellbore.

4.2.4. Reservoir Permeability

Unlike the case of increasing porosity, as shown in Figure 21, with the increase in permeability, the hydrate decomposition range gradually increases, and the effective plastic strain range around the well expands. This is mainly due to the radial expansion of the hydrate decomposition range increasing the range of weakened formation mechanical properties, so the wellbore is more prone to instability.

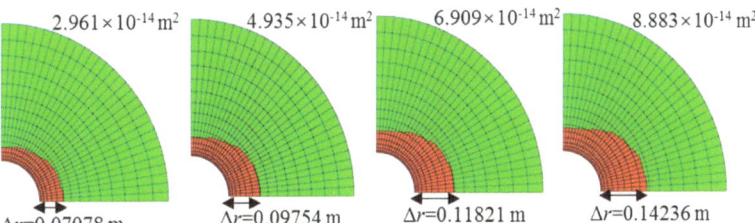

Figure 21. Variations of wellbore yield area at different permeabilities.

5. Conclusions

In this study, a comprehensive model for the stability analysis of horizontal wells in hydrate reservoirs was established, the coupling of multi-physical fields in the drilling process of hydrate reservoirs was realized, the stability of horizontal wells drilled in hydrate reservoirs was examined, and relevant sensitivity analyses were carried out. The conclusions are as follows:

(1) The intrusion of drilling fluid into the hydrate reservoir under differential pressures is the main factor leading to the decrease in the mechanical strength of the formation. Gases and liquids generated by hydrate decomposition in the formation increase the pore pressure in the formation, thereby reducing the effective stress in the formation.

The combination of the two causes the risk of wellbore destabilization to rise sharply in weakly cemented, poorly diagenetic hydrate reservoirs.
(2) Since the stress distribution around horizontal wells is different from that around vertical wells, the plastic yield zone of horizontal wells is no longer in the shape of a regular fan ring but is unevenly distributed. The closer the formation is to the wellbore, the more complete the decomposition of hydrates in the formation is, and the greater the effective plastic strain is.
(3) All other conditions being held constant, increases in drilling fluid intrusion pressure, temperature, and reservoir permeability will lead to an expansion of the range of decline in formation mechanical strength around the wellbore and the wellbore yield area. The wellbore production area will decrease with increasing reservoir porosity, which is beneficial for wellbore stability during drilling, but the risk of wellbore destabilization may increase dramatically once prolonged production occurs and the hydrates around the wellbore are fully decomposed.

Due to limited time and energy, further in-depth research should be conducted in the following areas in the future:

(1) Considering computational efficiency and accuracy, this article only established a two-dimensional numerical simulation model. In subsequent research, a three-dimensional physical model can be established to obtain more accurate research rules.
(2) This article simplifies the influence of formation heat transfer, and in the future, research on the effects of formation heat transfer on rock skeleton and formation fluid properties can be considered.

Author Contributions: Conceptualization, Z.S. and Z.L.; Methodology, Z.S., Z.G. and N.M.; Software, Z.S., X.W. and Z.G.; Validation, X.W. and Z.L.; Formal analysis, Z.L., N.M. and J.Z.; Resources, Z.G.; Data curation, J.Z.; Writing—original draft, Z.L. and N.M.; Writing—review & editing, Z.W.; Visualization, J.Z.; Supervision, Z.W. and B.S.; Project administration, Z.W. and B.S.; Funding acquisition, Z.W. and B.S. All authors have read and agreed to the published version of the manuscript.

Funding: This research was funded by [The Scientific and Technological Innovation Projects in Shandong Province] grant number [2022CXGC020407], [the National Natural Science Foundation of China] grant numbers [52288101, 52434002, 52304016], and [the Fundamental Research Funds for the Central Universities] grant number [24CX10004A].

Institutional Review Board Statement: Not applicable.

Informed Consent Statement: Not applicable.

Data Availability Statement: Data is contained within the article.

Conflicts of Interest: Authors Zhengfeng Shan and Xiansi Wang were employed by the company CNPC Offshore Engineering Co., Ltd. The remaining authors declare that the research was conducted in the absence of any commercial or financial relationships that could be construed as a potential conflict of interest.

References

1. Kvenvolden, K.A.; Ginsburg, G.D.; Soloviev, V.A. Worldwide distribution of subaquatic gas hydrates. *Geo-Mar. Lett.* **1993**, *13*, 32–40. [CrossRef]
2. Wang, W.; Zhang, C.; Wu, Y.; Zhang, S.; Xia, X. Prospects on technology for combining deep-sea geothermal energy with exploitation of natural gas hydrate. *Mod. Chem. Ind.* **2021**, *41*, 17–21.
3. Ran, Q.; Gu, X. Coupling Analysis of Multiphase Flow and Stress for Oil Reservoir. *Chin. J. Geotech. Eng.* **1998**, *20*, 69–73.
4. Hong, H.; Pooladi-Darvish, M.; Bishnoi, P.R. Analytical Modelling of Gas Production From Hydrates in Porous Media. *J. Can. Pet. Technol.* **2003**, *42*, 45–56. [CrossRef]
5. Xu, D.; Chen, Y.; Han, X.; Hao, G.; Lu, J.; Lan, Q. Prospects on combined technology for CO_2 replacement and exploitation of natural gas hydrate in sea. *Mod. Chem. Ind.* **2021**, *41*, 22–26, 32.
6. Zhang, R. *Study on Risk of Deepwater Gas Hydrate Formation Instability and Well Integrity during Drilling and Production*; China University of Petroleum (East China): Qingdao, China, 2017.

7. Li, G.; Moridis, G.J.; Zhang, K.; Li, X.-S. The use of huff and puff method in a single horizontal well in gas production from marine gas hydrate deposits in the Shenhu Area of South China Sea. *J. Pet. Sci. Eng.* **2011**, *11*, 49–68. [CrossRef]
8. Shen, H. *Fluid-Solid Coupling Numerical Simulation on Natural Gas Production from Hydrate Reservoirs by Depressurization*; China University of Petroleum (East China): Qingdao, China, 2009.
9. Liao, Y.; Wang, Z.; Chao, M.; Sun, X.; Wang, J.; Zhou, B.; Sun, B. Coupled wellbore–reservoir heat and mass transfer model for horizontal drilling through hydrate reservoir and application in wellbore stability analysis. *J. Nat. Gas Sci. Eng.* **2021**, *95*, 104216. [CrossRef]
10. Yun, T.S.; Santamarina, J.C.; Ruppel, C. Mechanical properties of sand, silt, and clay containing tetrahydrofuran hydrate. *J. Geophys. Res. Solid Earth* **2007**, *112*, 1–13. [CrossRef]
11. Hyodo, M.; Li, Y.; Yoneda, J. Mechanical behavior of gas-saturated methane hydrate-bearing sediments. *J. Geophys. Res. Solid Earth* **2013**, *118*, 5185–5194. [CrossRef]
12. Komatsu, Y.; Suzuki, K.; Fujii, T. Sedimentary facies and paleoenvironments of a gas-hydrate-bearing sediment core in the eastern Nankai Trough, Japan. *Mar. Pet. Geol.* **2015**, *66*, 358–367. [CrossRef]
13. Wang, X.H.; Sun, Y.F.; Wang, Y.F.; Li, N.; Sun, C.Y.; Chen, G.J.; Liu, B.; Yang, L.Y. Gas production from hydrates by CH_4-CO_2/H_2 replacement. *Appl. Energy* **2017**, *188*, 305–314. [CrossRef]
14. Sun, J.; Ning, F.; Lei, H.; Gai, X.; Sánchez, M.; Lu, J.; Li, Y.; Liu, L.; Liu, C.; Wu, N.; et al. Wellbore stability analysis during drilling through marine gas hydrate-bearing sediments in Shenhu area: A case study. *J. Pet. Sci. Eng.* **2018**, *170*, 345–367. [CrossRef]
15. Rutqvist, J.; Moridis, G.J.; Grover, T.; Silpngarmlert, S.; Collett, T.S.; Holdich, S.A. Coupled multiphase fluid flow and wellbore stability analysis associated with gas production from oceanic hydrate-bearing sediments. *J. Pet. Sci. Eng.* **2012**, *92*, 65–81. [CrossRef]
16. Yamini, O.A.; Movahedi, A.; Mousavi, S.H.; Kavianpour, M.R.; Kyriakopoulos, G.L. Hydraulic performance of seawater intake system using CFD modeling. *J. Mar. Sci. Eng.* **2022**, *10*, 988. [CrossRef]
17. Lei, L.; Santamarina, J.C. Laboratory strategies for hydrate formation in fine-grained sediments. *J. Geophys. Res. Solid Earth* **2018**, *123*, 2583–2596. [CrossRef]
18. Hyodo, M.; Li, Y.; Yoneda, J.; Nakata, Y.; Yoshimoto, N.; Nishimura, A. Effects of dissociation on the shear strength and deformation behavior of methane hydrate-bearing sediments. *Mar. Pet. Geol.* **2014**, *51*, 52–62. [CrossRef]
19. Freij-Ayoub, R.; Tan, C.; Clennell, B.; Tohidi, B.; Yang, J. A wellbore stability model for hydrate bearing sediments. *J. Pet. Sci. Eng.* **2007**, *57*, 209–220. [CrossRef]
20. Jin, G.; Lei, H.; Xu, T.; Xin, X.; Yuan, Y.; Xia, Y.; Juo, J. Simulated geomechanical responses to marine methane hydrate recovery using horizontal wells in the Shenhu area, South China Sea. *Mar. Pet. Geol.* **2017**, *92*, 424–436. [CrossRef]
21. Sun, J. *Characteristics of Reservoir Response to Drilling and Production in Gas Hydrate-Bearing Sediments in the South China Sea*; China University of Geosciences: Beijing, China, 2018.
22. Li, X.; Sun, B.; Ma, B.; Li, H.; Liu, H.; Cai, D.; Wang, X.; Li, X. Study on the Evolution Law of Wellbore Stability Interface during Drilling of Offshore Gas Hydrate Reservoirs. *Energies* **2023**, *16*, 7585. [CrossRef]
23. Wang, X.; Wang, Z.; Gong, Z.; Fu, W.; Liu, P.; Zhang, J. Study on Stability of Horizontal Wellbore Drilled in Marine Natural Gas Hydrate Reservoir. In *International Technical Symposium on Deepwater Oil and Gas Engineering*; Springer Nature: Singapore, 2023; pp. 523–535.
24. Wang, S.; Sun, Z. Current Status and Future Trends of Exploration and Pilot Production of Gas Hydrate in the World. *Mar. Geol. Front.* **2018**, *34*, 24–32.
25. Fung, L.S.K. A Coupled Geomechanic-Multiphase Flow Model For Analysis Of In Situ Recovery In Cohesionless Oil Sands. *J. Can. Pet. Technol.* **1992**, *31*, 56–67. [CrossRef]
26. Selim, M.S.; Sloan, E.D. Heat and mass transfer during the dissociation of hydrates in porous media. *AIChE J.* **1989**, *35*, 1049–1052. [CrossRef]
27. Nazridoust, K.; Ahmadi, G. Computational modeling of methane hydrate dissociation in a sandstone core. *Chem. Eng. Sci.* **2007**, *62*, 6155–6177. [CrossRef]
28. Cheng, Y.F.; Li, L.D.; Mahmood, S.; Cui, Q. Fluid-solid coupling model for studying wellbore instability in drilling of gas hydrate bearing sediments. *Appl. Math. Mech. (Engl. Ed.)* **2013**, *34*, 1421–1432. [CrossRef]
29. Wang, H.N.; Chen, X.P.; Jiang, M.J.; Guo, Z.Y. Analytical investigation of wellbore stability during drilling in marine methane hydrate-bearing sediments. *J. Nat. Gas Sci. Eng.* **2019**, *68*, 102885. [CrossRef]
30. Kim, H.C.; Bishnoi, P.R.; Heidemann, R.A.; Rizvi, S.S. Kinetics of methane hydrate decomposition. *Chem. Eng. Sci.* **1985**, *42*, 1645–1653. [CrossRef]
31. Fjaer, E.; Holt, R.M.; Horsrud, P.; Raaen, A.M.; Risnes, R. *Petroleum Related Rock Mechanics*; Elsevier: Amsterdam, The Netherlands, 2008.

Disclaimer/Publisher's Note: The statements, opinions and data contained in all publications are solely those of the individual author(s) and contributor(s) and not of MDPI and/or the editor(s). MDPI and/or the editor(s) disclaim responsibility for any injury to people or property resulting from any ideas, methods, instructions or products referred to in the content.

MDPI AG
Grosspeteranlage 5
4052 Basel
Switzerland
Tel.: +41 61 683 77 34

MDPI Books Editorial Office
E-mail: books@mdpi.com
www.mdpi.com/books

Disclaimer/Publisher's Note: The title and front matter of this reprint are at the discretion of the Topic Editors. The publisher is not responsible for their content or any associated concerns. The statements, opinions and data contained in all individual articles are solely those of the individual Editors and contributors and not of MDPI. MDPI disclaims responsibility for any injury to people or property resulting from any ideas, methods, instructions or products referred to in the content.